Lecture Notes in Bioinformatics 4115

Edited by S. Istrail, P. Pevzner, and M. Waterman

Subseries of Lecture Notes in Computer Science

De-Shuang Huang
Kang Li
George William Irwin (Eds.)

Computational Intelligence and Bioinformatics

International Conference on
Intelligent Computing, ICIC 2006
Kunming, China, August 16-19, 2006
Proceedings, Part III

Springer

Series Editors

Sorin Istrail, Brown University, Providence, RI, USA
Pavel Pevzner, University of California, San Diego, CA, USA
Michael Waterman, University of Southern California, Los Angeles, CA, USA

Volume Editors

De-Shuang Huang
Chinese Academy of Sciences
Institute of Intelligent Machines
Hefei, Anhui, China
E-mail: dshuang@iim.ac.cn

Kang Li
George William Irwin
Queen's University
Belfast, UK
E-mail: {K.Li,G.Irwin}@qub.ac.uk

Library of Congress Control Number: 2006930738

CR Subject Classification (1998): F.2.2, F.2, E.1, G.1, I.2, J.3

LNCS Sublibrary: SL 8 – Bioinformatics

ISSN 0302-9743
ISBN-10 3-540-37277-6 Springer Berlin Heidelberg New York
ISBN-13 978-3-540-37277-6 Springer Berlin Heidelberg New York

Springer is a part of Springer Science+Business Media

springer.com

Typesetting: Camera-ready by author, data conversion by Scientific Publishing Services, Chennai, India
Printed on acid-free paper SPIN: 11816102 06/3142 5 4 3 2 1 0

Preface

The International Conference on Intelligent Computing (ICIC) was formed to provide an annual forum dedicated to the emerging and challenging topics in artificial intelligence, machine learning, bioinformatics, and computational biology, etc. It aims to bring together researchers and practitioners from both academia and industry to share ideas, problems and solutions related to the multifaceted aspects of intelligent computing.

ICIC 2006 held in Kunming, Yunnan, China, August 16-19, 2006, was the second International Conference on Intelligent Computing, built upon the success of ICIC 2005 held in Hefei, China, 2005.

This year, the conference concentrated mainly on the theories and methodologies as well as the emerging applications of intelligent computing. It intended to unify the contemporary intelligent computing techniques within an integral framework that highlights the trends in advanced computational intelligence and bridges theoretical research with applications. In particular, bio-inspired computing emerged as having a key role in pursuing for novel technology in recent years. The resulting techniques vitalize life science engineering and daily life applications. In light of this trend, the theme for this conference was **"Emerging Intelligent Computing Technology and Applications"**. Papers related to this theme were especially solicited, including theories, methodologies, and applications in science and technology.

ICIC 2006 received over 3000 submissions from 36 countries and regions. All papers went through a rigorous peer review procedure and each paper received at least three review reports. Based on the review reports, the Program Committee finally selected 703 high-quality papers for presentation at ICIC 2006. These papers cover 29 topics and 16 special sessions, and are included in five volumes of proceedings published by Springer, including one volume of *Lecture Notes in Computer Science* (LNCS), one volume of *Lecture Notes in Artificial Intelligence* (LNAI), one volume of *Lecture Notes in Bioinformatics* (LNBI), and two volumes of *Lecture Notes in Control and Information Sciences* (LNCIS).

This volume of *Lecture Notes in Bioinformatics* (LNBI) includes 85 papers covering 10 relevant topics and 2 special session topics.

The organizers of ICIC 2006, including Yunan University, the Institute of Intelligent Machines of the Chinese Academy of Science, and Queen's University Belfast, have made enormous effort to ensure the success of ICIC 2006. We hereby would like to thank the members of the ICIC 2006 Advisory Committee for their guidance and advice, the members of the Program Committee and the referees for their collective effort in reviewing and soliciting the papers, and the members of the Publication Committee for their significant editorial work. We would like to thank Alfred Hofmann, executive editor from Springer, for his frank and helpful advice and guidance throughout and for his support in publishing the proceedings in the Lecture Notes series. In particular, we would like to thank all the authors for contributing their papers. Without the high-quality submissions from the authors, the success of the conference would not have been possible. Finally, we are especially grateful to the IEEE Computational Intelligence Society, The International Neural Network Society and the National Science Foundation of China for the their sponsorship.

June 2006

De-Shuang Huang
Institute of Intelligent Machines, Chinese Academy of Sciences, China
Kang Li
Queen's University Belfast, UK
George William Irwin
Queen's University Belfast, UK

ICIC 2006 Organization

General Chairs:

De-Shuang Huang, China
Song Wu, China
George W. Irwin, UK

International Advisory Committee

Aike Guo, China
Alfred Hofmann, Germany
DeLiang Wang, USA
Erke Mao, China
Fuchu He, China
George W. Irwin, UK
Guangjun Yang, China
Guanrong Chen, Hong Kong
Guoliang Chen, China
Harold Szu, USA
John L. Casti, USA
Marios M. Polycarpou, USA
Mengchu Zhou, USA
Michael R. Lyu, Hong Kong
MuDer Jeng, Taiwan
Nanning Zheng, China
Okyay Knynak, Turkey
Paul Werbos, USA
Qingshi Zhu, China
Ruwei Dai, China
Sam Shuzhi GE, Singapore
Sheng Zhang, China
Shoujue Wang, China
Songde Ma, China
Stephen Thompson, UK
Tom Heskes, Netherlands
Xiangfan He, China
Xingui He, China
Xueren Wang, China
Yanda Li, China
Yixin Zhong, China
Youshou Wu, China
Yuanyan Tang, Hong Kong
Yunyu Shi, China
Zheng Bao, China

Program Committee Chairs:

Kang Li, UK
Prashan Premaratne, Australia

Steering Committee Chairs:

Sheng Chen, UK
Xiaoyi Jiang, Germany
Xiao-Ping Zhang, Canada

Organizing Committee Chairs:

Yongkun Li, China
Hanchun Yang, China
Guanghua Hu, China

Special Session Chair:

Wen Yu, Mexico

Tutorial Chair:

Sudharman K. Jayaweera, USA

Publication Chair: Xiaoou Li, Mexico

International Liasion Chair: C. De Silva, Liyanage, New Zealand

Publicity Chairs: Simon X. Yang, Canada
Jun Zhang, China

Exhibition Chair: Cheng Peng, China

Program Committee

Aili Han, China
Arit Thammano, Thailand
Baogang Hu, China
Bin Luo, China
Bin Zhu, China
Bing Wang, China
Bo Yan, USA
Byoung-Tak Zhang, Korea
Caoan Wang, Canada
Chao Hai Zhang, Japan
Chao-Xue Wang, China
Cheng-Xiang Wang, UK
Cheol-Hong Moon, Korea
Chi-Cheng Cheng, Taiwan
Clement Leung, Australia
Daniel Coca, UK
Daqi Zhu, China
David Stirling, Australia
Dechang Chen, USA
Derong Liu, USA
Dewen Hu, China
Dianhui Wang, Australia
Dimitri Androutsos, Canada
Donald C. Wunsch, USA
Dong Chun Lee, Korea
Du-Wu Cui, China
Fengling Han, Australia
Fuchun Sun, China
Girijesh Prasad, UK
Guang-Bin Huang, Singapore
Guangrong Ji, China
Hairong Qi, USA
Hong Qiao, China
Hong Wang, China
Hongtao Lu, China
Hongyong Zhao, China
Huaguang Zhang, China
Hui Wang, China
Jiangtao Xi, Australia
Jianguo Zhu, Australia
Jianhua Xu, China
Jiankun Hu, Australia
Jian-Xun Peng, UK
Jiatao Song, China
Jie Tian, China
Jie Yang, China
Jin Li, UK
Jin Wu, UK
Jinde Cao, China
Jinwen Ma, China
Jochen Till, Germany
John Q. Gan, UK
Ju Liu, China
K. R. McMenemy, UK
Key-Sun Choi, Korea
Liangmin Li, UK
Luigi Piroddi, Italy
Maolin Tang, Australia
Marko Hočevar, Slovenia
Mehdi Shafiei, Canada
Mei-Ching Chen, Taiwan
Mian Muhammad Awais, Pakistan
Michael Granitzer, Austria
Michael J.Watts, New Zealand
Michiharu Maeda, Japan
Minrui Fei, China
Muhammad Jamil Anwas, Pakistan
Muhammad Khurram Khan, China
Naiqin Feng, China
Nuanwan Soonthornphisaj, Thailand
Paolo Lino, Italy
Peihua Li, China
Ping Guo, China
Qianchuan Zhao, China
Qiangfu Zhao, Japan
Qing Zhao, Canada
Roberto Tagliaferri, Italy
Rong-Chang Chen, Taiwan
RuiXiang Sun, China
Saeed Hashemi, Canada

Sanjay Sharma, UK
Seán McLoone, Ireland
Seong G. Kong, USA
Shaoning Pang, New Zealand
Shaoyuan Li, China
Shuang-Hua Yang, UK
Shunren Xia, China
Stefanie Lindstaedt, Austria
Sylvia Encheva, Norway
Tai-hoon Kim, Korea
Tai-Wen Yue, Taiwan
Takashi Kuremoto, Japan
Tarık Veli Mumcu, Turkey
Tian Xiang Mei, UK
Tim. B. Littler, UK
Tommy W. S. Chow, Hong Kong
Uwe Kruger, UK
Vitoantonio Bevilacqua, Italy
Wei Dong Chen, China
Wenming Cao, China
Wensheng Chen, China
Willi Richert, Germany
Worapoj Kreesuradej, Thailand
Xiao Zhi Gao, Finland
Xiaoguang Zhao, China
Xiaojun Wu, China
Xiaolong Shi, China
Xiaoou Li, Mexico
Xinge You, Hong Kong
Xiyuan Chen, China
Xiwen Zhang, China
Xun Wang, UK
Yanhong Zhou, China
Yi Shen, China
Yong Dong Wu, Singapore
Yuhua Peng, China
Zengguang Hou, China
Zhao-Hui Jiang, Japan
Zhen Liu, Japan
Zhi Wang, China
Zhi-Cheng Chen, China
Zhi-Cheng Ji, China
Zhigang Zeng, China
Ziping Chiang, Taiwa

Reviewers

Xiaodan Wang, Lei Wang, Arjun Chandra, Angelo Ciaramella, Adam Kalam, Arun Sathish, Ali Gunes, Jin Tang, Aiguo He, Arpad Kelemen, Andreas Koschan, Anis Koubaa, Alan Gupta, Alice Wang, Ali Ozen, Hong Fang, Muhammad Amir Yousuf , An-Min Zou, Andre Döring, Andreas Juffinger, Angel Sappa, Angelica Li, Anhua Wan, Bing Wang, Rong Fei, Antonio Pedone, Zhengqiang Liang , Qiusheng An, Alon Shalev Housfater, Siu-Yeung Cho, Atif Gulzar, Armin Ulbrich, Awhan Patnaik, Muhammad Babar, Costin Badica, Peng Bai, Banu Diri, Bin Cao, Riccardo Attimonelli, Baohua Wang, Guangguo Bi, Bin Zhu, Brendon Woodford, Haoran Feng, Bo Ma, Bojian Liang, Boris Bacic, Brane Sirok, Binrong Jin, Bin Tian, Christian Sonntag, Galip Cansever, Chun-Chi Lo, ErKui Chen, Chengguo Lv, Changwon Kim, Chaojin Fu, Anping Chen, Chen Chun , C.C. Cheng, Qiming Cheng, Guobin Chen, Chengxiang Wang, Hao Chen, Qiushuang Chen, Tianding Chen, Tierui Chen, Ying Chen, Mo-Yuen Chow, Christian Ritz, Chunmei Liu, Zhongyi Chu, Feipeng Da, Cigdem Turhan, Cihan Karakuzu, Chandana Jayasooriya, Nini Rao, Chuan-Min Zhai, Ching-Nung Yang, Quang Anh Nguyen, Roberto Cordone, Changqing Xu, Christian Schindler, Qijun Zhao, Wei Lu, Zhihua Cui, Changwen Zheng, David Antory, Dirk Lieftucht, Dedy Loebis, Kouichi Sakamoto, Lu Chuanfeng, Jun-Heng Yeh, Dacheng Tao, Shiang-Chun Liou, Ju Dai , Dan Yu, Jianwu Dang, Dayeh Tan, Yang Xiao, Dondong Cao, Denis Stajnko, Liya De Silva, Damien Coyle, Dian-Hui Wang, Dahai Zhang, Di Huang, Dikai Liu, D. Kumar, Dipak Lal Shrestha, Dan Lin, DongMyung Shin, Ning Ding, DongFeng Wang, Li Dong, Dou Wanchun, Dongqing Feng, Dingsheng Wan, Yongwen Du, Weiwei Du, Wei Deng, Dun-wei Gong, DaYong Xu, Dar-Ying Jan, Zhen Duan, Daniela Zaharie, ZhongQiang Wu, Esther Koller-Meier, Anding Zhu, Feng Pan, Neil Eklund, Kezhi

Mao, HaiYan Zhang, Sim-Heng Ong, Antonio Eleuteri, Bang Wang, Vincent Emanuele, Michael Emmerich, Hong Fu, Eduardo Hruschka, Erika Lino, Estevam Rafael Hruschka Jr, D.W. Cui, Fang Liu, Alessandro Farinelli, Fausto Acernese, Bin Fang, Chen Feng, Huimin Guo, Qing Hua, Fei Zhang, Fei Ge, Arnon Rungsawang, Feng Jing, Min Feng, Feiyi Wang, Fengfeng Zhou, Fuhai Li, Filippo Menolascina, Fengli Ren, Mei Guo, Andrés Ferreyra, Francesco Pappalardo, Chuleerat Charasskulchai, Siyao Fu, Wenpeng Ding, Fuzhen Huang, Amal Punchihewa, Geoffrey Macintyre, Xue Feng He, Gang Leng, Lijuan Gao, Ray Gao, Andrey Gaynulin, Gabriella Dellino, D.W. Ggenetic, Geoffrey Wang, YuRong Ge, Guohui He, Gwang Hyun Kim, Gianluca Cena, Giancarlo Raiconi, Ashutosh Goyal, Guan Luo, Guido Maione, Guido Maione, Grigorios Dimitriadis, Haijing Wang, Kayhan Gulez, Tiantai Guo, Chun-Hung Hsieh, Xuan Guo, Yuantao Gu, Huanhuan Chen, Hongwei Zhang, Jurgen Hahn, Qing Han, Aili Han, Dianfei Han, Fei Hao, Qing-Hua Ling, Hang-kon Kim, Han-Lin He, Yunjun Han, Li Zhang, Hathai Tanta-ngai, Hang-Bong Kang, Hsin-Chang Yang, Hongtao Du, Hazem Elbakry, Hao Mei, Zhao L, Yang Yun, Michael Hild, Heajo Kang, Hongjie Xing, Hailli Wang, Hoh In, Peng Bai, Hong-Ming Wang, Hongxing Bai, Hongyu Liu, Weiyan Hou, Huaping Liu, H.Q. Wang, Hyungsuck Cho, Hsun-Li Chang, Hua Zhang, Xia Huang, Hui Chen, Huiqing Liu, Heeun Park, Hong-Wei Ji, Haixian Wang, Hoyeal Kwon, H.Y. Shen, Jonghyuk Park, Turgay Ibrikci, Mary Martin, Pei-Chann Chang, Shouyi Yang, Xiaomin Mu, Melanie Ashley, Ismail Altas, Muhammad Usman Ilyas, Indrani Kar, Jinghui Zhong, Ian Mack, Il-Young Moon, J.X. Peng , Jochen Till, Jian Wang, Quan Xue, James Govindhasamy, José Andrés Moreno Pérez, Jorge Tavares, S. K. Jayaweera, Su Jay, Jeanne Chen, Jim Harkin, Yongji Jia, Li Jia, Zhao-Hui Jiang, Gangyi Jiang, Zhenran Jiang, Jianjun Ran, Jiankun Hu, Qing-Shan Jia, Hong Guo, Jin Liu, Jinling Liang, Jin Wu, Jing Jie, Jinkyung Ryeu, Jing Liu, Jiming Chen, Jiann-Ming Wu, James Niblock, Jianguo Zhu, Joel Pitt, Joe Zhu, John Thompson, Mingguang Shi, Joaquin Peralta, Si Bao Chen, Tinglong Pan, Juan Ramón González González, JingRu Zhang, Jianliang Tang, Joaquin Torres, Junaid Akhtar, Ratthachat Chatpatanasiri, Junpeng Yuan, Jun Zhang, Jianyong Sun, Junying Gan, Jyh-Tyng Yau, Junying Zhang, Jiayin Zhou, Karen Rosemary McMenemy, Kai Yu, Akimoto Kamiya, Xin Kang, Ya-Li Ji, Guo-Shiang Lin, Muhammad Khurram, Kevin Curran, Karl Neuhold, Kyongnam Jeon, Kunikazu Kobayashi, Nagahisa Kogawa, Fanwei Kong, Kyu-Sik Park, Lily D. Li, Lara Giordano, Laxmidhar Behera, Luca Cernuzzi, Luis Almeida, Agostino Lecci, Yan Zuo, Lei Li, Alberto Leva, Feng Liang, Bin Li, Jinmei Liao, Liang Tang, Bo Lee, Chuandong Li, Lidija Janezic, Jian Li, Jiang-Hai Li, Jianxun Li, Limei Song, Ping Li, Jie Liu, Fei Liu, Jianfeng Liu, Jianwei Liu, Jihong Liu, Lin Liu, Manxi Liu, Yi Liu, Xiaoou Li, Zhu Li, Kun-hong Liu, Li Min Cui, Lidan Miao, Long Cheng , Huaizhong Zhang, Marco Lovera, Liam Maguire, Liping Liu, Liping Zhang, Feng Lu, Luo Xiaobin, Xin-ping Xie, Wanlong Li, Liwei Yang, Xinrui Liu, Xiao Wei Li, Ying Li, Yongquan Liang, Yang Bai, Margherita Bresco, Mingxing Hu, Ming Li, Runnian Ma, Meta-Montero Manrique, Zheng Gao, Mingyi Mao, Mario Vigliar, Marios Savvides, Masahiro Takatsuka, Matevz Dular, Mathias Lux, Mutlu Avci, Zhifeng Hao, Zhifeng Hao, Ming-Bin Li, Tao Mei, Carlo Meloni, Gennaro Miele, Mike Watts, Ming Yang, Jia Ma, Myong K. Jeong, Michael Watts, Markus Koch, Markus Koch, Mario

Koeppen, Mark Kröll, Hui Wang, Haigeng Luo, Malrey Lee, Tiedong Ma, Mingqiang Yang, Yang Ming, Rick Chang, Nihat Adar, Natalie Schellenberg, Naveed Iqbal, Nur Bekiroglu, Jinsong Hu, Nesan Aluha, Nesan K Aluha, Natascha Esau, Yanhong Luo, N.H. Siddique, Rui Nian, Kai Nickel, Nihat Adar, Ben Niu, Yifeng Niu, Nizar Tayem, Nanlin Jin, Hong-Wei Ji, Dongjun Yu, Norton Abrew, Ronghua Yao, Marco Moreno-Armendariz, Osman Kaan Erol, Oh Kyu Kwon, Ahmet Onat, Pawel Herman, Peter Hung, Ping Sun, Parag Kulkarni, Patrick Connally, Paul Gillard, Yehu Shen, Paul Conilione, Pi-Chung Wang, Panfeng Huang, Peter Hung, Massimo Pica Ciamarra, Ping Fang, Pingkang Li, Peiming Bao, Pedro Melo-Pinto, Maria Prandini, Serguei Primak, Peter Scheir, Shaoning Pang, Qian Chen, Qinghao Rong, QingXiang Wu, Quanbing Zhang, Qifu Fan, Qian Liu, Qinglai Wei, Shiqun Yin, Jianlong Qiu, Qingshan Liu, Quang Ha, SangWoon Lee , Huaijing Qu, Quanxiong Zhou , Qingxian Gong, Qingyuan He, M.K.M. Rahman, Fengyuan Ren, Guang Ren, Qingsheng Ren, Wei Zhang, Rasoul Milasi, Rasoul Milasi, Roberto Amato, Roberto Marmo, P. Chen, Roderick Bloem, Hai-Jun Rong, Ron Von Schyndel, Robin Ferguson, Runhe Huang, Rui Zhang, Robin Ferguson, Simon Johnston, Sina Rezvani, Siang Yew Chong, Cristiano Cucco, Dar-Ying Jan, Sonya Coleman, Samuel Rodman, Sancho Salcedo-Sanz, Sangyiel Baik, Sangmin Lee, Savitri Bevinakoppa, Chengyi Sun, Hua Li, Seamus McLoone, Sean McLoone, Shafayat Abrar, Aamir Shahzad, Shangmin Luan, Xiaowei Shao, Shen Yanxia, Zhen Shen, Seung Ho Hong, Hayaru Shouno, Shujuan Li, Si Eng Ling, Anonymous, Shiliang Guo, Guiyu Feng, Serafin Martinez Jaramillo, Sangwoo Moon, Xuefeng Liu, Yinglei Song, Songul Albayrak, Shwu-Ping Guo, Chunyan Zhang, Sheng Chen, Qiankun Song, Seok-soo Kim, Antonino Staiano, Steven Su, Sitao Wu, Lei Huang, Feng Su, Jie Su, Sukree Sinthupinyo, Sulan Zhai, Jin Sun, Limin Sun, Zengshun Zhao, Tao Sun, Wenhong Sun, Yonghui Sun, Supakpong Jinarat, Srinivas Rao Vadali, Sven Meyer zu Eissen, Xiaohong Su , Xinghua Sun, Zongying Shi, Tony Abou-Assaleh, Youngsu Park, Tai Yang, Yeongtak Jo, Chunming Tang, Jiufei Tang, Taizhe Tan, Tao Xu, Liang Tao, Xiaofeng Tao, Weidong Xu, Yueh-Tsun Chang, Fang Wang, Timo Lindemann, Tina Yu, Ting Hu, Tung-Kuan Liu, Tianming Liu, Tin Lay Nwe, Thomas Neidhart, Tony Chan, Toon Calders, Yi Wang, Thao Tran, Kyungjin Hong, Tariq Qureshi, Tung-Shou Chen, Tsz Kin Tsui, Tiantian Sun, Guoyu Tu, Tulay Yildirim, Dandan Zhang, Xuqing Tang, Yuangang Tang, Uday Chakraborty, Luciana Cariello, Vasily Aristarkhov, Jose-Luis Verdegay, Vijanth Sagayan Asirvadam, Vincent Lee, Markus Vincze, Duo Chen, Viktoria Pammer, Vedran Sabol, Wajeeha Akram, Cao Wang , Xutao Wang, Winlen Wang, Zhuang Znuang, Feng Wang, Haifeng Wang, Le Wang, Wang Linkun, Meng Wang, Rongbo Wang, Xin Wang, Xue Wang, Yan-Feng Wang, Yong Wang, Yongcai Wang, Yongquan Wang, Xu-Qin Li, Wenbin Liu, Wudai Liao, Weidong Zhou, Wei Li, Wei Zhang, Wei Liang, Weiwei Zhang, Wen Xu, Wenbing Yao, Xiaojun Ban, Fengge Wu, Weihua Mao, Shaoming Li, Qing Wu, Jie Wang, Wei Jiang, W Jiang, Wolfgang Kienreich, Linshan Wang, Wasif Naeem, Worasait Suwannik, Wolfgang Slany, Shijun Wang , Wooyoung Soh, Teng Wang, Takashi Kuremoto, Hanguang Wu, Licheng Wu, Xugang Wang, Xiaopei Wu, ZhengDao Zhang, Wei Yen, Yan-Guo Wang, Daoud Ait-Kadi, Xiaolin Hu, Xiaoli Li, Xun Wang, Xingqi Wang, Yong Feng, Xiucui Guan, Xiao-Dong Li, Xingfa Shen, Xuemin Hong, Xiaodi Huang, Xi Yang, Li

Table of Contents

Ant Colony Optimisation

Particle Swarm Optimisation

Swarm Intelligence

Autonomy-Oriented Computing

Quantum and Molecular Computations

Biological and DNA Computing

Intelligent Computing in Bioinformatics

Intelligent Computing in Computational Biology and Drug Design

Computational Genomics and Proteomics

RNA Secondary Structure Prediction with Simple Pseudoknots Based on Dynamic Programming

Oyun-Erdene Namsrai, Kwang Su Jung, Sunshin Kim, and Keun Ho Ryu*

Database/BioInformatics lab,
School of Electrical & Computer Engineering,
Chungbuk National University
Cheongju, Chungbuk 361-763, Korea
Tel.: +82-43-261-2254, Fax: +82-43-275-2254
{oyunerdene, ksjung, sskim04, khryu}@dblab.chungbuk.ac.kr

Abstract. RNA molecules are sequences of nucleotides that serve as more than mere intermediaries between DNA and proteins, e.g. as catalytic molecules. The sequence of nucleotides of an RNA molecule constitutes its primary structure, and the pattern of pairing between nucleotides determines the secondary structure of an RNA. Computational prediction of RNA secondary structure is among the few structure prediction problems that can be solved satisfactory in polynomial time. Most work has been done to predict structures that do not contain pseudoknots. Pseudoknots have generally been excluded from the prediction of RNA secondary structures due to its difficulty in modelling. In this paper, we present a computation the maximum number of base pairs of an RNA sequence with simple pseudoknots. Our approach is based on pseudoknot technique proposed by Akutsu. We show that a structure with the maximum possible number of base pairs could be deduced by a improved Nussinov's trace-back procedure. In our approach we also considered wobble base pairings (G·U). We introduce an implementation of RNA secondary structure prediction with simple pseudoknots based on dynamic programming algorithm. To evaluate our method we use the 15 sequences with simple pseudoknots of variable size from 19 to 25 nucleotides. We get our experimental data set from PseudoBase. Our program predicts simple pseudoknots with correct or almost correct structure for 53% sequences.

1 General Introduction

A ribonucleic acid (RNA) is one of the two types of nucleic acids found in living organisms (the other one is deoxyribonucleic acid—DNA). An RNA molecule represents a long chain of monomers called nucleotides. RNAs contain four different nucleotides, adenine (A), guanine (G), cytosine (C), and uracil (U). RNA differs from DNA by being single stranded and by having the nucleic acid Uracil (U) replace thymine.

The sequence of nucleotides of an RNA molecule constitutes its primary structure, and the pattern of pairing between nucleotides determines the secondary structure of

* Corresponding author.

D.-S. Huang, K. Li, and G.W. Irwin (Eds.): ICIC 2006, LNBI 4115, pp. 303–311, 2006.

RNA. In other words primary structure is the linear sequence of nucleotides and secondary structure refers to strong intramolecular contacts between pairs of non-adjacent nucleotides, each nucleotide being paired with either none or one other nucleotide.

The most stable pairs form between GC, AU, and GU and their mirrors, CG, UA, and UG. Their stability also makes them the most common pairs. These pairs are called canonical base pairs. One of property of secondary structure is that pairs tend to form in groups yielding higher order structure. Common RNA substructures are hairpin loops, internal loops and bulges, multi-branch loops, and dangling ends. Forming adjacent pairs tends to increase the stability. These stacked pairs are also called stems, or helices.

RNA serves several functions: coding, catalysis, structure. Messenger RNA (mRNA) carries to the ribosome the coding of a gene with the introns spliced out. The ribosome translates the mRNA codon by codon to a string of amino acids. Each amino acid is brought to the ribosome by a transfer RNA (tRNA) that corresponds to the codon currently being translated. The ribosome itself is a complex consisting of proteins and ribosomal RNA (rRNA). The function of the RNA within the cell is determined in large part by the three-dimensional structure of the RNA molecule, when it folds. In turn, the three-dimensional structure is partly determined by the secondary structure of the molecule.

Determining the secondary structure of an RNA molecule is an integral part of understanding the function of the RNA molecules. Methods for determining the structure of an RNA molecule in a biology lab are expensive, and so a computational means for reliably predicting the secondary structure directly from the base sequence would be very valuable.

Thus, the knowledge of secondary and tertiary structures of an RNA molecule is highly desirable when investigating its role in a cell. Since experimental determination of RNA structure is time-consuming and expensive, its computational prediction is of great interest, and some efficient solutions are known.

2 Related Work

Conventionally, most methods developed so far for predicting the secondary structure of RNAs might be roughly classified into two categories: energy minimization and phylogenetic comparisons.[1,2,3]

The energy minimization technique comprises combinatorial and dynamic programming approaches, and is based on computation of the lowest free energy structure(s) of a sequence. [3,4]

The combinatorial approach first generates all the possible helices of a sequence; and in a second step a branch and bound algorithm combines compatible helices until optimal or suboptimal structures are formed [5,6]. This technique, that exhaustively search the solution space, cannot deal with sequences much longer than 200 nucleotides.

Dynamic programming approach computes the lowest free energy structure by mean of an energy base-pairing optimization based on a recursive relation between the best structures of length k and the best structures of length k-1 [7,8,9,10,11].

Dynamic programming allows to treat sequences containing up to 2000 nucleotides, however such methods method neither consider pseudo-knots nor find sub-optimal solutions.

A more recent method partially solves the last problem [12]. The phylogenetic methods use covariation analysis to identify conserved paired bases among a set of homologous sequences. This is a satisfying procedure that gives excellent results, including pseudo-knots identification. [2,3,11,13]

However the procedure requires a prior alignment of sequences and multiple alignment is, in turn, a difficult problem. One must also quote miscellaneous methods based either on parallel algorithms, or formal grammar, graph theory and simulation of the RNA folding process.

Recent methods intend both to align RNA sequences and to predict their consensus secondary structure[14,15,16,17,18].

Finally few recent methods have been proposed in order to predict the secondary structure, possibly included pseudoknots, of a single RNA sequence:

Rivas and Eddy [1,4,19,20] described a dynamic programming algorithm for predicting optimal RNA secondary structure, including pseudoknots. The algorithm has a worst case complexity of $O(N^6)$ in time and $O(N^4)$ in storage. They presented an implementation of the algorithm that generates the optimal minimum energy structure for a single RNA sequence, using standard RNA folding thermodynamic parameters augmented by a few parameters describing the thermodynamic stability of pseudoknots.

Akutsu[1] and Uemura[22] also describe dynamic programming algorithms that can handle a more general class of structures than Dirks and Pierce [23], but uses a more simplistic energy model.

Dirks and Pierce [23] described dynamic programming algorithm for computing the minimum energy structure over a more restricted class of secondary structures. This algorithm uses $O(n^5)$ time and $O(n^4)$ space.

3 Preliminaries

3.1 RNA Secondary Structure

A secondary structure S is formally defined as the set of all base pairs (i, j) with $i < j$ such that for any two base pairs (i, j) and (k, l) with $i \leq k$ the two following conditions:

1. $i = k$ if and only if $j = l$.
2. There are no knots or pseudo knots allowed. For any two base pairs (i, j) and (k, l) the condition $i < k < l < j$ or $k < i < j < l$ must be satisfied.

3.2 RNA Secondary Structure with Pseudoknots

A folding of RNA secondary structure such as that shown in Figure 2 is a pseudoknotted structure.

To define a pseudoknot, let us consider a RNA sequence which consists of bases with index 0 to n. We also define a base-pair to be (i, j) if a base with index i form a pair with another base with index j in the same sequence. A pseudoknot is then defined as two base-pairs (i, j) and (i', j') such that $0 < i < i' < j < j' < n$ where 0 and n

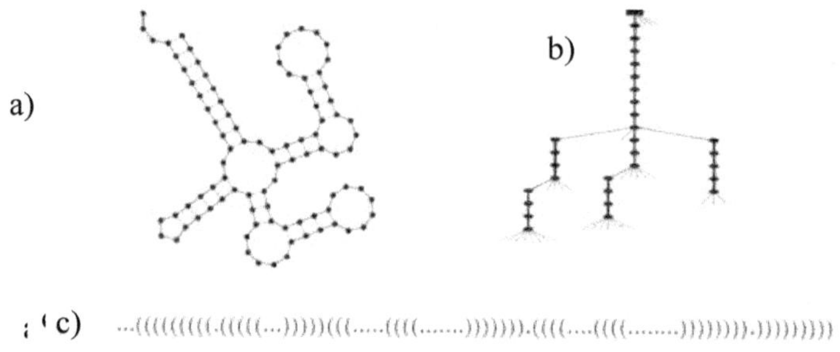

Fig. 1. Different representations of one RNA secondary structure. Notation a) shows the secondary structure as a planar graph, b) is the corresponding tree, and c) gives the linear string encoding for the structure. There are several other representations are well known including mountain plot, circle and arc representation.

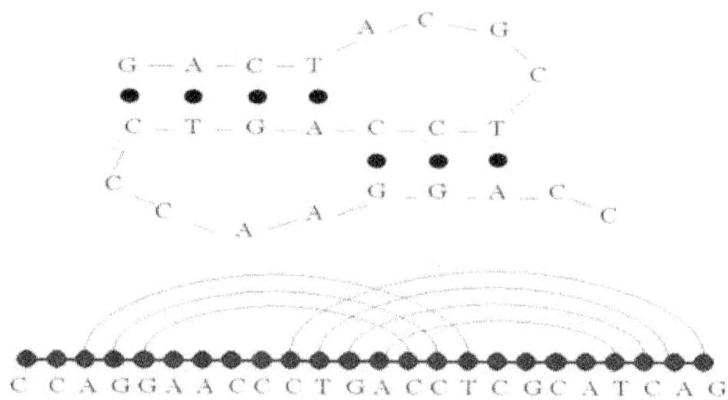

Fig. 2. Pseudoknotted RNA secondary structure

are the indices to the two ends of the RNA. In other words, if the base-pairs overlap each other, we have a pseudoknot as shown in Figure 2. In an arc diagram of a pseudoknotted secondary structure, at least one arc crosses another arc in the structure.

Pseudoknots can be classified into 3 main types:

- I-type pseudoknot (interior loop)
- B-type pseudoknot (bulge loop)
- H-type pseudoknot (hairpin loop)

H-type pseudoknots are the most abundant of all known pseudoknots, furthermore they are the only type of pseudoknot for which an energy model exists.

4 Proposed Algorithm

An RNA secondary structure prediction with simple pseudoknots which has the maximum number of base pairs can be computed by the dynamic programming algorithm in $O(N^4)$ time and $O(N^3)$ space.

Generally all dynamic programming algorithms have to pass three stages:

- Initialization of the matrices
- Matrix filling - scoring
- Traceback/Backtracking

Depending on the scoring function in step two several problems with different time and space complexity can be solved for a given RNA sequence.

Our approach is based on combining a dynamic programming formulation for the basic version of RNA secondary structure prediction with simple pseudoknots presented in [21] with earlier Nussinov's [8] RNA secondary structure prediction method.

We used Akutsu's algorithm to calculate Optimal score matrix and score matrix with Pseudoknotted case. In order to describe the pseudoknot algorithm we need to introduce three three-dimensional (N × N × N) matrices, to be called Sl, Sm, and Sr and one triangular N × N matrix to be called PS.

For finding a simple pseudoknot substructure whose endpoints are Ith and Kth residues, the algorithm considers three types of triplets Sl(i, j, k), Sm(i, j, k), and Sr(i, j, k) for each i, j, and k such that $(I \le i < j < k \le K)$.

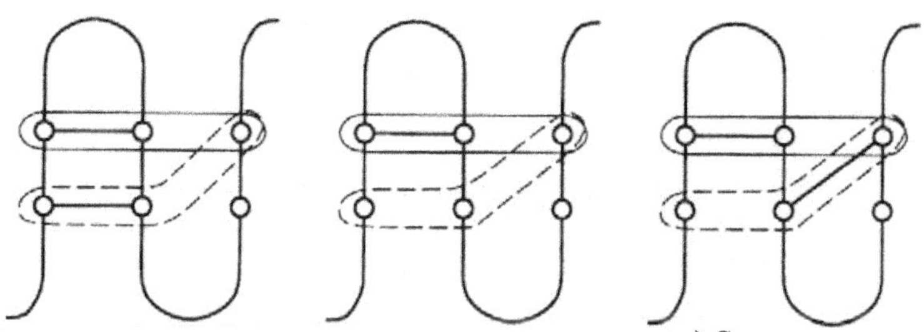

Fig. 3. Illustration of the recurrence used in the dynamic programming procedure

These matrices are defined in the following way: Sl(i, j, k) is the score of the best folding between positions I and i, and j and k, provided that ith and jth residues make a base pair; Sr(i, j, k) is the score of the best folding between positions I and i, and j and k, provided that jth and kth residues make a base pair; Sm(i, j, k) is the score of the best folding between positions I and i, and j and k, provided that neither i pairs with j, no j pairs with k. Sp(i, j) is the score of the best pseudoknot fold with ending points i and j. Recurrence relations for computing these triplet can be found in (Akutsu 2000). For each pair (I,K) we compute the above three matrices and obtain the score of a pseudoknot Sp(I,K) by

$$Sp(I,K) = \max_{I<i<K} \begin{cases} SL(i, i+1, K) \\ SM(i, i+1, K) \\ SR(i, i+1, K) \end{cases} \qquad (1)$$

Finally, the optimal score for each pair (i, j) can be computed by the following recurrence:

$$S(i,j)=\max\begin{cases} PS(i,j) \\ \mu(i,j)+S(i+1,j-1) \\ \max\limits_{i\le m<j}\{\,S(i,m)+S(m+1,j)\,\} \end{cases} \tag{2}$$

One drawback of the original Akutsu's algorithm is, it considers only regular base pairs. RNA base pairs are the canonical Watson-Crick A·U and G·C pairs. Crick proposed, after examining how tRNAs might recognize the genetic code, that G·U is also a valid base pair in RNA secondary structure [24]. The scoring table used by the Akutsu's dynamic programming algorithm, however, only counts regular base pairs and equate A-U with G-C.

In our RNAStruct software we compensate above problem to get more good result when predicting RNA secondary structures. Our sub function basepair begins with checking regular base pairs and it returns true if i-th element paired with j-th element, also it checks given two elements is paired irregularly (G-U). Otherwise our function returns NULL value.

Our matrix for score of pseudoknot Sp and matrix for optimal score S are logical matrix defined by define preprocessor command and originally we are using P pointer with integer type to save space.

Akutsu only described the procedures for computing scores or free energies in [21], but didn't described how to trace back that the optimal scoring matrix S and, pseudoknot Sp matrix.

Trace back method of Nussinov's algorithm is valid for only pseudoknot free structures of RNA. We considered three types of triplets Sl(i, j, k), Sm(i, j, k), and Sr(i, j, k) when predict simple pseudoknotted RNA secondary structure. Our recursive function traceback computes traceback starting from position (i, j) and it prints matched pairs to output.txt file when it found.

Our program input is simple pseudoknotted RNA sequence without any description and comment.

And our program **RNAStruct** outputs Optimal scoring matrix S, Score of Pseudoknot Sp and matched pairs.

5 Experimental Review

5.1 Experimental Settings and Results

We predicted the structure of 15 pseudoknotted RNA sequences taken from PseudoBase. PseudoBase is a database containing structural, functional and sequence data related to RNA pseudoknots. It can be reached by its central page at http://wwwbio.LeidenUniv.nl/~Batenburg/PKB.html. From here one can retrieve pseudoknot data as well as submit data for pseudoknots that are not yet in the database.

Average lengths of our experimental sequences taken from PseudoBase were around 20 nucleotides.

To evaluate our software we compared our result to that of the currently best known software for globally optimal RNA pseudoknot prediction developed by Elena Rivas [19]. That algorithm has a worst case complexity of $O(N^6)$ in time and $O(N^4)$ in storage.

Table 1. Some output of our RNAStruct program, PKB # is the Sequence ID on PeudoBase database, PBase is denotes the bracket view of structure on PseudoBase

1.	PKB #	PKB86
	Sequence	UAGGGGCUUACCGAAAUAAGCC
	PBase	:(((:[[[[[))) : : :]]]]]:
	Prediction	::((:[[[[[)): : : :]]]]]:
2.	PKB #	PKB96
	Sequence	CCGUGGCGAGUACGAUAACUCGUA
	PBase	: (((:[[[[[[))) :::]]]]]]:
	Prediction	: :((::[[[[[))::::]]]]]::
3.	PKB #	PKB63
	Sequence	CGUGGUGCAUACGAUAAUGCAU
	PBase	(((:[[[[[[))):::]]]]]]
	Prediction	::(:[[[[[[::):::]]]]]]
4.	PKB #	PKB102
	Sequence	GCGUAUCUCUUAUGUCAACAGAGA
	PBase	:((((:[[[[)))) :::::]]]] :
	Prediction	:((((::::[]))) ::::: :::) :

Our approach is based on pseudoknot technique for maximizing the number of base pairs proposed by Akutsu[21]. The algorithm has worst case time and space complexities of $O(N^4)$ and $O(N^3)$, respectively.

RNAStruct folds 14 pseudoknots out of 15 in the Pseudobase database. Our program predicts simple pseudoknots with correct or almost correct structure for 53% sequences. From 130 pairs which are included in our total 15 pseudoknotted sequences, we predicted total 62 base pairs.

6 Conclusion

We presented an implementation of RNA secondary structure prediction with simple pseudoknots based on dynamic programming algorithm. Our experimental results indicate that our approach works well.

But due to the space complexity of dynamic programming technique for maximizing the number of base pairs, our algorithm cannot deal with sequences more than 100 nucleotides.

In the future work we will consider graphical representation of RNA secondary structure prediction and we will check the possibility of new solution for predicting more long sequences (we will consider space and time complexity) of pseudoknotted RNA secondary structure.

Acknowledgement

This work was supported by the Regional Research Centers Program of Ministry of Education & Human Resources Development and by Korea Science and Engineering Foundation (# 1999-2-303-006-3) in Korea.

References

1. Eddy, S. R.: What is Dynamic Programming? Nature BioTechnology Vol. 22. Number 7. Nature Publishing Group, November, July (2004)
2. Robin, D. D.: Prepared under the Direction of Sean R. Eddy.: RNA Structural Alignment Using Stochastic Context-free Grammars. Ph.D thesis presented to the Sever Institute of Washington University, December, (2004)
3. Stephen McCauley advised by Ian Holmes.: An Analysis of the Relative Efficacy of the Nussinov-Felsenstein, and the Knudsen-Hein, RNA Secondary Structure Prediction, Ph.D thesis presented in October 6, (2003)
4. Eddy, S. R.: How do RNA Folding Algorithms Works? Nature BioTechnology Vol. 22. Nature Publishing Group (2004)
5. Pipas, J., McMahon, J.: A Method for Predicting RNA Secondary Structure. Proc Natl Acad Sci USA., Vol. 72. (1975) 2017-2021
6. Sankoff, D.: Simultaneous Solution of the RNA Folding, Alignment, and Protosequence Problems. SIAM J Appl Math Vol. 45. (1985), 810-825
7. Nussinov, R.P., Eddy, S.R., Griggs, J. R., Kleitman, D. J.: Algorithms for Loop Matching. SIAM Journal of Applied Mathematics, Vol. 35. (1978) 68–82
8. Nussinov, R., Jacobson, A.: Fast Algorithm for Predicting the Secondary Structure of Single-stranded RNA. Proc Natl Acad Sci USA,Vol. 77.(1980) 6309-6313
9. Waterman, M.S., Smith, T.F.: RNA Secondary Structure: A Complete Mathematical Analysis, Mathematical Bioscience, Vol. 42,(1978) 257-266
10. Zuker, M., Stiegler, P.: Optimal Computer Folding of Large RNA Sequences Using Thermodynamics and Auxiliary Information. Nucleic Acids Res.,Vol. 9. No. 133. (1981)
11. Zuker, M., Sankoff, D.: RNA Secondary Structures and Their Prediction, Mathematical Bioscience, Vol. 46, (1984) 591-621
12. Zuker, M.: On Finding All Suboptimal Foldings of an RNA Molecule, Science, Vol. 244. (1989) 48-52
13. Zuker, M., Mathews, D. H., Turner, D. H.: Algorithms and Thermodynamics for RNA Secondary Structure Prediction: A Practical Guide in RNA Biochemistry and Biotechnology, ser. NATO ASI Series, J. Barciszewski and B. Clark, Eds. Kluwer Academic Publishers (1999)
14. Eddy, S.R., Durbin R.: RNA Sequence Analysis Using Covariance Models. Nucl Acids Res Vol. 22. (1994), 2079-2088
15. Gorodkin, J., Heyer, L.J., Stormo, G.D.: Finding the Most Significant Common Sequence and Structure Motifs in a set of RNA Sequences. Nucl Acids Res Vol. 25. (1997) 3724-3732
16. Samuel, I., Ming-Yang, K.: Predicting RNA Secondary Structures with Arbitrary Pseudoknots by Maximizing the Number of Stacking Pairs, Journal of Computational biology, Vol. 10. No 6. (2003) 981-995

17. Tabaska, J. and Stormo, G.: Automated Alignment of RNA Sequences to Pseudoknotted Structures, Fifth International Conference on Intelligent Systems for Molecular Biology, The AAAI Press, Menlo Park, California (USA), (1997) 311-318
18. Tabaska, J.E., Cary, R.B., Gabow, H.N., Stormo, G.D.: An RNA Folding Method Capable of Identifying Pseudoknots and Base Triples. Bioinformatics Vol. 14. (1998), 691-699
19. Rivas, E., Eddy, S.: A Dynamic Programming Algorithm for RNA Structure Prediction Including Pseudoknots, Journal of Molecular Biology, Vol. 285. (1999) 285, 2053
20. Rivas,E., Sean R.Eddy.: Noncoding RNA Gene Detection Using Comparative Sequence Analysis, BioMedCentral 2:8, Bioinformatics (2001)
21. Akutsu T.: Dynamic Programming Algorithm for RNA Secondary Structure Prediction with Pseudoknots, Discrete Appl. Math, Vol. 104. (2000) 45-62
22. Uemura, Y., Hasegawa, A., Kobayashi, S., Yokomori, T.: Tree Adjoining Grammars for RNA Structure Prediction. Theoretical Computer Science Vol. 210. (1999) 277–303
23. Dirks, R. M., Pierce, N. A.: A Partition Function Algorithm for Nucleic Acid Secondary Structure Including Pseudoknots, J. Comput. Chem.,Vol. 24. (2003) 1664-1677
24. Crick, F.H.: Codon–anticodon Pairing: The Wobble Hypothesis. J. Mol.Biol. Vol. 19. (1996) 548–55

Template Frame for DNA Computing

Wenbin Liu[1,2] and Xiangou Zhu[1]

[1] School of Computer Science and Engineering, Wenzhou University, Wenzhou City 325027, China
[2] Department of Control Science and Engineering, Huazhong University of Science and Technology, Wuhan City 430074, China
wbliu6910@126.com

Abstract. The encoding problem is very crucial for the information processing based on DNA computing. The main difficulty lies in that it is hard to control the shift hybridization between DNA sequences. In this paper, a new data structure, named Template Frame, is proposed based on the template strategy. As the Template Frame provides an effective tool to map the various concatenations of code sequences with the template set, thus it can be used to tackle the shift distance problem.

1 Introduction

DNA computing is the first paradigm trying to exploits the parallelism inherent in biochemical reactions. As information is encoded with specific DNA sequences, its retrieval is mainly accomplished through specific hybridization between complementary sequences [1]. Unfortunately, hybridization between DNA sequences may take place even in imperfect matching due to the false positive and the false negative. Up to now, most efforts have been focused on designing non-interacting code set by some combinatorial constraints [2][3][4] [5][6][7][8]. Though this could help to prevent some undesired hybridizations, but they are not capable of taking into account the overlapping region between two concatenated code sequences. For example, assuming two code sequences s_1 and s_2 hold enough unlikeness, the concatenation of them, s_1s_2, may contain some subsequence s which still keeps a high similarity with one of the code sequences. In this case, the complementary of some sequence s_i (i=1,2) may hybrid with s instead of its correspondent counterpart. As a matter of fact, this is the reason that makes the encoding problem hard to deal with in DNA based computing.

The template strategy proposed by Frutos et al. intends to reduce the encoding problem in alphabet $\Sigma_{DNA}=\{A,G,C,T\}$ to two sub-problems in alphabet $\Sigma_{01}=\{0,1\}$ [10]. Thus it is convenient for us to take further theoretical analysis on the encoding problem. In this paper, a new data structure, named Template Frame, is proposed based on the template strategy to tackle the shift hybridization in DNA computing. In section 2, we briefly review the template strategy, and then give the definition of the

D.-S. Huang, K. Li, and G.W. Irwin (Eds.): ICIC 2006, LNBI 4115, pp. 312–319, 2006.

template frame and some theoretical analysis. Followed by, in section 3, we present some calculation results based on several template sets and discuss the optimization of template frame. Finally, we conclude in section 4 with some discussions and some future studying directions.

2 The Template Frame

2.1 The Template Strategy

In template strategy, DNA sequence is produced as the following steps: First, positions for [AT] and [GC] are determined by a template string $t_i = t_{i1}t_{i2}\cdots t_{in}$ ($t_{ij} \in \{0,1\}$), where 1 indicates [AT] and 0 indicates [GC]. Second, either A or T is chosen for template positions $t_{ij} = 1$, and either G or C for positions $t_{ij} = 0$ ($1 \le j \le n$) by a map string $m_i = m_{i1}m_{i2}\cdots m_{in}$ ($m_{ij} \in \{0,1\}$). Thus the production rule can be formulated as: "A" is derived from 1×0, "T" from 1×1, "C" from 0×0, and "G" from 0×1. Fig.1 shows the production of a DNA sequence through its template strand and map strand.

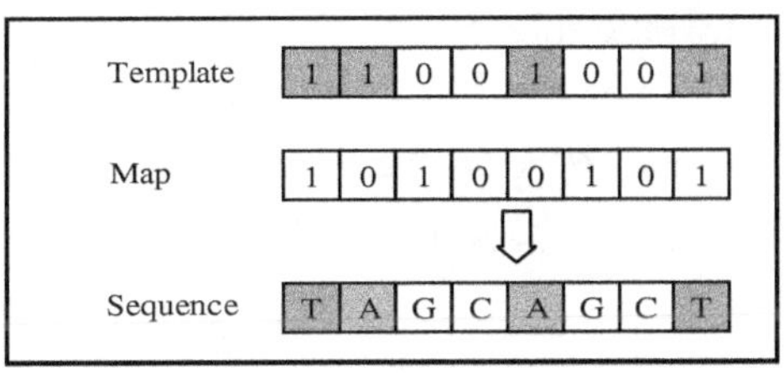

Fig. 1. A 8-mer sequence produced by a template strand and a map strand

As the likelihood of hybridization between DNA sequences decreases as the mismatches between them increases, it is usually required that the code set S should hold the following conditions:

1. In order to keep similar binding strength for all perfected matches, the GC content of the code sequences is fixed at around 50%;
2. The Hamming distance $H(s_i, s_j) \ge d$ ($s_i, s_j \in S$);
3. The Hamming distance $H(s_i^r, s_j^c) \ge d$ ($s_i, s_j \in S$);

where r denotes the reversal of a sequence, c denotes its Waston-Crick (in short WC) complement, and d is an integer equals to $\lfloor n/2 \rfloor$ ($\lfloor * \rfloor$ denotes the floor integer of a real number). The production of the code set S can be denoted as:

$$T \times M \rightarrow S \tag{1}$$

To get a further insight on the template strategy, the production of a code sequence s, its reversal s^r and its WC complementary s^c can be written as:

$$t \times m \rightarrow s \;\; (5' \rightarrow 3') \tag{2}$$

$$t^r \times m^{rc} \rightarrow s^c \;\; (3' \rightarrow 5') \tag{3}$$

$$t^r \times m^r \rightarrow s^r \;\; (3' \rightarrow 5') \tag{4}$$

Because of the symmetry strings appeared in the template set T and the map set M, we decompose them as:

$$T = T_n \cup T_s \tag{5}$$

$$M = M_n \cup M_s \tag{6}$$

where T_n and T_s represent the non-palindrome subset and the palindrome subset respectively, and M_n and M_s are the corresponding map subset to them respectively.

Lemma 2.1 (**the non-palindrome subset**)
For any two template strings $t_i, t_j \in T_n$ and any two map strings $m_p, m_q \in M_n$, if $H(t_i, t_j) \geq d$, $H(t_i, t_j^r) \geq d$, and $H(m_p, m_q) \geq d$, then all the DNA sequences $s \in S_n$, produced by $T_n \times M_n \rightarrow S_n$, will satisfy both condition 2 and 3 simultaneously.

Lemma 2.2 (**the palindrome subset**)
For any two template strings $t_i, t_j \in T_s$ and any two map strings $m_p, m_q \in M_s$, if $H(t_i, t_j) \geq d$, $H(m_p, m_q) \geq d$, and $H(m_p, m_q^c) \geq d$, then all the DNA sequences $s \in S_s$, produced by $T_s \times M_s \rightarrow S_s$, will satisfy both condition 2 and 3 simultaneously.

Theorem 2.1
For any two template strings $t_i \in T_n$ and $t_j \in T_s$, if $H(t_i, t_j) \geq d$, then the DNA sequences $S = S_n \cup S_s$ will satisfy both condition 2 and 3 simultaneously.

Note: constraint $H(t_i, t_j^r) \geq d$ in lemma 2.1 also holds for a template string itself and $H(m_i, m_j^c) \geq d$ in lemma 2.2 also holds for a map string itself.

Then the basic idea of the template strategy can be formulated as: if the binary strings both in the template set T and the map set M hold some distance measure, then the DNA code sequences produced by them must hold some distance measures. Recently, we introduced the H–measure, proposed by Garzon [4], to the template strategy, and our result showed that it could be used to optimize the quality of sequences significantly. Detailed descriptions are recommended to refer [10][11].

2.2 The Template Frame

Uunder the template strategy, our previous work shows that the template set T plays a dominant role on the prosperities of the code set S. If we can find some data structure of the template set T to map it with the various concatenations of the code sequences, then the shift hybridization problem can be solved.

Definition 2.2.1
For a template set $T=\{t_1,t_2,\cdots t_m\}$, where t_i ($1\le i\le m$) is a binary string with length n. A template frame P of T is defined as a permutation of the m template strings

$$P=t_{i_1}t_{i_2}\cdots t_{i_m}t_{i_1}\quad(1\le i_1,i_2,\cdots,i_m\le m) \tag{7}$$

As the first template sequence appears twice in a template frame P, then its length becomes $|P|=(m+1)n$ ($|*|$ represent the length of a binary string). Obviously, there are $m!$ template frames in total for template set T.

In order to count the mismatches of two strings in different length, Arita introduced a notation $H_M(y,x)$ ($|y|\ge|x|$), named the mass of string y to string x, which is defined as the minimal Hamming distance between $|y|-|x|+1$ substrings of length $|x|$ in string y and string x [8]. In the following, this definition of mass is also adopted.

Definition 2.2.2
To a template string t, the mass of template frame P is defined as:

$$\|P\|_t=\begin{cases}H_M(P',t) & t=t_{i_1}\\ \min[H_M(P_1',t),H_M(P_2',t)] & t\ne t_{i_1}\end{cases} \tag{8}$$

where P' is the substring of P without the terminal letters at either end; P_1' is the substring obtained by deleting the substring started at the last letter of t; P_2' is the substring obtained by deleting the substring ended at the first letter of t (see Figure 2).

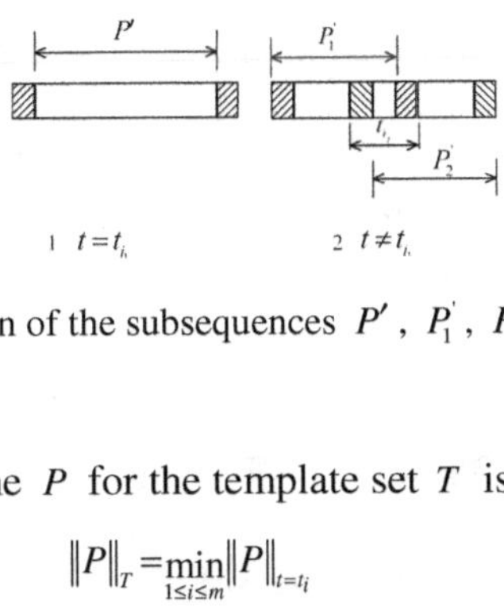

Fig. 2. A demonstration of the subsequences P', P_1', P_2' in template frame P

Definition 2.2.3
The mass of a template frame P for the template set T is defined as:

$$\|P\|_T=\min_{1\le i\le m}\|P\|_{t=t_i} \tag{9}$$

Definition 2.2.4
The average mass of a template frame P for the template set T is defined as:

$$Q=\sum_{i=1}^{m}\|P\|_{t_i}/m \tag{10}$$

Given a code set S and its template set $T=\{t_1,t_2,\cdots t_m\}$, strand $W=W_1W_2\cdots W_k$ ($k\ge1$) is encoded under template frame $P=t_{i_1}t_{i_2}\cdots t_{i_m}t_{i_1}$ ($1\le i_1,i_2,\cdots,i_m\le m$) if each of its subsequence

$W_i = s_{i_1} s_{i_2} \cdots s_{i_m}$ ($1 \le i \le k$) is composed of m code sequences whose template strings are $t_{i_1}, t_{i_2}, \cdots t_{i_m}$ accordingly, and any code sequence $s \in S$ is used at most once in .

Theorem 2.2
Under template strategy, if strand W is encoded under template frame P, then $\|W\|_s \ge \|P\|_T$ holds. (proof is omitted)

Proof
For strand W encoded under template frame P, we can rewrite W and its corresponding template string $(T)_W$ as follows:

$$W = s_{1_1} s_{1_2} \cdot s_{1_j} \cdot s_{1_m} s_{2_1} s_{2_2} \cdot s_{2_j} \cdot s_{2_m} \cdots s_{k_1} s_{k_2} \cdot s_{k_j} \cdot s_{k_m} \tag{11}$$

$$(T)_W = \underbrace{t_{i_1} t_{i_2} \cdot t_{i_j} \cdot t_{i_m}}_{1} \underbrace{t_{i_1} t_{i_2} \cdot t_{i_j} \cdot t_{i_m}}_{2} \cdots \underbrace{t_{i_1} t_{i_2} \cdot t_{i_j} \cdot t_{i_m}}_{k} \tag{12}$$

Under template strategy, it is easy to see that $\|P\|_T \le d$ always holds. Considering a code sequence $s \in S$, there are only two situations about its relation with strand W:

1. The code sequence s is not appeared in W, then $\|W\|_s = H_M(W, s)$. No loss of generality, we can assume its template string $t = t_{i_j}$. For clarity, we divide the subsequence in W into two groups: those share the same template string with t and the others. For the first case, the Hamming distance $H(s_{i_j}, s) \ge d \ge \|P\|_T$ ($1 \le i \le k$). For the second case, we have $\|(T)_W\|_t = \|P\|_t \ge \|P\|_T$ by the equation (8), and then $H(s', s) \ge \|P\|_T$ (s' is other subsequences in W where $s' \ne s_{i_j}$). Then we have $\|W\|_s = H_M(W, s) \ge \|P\|_T$.
2. The code sequence s is appeared in a subsequence W_i ($1 \le i \le k$). In this case, we have to calculate the mass of two substrands in W as in Figure 2.. By same method in used in 1, we can also reach $\|W\|_s \ge \|P\|_T$. Then theorem 2.2 is proved.

3 Optimization of the Template Frame

Theorem 2.2 shows that if the possible solution strands in DNA computing are encoded under a template frame P as that in W, then the Hamming distance between the code sequences and other subsequence in each solution strand can be held to be no less than $\|P\|_T$ (except for the correspondent places for each code). Therefore it provides a useful data structure to improve the reliability of DNA computing. Now the problem is how to find the template frame with the largest mass $\|P_{best}\|_T$ and average mass Q_{best}.

Table 1. The maximal mass $\|P\|_T$ and mQ for different template sets with code length $n = 8, 12$

	8					12		
Label	T1	T2	T3	T4	T5	T1	T2	T3
0011	3(31)	4(32)	2(28)	3(30)	2(27)	4(38)	4(41)	4(39)
1100	3(31)	4(32)	2(28)	3(30)	2(27)	4(43)	4(43)	4(36)
0101	3(28)	3(31)	3(29)	3(30)	3(30)	4(39)	4(40)	2(36)
1010	3(28)	3(31)	3(29)	3(30)	3(30)	4(36)	3(35)	2(36)
0110	2(28)	0(28)	0(27)	0(24)	2(28)	4(39)	2(30)	4(38)
1001	2(26)	0(28)	0(27)	0(24)	2(28)	4(42)	4(38)	4(40)
00011						4(44)	4(43)	4(44)
11000						5(46)	4(45)	3(37)
01100						4(44)	4(39)	4(44)
00110						4(44)	4(42)	5(44)
10001						4(44)	4(38)	4(36)
none	1(12)	1(11)	1(10)	0(9)	0(8)	2(19)	2(22)	2(21)
Average H-measure	3.6	3.5	3.4	3.3	3.2	5	5.03	4.9

Table 1 shows the largest mass $\|P\|_T$ and its corresponding mQ (in parentheses) for several template sets with code length 8 and 12. The results show:

1. The addition of suitable labels can greatly improve the quality of a template frame. The largest mass $\|P\|_T$ and mQ are 1(12) and 2(22) respectively as the template strings are directly concatenated. In [10], the authors added a pair of labels at both ends of the code sequences to reduce the slide match. Here we introduce it to the template frame, the calculation results show that the property of a template frame can be improved significantly by using suitable labels, and the largest mass $\|P\|_T$ and mQ are 4(32) and 5(46) respectively. In order to keep the uniform of GC content, we require the labels should also have a "01" content about 50%.
2. The average H-measure has a great influence on the property of a template frame. Generally speaking, if a template set has a larger average H-measure, it tends to have a larger mass $\|P\|_T$ and parameter mQ.
3. The increase of the labels' length can also improve the mass $\|P\|_T$ and mQ of a template frame. The drawback of increasing the label length is that it also leads to the increase of the code length. So we should take a trade off between the quality of a template frame and the code length in practice. A parameter σ can be used to take an evaluation on this aspect:

$$\sigma = \frac{\|P_l\|_T}{n+l} \Big/ \frac{\|P\|_T}{n} = \frac{n\|P_l\|_T}{(n+l)\|P\|_T} \tag{13}$$

where P_l denotes the template frame with labels, and l denotes the length of it. For example in code length n=8,12, parameter σ can be calculated respectively as: $\sigma_8 = \frac{n\|P_l\|_T}{(n+l)\|P_l\|_T} = \frac{8\times 4}{(8+4)\times 1} = \frac{8}{3}$ and $\sigma_{12} = \frac{n\|P_l\|_T}{(n+l)\|P_l\|_T} = \frac{12\times 5}{(12+5)\times 2} = \frac{30}{17}$. By its definition, it is easy to see that σ is actually a ratio of the relative mass for template frame P. Theoretically, the length of labels can be increased until the maximal σ is achieved.

Under template strategy, we can get lots of template sets satisfying the constraints in section 2.1 through a random search algorithm. So it is a hard problem for us to find a template set that has the largest mass $\|P\|_T$ and largest average mass Q. However we find a template with relative good property by two steps:

1. Find the template sets with a larger average H-measure;
2. As the size of the template set T is relatively small, take an exhaustive search for the $m!$ template frames for each template set with different labels.

4 Conclusions

One of the challenges in the encoding problem of DNA computing is the shift hybridization between DNA sequences. In order to tackle this problem, a new concept– Template Frame is proposed. As the template frame provides a useful data structure that set up a map between the combination of code sequences and its template set, then the shift hybridization in DNA computing can be effectively avoid by a suitable template frame. The template frame proposed by this paper can be easily adopted to models which has no special requirement on the arrangement of the code sequences, such as the DNA algorithms for the Graph Coloring Problem, the Minimal Set Covering Problem, the SAT problem, the construction of a DNA database proposed by Reif [12], and the design of DNA chips.

Although our result is exciting, there are still some open problem needed further research: First, theoretical analysis of the mass of the template frame and its optimization; Second, how to apply the template frame in those algorithms where the possible solution space are concatenated according to some constraints, for example the Directed Hamiltonian Path Problem in Adleman's paper[1].

Acknowledgment

This paper is supported by the NSFC under No.60403002 and 30570431, the CPSF under No.2004036130, the ZJNSF under No.Y106654 & Y405553, Excellent Youth Science and Technology Foundation of Anhui Province of China under No. 06042088.

References

1. Adleman. L.: Molecular Computations to Combinatorial Problems. Science, 266(1994) 1021-1024
2. Deaton, R. et al.: Genetic Search of Reliable Encodings for DNA-Based Computation. 1st Genetic Programming Conference, Stanford University, (1996) 9-15

3. Deaton, R. et al.: Good Encodings for DNA-Based Solutions to Combinatorial Problems. DIMACS Series in Discrete Mathematics and Theoretical Computer Science, 44(1999)247-258
4. Garzon, M. et al.: A New Metric for DNA Computing. Proceedings of the 2nd Annual Genetic Programming Conference,42(1997)472-487
5. Marathe, A. et al.: On Combinatorial DNA Code Design. DIMACS Series in Discrete Mathematics and Theoretical Computer Science, 44(1999):75-87
6. Li, M. et al.: DNA Word Design Strategy for Creating Sets of Noninteracting Oligonucleotides for DNA Microarrays. Langmuir,18(2002)805-812
7. Arita, M. et al.: Improving Sequence Design for DNA computing. Proceedings of the Genetic and Evolutionary Computation Conference, Las Vegas(2000) 875-882
8. Arita, M., Kobayashi, S.: DNA Sequence Design Using Template. *New Generation Comput.* 20(2002): 263-277
9. Arita, M., Kobayashi, S.: The Power of Sequence Design in DNA Computing. 4th International Conference on Computational Intelligence and Multimedia Applications, (2001)163-166
10. Frutos, A. G. et al.: Demonstration of a Word Design Strategy for DNA Computing on Surface. *Nucleic Acids Res.*, 25(1997)4748-4757
11. Liu, W. et al.: DNA Sequence Design Based on Template Strategy. J. Chem. Inf. Comput. Sci, 43(2003)2014－2018
12. Reif J.H. et al.: Experimental Construction of Very Large Scale DNA Databases with Associative Search Capability. Lecture Notes in Computer Science, Springer(2002) 231-247

A New DNA-Based Approach to Solve the Maximum Weight Clique Problem*

Aili Han[1,2] and Daming Zhu[2]

[1] Department of Computer Science and Technology, Shandong University, Weihai 264209, China
[2] School of Computer Science and Technology, Shandong University, Jinan 250061, China
hanal@sdu.edu.cn

Abstract. Given an undirected graph with weights on the vertices, the maximum weight clique problem is to find a subset of mutually adjacent vertices, i.e. a clique, which have the largest total weight. We devised a new *DNA* encoding method to solve the maximum weight clique problem whose basic idea is that each vertex on weighted graph is encoded by two *DNA* strands of different length and each edge is encoded by one *DNA* strand with a length of 20. The longer *DNA* strand corresponding to vertex v_i consists of three parts and its center part is with a length of w_i; the shorter *DNA* strand is the reverse complementation of the longer one's center part. We also gave the correspond- ing molecule algorithm and its biological implementation. The proposed *DNA* computing method can be expanded to solve other *NP*-hard problems, and it provides further evidence for the ability of *DNA* computing to solve numerical optimization problems.

1 Introduction

Based on the massive parallelism of *DNA* computing, many researchers tried to solve a host of difficult problems, especially *NP* problems. In 1994, Adleman[1] solved a 7-vertex instance of the *Hamiltonian path* problem by means of the molecular biology techniques, which was published in *Science*. The creative research opened up a new way to computation using *DNA* molecules.

A major goal of subsequent research in this area is to understand how to solve *NP*-complete problems. To address this goal, Lipton[2] abstracted a parallel model of molecular computation in 1995 on the basis of Adleman's experiment, and used it to solve the 3-*SAT* problem; Ouyang et al[3] solved the *maximal clique* problem(MCP) in 1997; Head et al[4] solved the *maximal independent set* problem using operations on *DNA* plasmids in 2000; Sakamato et al[5] presented a molecular algorithm of Boolean calculation by *DNA* hairpin formation. These previous researches in *DNA*

* This work was supported by the Science and Technology Development Foundation from Shandong University at Weihai under Grant No.XZ2005005; the National Natural Science Foundation of China under Grant No.60573024; the National Grand Fundamental Research 973 Program of China under Grant No.2005CCA04500.

D.-S. Huang, K. Li, and G.W. Irwin (Eds.): ICIC 2006, LNBI 4115, pp. 320 – 327, 2006.

computing do not require the consideration of weight representation in *DNA* strands. However, there are many practical applications related to weights, such as the *shortest path* problem, the *traveling salesman* problem, the *minimum spanning tree* problem, and the *maximum weight clique* problem (MWCP).

To represent weight in the *DNA* computing model, Narayanan et al[6] presented a conceptual weight encoding method for the *shortest path* problem; Shin et al[7] proposed a weight representation method of varying the number of hydrogen bonds in fixed-length *DNA* strands for the *traveling salesman* problem; Yamamura et al[8] gave a concentration control method for the *shortest path* problem; Lee et al[9] proposed a melting temperature control method for the *traveling salesman* problem; Yin[10] introduced two weight encoding methods for the *minimum spanning tree* problem and the Chinese postman problem; Ma et al[11] introduced a *DNA* computing model for MWCP which is based on plasmids. The existing weight encoding methods are suitable for the specific instances, but they cannot be easily generalized. So we need to do further research on weight encoding method.

In this paper, we devised a new weight encoding method and a corresponding *DNA* algorithm for MWCP. It is organized as follows. In Section 2, we describe a new *DNA* encoding method for MWCP. In Section 3, we present a *DNA* algorithm based on the proposed encoding method and analyze its time complexity and space complexity. Section 4 gives the implementation of the *DNA* algorithm using biological operations. Section 5 compares the new *DNA* computing model with Ouyang's algorithm[3]. Finally, in Section 6, we draw some conclusions.

2 DNA Encoding Method for MWCP

In order to be convenient for describing the new *DNA* encoding method, we firstly give the representation method of orientation of *DNA* strand.

2.1 Orientation Representation of *DNA* Strand

When one *DNA* molecule combines with another to form a *DNA* strand, 5′-phosphate group of one nucleotide always combine with 3′-hydroxyl group of another nucleotide by phosphodiester bonds, shown in Fig.1. This is called as 5′-3′ orientation or 3′-5′ orientation[12]. In Fig. 1(*a*), the nucleotide with 5′-free end is located at the most left end and the nucleotide with 3′-free end is located at the most right end, which is marked as 5′-*AGC*-3′ or 5′-*AGC*; In Fig. 1(*b*), the nucleotide with 3′-free end is located at the most left end and the nucleotide with 5′-free end is located at the most right end, which is marked as 3′-*CGA*-5′ or 3′-*CGA*. In fact, 5′-*AGC* is the same as 3′-*CGA*, which can be seen by overturning Fig. 1(*b*). In order to be convenient for description, 5′-*AGC* is written as *AGC*, and 3′-*CGA* is written as -*CGA* in this paper. Note that -*CGA*=*AGC*.

Definition 1. For any a *DNA* strand *s*, let *h* represent a mapping function from each base to its Watson-Crick complementary base, or *h*(*A*)=*T*, *h*(*G*)=*C*, *h*(*C*)=*G*, *h*(*T*)=*A*. The obtained *DNA* strand is called the *complementary strand* of *s*, and its reversal is called the *reverse complementation* of *s*.

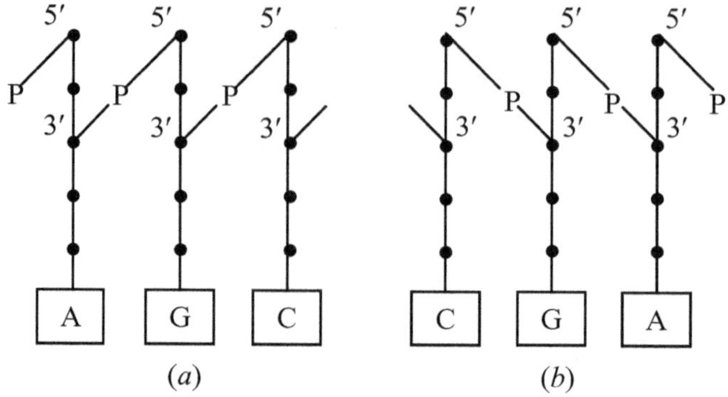

Fig. 1. Orientation of *DNA* strand

Let *s=AGC*. Its *complementary strand* is *h*(*s*)=*TCG*, and its *reverse complement-ation* is -*h*(*s*)=-*TCG*=*GCT*. Obviously, the *DNA* strand *AGC* can combine with *GCT* to form a double-stranded *DNA* (ds*DNA*) through hydrogen bonds. In general, the *DNA* strand *s* can combine with its *reverse complementation* -*h*(*s*) to form a ds*DNA* through hydrogen bonds.

2.2 DNA Encoding Method for MWCP

MWCP can be described as follows: For an undirected graph $G=(V,E)$, $v_i \in V$, $e_{ij} \in E$, where the weight on vertex v_i is w_i, $w_i \geq 0$, $w_i \in Z$[1], it is to find a subset of mutually adjacent vertices which has the largest total weight. We suppose that *G* is a simple graph, or there aren't parallel edges on it. The *DNA* encoding method is as follows.

1) For any vertex $v_i \in V$, we use two *DNA* strands, s_{i1} and s_{i2}, to encode it. The longer *DNA* strand s_{i1} consists of three parts, which are respectively marked as s_{i1}', s_{wi} and s_{i1}'', where s_{wi} is with a length of w_i, and s_{i1}' or s_{i1}'' is with a length of 10, that is, the longer *DNA* strand $s_{i1}=s_i' s_{wi} s_i''$ is with a length of $20+w_i$. The shorter strand s_{i2} is the *reverse complementation* of s_{wi}, or $s_{i2}=-h(s_{wi})$. Obviously, the code s_{i2} can combine with the center part of s_{i1} to form a fragment of ds*DNA*. After each vertex is encoded, we embedded restriction sequences at both sides of s_{wi} and s_{i2}. The codes of vertex v_i are shown in Fig. 2(*a*).

2) For any edge $e_{ij} \in E$, we encode it using the *DNA* strand se_{ij} which is the *reverse complementation* of the last part of s_{i1} and the first part of s_{j1}, or $-h(s_{i1}''s_{j1}')$. Obviously, the *DNA* strand se_{ij} is with a length of 20. Thus, the *DNA* strands corresponding to vertices v_i,v_j can combine with the *DNA* strand corresponding to edge e_{ij} to form a stable ds*DNA*, as shown in Fig. 2(*b*).

Take the graph shown in Fig. 3(*a*) as an example, where the number in each bracket represents the weight of the corresponding vertex. We illustrate the proposed *DNA* encoding method in the following.

---

[1] If w_i is a real number, all the weights multiplied by a certain integer (i.e. 10) and then they are rounded into integers.

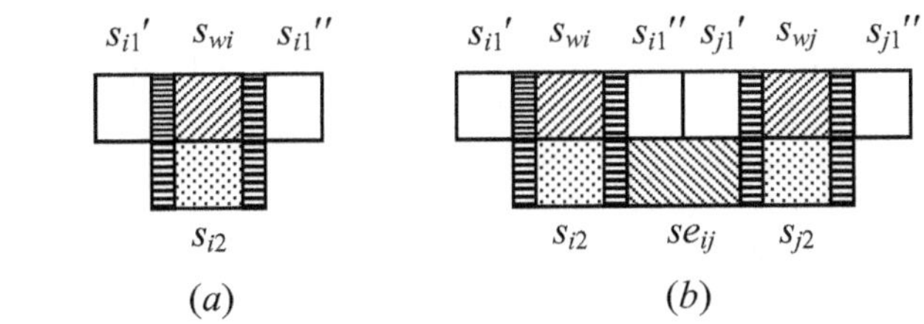

Fig. 2. The *DNA* strands s_{i1} and s_{i2} to encode vertex v_i and the joint of *DNA* strands s_{i1}, s_{i2}, s_{j1}, s_{j2} and se_{ij} to encode vertices v_i, v_j and edge e_{ij}

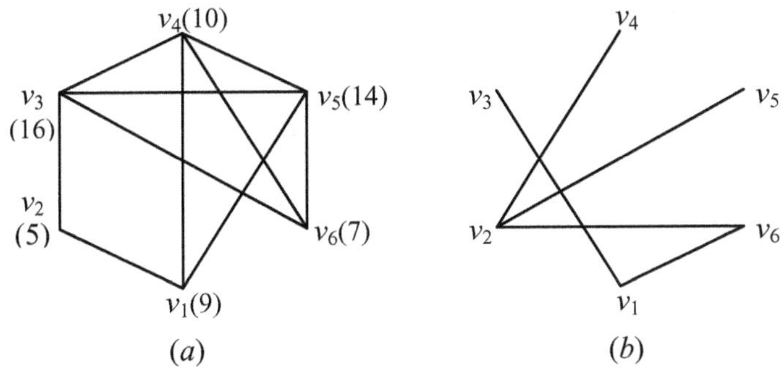

Fig. 3. A weighted graph and its complementary graph

For the vertices v_1, v_2, … v_6 in Fig. 3(*a*), the following *DNA* strands s_{11}, s_{12}, s_{21}, s_{22}, … s_{61}, s_{62} are respectively selected to encode them.

s_{11}=*ATATCGCGGG-TTCAACGTG-CTCAACGTGC*
s_{12}=*-h(TTCAACGTG)=CACGTTGAA*
s_{21}=*CAGTTGACAT-GCAGG-ATCGAATCGA*
s_{22}=*-h(GCAGG)=CCTGC*
s_{31}=*TATAGTACTG-ATCGTAGTCAGTACTG-ATCGTAGTCA*
s_{32}=*-h(ATCGTAGTCAGTACTG)=CAGTACTGACTCCGAT*
s_{41}=*AACCTGGTAC-CAAGCTTGAC-CAAGCTTGAC*
s_{42}=*-h(CAAGCTTGAC)=GTCAAGCTTG*
s_{51}=*TGGTTTGGAC-TGGTCAAGTTGGAC-TGGTCAAGTT*
s_{52}=*-h(TGGTCAAGTTGGAC)=GTCCAACTTGACCA*
s_{61}=*TTTAGCGCAT-GCAGGAT-CGAGGATCGA*
s_{62}=*-h(GCAGGAT)=ATCCTGC*

For any edge e_{ij} in Fig. 3(*a*), we use one *DNA* strand $se_{ij}=-h(s_i''s_j')$ to encode it. Consider the edges linked to vertex v_1. The corresponding codes are as follows:

se_{12}=*ATGTCAACTGGCACGTTGAG*
se_{14}=*GTACCAGGTTGCACGTTGAG*
se_{15}=*GTCCAAACCAGCACGTTGAG*

When we select the above *DNA* strands to encode the vertices and the edges, we must avoid the following situations[13]: (1) one part of a *DNA* strand may combine with another part of it through intercrossing in a molecule; (2) one *DNA* strand may combine with another *DNA* strand through intercrossing between molecules; (3) one *DNA* strand may combine with other two or more *DNA* strands through intercrossing between molecules.

With the help of the property of *reverse complementation* between the *DNA* strands, the codes corresponding to vertices can combine with the codes corresponding to edges to form many ds*DNAs*, some of which are given in Fig. 4.

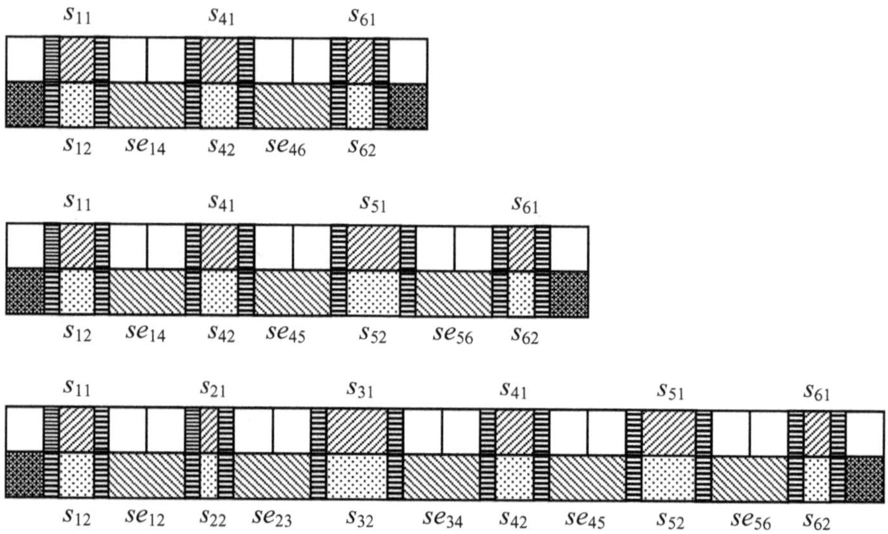

Fig. 4. Joint of the vertex codes and the edge codes

3 DNA Algorithm for MWCP

MWCP has been proved to be *NP*-complete. The determinate algorithms of solving it are with exponential time complexity. By means of massive parallel computation of *DNA* molecules and the principle of Watson-Crick complementary bases, the optimal solution of MWCP can be found in polynomial time. In order to be convenient for describing the *DNA* algorithm, we firstly give the definition of *complementary graph* in the following.

Definition 2. Given an undirected graph G, the graph G' containing all edges missing in G is called the *complementary graph* of G. For example, Fig. 3(*b*) is the *complementary graph* of Fig. 3(*a*). In this paper, the number of edges in the *complementary graph* is marked as *nc*.

Obviously, any two vertices connected in the *complementary graph* are disconnected in the original graph, so they cannot be members of the same clique. Thus, we can give the new *DNA* algorithm of solving MWCP as follows:

(1) For any vertex $v_i \in V$, two *DNA* strands, s_{i1} and s_{i2}, are used to encode it. The *DNA* strand $s_{i1}=s_{i1}'s_{wi}s_{i1}''$ is with a length of $20+w_i$, where the length of s_{wi} is equal to w_i and the length of s_{i1}' and s_{i1}'' is 10 respectively. The strand $s_{i2}=-h(s_{wi})$;

(2) For any edge $e_{ij} \in E$, the *DNA* strand se_{ij}=-$h(s_{i1}''s_{j1}')$ is used to encode it. Obviously, se_{ij} is with a length of 20;

(3) By means of *merge* operation, sufficient *DNA* strands s_{i1}, s_{i2} and se_{ij} ($1 \le i,j \le n$) are mixed together in a single ligation reaction;

(4) Let $sumw = \sum_{i=1\sim n} w_i$. Based on the principle of Watson-Crick complementary bases, the various ds*DNA*s are generated randomly, and the ds*DNA*s starting with v_1 and ending at v_n are amplified by polymerase chain reaction (PCR) with the primers –$h(v_1)$ and $-h(v_n)$. Only those amplified ds*DNA*s whose length is equal to or less than *sumw* are saved. The set of the saved ds*DNA*s is called the complete data pool;

(5) Digesting the ds*DNA*s in the complete data pool with restriction enzymes. The enzymes break *DNA* at specific restriction sites, which were embedded within the sequences for v_i($1 \le i \le n$);

(6) For k:=1 to *nc* do {For any edge e_{ij} in the complementary graph, let $sumw=sumw-min\{w_i,w_j\}$}. Separate away all the ds*DNA* whose length is above to the updated *sumw*;

(7) Separate the longest ds*DNA*s in the remaining data pool by means of gel electrophoresis;

(8) Sequencing the longest ds*DNA*s to determinate its nucleotides sequence. The sequence corresponds to the optimal solution.

In the following, the time complexity of the *DNA* algorithm is analyzed according to the data structure used in the algorithm[14,15]. The steps 1 and 2 are preliminaries for *DNA* computing. The steps 3 and 4 are irrelevant to the scale of problem, so their time complexity is $O(1)$. In steps 5 and 6, the digestion operation need to do *nc* times which is at most $n(n-1)/2$, so its time complexity is $O(n^2)$. In steps 7 and 8, separating and sequencing the paths are irrelevant to the scale of problem, so their time complexity is $O(1)$. Therefore, the total time complexity of our *DNA* algorithm is $O(n^2)$, where n is the number of vertices in graph G.

The space complexity of the *DNA* algorithm is analyzed as follows. In step 1, the number of the vertex codes is $2n$. In step 2, the number of the edge codes is m, $m=O(n^2)$. Let d_i denote the degree of vertex v_i in graph G, and d_{max} denotes the maximum of all the vertex degrees. In step 3 and 4, the number of the ds*DNA*s in the complete data pool is $O(C_{d1}^1 C_{d2}^1 C_{d3}^1 \ldots C_{dn-1}^1) = O(d_{max}{}^n)$. In steps 5 and 6, the digestion operation doesn't increase the number of storage unit, so its space complexity is still $O(d_{max}{}^n)$. In steps 7 and 8, the separating and sequencing operations doesn't increase the number of storage unit either, so its space complexity is also $O(d_{max}{}^n)$. Therefore, the total space complexity of our *DNA* algorithm is $O(d_{max}{}^n)$.

4 Biological Implementation of DNA Algorithm

By means of the basic molecular biological operations, the steps 1~4 and 6~8 in the *DNA* algorithm can be easily carried on. In the following, we only give the implementation of step 5. In step 5, the ds*DNA*s are digested by restriction enzymes. The restriction enzymes break the ds*DNA*s at specific restriction sites, which were embedded within the sequences for v_i($1 \le i \le n$).

To cut a connection in the *complementary graph*, for example, the 1-3 connection, we first divided the ds*DNA*s into two test tubes, t_0 and t_1. In t_0 we cut the *DNA* fragments located at the center part of the codes corresponding to v_1 with specific restriction enzymes; in t_1 we cut the *DNA* fragments located at the center part of the codes corresponding to v_3 with other specific restriction enzymes. And then we combined t_0 and t_1 into test tube t, which didn't contain the codes corresponding to v_1 and v_3 at the same time. Let $sumw$=$sumw$-min{w_1, w_3}.

By nc sequential restriction operations with different enzymes, all the *DNA* fragments connected by the edges in the *complementary graph* are digested. Each time of digesting the *DNA* fragments connected by edges e_{ij} in the *complementary graph*, let $sumw$=$sumw$-min{w_i, w_j}. Thus, through n-1 times digestion operations, the total weight value of the remaining ds*DNA*s is less than or equal to the updated $sumw$.

5 Comparison with Ouyang's DNA Algorithm

At present, there exists some previous works[1-9,11,16-20] to compute *NP* problems with *DNA* molecules. Our *DNA* computing method for MWCP is an improvement on Ouyang's algorithm[3] of solving MCP. The main improvements are as follows. (1) On the basis of Ouyang's *DNA* computing model, we added weight representation in *DNA* strands. Thus, it can be expanded to solve *NP*-hard problems. (2) In Ouyang's algorithm of solving MCP, the number of elements in the complete data pool is $O(n^n)$, or its space complexity is $O(n^n)$. And in our *DNA* algorithm for MWCP, the number of elements in the complete data pool is $O(d_{max}{}^n)$, or its space complexity is $O(d_{max}{}^n)$. Obviously, the number of storage unit used in our *DNA* algorithm is less than that in Ouyang's algorithm because the degree of any vertex is always less than the number of vertices in graph. So our *DNA* algorithm is with less space complexity than Ouyang's algorithm. (3) In Ouyang's algorithm, all the combinations of vertices are in the complete data pool, whereas in our algorithm, those vertices disconnected by edge in graph G aren't in the complete data pool. So the number of *DNA* molecules being cut in each digestion operation is reduced in our algorithm.

6 Conclusion

We presented a new weight encoding method and the corresponding molecular algorithm for MWCP. The results show that a series of biological experiment operations successfully find the optimal solution of the MWCP instance shown in Fig.3(a), which consists of 10 edges and 6 vertices with six different costs. One of the most important features of our weight encoding method is that it uses two *DNA* strands of different length to encode one vertex, or it can handle quantitative expression of numerical values using *DNA* strands of different length. The ability of representing and manipulating numerical data opens up the possibility of solving a wide range of numerical optimization problems. Our work provides further evidence for the ability of *DNA* computing to solve numerical optimization problems.

References

1. Adleman, L.M.: Molecular Computation of Solutions to Combinatorial problems. Science, Vol. 266 (1994) 1021-1024
2. Lipton, R.J.: DNA solution of Hard Computational Problems. Science, Vol. 268 (1995) 542-545
3. Ouyang, Q., Kaplan, P.D., Liu, S., et al.: DNA Solution of the Maximal Clique Problem. Science, Vol. 278 (1997) 446-449
4. Head, T., Rozenberg, G., Bladergroen, R.S., et al.: Computing with DNA by Operating on Plasmids. Biosystems, Vol. 57 (2000) 87-93
5. Sakamoto, K., Gouzu, H., Komiya, K., et al.: Molecular Computation by DNA Hairpin Forma-tion. Science, Vol. 288 (2000) 1223-1226
6. Narayanan, A., Zorbalas, S., et al.: DNA Algorithms for Computing Shortest Paths. Proceed-ings of the Genetic Programming, Morgan Kaufmann (1998) 718-723
7. Shin, S.Y., Zhang, B.T., Jun, S.S., et al.: Solving Traveling Salesman Problems Using Molecu-lar Programming. Proceedings of the Congress on Evolutionary Computation. IEEE Press (1999) 994-1000
8. Yamamura, M., Hiroto, Y., Matoba, T.: Solutions of Shortest Path Problems by Concentration Control. Lecture Notes in Computer Science, 2340 (2002) 231-240
9. Lee, J.Y., Shin, S.Y., Park, T.H., et al.: Solving Traveling Salesman Problems with DNA Molecules Encoding Numerical Values. BioSystems, Vol. 78 (2004) 39-47
10. Yin, Z.: DNA Computing in Graph and Combination Optimization. Science Press (2004) 57-72 (in Chinese)
11. Ma, R., Zhang, Q., Gao, L., Xu, J.: Using DNA to Solve the Maximum Weight Clique of Graphs. ACTA Electronica Sinica, Vol. 32 (2004) 13-16
12. Xu, J., Wang, S., Pan, Q., translate., Paun, G., Rozenberg, G., Salomaa, A. write: DNA Computing: New Computing Paradigms. Tsinghua University publishing company, Vol. 1(2004) 3-54 (in Chinese)
13. Liu, W., Wang, S., Xu, J.: Research on the Encoding Method of DNA Computing. Computer Engineering and Applications, Vol. 27 (2003) 118~121 (in Chinese)
14. Han, A., Yang, Z., et al.: Complexity Analysis for HEWN Algorithm. Journal of Software, Vol. 13 (2002) 2337-2342 (in Chinese)
15. Han, A.: Complexity Research for *B* Algorithm. Proceedings of the Tenth Joint Internation-al Computer Conference, International Academic Publishers (2004) 188-192
16. Wang, L., Lin, Y., Li, Z.: DNA Computation for a Category of Special Integer Planning Problem. Computer Research and Development, Vol. 42 (2005) 1431-1437 (in Chinese)
17. Chen, Z., Li, X., Wang, L., et al.: A Surface-Based DNA Algorithm for the Perfect Matching Problem. Computer Research and Development, Vol. 42 (2005) 1241-1246 (in Chinese)
18. Lancia, G.: Integer Programming Models for Computional Biology Problems. Journal of Computer Science and Technology, Vol. 19 (2004) 60-77
19. Ibrahim, Z., Tsuboi, Y., Muhammad, M.S., et al.: DNA Implementation of k-shortest Paths Computation. IEEE Congress on Evolutionary Computation. IEEE CEC 2005 Proceedings, Vol. 1 (2005) 707-713
20. Liu, Q., Wang, L., Frutos, A.G. et al.: DNA Computing on Surfaces. Nature, Vol. 403 (2000) 175-179

A New DNA Encoding Method for Traveling Salesman Problem*

Aili Han[1, 2] and Daming Zhu[2]

[1] Department of Computer Science and Technology, Shandong University, Weihai 264209, China
[2] School of Computer Science and Technology, Shandong University, Jinan 250061, China
hanal@sdu.edu.cn

Abstract. We have devised a new *DNA* encoding method to represent weight and apply it to solve the traveling salesman problem, an instance of optimization problems on weighted graphs. For any weighted graph $G=(V,E)$, $v_i \in V$, $1 \le i \le n$, where exists weight w_{ij} on edge v_iv_j, we use two *DNA* strands with different lengths to encode each of the edges. The longer *DNA* strand consists of three parts: one for the departure vertex, another for the weight, and the last for the arrival vertex. The shorter *DNA* strand is the *reverse complementation* of the center part of the longer one. The proposed weight encoding method is an improvement on the previous weight encoding methods, and it can more easily find the optimal solutions than the former ones. This work extends the capability of *DNA* computing to solving numerical optimization problems, which is contrasted with other *DNA* computing methods focusing on decision problems.

1 Introduction

DNA computing is a computational paradigm that uses synthetic or natural *DNA* molecules as information storage media, in which the techniques of molecular biology, such as polymerase chain reaction(*PCR*), gel electrophoresis, and enzymatic reactions, are used as computational operators for copying, sorting, and splitting/ concatenating the information respectively.

Based on the massive parallelism of *DNA* computing, many researchers tried to solve a host of difficult problems. In 1994, Adleman [1] solved a 7-vertex instance of the *Hamiltonian path* problem by means of the molecular biology techniques. The creative research opened up a new way to computation using *DNA* molecules. A major goal of subsequent research in this area is to understand how to solve *NP*-complete problems. To address this goal, Lipton [2] abstracted a parallel model on the basis of Adleman's experiment, and applied it to solve the 3-*SAT* problem; Ouyang *et al* [3] solved the *maximal clique* problem; Head *et al* [4] solved the *maximal independent set* problem based on *DNA* plasmids; Sakamato *et al* [5] presented a *DNA* algorithm of Boolean calculation by *DNA* hairpin formation. These previous

* This work was supported by the Science and Technology Development Foundation from Shandong University at Weihai under Grant No.XZ2005005; the National Natural Science Foundation of China under Grant No.60573024; the National Grand Fundamental Research 973 Program of China under Grant No.2005CCA04500.

D.-S. Huang, K. Li, and G.W. Irwin (Eds.): ICIC 2006, LNBI 4115, pp. 328 – 335, 2006.

researches do not require the consideration of the representation of weight in *DNA* strands. However, many practical applications involve weighed graphs. The representation of weight in *DNA* strands is an important issue toward expanding the capability of *DNA* computing to solve numerical optimization problems. There exists previous work to represent weight with *DNA*. Narayanan *et al* [6] presented a conceptual encoding method that represents costs with the lengths of *DNA* strands. Shin *et al* [7] proposed a method to represent real numbers by varying the number of hydrogen bonds. Yamamura *et a l*[8] gave a concentration control method which encodes weight by the concentrations of *DNA* strands. Lee *et al* [9] introduced a melting temperature control method which represents costs by melting temperatures of fixed-length *DNA* strands. Yin [10] presented a kind of weight encoding method which can be easily generalized.

In this paper, we introduce a new weight encoding method in which two *DNA* strands with different lengths are designed to encode one edge to improve the previous work. The *traveling salesman* problem (*TSP*) is used as a test problem for the proposed weight encoding method. This paper is organized as follows. In Section 2, we describe a new *DNA* encoding method to represent weight. Section 3 describes a *DNA* algorithm of solving *TSP* based on the proposed encoding method. Section 4 gives the implementation of the *DNA* algorithm using biological operations. Section 5 compares the new *DNA* encoding method with the previous encoding methods. Finally, in Section 6, we draw some conclusions.

2 DNA Encoding Method for TSP

Representation of weight for weighted-graph problems is one of the most important but challenging problems. This paper presents a new weight encoding method that can more easily find the optimal solution than the former ones. In order to be convenient for description, we firstly give the orientation representation of *DNA* strand.

2.1 Representation of DNA Orientation

When one *DNA* molecule combines with another to form a *DNA* strand, 5′-phosphate group of one nucleotide always combine with 3′-hydroxyl group of another nucleotide by phosphodiester bonds. This is called as 5′-3′ orientation or 3′-5′ orientation. The nucleotide with 5′ free-end being located at the most left end and 3′ free-end being located at the most right end is marked as 5′-*XX*...*X*-3′, and the nucleotide with 3′ free-end being located at the most left end and 5′ free-end being located at the most right end is marked as 3′-*XX*...*X*-5′, where *X* denotes a letter in the set {*A*,*G*,*C*,*T*}. In this paper, 5′-*AGTC*-3′ is briefly written as *AGTC*, and 3′-*CTGA*-5′ is written as -*CTGA*. Note that -*CTGA*=*AGTC*.

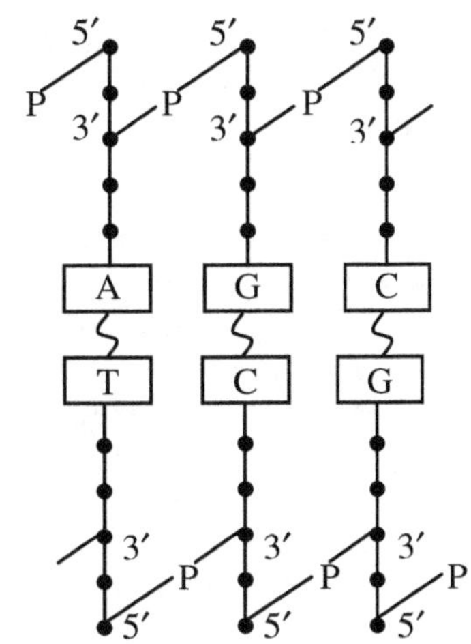

Fig. 1. A double-stranded *DNA*

Definition 1. For any *DNA* strand *s*, let *h* represent a mapping function from each base to its Watson-Crick complementary base, or $h(A)=T$, $h(G)=C$, $h(C)=G$, $h(T)=A$. The obtained *DNA* strand is called as *complementary strand* of *s*, and its reversal is called as *reverse complementation* of *s*. The mapping function *h* is called as *complementary mapping*.

Take the double-stranded *DNA* (ds*DNA*) shown in Fig. 1 as an example. Let $s=AGC$. Its *complemen-tary strand* is $h(s)=TCG$, and its *reverse comple-menttation* is $-h(s)=-TCG=GCT$. Obviously, the *DNA* strands *AGC* and *GCT* can combine with each other to form a ds*DNA* through hydrogen bonds. In general, *DNA* strand *s* and its *reverse comple-mentation* $-h(s)$ can combine with each other to form a ds*DNA* through hydrogen bonds.

2.2 DNA Encoding Method for TSP

TSP is to find a minimum cost path for a given set of cities and roads. The path must begin at a specified city and end there after passing through all the given cities. *TSP* can be abstracted as follows: for a connected, weighted, undirected graph $G=(V,E)$, $v_i \in V$, $1 \le i \le n$, where the weight on edge v_iv_j is w_{ij}, $w_{ij} \ge 0$, $w_{ij} \in Z$[1], it is to find a circuit with minimum weight which begins at a specified vertex and ends there after through all the given vertexes. We suppose that $G=(V,E)$ is a simple graph just as well, or there aren't parallel edges on it. The *DNA* encoding method for *TSP* is as follows.

1) For any vertex $v_i (1 \le i \le n)$, *DNA* stand s_i of length 20 is used to encode it.

2) For any edge $v_iv_j (1 \le i,j \le n)$, we encode it using two different length *DNA* strands. The longer strand s_{ij1} consists of three parts, which are respectively marked as s_{ij1}', s_{ij1}'', s_{ij1}''', where s_{ij1}' is the *reverse complementation* of the second half of s_i, s_{ij1}'' is the *reverse complementation* of s_{wij} whose length is equal to w_{ij}, and s_{ij1}''' is the *reverse complementation* of the first half of s_j. The shorter strand s_{ij2} is the *reverse complementation* of s_{ij1}''. Specifically, s_i and s_j are respectively divided into two substrands with equal length, or $s_i=s_i's_i''$, $s_j=s_j's_j''$. The longer *DNA* strand $s_{ij1}=-h(s_i''s_{wij}s_j')$, whose length is $20+w_{ij}$, and the shorter *DNA* strand $s_{ij2}=-h(s_{ij1}'')=s_{wij}$.

Note that, for any undirected edge v_iv_j, when walking from v_i to v_j, the longer code s_{ij1} is the *reverse complementation* of $s_i''s_{wij}s_j'$, or $s_{ij1}=-h(s_i''s_{wij}s_j')$, and the shorter code $s_{ij2}=-h(s_{ij1}'')=s_{wij}$; when walking from v_j to v_i, the longer code s_{ji1} is the *reverse complementation* of $(-s_j)''(-s_{wij})(-s_i)'$, or $s_{ji1}=-h((-s_j)''(-s_{wij})(-s_i)')= h(s_i''s_{wij}s_j')=-s_{ij1}$, and the shorter code $s_{ji2}=-h(s_{ji1}'') =-s_{wij}=-s_{ij2}$. That is, only need $s_{ij1}=-h(s_i''s_{wij}\ s_j')$ and $s_{ij2}=s_{wij}$ to encode edge v_iv_j.

Take the weighted graph *G* shown in Fig. 2 as an example. We specifically analyze the pro-posed *DNA* encoding method. For the vertices $v_1,v_2,\ldots,v_7$, the following *DNA* strands $s_1,s_2,\ldots,s_7$ are respectively selected to encode them.

---

[1] If w_{ij} is a real number, all the weights multiplied by a certain integer (i.e. 10) and then they are rounded into integers.

s_1=*ATATCGCGGGTTCAACGTGC*,
s_2=*CAGTTGACATGCAGGATCGA*,
s_3=*TATAGTACTGATCGTAGTCA*,
s_4=*AACCTGGTACCAAGCTTGAC*,
s_5=*TGGTTTGGACTGGTCAAGTT*,
s_6=*TTTAGCGCATGCAGGATCGA*,
s_7=*GTTACGTGAGTCTGACTGAC*.

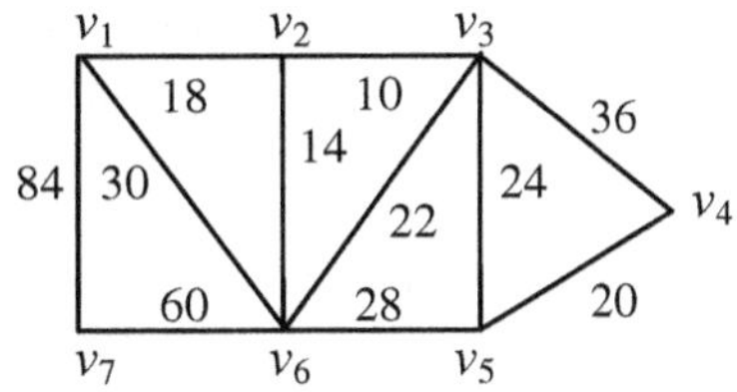

Fig. 2. A weighted graph

For any edge v_iv_j, we use two different length *DNA* strands $s_{ij1}=-h(s_i''s_{wij}s_j')$ and $s_{ij2}=s_{wij}$ to encode it. For example, the codes of edge v_1v_2 are as follows:

$s_{121}=-h(s_1''s_{w12}s_2')$=-*h*(*TTCAACGTGCAGTGCATGCTGATCGATGCAGTTGACAT*)
=-*AAGTTGCACGTCACGTACGACTAGCTACGTCAACTGTA*

$s_{122}=s_{w12}$=*AGTGCATGCTGATCGATG*

The *DNA* strands s_1,s_2,s_{121} and s_{122} can combine with each other, as shown in Fig.3.

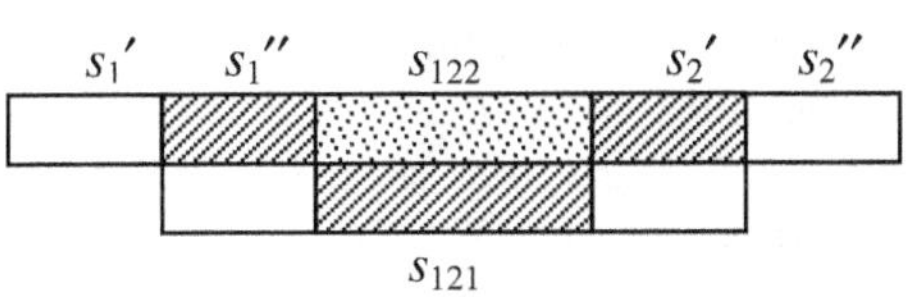

Fig. 3. Joint of *DNA* strands s_1, s_2, s_{121} and s_{122}

Based on Fig.3, the next is to extend forward right to s_3 or s_6 and to extend forward left to s_7 or s_6. The right extension is shown in Fig.4, and its left extension can be similarly drawn.

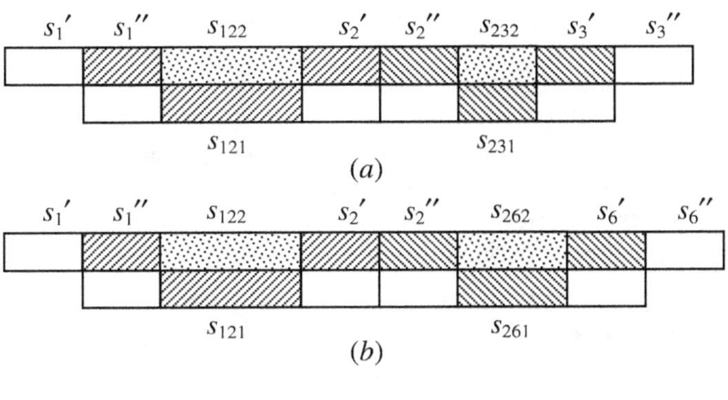

Fig. 4. The right extension of Fig.3

On the basis of Fig. 4(*a*), the next is to extend forward right to s_4, s_5 or s_6, and to extend forward left to s_7 or s_6, and so on. Thus, with the help of the property of *reverse complementation* between the vertex codes and edge codes, the ds*DNA*s may extend continually to form various random paths including the optimal solution.

3 DNA Algorithm for TSP

TSP has been proved to be *NP*-hard. The determinate algorithms of solving it so far have only complete enumeration, which is with exponential time complexity. By means of massive parallel computation of *DNA* molecules and the principle of Watson-Crick complementary bases, the optimal solution of *TSP* can be found in polynomial time. We suppose that v_1 is the original vertex just as well. This is because that, for any a shortest circuit *C* going through all the given vertexes, the length of the route which begins at v_1 and ends there along *C* is equal to that of the route which begins at v_i ($i \neq 1$) and ends there along *C*.

Note that our *DNA* algorithm searches for the shortest *path* instead of the shortest *circuit*. This is because that the length of a shortest path which begins at v_1 and end there after going through all the given vertexes is equal to that of a shortest circuit which begins at v_1 and end there after going through all the given vertexes. Moreover, *PCR* in the biological techniques is generally carried out on a linear template, and there isn't circular template so far.

In order to generate more molecules which correspond to the optimal solution, we add the primer $-h(s_1')$ to one end of the *PCR* template and add the primer $-h(s_1'')$ to another end. For the *DNA* strand s_1 given in section 2, the primers are as follows:

$-h(s_1')=-h(ATATCGCGGG)=-TATAGCGCCC$

$-h(s_1'')=-h(TTCAACGTGC)=-AAGTTGCACG$

The *DNA* algorithm of *TSP* is given as follows:

(1) By means of *merge*[2] operation, sufficient *DNA* strands s_i, s_{ij1} and s_{ij2}, $1 \leq i,j \leq n$, are mixed together in a single ligation reaction, and the primers $-h(s_1')$ and $-h(s_1'')$ are respectively added to the two ends of the *PCR* template;

(2) Based on the principle of Watson-Crick complementary bases, randomly generate various *DNA* molecules corresponding to the random paths;

(3) Separate away all the paths without the departure vertex v_1, or separate away all the *DNA* molecules without 5′-end being s_1;

(4) Separate away all the paths without the arrival vertex v_1, or separate away all the *DNA* molecules without 3′-end being s_1;

(5) For each vertex v_i ($2 \leq i \leq n$), separate away all the paths without v_i;

(6) Separate the shortest path by means of gel electrophoresis;

(7) Sequencing the shortest path to determinate its nucleotides sequence. The sequence corresponds to the optimal solution.

In the *DNA* algorithm of *TSP*, the steps of inputting *DNA* strands, generating various random paths, and separating away all the paths without v_1 are irrelevant to the scale of problem, so the time complexity is $O(1)$; for each vertex v_i ($2 \leq i \leq n$), separating away all the paths without s_i is relevant to the number of vertexes in *G*, so its time complexity is $O(n)$; separating the shortest path through agarose gel electrophoresis and sequencing the shortest paths are irrelevant to the scale of problem, so the time complexity is $O(1)$. Therefore, the total time complexity of *DNA* algorithm is $O(n)$, where *n* is the number of vertexes in *G*.

---

[2] The function and implementation of *merge* is given in section 4.

4 Biological Implementation of DNA Algorithm

By means of the molecular biology techniques, the steps 1, 2 and 6 can be easily carried on. In the following, we only give the implementation of steps 3~5 and 7.

4.1 Separation of Random Paths

The various random paths generated in our *DNA* algorithm are put in a tube *N*, and then heat the tube to make the ds*DNA*s become *DNA* strands. Through *separate* operation, all the paths without the departure vertex v_1 are separated away, or remain the *DNA* molecules with 5′-free end being s_1; and then separate away all the paths without the arrival vertex v_1 from the *DNA* molecules with 5′-free end being s_1, or remain the *DNA* molecules with 5′-free end and 3′-free end being s_1. Finally, through a *n*-1 times circle of *separate* operations, separate away all the paths without $s_i(2\leq i\leq n)$. The *DNA* program is as follows:

Step 1. $N \leftarrow B(N, s_1)$;
Step 2. $N \leftarrow E(N, s_1)$;
Step 3. FOR i:=2 TO n DO $N \leftarrow +(N, s_i)$;

4.2 Sequencing the Shortest Paths

The *DNA* molecules with s_1 being at the two ends and including $s_i(2\leq i\leq n)$ are separated through agarose gel electrophoresis, and the furthest *DNA* molecules correspond to the shortest paths. By means of *PCR* amplification and purification, it is carried out several times to increase its purity. Sequencing the shortest paths to determine its nucleotide sequence. The sequence corresponds to the shortest circuit.

5 Comparison with Other DNA Encoding Methods

For the classical *NP*-hard problems, there exists previous work to represent the numerical values with *DNA*. Narayanan *et al*[6] presented a conceptual encoding method which represents costs with the lengths of *DNA* strands. This method requires very long *DNA* strands to solve an instance containing a wide range of costs, and this makes experimental implementation impossible. Shin *et al*[7] proposed a method to represent weight by varying the number of hydrogen bonds. Although the number of hydrogen bonds is an important factor in deciding the thermal stability of *DNA* strands, it is not a sufficient factor. Yamamura *et al*[8] proposed a concentration control method which encodes the weights using the concentrations of *DNA* strands. Though they showed the possibility of concentration control method, they failed in finding the final solution. Lee *et al*[9] proposed a melting temperature control method which represents costs by melting temperatures of fixed-length *DNA* strands. All of the above methods are suitable for some specific instances, but they cannot easily be generalized. Yin[10] introduced a kind of weight encoding methods which can easily be generalized whose feature is that the edge code is divided into three parts, and the center part is used to represent weight. This makes the generated paths in a single ligation reaction are alternant *DNA* strands or double strand, as shown in Fig.5. Since the stable structure of *DNA* molecules is ds*DNA*, the part of *DNA* strand easily combined with other molecules through hydrogen bonds based on the principle of Watson-Crick complementary bases. Therefore, the methods affected the generation of the optimal solution when the difference between the biggest weight and the smallest weight is large or when the lengths of *DNA* strands are large.

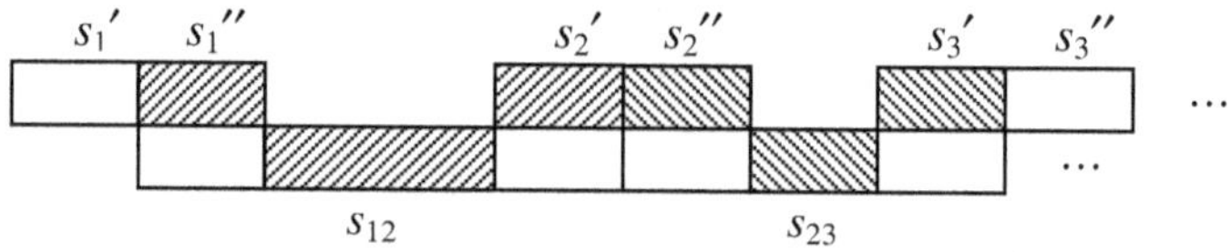

Fig. 5. Alternant *DNA* strand or double strand

Our weight encoding method is an improvement on the previous encoding methods, and it can more easily generate the optimal solutions than the former ones. In our encoding method, each vertex v_i ($1 \le i \le n$) is encoded using *DNA* strand s_i of length 20, and each edge is encoded using two different length *DNA* strands: the longer strand s_{ij1} consists of three parts, which are respectively marked as s_{ij1}', s_{ij1}'', s_{ij1}''', where s_{ij1}' is the *reverse complementation* of the second half of s_i, s_{ij1}'' is the *reverse complementation* of s_{wij} whose length is equal to w_{ij}, and s_{ij1}''' is the *reverse complementation* of the first half of s_j; the shorter strand s_{ij2} is the *reverse complementation* of s_{ij1}''. Thus, the generated random paths in a single ligation reaction are ds*DNA*s, as shown in Fig.4. The generated ds*DNA*s are more stable than alternant *DNA* strand or double strand, so our encoding method can generate the optimal solution even when the difference between the biggest weight and the smallest weight is large or when the lengths of *DNA* strands are large. Moreover, the proposed weight encoding method is suitable for the larger scale graph at least in theory, and it can be generalized and implemented automatically. Our weight encoding method also has characteristics of easy encoding and low error rate.

6 Conclusion

We presented a new weight encoding method and an associated molecular algorithm for *TSP*. The results show that a series of biological experiment operations successfully find the optimal solution of the *TSP* instance shown in Fig. 2, which consists of 7 vertexes and 11 edges of eleven differing cost values. One of the most important features of our weight encoding method is that it uses two different length *DNA* strands to encode one edge, or it can handle quantitative expression of numerical values using *DNA* strands with different lengths. The ability of representing and manipulating numerical data in *DNA* computing opens up the possibility of solving a wide range of numerical optimization problems. In items of efficiency, it makes optimization problems be dealt with in the same efficiency as decision problems.

References

1. Adleman, L.M.: Molecular Computation of Solutions to Combinatorial problems. Science, 266 (1994) 1021-1024
2. Lipton, R.J.: DNA Solution of Hard Computational Problems. Science, 268 (1995) 542-545
3. Ouyang, Q., Kaplan, P.D., Liu, S., *et al*: DNA Solution of the Maximal Clique Problem. Science, 278 (1997) 446-449
4. Head, T., Rozenberg, G., Bladergroen, R.S., *et al*: Computing with DNA by Operating on Plasmids. Biosystems, 57 (2000) 87-93
5. Sakamoto, K., Gouzu, H., Komiya, K., *et al*: Molecular Computation by DNA Hairpin Formation. Science, 288 (2000) 1223-1226
6. Narayanan, A., Zorbalas, S., *et al*: DNA Algorithms for Computing Shortest Paths. Proceed-ings of the Genetic Programming, Morgan Kaufmann (1998) 718-723
7. Shin, S.Y., Zhang, B.T., Jun, S.S., *et al*: Solving Traveling Salesman Problems Using Molecular Programming. Proceedings of the Congress on Evolutionary Computation. IEEE Press (1999) 994-1000
8. Yamamura, M., Hiroto, Y., Matoba, T.: Solutions of Shortest Path Problems by Concentration Control. Lecture Notes in Computer Science, 2340 (2002) 231-240
9. Lee, J.Y., Shin, S.Y., Park, T.H., *et al*: Solving Traveling Salesman Problems with DNA Molecules Encoding Numerical Values. BioSystems, 78 (2004) 39-47
10. Yin, Z.: DNA Computing in Graph and Combination Optimization. Science Press (2004) 57-72 (in Chinese)
11. Xu, J., Dong, Y. *et al*: Sticker DNA Computer Model-Part I: Theory. Chinese Science Bulletin, 49 (2004) 772-780
12. Zhang, Z., Zhao, J., He, L.: Molecular Biology Development of DNA Computer. Journal of Genetics, 30 (2003) 886~892 (in Chinese)
13. Wang, L., Lin, Y., Li, Z.: DNA Computation for a Category of Special Integer Planning Problem. Computer Research and Development, 42 (2005) 1431-1437 (in Chinese)
14. Chen, Z., Li, X., Wang, L., *et al*: A Surface-Based DNA Algorithm for the Perfect Matching Problem. Computer Research and Development, 42 (2005) 1241-1246 (in Chinese)
15. Braich, R.S., Chelyapov, N., Johnson, C., *et al*: Solution of a 20-variable 3-SAT Problem on a DNA Computer. Science, 296 (2002) 499~502
16. Lancia, G.: Integer Programming Models for Computional Biology Problems. Journal of Computer Science and Technology, 19 (2004) 60-77
17. Han, A., Yang, Z. *et al*: Complexity Analysis for HEWN Algorithm. Journal of Software, 13 (2002) 2337-2342 (in Chinese)
18. Han, A.: Complexity Research for *B* Algorithm. Proceedings of the Tenth Joint Internation-al Computer Conference, International Academic Publishers (2004) 188-192
19. Han, A.: A Study on the Solution of 9-room Diagram by State Space Method. Journal of Shandong University (Engineering Science), 34 (2004) 51-54 (in Chinese)
20. Han, A., Pan, J.: A Network Layout Algorithm Based on the Principle of Regular Hexagons Covering a Plane. Proceedings of the 8th International Conference for Young Computer Scientist. International Academic Publishers (2005) 223-227
21. Liu, Q., Wang, L., Frutos, A.G. *et al*: DNA Computing on Surfaces. Nature, 403 (2000) 175-179
22. Ibrahim, Z., Tsuboi, Y., Muhammad, M.S. *et al*: DNA Implementation of K-shortest Paths Computation. IEEE Congress on Evolutionary Computation. IEEE CEC 2005 Proceedings, 1 (2005) 707-713

Computational Design Approach to Hydrodynamic Focusing in a Flow Cytometer

An-Shik Yang and Chun-Yao Wu

Department of Mechanical and Automation Engineering, Da Yeh University, Chang Hwa, Taiwan 515
asyang@mail.dyu.edu.tw

Abstract. A two-fluid theoretical model was developed to simulate the hydrodynamic focusing process formed by the coflowing sample and sheath fluids in a flow cytometer. The analysis consists of two groups of time-dependent three-dimensional conservation equations of mass and momentum. To validate the computer code, the predicted focused width in the two-dimensional test configuration was compared with Lee et al.'s measured data. The present study also examines the pressure distribution of the three-dimensional hydrodynamic focusing flowfield. For the u_{Sh}/u_S ratio ranging from 10 to 80, the focused width was determined to explore the applicability of the proposed flow cytometer.

1 Introduction

The application of flow cytometers for cell counting and sorting is one of the most widely used analytical techniques in the biological, biomedical, and forensic sciences [1, 2]. A flow cytometer is designed to transport cells individually into a certain location intersected by the illuminating beams. In practice, the sample cells are pushed through a coaxial liquid stream encircled by the sheath stream. The sheath stream is essentially a tube of cell-free fluid surrounding the core sample stream containing the cells aligned in a single file. The associated phenomenon is described as hydrodynamic focusing [3, 4]. The sheath flow can be regulated to position the cells into the proper observation area in the center of the core stream under a stable laminar flow environment.

Considerable efforts have been made to study the manipulations of fluid flows in cytometers for counting and sorting particles or cells [5-8]. To study the hydrodynamic focusing behavior, Lee et al. [9, 10] designed and fabricated a novel flow cytometer to measure the cellular characteristics with the magnified visualized images of coflowing dye-containing sample and sheath water flows. Chung et al. [11] also conducted a lumped analysis with the Poiseuille-flow assumption to calculate the shape of the focused flow in a microchip flow cytometer.

To assist the design process of flow cytometers, a computational fluid dynamics (CFD) computer code is useful to probe the complicated focusing flows. The purpose of this paper is to study the hydrodynamic focusing behavior in a cytometer. The predictions were compared with the measured results of the focused width from Lee et al. in Ref. 10 to validate the theoretical model for the two-dimensional test

D.-S. Huang, K. Li, and G.W. Irwin (Eds.): ICIC 2006, LNBI 4115, pp. 336 – 341, 2006.

configuration. Numerical experiments were then extended to the three-dimensional focusing flow by varying the ratios of the sheath to the sample fluid velocity for determining the focused width

2 Theoretical Approach

In the analysis, the sample and the sheath fluids are considered to be incompressible, laminar, isothermal flows. A three-dimensional two-fluid model is developed using two groups of transient conservation equations of mass and momentum, with considering the interfacial momentum exchange between two fluids. The mean flow properties are evaluated through the volume fraction [12-14]. The governing equations are given as below.

$$\frac{\partial(\alpha_k)}{\partial t} + \nabla \bullet (\alpha_k \vec{u}_k) = 0. \tag{1}$$

$$\frac{\partial(\alpha_k \rho_k \vec{u}_k)}{\partial t} + \nabla \bullet (\alpha_k \rho_k \vec{u}_k \vec{u}_k) = -\alpha_k \nabla P + \nabla \bullet \alpha_k \tau_k + \alpha_k \rho_k \vec{g} + \overrightarrow{M}_k . \tag{2}$$

In this study, the subscript k is the phase indicator, which can be S or Sh to denote the sample and sheath fluids, while the variables ρ, α, $\vec{u}$, P, τ, $\vec{g}$, and $\overrightarrow{M}_k$ stand for density, volume fraction, velocity vector, pressure, stress tensor, gravitational acceleration, and interfacial momentum exchange rate. For instance, the variable α_S defines the volume fraction of each computational cell taken by the sample fluid. And the sum of the volumetric fractions for the sample and the sheath fluids is equal to unity ($\alpha_S + \alpha_{Sh} = \mathbf{1}$). In Eq. (2), the $\overrightarrow{M}_k$ term represents the interfacial momentum transfer due to the effects of drag and virtual mass ($\overrightarrow{M}_k = \overrightarrow{M}_{D,k} + \overrightarrow{M}_{V,k}$). In the momentum equation of the sample fluid, the terms $\overrightarrow{M}_{D,S}$ and $\overrightarrow{M}_{V,S}$ can be described as below.

$$\overrightarrow{M}_{D,S} = \frac{\mathbf{3}}{\mathbf{4}} \frac{C_D}{D} \alpha_{Sh} \rho_S \left| \vec{u}_S - \vec{u}_{Sh} \right| (\vec{u}_S - \vec{u}_{Sh}), \tag{3}$$

$$\overrightarrow{M}_{V,S} = -\alpha_{Sh} \rho_S C_{vm} \left[\left(\frac{\partial \vec{u}_S}{\partial t} + \vec{u}_S \cdot \nabla \vec{u}_S \right) - \left(\frac{\partial \vec{u}_{Sh}}{\partial t} + \vec{u}_{Sh} \cdot \nabla \vec{u}_{Sh} \right) \right]. \tag{4}$$

Where D and C_D are the interfacial characteristic length and the drag coefficient [15, 16]. The symbol C_{vm} is the virtual mass coefficient with a typical value of 0.5 [17]. In addition, the $\overrightarrow{M}_{Sh}$ term on the sheath-fluid part is given as $\overrightarrow{M}_{Sh} = -\overrightarrow{M}_S$.

In the setting of the boundary equations, the velocity and volumetric fraction of the incoming sample flow are specified to be uniform profiles with a given pressure at the inlet. The zero normal gradient of pressure and the no-slip condition are imposed along the wall. At the exit, the flowfield becomes fully developed and all primitive variables are extrapolated from the interior values. The theoretical model is solved using a numerical procedure unifying two sets of conservation equations for the sample and the sheath fluids with the consideration of inter-fluid interactions. An iterative semi-implicit method for pressure-linked equations consistent (SIMPLEC) algorithm is employed for treating the velocity-pressure coupling [18- 20]. The velocities for the sample and the sheath fluids as well as the distributions of pressure and void fractions are solved after completing the calculation procedure.

3 Results and Discussion

The current theoretical model was validated by comparing the predictions with Lee et al.'s experimental observations for a cytometer [10]. The flow cytometer was fabricated on plastic substrates with the sample fluid focused in a quasi-two-dimensional manner. Figure 1 shows the schematic diagram of a flow cytometer and the comparison of the predicted focused width with measured data at different u_{Sh}/u_S ratios. For computation and visualization purposes, the iso-contour of $\alpha_s = 0.5$ was employed to identify the boundary of the sample stream, which in turn determines the focused width. When the u_{Sh}/u_S ratio increases, the calculated focused width tends to be reduced and agrees well with experimental results. The discrepancy of the prediction with the measurement of the focused width is 2.6 % at $u_{Sh}/u_S = 35$.

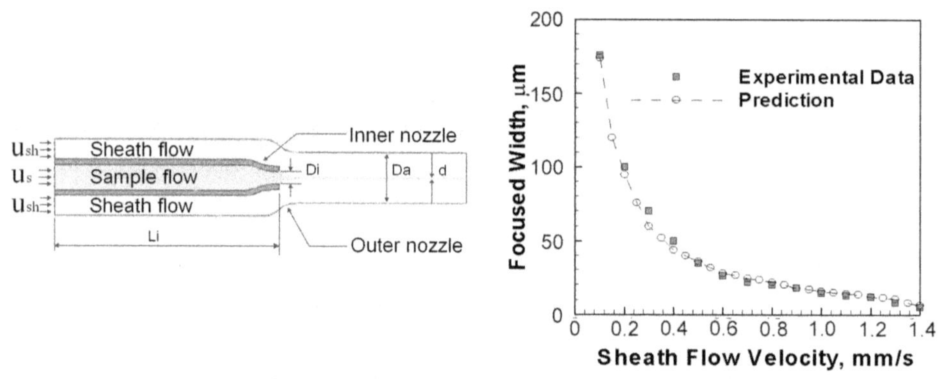

(a) (b)

Fig. 1. (a) Schematic diagram of a flow cytometer; (b) Comparison of the predicted focused width with Lee et al.'s measured data at different u_{Sh}/u_S ratios ($u_S = 0.02$ mm/s, $u_{Sh} = 0.1$-1.4 mm/s)

For three-dimensional calculations, the numerical grid setup of the flow cytometer is displayed in Fig. 2. The dimensions of the proposed cytometer in an axisymmetric configuration are 20 mm in length and 3.6 mm/2.4 mm in diameter for the inflow/ outflow section. The inner nozzle length is 11 mm, while the wall thickness and exit diameter are 0.21 mm and 0.6 mm, respectively. The geometries and design profiles of the inner and outer nozzles can be found in Ref. 10. To capture the sharp variations of the flow properties associated with the flow focusing progress, finer grids have been arranged in the convergent regions of the nozzles and near wall boundaries. The

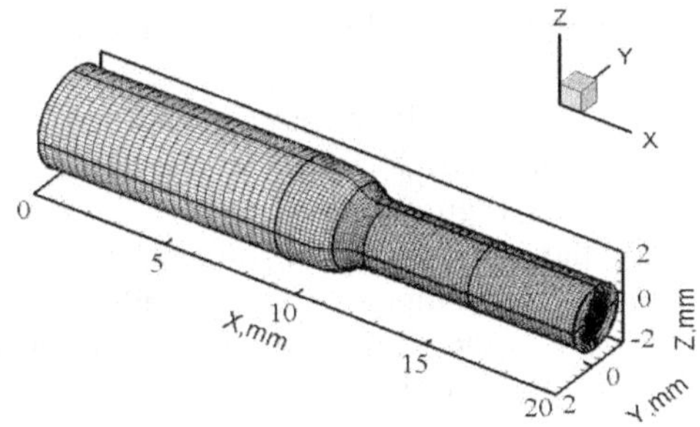

Fig. 2. Numerical grid setup of the proposed flow cytometer

smallest spacing of the grids is around 1.8μm. Using water with dye as the testing liquid, the density and viscosity of the dye-containing sample fluid and the sheath water are 998 kg/m^3 and 0.001 N-s/m^2 for the baseline case.

Figure 3 presents (a) the centerline profile and (b) the transverse profile (at the axial location of 11.2 mm) of gage pressure (P-P_{atm}) in the hydrodynamic focusing flow field for the case of u_{Sh}/u_S= 50. The inlet velocities of the sample and the sheath flows are 0.02 mm/s and 1.0 mm/s, with the annular sheath flow surrounding the core sample fluid issued from the inner nozzle. In general, the pressure decreases with the axial distance from 0.074 N/m^2 at the inlet to 0.068 N/m^2 at the outlet of the flow cytometer. When two streams merge near the exit area of the inner nozzle, the high-velocity sheath water tends to flow inwards and to develop a pressure bulge for compressing and stretching the sample fluid. Hence, the diameter of the sample stream is significantly narrowed when the flow exhibits the phenomenon of hydrodynamic focusing.

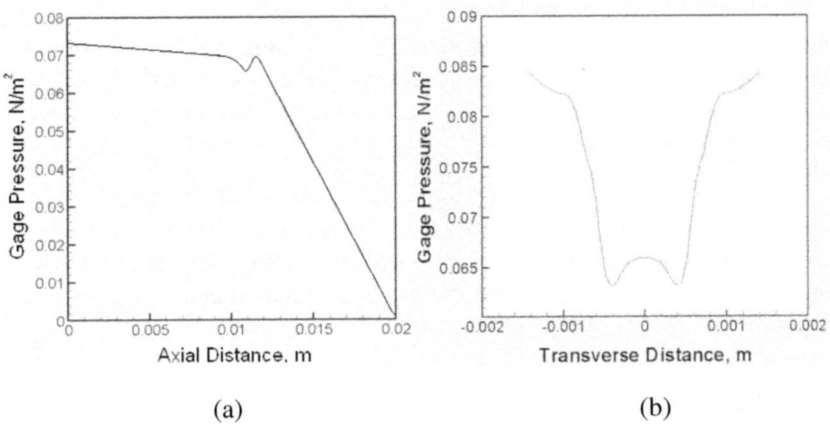

Fig. 3. (a) Centerline pressure profile and (b) transverse pressure profile at the axial location of 11.2 mm of the proposed flow cytometer (u_S = 0.02 mm/s, u_{Sh} =1.0 mm/s)

A lower pressure can also pull the cells to stay in the core region and continue to move individually along the focused stream.

In effect, the human red/white cells have a typical size of 7 μm/10 μm [21]. Accordingly, the required width of the focused sample stream should be around 15 μm to sort and count sample cells. Figure 4 shows the predicted focused width at different u_{Sh}/u_S ratios. In operating the proposed flow cytometer, the focused width decreases as the u_{Sh}/u_S ratio is increased. The computed focused width can lower to 15 μm at u_{Sh}/u_S = 80. In this study, the u_{Sh}/u_S ratio should be regulated to 80 in order to obtain a 15-μm width for holding cells in single file.

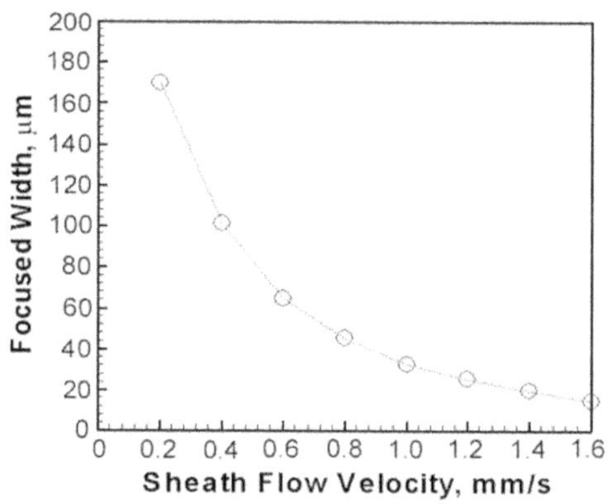

Fig. 4. Focused width at different u_{Sh}/u_S ratios (u_S = 0.02 mm/s, u_{Sh} =0.2-1.6 mm/s)

4 Conclusions

Numerical calculations were performed using a two-fluid theoretical model to simulate the hydrodynamic focusing flow process in a flow cytometer. For the computer code validation in the two-dimensional configuration, the calculated focused width agrees well with the Lee et al.'s measured data for the u_{Sh}/u_S ratio ranging from 5 to 70. The predictions indicate a pressure gap occurred at the merging layer of the sample and sheath streams associated with the hydrodynamic focusing, which can squeeze the sample fluid to form a focused sample stream. The focused width is substantially reduced when varying the u_{Sh}/u_S ratio from 10 to 80. The focused width can be narrowed to 15 μm at u_{Sh}/u_S = 80, indicating that the proposed flow cytometer can achieve a proper focused width to sort and count cells.

Acknowledgments

This study represents part of the results obtained under Contract NSC92-2218-E-194-021, sponsored by the National Science Council of Taiwan, R.O.C.

References

1. Ormerod, M.: Flow Cytometry: A Practical Approach. 3rd Edn. Oxford University Press, Surrey, UK (2000)
2. Shapiro, H. M.: Practical Flow Cytometry: Parameters and Probes. 3rd Edn. Wiley-Liss, New York (1994)
3. Mandy, F. F., Bergeron, M., Minkus, T.: Principles of Flow Cytometry. Transfusion Science, 16 (4) (1995) 303-314
4. Schrum, D., Culbertson, P. T., Jacobson, S. C., Ramsey, J. M.: Microchip Flow Cytometry Using Electrokinetic Focusing. Analytical Chemistry, 71(19) (1999) 4173- 4177
5. Miyake, R., Ohki, H., Yamazaki, I., Yabe, R.: A Development of Micro Sheath Flow Chamber. Proceedings of 4th IEEE MEMS, (1991) 265-270
6. Dietz, L. J., Dubrow, R. S., Manian, B. S., Sizto, N. L.: Volumetric Capillary Cytometry: A New Method for Absolute Cell Enumeration. Cytometry, 23 (1996) 177-186
7. Van, O. A., Keller, R. A., Ambrose, W. P.: High Throughput Flow Cytometric DNA Fragment Sizing. Analytical Chemistry, 72(1) (2000) 37-41
8. McClain, M. A., Culbertson, C. T., Jacobson, S. C., Ramsey, J. M.: Flow Cytometry of Escherichia coli on Microfluidic Devices. Analytical Chemistry, 73 (21) (2001) 5334-5338
9. Lee, G. B., Hwei, B. H., Huang, G.: Micromachined Pre-focused M×N Flow Switches Continuous Multi-sample Injection. Journal of Micromechanics and Microengineering, 11 (2001) 1- 8
10. Lee, G. B., Hung, C. I., Ke, B. J., Huang, G. R., Hwei, B. H., Lai, H. F.: Hydrodynamic Focusing for a Micromachined Flow Cytometer. Journal of Fluids Engineering, 123 (2001) 672- 679
11. Chung, S., Park, S. J., Kim, J. K., Chung, C., Han, D. C., Chang, J. K.: Plastic Microchip Flow Cytometer Based on 2- and 3-Dimensional Hydrodynamic Flow Focusing. Microsystem Technologies, 9 (2003) 525- 533
12. Drew, D. A., Lahey, R. T.: Phase Distribution Mechanisms in Two-phase Flow in a Circular Pipe. J. Fluid Mechanics, 117 (1982) 91-106
13. Park, J. W., Drew, D. A., Lahey, R. T.: The Analysis of Void Wave Propagation in Adiabatic Mono-dispersed Bubbly Two-phase Flows Using an Ensemble-averaged Two-fluid Model. International Journal of Multiphase Flow, 24 (1998) 1205-1244
14. Lahey, R.T., Drew, D. A.: The Analysis of Two-Phase Flow and Heat Transfer Using a Multidimensional, Four Field, Two-fluid Model. Nuclear Engineering and Design, 204 (2001) 29-44
15. Ishii, M., Mishima, K.: Two-fluid Model Hydrodynamic Constitutive Relation. Nuclear Engineering and Design, 82 (1984) 107-126
16. Ishii, M., Zuber, N.: Drag Coefficient and Relative Velocity in Bubbly, Droplet or Particulate Flows. AIChE Journal, 25 (1979) 843-855
17. Lahey, R. T.: The Analysis of Phase Separation and Phase Distribution Phenomena Using Two-fluid models. Nuclear Engineering and Design, 122 (1990) 17-40
18. Van, D. J. P., Raithby, G. D.: Enhancements of the SIMPLE Method for Predicting Incompressible Fluid Flows. Numerical Heat Transfer, 7 (1984) 147-163
19. Jang, D. S., Jetli, R., Acharya, S.: Comparison of the PISO, SIMPLER, and SIMPLEC Algorithms for the Treatment of the Pressure-Velocity Coupling in Steady Flow Problems. Numerical Heat Transfer, 10 (1986) 209-228
20. Patankar, S. V.: Numerical Heat Transfer and Fluid Flow, Hemisphere Publishing, Washington, DC (1980)
21. Mazumdar, J. N.: Biofluid Mechanics, World Scientific Publishing, Singapore (1998)

A Biometric Encryption Approach Incorporating Fingerprint Indexing in Key Generation*

Fengling Han, Jiankun Hu, and Xinghuo Yu

Science, Engineering and Technology Portfolio, Royal Melbourne Institute of Technology, Melbourne, VIC 3001, Australia
{fengling, jiankun}@cs.rmit.edu.au x.yu@rmit.edu.au

Abstract. This paper presents a new biometric encryption protocol in which encryption key is incorporated with fingerprint indexing. Based on the extended chaotic Baker map, the pixel permutation and gray level value substitution are performed to shuffle pixel positions in the original fingerprint images. The encryption key of the pixel permutation is generated by the combination of the random pixel distribution of a fingerprint imprint as well as features used for the fingerprint indexing. In addition to the advantage enjoyed by biometric keys over traditional password/PINs, the proposed biometric encryption approach is very efficient in identity identification within a large database due to the pre-filtering feature of the fingerprint indexing incorporated in the keys. This approach is applicable to the centralized matching scenario where fingerprints need to be encrypted before transmitted. Simulation results have validated the proposed schemes.

1 Introduction

Fingerprint images have long been used for person recognition due to their uniqueness and immutability. Large volumes of fingerprints are collected and used in a wide range of applications, including forensics, access control, and driver license registration [1]. In a centralized system, when used in personal identification, biometric images should be encrypted before transmitted to the central server. Fingerprint representations are of two types: local and global. The uniqueness of a fingerprint can be determined by the global pattern of ridges and valleys as well as the local anomalies [a ridge bifurcation or a ridge ending, called minutiae points] [2].

There are two general ways in which fingerprint based biometric systems are used: verification and identification. In verification, the user inputs a fingerprint image and claims an identity (ID), the system then verifies whether the input image is consistent with the input ID. In an identification system, the user only inputs a fingerprint image, and the system identifies potential corresponding fingerprints in the database [3]. Identification has found most popular applications in the forensic field. Identification is more complex than verification as it includes an additional task of searching the target. This will be a time consuming huge work as the normal biometric database is

* The work is financially supported by the ARC Linkage project LP0455324.

D.-S. Huang, K. Li, and G.W. Irwin (Eds.): ICIC 2006, LNBI 4115, pp. 342 – 351, 2006.

of prohibitive large. Fingerprint classification helps to reduce the workload. The Henry system of fingerprint classification that classifies fingerprints into five classes: Right Loop (R), Left Loop (L), Whorl (W), Arch (A), and Tented Arch (T), is based on the global patterns of ridges, and is commonly used by most of the developers and users [4]. However, the problem with this classification technique is that the number of principal class is small and the fingerprints are unevenly distributed (31.7%, 33.8%, 27.9%, 3.7%, and 2.9% for class R, L, W, A, and T). The classification approach does not narrow down enough search space in the database for efficient identification of a fingerprint. Continuous classification schemes slightly increases the performance, but is still not acceptable for commercial use in large database [9]. To reduce the search time and computational complexity, it is desirable to classify these fingerprints in an accurate and consistent manner such that the input fingerprint needs to be matched only with a subset of the fingerprints in the database [5]. These approaches are called indexing techniques in the fingerprint recognition area, which is achieved by extracting features from the fingerprints and first matching the fingerprints that have the smallest feature distance to the query fingerprint [9]. Some indexing algorithms have been proposed recently [2, 3, 6-11], which are mainly based on three possible indexing features: the registered directional field estimate [10, 11], FingerCode [2] and minutiae triplets [3, 6-9]. Further effort is the combination of above three methods [8, 9]. The global features of a fingerprint are typically used for indexing, but they do not offer very good individual discrimination [2]. Automatic fingerprint identification (AFI) systems, that match a query print against a large database of prints, rely on the pattern of ridges in the query image to narrow their search in the database (fingerprint indexing), and then use the minutiae points to determine an exact fingerprint matching.

AFI systems normally need to send the fingerprint to the back-end server for processing which requires encryption at the sender. A conventional approach is to encode the fingerprint at the pixel level and treat it as a normal text using symmetric encryption scheme. At the receiver side, decryption is done and then global features are retrieved to go through the stages of indexing and matching. There are several issues with such schemes. First, distribution and maintenance of a shared secrete key is a well known hard problem. Secondly it is expensive to encrypt images using conventional cryptography. Researchers are set out to find a better way of combing biometrics and cryptography. A biometric system may be viewed as a pattern recognition system. The hardest problem with biometrics is the unreliability of biometric features used to encode cryptographic keys in the template, while cryptography demands correctness in keys. There have been a number of attempts to bridge the gap between the fuzziness of biometrics and the exactitude of cryptography [16].

To avoid the contradiction of unreliability of biometric template and the exactitude of cryptography key, this paper proposes a new protocol for scenarios where fingerprint images are encrypted at the sender. In this protocol, the parameters of fingerprint indexing with features of minutiae triplet [3] and the random pixel distribution are suggested to be used as the encryption key; and extended chaotic Baker map is employed to permute the pixels position. The advantage of the proposed scheme is that the encryption key is extracted from the query fingerprint image and is generated randomly at each fingerprint impression. Furthermore, the generation of encryption is the byproduct of fingerprint image processing which will save workloads in large fingerprint database. In addition, the encryption key is separated from the fingerprint template which located at central server.

The rest of the paper is organized as follows: Section 2 introduces the extended chaotic Baker map. Incorporation of cipher key with fingerprint indexing is presented in Section 3. Simulations and evaluations are presented in Section 4. Conclusions are given in Section 5.

2 Improved Baker Map for Fingerprint Image Encryption

In order to protect the user's biometric templates in centralized matching circumstance, the fingerprint samples/templates must be encrypted before transmitted. Generally, there are two major approaches that are used to protect digital image/video from attacker. One is information hiding such as digital watermarking of image/video. The other is encryption, which includes conventional encryption and others such as chaotic encryption. There are two classes of key-based cryptographic algorithms, which are symmetric (private-key) and asymmetric (public-key) cryptographic algorithms. Symmetric cryptographic algorithms use the same key for both encryption and decryption, such that the decryption key can easily be derived from the encryption key. Asymmetric cryptographic algorithms use different keys for encryption and decryption. The decryption key cannot be derived from the encryption key. The problem with public-key system is its lower speed compared to that of private-key system. Private-key cryptographic systems, which are designed for text-based encryption, can be used to encrypt image. But the image data must be converted to bit-stream first in order up to match the standard encryption system. With the development of nonlinear dynamics and chaos theory, some common features between chaos and cryptography such as sensitivity to variables and parameters, were revealed, and a number of digital chaotic cryptographic schemes have been proposed [12-15]. But most of chaotic ciphers are vulnerable to attacks for their small key space. To encrypt a fingerprint image, this section introduces the extended chaotic Baker map which is efficiency and has a large key space. A few rounds of iterations will make the original images being permuted thoroughly.

Assume a square image consists of $N \times N$ pixels with L levels of gray, the two-dimensional Baker map B is described as:

$$\begin{cases} B(x,y) = (2x, 0.5y) & 0 \le x < 0.5 \\ B(x,y) = (2x-1, 0.5y+0.5) & 0.5 \le x \le 1 \end{cases} \tag{1}$$

The discretized Baker map is denoted as $B_{(n_1,\ldots,n_k)}$, where the sequence of k integers, $n_1,\ldots, n_k$ is chosen such that each integer n_i divides N, and $n_1+\ldots+n_k = N$. Denoting $N_i = n_1+\ldots+n_{i-1}$, the pixel (r, s), with $N_i \le r < N_i+n_i$, and $0 \le s < N$ is mapped to

$$B_{(n_1,\cdots,n_k)}(r,s) = \left(\frac{N}{n_i}(r-N_i) + s \bmod \frac{N}{n_i}, \frac{n_i}{N}(s - s \bmod \frac{N}{n_i}) + N_i\right). \tag{2}$$

Eq.(2) is used to permute pixels in a square image. In addition to the pixel permutation, a gray level value substitution which can create a random-looking image with a uniform histogram appearing in only one round of iteration should be used. The gray level value substitution is performed as: Consider an $N \times N$ square image with 256 gray levels. Let g_{ij} denote the gray level of the pixel (i, j). A map $\boldsymbol{h}$: $\boldsymbol{N}^N{}_0\times \boldsymbol{N}^N{}_0\times \boldsymbol{N}^L{}_0\rightarrow\boldsymbol{N}^L{}_0$ needs to be found such that the pixel (i, j) with gray level g_{ij} is mapped to $\boldsymbol{B}(i, j)$ with a gray level $h(i, j, g_{ij})$. The three-dimensional map $\boldsymbol{B}_3$: $\boldsymbol{N}^N{}_0\times \boldsymbol{N}^N{}_0\times \boldsymbol{N}^L{}_0 \rightarrow \boldsymbol{N}^N{}_0\times \boldsymbol{N}^N{}_0\times\boldsymbol{N}^L{}_0$ should be invertible to make deciphering possible. This means that $\boldsymbol{B}_3(i, j, g) \neq \boldsymbol{B}_3(i, j, g')$ for each $(i, j)\in \boldsymbol{N}^N{}_0\times \boldsymbol{N}^N{}_0$ and $g, g'\in \boldsymbol{N}^L{}_0$. For example this gray level value permutation can be done with: $\boldsymbol{h}(i, j, g_{ij})= g_{ij} + \hat{\boldsymbol{h}}(i, j) \bmod L$, where $\hat{\boldsymbol{h}}$ is any (possible not one-to-one) function of i and j, and it produces an acceptable map $\boldsymbol{h}$. In this case, $\boldsymbol{h}$ can be interpreted as a simple shift cipher with the shift size $\hat{\boldsymbol{h}}(i, j)$ dependent on the position of the pixel (i, j). The shift size $\hat{\boldsymbol{h}}(i, j)$ could be computed quickly using some bit operations on i and j, or it could be stored in a look up table. The details are referenced to [14].

In [14], the author only presented formula when integer n_i divides N, where a small key space is generated which is weak in security. The improvement proposed in this section is a pixel permutation. With this improvement, a pixel permutation formula that integer n_i does not divide N is presented.

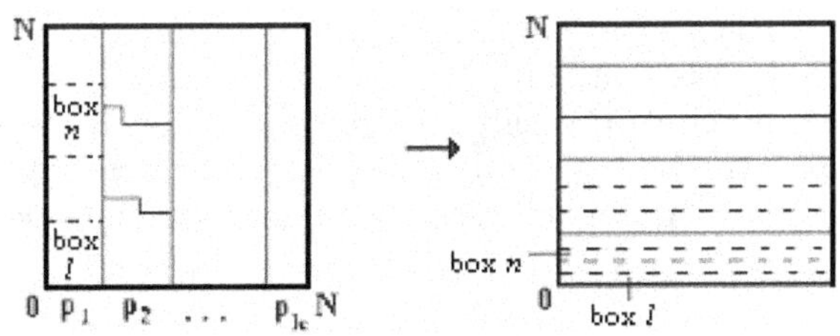

Fig. 1. Permutation of pixels in a square image

In Fig.1, a sequence of k integer, $p_1, \ldots, p_k$, are employed as the encryption key for the permutation of pixels in an $N\times N$ square image, where $p_1 + \ldots + p_k = N$. The image is divided into k vertical rectangles, and there are p_i ($i = 1, \ldots, k$) pixels in one row. The pixels in the original images are represented as $P(r, s)$, $r, s = 1, 2, \ldots N$ are the vertical and pixel horizontal coordinate. After permutation, the pixels are represented as $Q(\cdot,\cdot)$

If p_i divides N, the pixels permutation is similar to formula (2). If p_i does not divide N, inside each column, the pixels are divided into p_i boxes, each box containing exactly N pixels. In this case, the permutation formulae are as follows:

$$Q(N - N_i - j + 1, j) = P(N - (s - 1) - m_2(j), N_i + r) \tag{3}$$

$$r = 1, 2, \ldots, n_i;\ s = 1 - l_{dd};\ \ldots;\ d - l_{uu};\ j = 1, 2, \ldots, n_i;\ i = 1, 2, \ldots, k$$

with

$$d = fix\left(\frac{N - l_d(j) + p_i - 1}{p_i}\right)$$

$$l_{dd} = fix\left(\frac{p_i - j - l_d(j)}{p_i - j - l_d(j) + 0.1}\right)$$

$$l_{uu} = fix\left(\frac{p_i - j - l_u(j)}{p_i - j - l_u(j) + 0.1}\right)$$

and

$$l_u(j) = d - p_i + l_d(j) - N \text{, } l_d(j+1) = l_u(j) \text{ , } l_d(1) = 0$$

$$m_2(j+1) = m_2(j) + d \text{, } m_2(1) = 0$$

where *fix*(*X*) is a function which rounds the elements of *X* to the nearest integers towards zero.

The decryption key is the same with that of encryption key. Only with valid fingerprint, can the fingerprint image be recovered correctly at central server. Then it is possible to get a successful fingerprint matching.

The proposed fingerprint image encryption scheme is more applicable and effective. The encryption key can be an arbitrary combination of integers. It can be designed purposely, for example, extracted from the features of fingerprint images, from random pixels distribution in the fingerprint impressions, or from pseudo random number generator, et al. The only requirement for key is the sum of all digits equals to the pixels number in a row/column (for square image). This chaotic encryption approach has a large key space, and the fingerprint images can be fully recovered.

3 Key Generation by Incorporating Fingerprint Indexing

In this section, we propose a protocol in a fingerprint identification system, in which the encryption key is integrated with fingerprint indexing.

When transmitted over insecure channel, fingerprint images should not be disclosed to any unauthorized persons. They need to be encrypted before transmitted. In addition, a completely loss-free image should be recovered for further matching. To save the computational complexity and meet the above two requirements, we proposed a protocol to transmit the images over public channel:

- Private-key cryptography system, such as extended Baker map, is used in transmitting the encrypted images, while the public key cryptosystem is employed to encrypt the private key.
- The encryption keys are directly derived from the query fingerprint images.

The extended Baker map is used to encrypt the query fingerprint images in this protocol. The pixels permutation, together with gray level values substitution result in a fully recoverable encrypted image with high security and low computation complexity.

The block diagram of incorporating encryption key with fingerprint indexing is as shown in Fig.2.

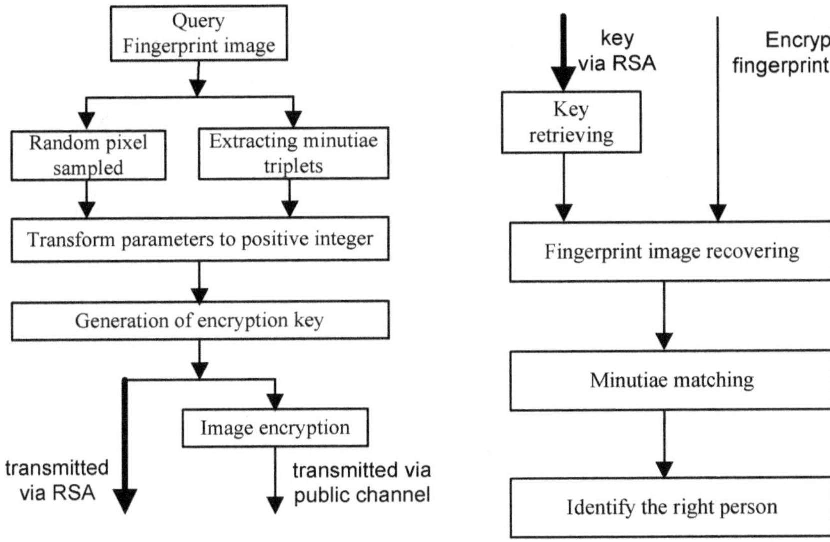

(a) (b)

Fig. 2. Block diagram of incorporating private key with fingerprint indexing; (a) At a terminal. (b) At central server.

As shown in Fig.2(a), the encryption key is directly derived from query fingerprint imprint. After a qualified image is captured, some random pixels are picked up; their coordinates and gray values are processed as part of the private key. Then the image is being pre-processed. The fingerprint indexing method proposed in [3] is suggested to be used here. The features of the query fingerprint $\boldsymbol{H}(\alpha_{min}, \alpha_{med}, \varphi, \eta, \lambda, \chi, \xi)$ (details are referenced to [3]) are extracted and computed, which are transformed into positive integer and used as part of key. After that, improved Baker map is used to encrypt the fingerprint images. The encrypted image is transmitted to the central server via private key cryptosystems, while the key is transmitted to the central server via public key cryptosystems such as RSA. It can be seen from Fig.2(b), after encryption key being retrieved, the minutiae matching is used at last to identify the query fingerprint image. While part of the encryption key is used as fingerprint indexing directly in order to narrow the candidates to be matched.

4 Simulation and Security Analysis

In this section, the proposed fingerprint image encryption which key being derived from features of fingerprint indexing is tested. Simulation results are presented.

Fig.3 shows a 380× 380 gray level fingerprint image and its histogram. Using the proposed extended chaotic Baker map and encryption key being incorporated with fingerprint index scheme, the encryption key is composed of minutiae triplets and random pixels distribution. For the random pixel distribution, for example, coordinates and gray values of ten points in the query fingerprint imprint can be picked up.

For the features of minutiae triplets, twenty minutiae triplets are found, the feature values are computed. These parameters can be divided into five groups, which are the iteration times of extended Baker map. Some transformation and mod operations are conducted to guarantee each digit is a positive integer. And the last digit in each group is the supplementary digit to make the sum of key equals to 380 (pixel number in a row of the square image).

For example, 180 parameters are obtained from the processed minutiae triplets and random pixels distribution in a query fingerprint image. These parameters are further divided into five groups. After processing, the first group contains 25 positive integers: [16,17,6,28,27,25,5,35,15,8,9,5,3,12,6,11,4,33,10,7,16,20,50,4,8], which are used as the encryption key to perform the first round extended Baker map. The encrypted fingerprint image and histogram are shown in Fig.4(a), (b). There are 32, 40, 27, 56 positive integers in the other four groups, which are used as encryption keys in the four iterations followed. After five times extended Baker map with different key each time, the encrypted image and histogram are shown in Fig.4 (c) and (d), respectively.

(a)

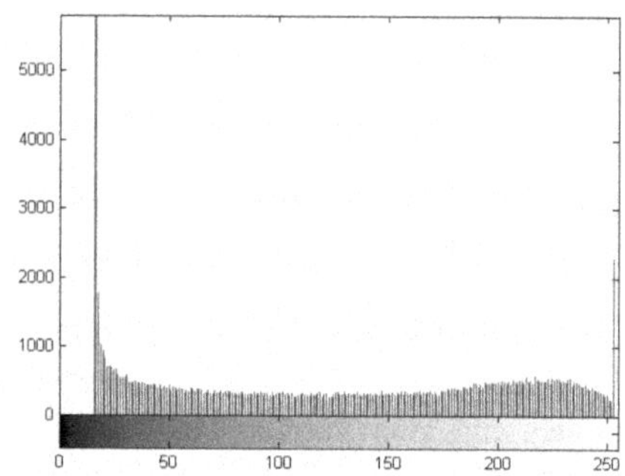

(b)

Fig. 3. Gray level fingerprint image and histogram. (a) Fingerprint image; (b) Histogram.

From Fig.4, one can see that after one round of iteration, the pixels have been permutated significantly, and the histogram has been uniformed. After five rounds of iteration with different encryption key each time, the original image has been scrambled perfectly. One can also see that the histogram in Fig.4(d) has much better statistic character than that in Fig.3(b). So the image can be well hidden.

The decryption is the inverse of the encryption. Only with the right key can the fingerprint image being fully recovered.

In this example, without taking diffusion into consideration, there are 180 elements in the key, and the value of each element can be processed between 0 to 99. Preliminary analysis indicates that the keyspace for five rounds of iteration can be as large as $100^{180} = 10^{360}$. For this fingerprint image, five rounds of pixel permutation should yield a secure cipher which would mean that an exhaustive search of key space is infeasible. The randomness of parameters in individual fingerprint query image can make contributions to the key entropy.

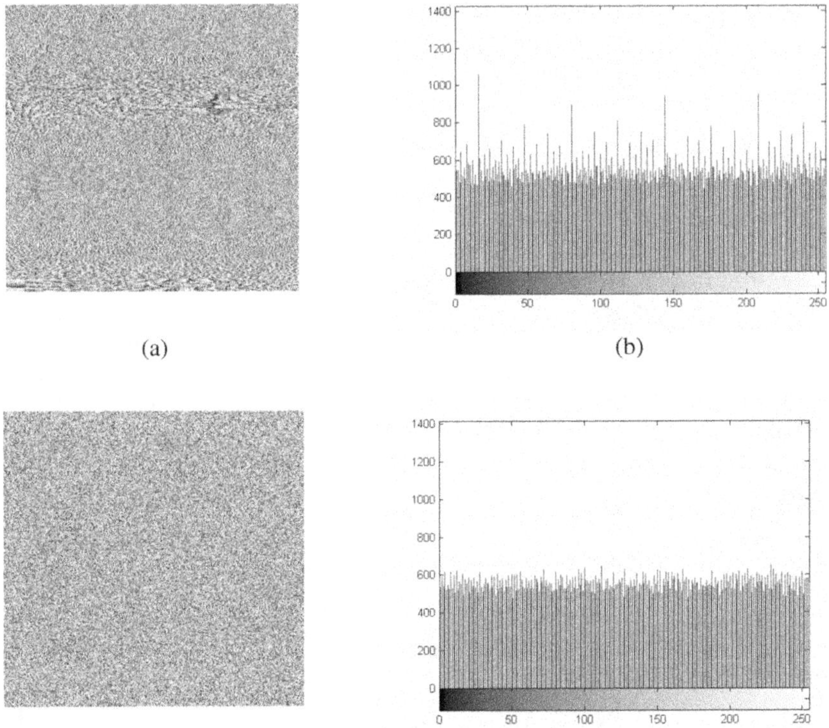

(a) (b)

(c) (d)

Fig. 4. The encrypted images. (a) After one round of iteration; (b) Histogram of Fig.4(a); (c) After five rounds of iterations; (d) Histogram of Fig.4(c).

Remark:

The advantages of the proposed protocol are:

1) The protocol proposed in this paper aiming at generating cryptography key directly from the query fingerprint images. While the objective of Biometric Encryption algorithm in literatures is to provide a mechanism for the linking and subsequent retrieval of a digital key using a biometric such as a fingerprint. This digital key can then be used as a cryptographic key [17].
2) With the proposed encryption key being incorporated with fingerprint indexing scheme, part of the fingerprint matching load in the fingerprint matching stage is performed in the processing of private key generation. This can reduces the computation load of fingerprint matching greatly.
3) The length of the encryption key can be as long as one desired with key being generated from the coordinates and gray values of random picked pixels and from features of minutiae triplets. This could result in a large key space which is vital to the strength of cryptography.

4) At central server, the encrypted fingerprint images are fully recovered, which guarantee the successive fingerprint matching for identify the owner of the query fingerprint image.

5 Conclusions

A biometric encryption protocol which incorporates encryption key with fingerprint indexing has been proposed. The extended chaotic Baker map is employed to encrypt the fingerprint image. The encryption key is directly generated from the query fingerprint image: from features of minutiae triplets and from some random pixels distribution. At the central server, after the key being retrieved and the fingerprint image is fully recovered, then fingerprint image pre-processing are employed as the fingerprint indexing, which will reduce computation load in fingerprint minutiae matching stage significantly. The successful fingerprint minutiae matching at last identify the identity. As the key has included a random pixel distribution at each imprint and it has a large space, this biometric encryption scheme is strong in security.

References

1. Maltoni, D., Maio, D., Jain, A. K., Prabhakar, S.: Handbook of Fingerprint Recognition, Springer, New York (2003)
2. Jain, A. K., Prabhakar, S., Hong, L., Pankanti, S.: Filterbank-based Fingerprint Matching. IEEE Trans. on Image Processing, 9(5) (2000) 846–859
3. Bhanu, B., Tan, X.: Fingerprint Indexing Based on Novel Features of Minutiae Triplet. IEEE Trans. on Pattern Anal. Mach. Intell., 25(5): (2003) 616–622
4. Tan, X., Bhanu, B., Lin, Y.: Fingerprint Classification Based on Learned Features. IEEE Trans. on Systems, Man, and Cyber.–part C: Appl. Reviews, 35(3): (2005) 287–300
5. Jain, A. K., Prabhakar, S., Hong, L., A Multi-channel Approach to Fingerprint Classification. IEEE Trans. on Pattern Anal. Mach. Intell., 21(4): (1999) 348–359
6. Tan, X., Bhanu, B., Lin, Y.: Fingerprint Identification: Classification vs. Indexing. Proc. of the IEEE Conf. on Advanced Video and Signal Based Surveillance, (2003) 151-156
7. Ratha, N. K., Karu, K., Chen, S., Jain, A. K.: A Real-time Matching System for Large Fingerprint Databases. IEEE Trans. on Pattern Anal. Mach. Intell., 18(8): (1996) 799–813
8. Liu, T., Zhu, G., Zhang, C., Hao, P.: Fingerprint Indexing Based on Singular Point Correlation. Proc. of Intern. Conf. on Image Processing. vol.3: (2005) 293-296
9. Boer, J., Bazen, A. M., Gerez,S. H.: Indexing Fingerprint Databases Based on Multiple Features. Workshop on Circuits, Systems and Signal Processing, Veldhoven, The Netherlands. (2001)
10. Nandakumar, K., Jain, A. K.: Local Correlation-based Fingerprint Matching. Proc. of 4th Indian Conf. on Computer Vision, Graphics and Image Processing, Kolkata, (2004)
11. Wang, S., Zhang, W.,Wang, Y.: Fingerprint Classification by Directional Fields. Proc. of 4th IEEE Conf. on Multimodal Interfaces, Pittsburgh, PA, USA. (2002)
12. Yen, J. C., Guo, J. I.: A New Key-based Design for Image Encryption and Decryption, Proc. of the IEEE Circuits and Systems, 4, (2000) 49–52
13. Jakimoski, G., Kocarev, L.: Chaos and Cryptography: Block Encryption Ciphers Based on Chaotic Map, IEEE Trans. on Circuits and Systems - I, (2001) 163–169

14. Fridrich, J.: Symmetric Ciphers Based on Two-dimensional Chaotic Maps, Int. J. Bifurcation and Chaos, 8(6): (1998) 1259–1284
15. Chen, G., Mao, Y., Chui, C.: A Symmetric Encryption Scheme Based on 3D Chaotic Cat Map, Chaos, Solitons & Fractals, 21: (2004) 749–761
16. Hao, F., Anderson, R., Daugman, J.: Combining Cryptography with Biometrics Effectively, U. Cambridge Tech report UCAM-CL-TR-640
17. Biometric Encryption, http://www.bioscrypt.com/assets/biometric_encryption.pdf

A General Solution for the Optimal Superimposition of Protein Structures

Qishen Li, Jian Shu, Zhaojun Shi, and Dandan Zhang

Faculty of Computing, Nanchang Institute of Aeronautical Technology,
Nanchang, Jiangxi 330063, P.R. China
li_qishen@yahoo.com

Abstract. The theorem presented in this paper is a general solution for the optimal superimposition between two protein structures, which is actually the problem of the weighted optimal rigid superimposition between two vector sets with the same size. The theorem gives not only the rotational and translational parameters, but also the minimal value of the mean squared deviation of the optimal superimposition.

1 Introduction

Protein structure alignment is to establish structural equivalences between three-dimensional (3D) structures of two protein folds based on their amino acid positions, which is an indispensable bioinformatics tool for studying protein function and evolution [1, 2], due to the rapid accumulation of protein 3D structures in the Protein Data Bank [3]. Most alignment methods rely on a structure superimposition procedure, which finds the transformation (i.e. rotation and translation) to optimally match the aligned residue pairs of two proteins, in terms of their RMSD (Root Mean Squared Deviation) [4].

The problem of the optimal superimposition between two protein structures is actually the weighted optimal rigid superimposition between two vector sets, which are the 3D coordinates of atoms in the protein molecules. The solution of superimposition has attracted attentions of a number of researchers, notably Mclachlan [5, 7], Kabsch [8, 9], Diamond [10, 11], Ferro & Hermans [12], Lesk [13], Kaindl & Steipe [14] and Steipe [15]. All of the above methods achieve the optimal superimposition by obtaining the optimal rotation matrix under the precondition that the centroids of two vector sets are coincident.

Without the coincidence of both centroids as the precondition, the presented theorem is a general solution for the optimal superimposition, which gives not only the transformational parameters but also the minimal value of the mean squared deviation. This work is inspired by Umeyama's work [16], which is applied in the field of computer vision.

2 Representation of the Superimposition Problem

The structure of a protein molecular that consists of n atoms can be described by n 3D coordinate vectors such as (x_1, y_1, z_1), (x_2, y_2, z_2), …, (x_n, y_n, z_n). Therefore, the

D.-S. Huang, K. Li, and G.W. Irwin (Eds.): ICIC 2006, LNBI 4115, pp. 352–361, 2006.

problem of protein structure optimal superimposition can be stated in mathematics as follows.

Given two sets X_0 and Y_0 of n 3D column vectors x_k and y_k (k = 1, 2, …, n), find a 3×3 orthogonal rotation matrix R with determinant +1 and a 3×1 translation vector t which convert the coordinates x_{ik} (i = 1, 2, 3) to

$$x'_{ik} = \sum_j R_{ij} x_{jk} + t_i , \tag{1}$$

and minimize the function

$$e^2(R,t) = \frac{1}{n} \sum_{ik} w_k (y_{ik} - x'_{ik})^2 , \tag{2}$$

here w_k is a weighting factor assigned to the kth atom. The Eq.(2) can be converted to

$$\begin{aligned} e^2(R,t) &= \frac{1}{n} \sum_k \left\| \sqrt{w_k}\, y_k - \sqrt{w_k}\,(Rx_k + t) \right\|^2 \\ &= \frac{1}{n} \sum_k \left\| \tilde{y}_k - (R\tilde{x}_k + \sqrt{w_k}\, t) \right\|^2 \\ &= \frac{1}{n} \left\| Y - (RX + tw^T) \right\|^2 , \end{aligned} \tag{3}$$

where $\tilde{x}_k = \sqrt{w_k}\, x_k$, $\tilde{y}_k = \sqrt{w_k}\, y_k$, X and Y are two sets consisting of n column vectors of $\tilde{x}_k$ and $\tilde{y}_k$ respectively, $w = (\sqrt{w_1}, \sqrt{w_2}, \cdots, \sqrt{w_n})^T$. Thus, the problem of the optimal superimposition between two protein structures is converted to obtain the rotational and translational parameters R and t of the transformation between two vector sets X and Y.

3 Theorem of a General Solution for Optimal Superimposition

Umeyama proposed a very useful theorem for finding the similarity transformational parameters (rotation, translation, and scaling) between two point sets that give the least mean squared error [16]. In this section, we derive a valuable theorem, which gives a general solution for the considered problem of optimal superimposition by introducing weighting factors to the vector sets and canceling the scaling component of the transformation considered in Umeyama's theorem. Not only the rotational and translational parameters, but also the minimal value of the mean squared deviation of the optimal superimposition is obtained from the datum of the superimposed vector sets in the presented theorem. Before showing the theorem, we introduce a lemma presented and proved in Umeyama's correspondence [16], in which the rotational parameters of least-squares residual are given. The proof of the Lemma is shown in Appendix.

Lemma
Let A and B be $m \times n$ matrices, and R an $m \times m$ rotation matrix, and UDV^T a singular value decomposition of AB^T ($UU^T = VV^T = I$, $D = \mathrm{diag}(d_i)$, $d_1 \geq d_2 \geq \cdots \geq d_m \geq 0$). Then the minimum value of $\|A - RB\|^2$ with respect to R can written as

$$\min_R \|A - RB\|^2 = \|A\|^2 + \|B\|^2 - 2\mathrm{tr}(DS), \tag{4}$$

where

$$S = \begin{cases} I, & \text{if } \det(AB^T) \geq 0; \\ \mathrm{diag}(1,1,\cdots,1,-1), & \text{if } \det(AB^T) < 0. \end{cases} \tag{5}$$

When rank (AB^T) $\geq m-1$, the optimum rotation matrix R which achieves the above minimum value is uniquely determined as

$$R = USV^T, \tag{6}$$

where S in Eq.(6) must be chosen as

$$S = \begin{cases} I, & \text{if } \det(U)\det(V) = 1; \\ \mathrm{diag}(1,1,\cdots,1,-1), & \text{if } \det(U)\det(V) = -1, \end{cases} \tag{7}$$

when det (AB^T) = 0 (rank (AB^T) = $m-1$).

We can derive the following theorem using this lemma.

Theorem
Let $X = \{x_1, x_2, \ldots, x_n\}$ and $Y = \{y_1, y_2, \ldots, y_n\}$ be corresponding point sets in m-dimensional space. The mean squared deviation of the optimal rigid superimposition between these two point sets with respect to the transformational parameters [R (rotation) and t (translation)] is expressed by

$$e^2(R,t) = \frac{1}{n}\sum_{i=1}^{n} \left\| Y - (RX + tw^T) \right\|^2. \tag{8}$$

The minimum value ε^2 of the mean squared deviation $e^2(R,t)$ is given as the following:

$$\varepsilon^2 = \frac{1}{n}\left\{ \|YK\|^2 + \|XK\|^2 - 2\mathrm{tr}(DS) \right\}, \tag{9}$$

where

$$K = I - \frac{1}{w_0} ww^T, \tag{10}$$

here, I is the identity matrix, $w_0 = \sum_k w_k$, D is derived by letting a singular value decomposition of $YKK^T X^T$ be UDV^T, the expression of D is $D = \mathrm{diag}(d_i)$, $d_1 \geq d_2 \geq d_m \geq 0$, and

$$S = \begin{cases} I, & \text{if } \det(YKK^T X^T) \geq 0; \\ \mathrm{diag}(1,1,\cdots,1,-1), & \text{if } \det(YKK^T X^T) < 0. \end{cases} \tag{11}$$

When $\mathrm{rank}(YKK^T X^T) \geq m-1$, the optimum transformational parameters are determined uniquely as the listed below:

$$R = USV^T, \tag{12}$$

$$t = \frac{1}{w_0}(Yw - RXw), \tag{13}$$

where S in Eq.(12) must be chosen as

$$S = \begin{cases} I, & \text{if } \det(U)\det(V) = 1; \\ \mathrm{diag}(1,1,\cdots,1,-1), & \text{if } \det(U)\det(V) = -1, \end{cases} \tag{14}$$

when $\mathrm{rank}(YKK^T X^T) = m-1$.

Proof of the Theorem

To prove the theorem, we introduce an $n \times n$ symmetry matrix $K = I - \frac{1}{w_0} ww^T$ ($K = K^T$). Here, $w_0 = \sum_k w_k$. Using the following equations,

$$X = XK + \frac{1}{w_0} Xww^T, \tag{15}$$

$$Y = YK + \frac{1}{w_0} Yww^T, \tag{16}$$

the mean squared deviation $e^2(R,t)$ is further reformulated as

$$\begin{aligned} e^2(R,t) &= \frac{1}{n} \| YK + \frac{1}{w_0} Yww^T - RXK - \frac{1}{w_0} RXww^T - tw^T \|^2 \\ &= \frac{1}{n} \| YK - RXK + (\frac{1}{w_0} Yw - \frac{1}{w_0} RXw - t)w^T \|^2 \\ &= \frac{1}{n} \| YK - RXK - t'w^T \|^2 \\ &= \frac{1}{n} \{ \| YK - RXK \|^2 + \| t'w^T \|^2 - 2\mathrm{tr}(K^T(Y^T - X^T R^T)t'w^T \}, \end{aligned} \tag{17}$$

where

$$t' = -\frac{1}{w_0} Yw + \frac{1}{w_0} RXw + t \cdot \tag{18}$$

Since the following equations

$$\begin{aligned} \mathrm{tr}(K^T(Y^T - X^T R^T)t'w^T) &= \mathrm{tr}(w^T(I - \frac{1}{w_0} ww^T)(Y^T - X^T R^T)t') \\ &= \mathrm{tr}((w^T - \frac{1}{w_0} w^T ww^T)(Y^T - X^T R^T)t') \\ &= \mathrm{tr}((w^T - w^T)(Y^T - X^T R^T)t') \\ &= 0, \end{aligned} \tag{19}$$

and

$$\left\|t'w^T\right\|^2 = w_0 \left\|t'\right\|^2, \tag{20}$$

are derived, we have

$$e^2(R,t) = \frac{1}{n}\left\|YK - RXK\right\|^2 + \frac{w_0}{n}\left\|t'\right\|^2. \tag{21}$$

From Eq.(21), t' must be equal to 0 in order to minimize $e^2(R,t)$, that is,

$$t = \frac{1}{w_0} Yw - \frac{1}{w_0} RXw = \frac{1}{w_0}(Yw - RXw) \cdot \tag{22}$$

Next, let UDV^T be a singular value decomposition of $YKK^T X^T$. Then, the minimum value e^2 of $\frac{1}{n}\left\|YK - RXK\right\|^2$ with respect to R is given from the above lemma as follows:

$$\varepsilon^2 = \frac{1}{n}\{\left\|YK\right\|^2 + \left\|XK\right\|^2 - 2\mathrm{tr}(DS)\}, \tag{23}$$

where

$$S = \begin{cases} I, & \text{if } \det(YKK^T X^T) \geq 0; \\ \mathrm{diag}(1,1,\cdots,1,-1), & \text{if } \det(YKK^T X^T) < 0. \end{cases} \tag{24}$$

Also from the lemma, if $\mathrm{rank}(YKK^T X^T) \geq m-1$,

$$R = USV^T, \tag{25}$$

where S must be chosen as

$$S = \begin{cases} I, & \text{if } \det(U)\det(V) = 1; \\ \mathrm{diag}\,(1,1,\cdots,1,-1), & \text{if } \det(U)\det(V) = -1, \end{cases} \tag{26}$$

when $\mathrm{rank}(YKK^T X^T) = m-1$.

The theorem is proved. **Q.E.D.**

The presented theorem is an extension of Umeyama's theorem [16]. It is proved by replacing c and h in Umeyama's formulae with 1 and w, respectively. Accordingly, n is replaced with w_0 in the computation of K. Here c is the scaling parameter of transformation and $h = (1, 1, \ldots, 1)^T$.

4 Examples

In order to elucidate the validity of the proposed theorem, some numerical results of RMSD are shown in Table 1. Since the superimposition method requires the same number of pairs of vector sets, N residues are selected to be superimposed from the beginning of the N-terminal of the protein. Generally, according to the atoms of protein molecules used in the superimposition process, researchers adopt two kinds of superimposition. One is when only the alpha-carbon atom; the other is where four atoms, nitrogen, alpha-carbon, carbon and oxygen, of the peptide backbone are superimposed for per residue. The weighting factors assigned to these four kinds of atoms in this paper are 0.6, 1, 0.8 and 0.9, which may not have any biological significance but demonstrates the RMSD results. When only alpha-carbon atom of each residue is used, the number of atom pairs in the superimposition sequence is N. When four atoms (nitrogen, alpha-carbon, carbon and oxygen) of each residue are used, the number of atom pairs is $4 \times N$. In this experiment, three pairs of protein molecules are superimposed. The values of RMSD listed in table 1 are then calculated with pairs of the transformed vector sets. It is indicated that the same RMSD values are derived as applying Kabsch's method [8]-[9].

Table 1. Some results of RMSD between two sets of atom pairs

Atoms used	PDB pair IDs	RMSD (Å)		
		N=10	N=50	N=100
alpha-carbon only	1LYZ .vs. 2LZM	1.26	1.69	1.48
	1HFC .vs. 1MNC	1.03	0.52	0.37
	1ULB .vs. 2CTC	1.34	1.45	1.25
nitrogen, alpha-carbon, carbon and oxygen	1LYZ .vs. 2LZM	0.58	0.83	0.73
	1HFC .vs. 1MNC	0.49	0.25	0.18
	1ULB .vs. 2CTC	0.64	0.72	0.62

Formally, given N atom positions from protein structures x and y respectively with the weighting factors $w(i)$, the RMSD is defined as:

$$\mathrm{RMSD}(N;\, x,\, y) = \left[\frac{\sum_{i=1}^{N} w_i \left\| x_i - y_i \right\|^2}{N \sum_{i=1}^{N} w_i} \right]^{\frac{1}{2}} . \tag{27}$$

5 Conclusion

A general solution for the optimal superimposition between two protein structures, which is actually the weighted optimal rigid superimposition between two 3D vector sets, is presented in this paper. Not only the rotational and translational parameters, but also the minimal value of the mean squared deviation of the optimal superimposition is obtained from the datum of the superimposed vector sets in the presented theorem.

Acknowledgments

This work was partially supported by the doctoral research project through Nanchang Institute of Aeronautical Technology.

References

1. Holm L., and Sander C.: Searching Protein Structure Databases has come of Age. Proteins, 19(3) (1994) 165–173
2. Gibrat J. F., Madej T., and Bryant S. H.: Surprising Similarities in Structure Comparison. Curr. Opin. Struct. Biol., 6(3) (1996) 377–385
3. Berman H. M., Westbrook J., Feng Z., Gilliland G., Bhat T. N., Weissig H., Shindyalov I. N., and Bourne P. E.: The Protein Data Bank. Nucleic Acids Res., 28(1) (2000) 235–242
4. Kolodny R., Koehl P. and Levitt M.: Comprehensive Evaluation of Protein Structure Alignment Methods: Scoring by Geometric Measures. J. Mol. Biol., 346(4) (2005) 1173–1188
5. Mcachlan A. D.: A Mathematical Procedure for Superimposing Atomic Coordinates of Proteins. Acta Crystallogr. Sect. A, 28(6) (1972) 656–657
6. Mcachlan A. D.: Gene Duplications in the Structural Evolution of Chymotrypsin. J. Mol. Biol., 128(1) (1979) 49–79
7. Mcachlan A. D.: Rapid Comparison of Protein Structures. Acta Crystallogr. Sect. A, 38(6) (1982) 871–873
8. Kabsch W.: A Solution for the best Rotation to Relate two Sets of Vectors. Acta Crystallogr. Sect. A, 32(5) (1976) 922–923
9. Kabsch W.: A Discussion of the Solution for the Best Rotation to Relate two Sets of Vectors. Acta Crystallogr. Sect. A, 34(5) (1978) 827−828
10. Diamond R.: On the Comparison of Conformations Using Linear and Quadratic Transformations. Acta Crystallogr. Sect. A, 32(1) (1976) 1–10
11. Diamond R.: A Note on the Rotational Superposition Problem. Acta Crystallogr. Sect. A, 44(2) (1988) 211–216
12. Ferro D. R., and Hermans J.: A Different Best Rigid-body Molecular Fit Routine. Acta Crystallogr. Sect. A, 33(2) (1977) 345–347
13. Lesk A. M.: A Toolkit for Computational Molecular Biology. II. On the Optimal Superpose- tion of two Sets of Coordinates. Acta Crystallogr. Sect. A, 42(2) (1986) 110–113
14. Kaindl K. and Steipe B.: Metric Properties of the Root-Mean-Square Deviation of Vector Sets. Acta Crystallogr. Sect. A, 53(6) (1997) 809

15. Steipe B.: A Revised Proof of the Metric Properties of Optimally Superimposed Vector Sets. Acta Crystallogr. Sect. A, 58(5) (2002) 506
16. Umeyama S.: Least-squares Estimation of Transformation Parameters Between two Point Patterns. IEEE Trans. on Pattern Anal. Mach. Intell., 13(4) (1991) 376–386

Appendix: Proof of the Lemma

Define an objective function F as

$$F = \|A - RB\|^2 + \mathrm{tr}(L(R^T R - I)) + g\{\det(R) - 1\}, \tag{28}$$

where g is a Lagrange multiplier and L is a symmetric matrix of Lagrange multipliers. The second and third term of F represent the conditions for R to be an orthogonal and proper rotation matrix respectively. Partial differentiations of F with respect to R, L and g lead to the following system of equations:

$$\frac{\partial F}{\partial R} = -2AB^T + 2RBB^T + 2RL + gR = 0, \tag{29}$$

$$\frac{\partial F}{\partial L} = R^T R - I = 0, \tag{30}$$

$$\frac{\partial F}{\partial g} = \det(R) - 1 = 0, \tag{31}$$

where we used

$$\frac{\partial}{\partial L}\det(R) = \mathrm{adj}(R^T) = \det(R^T)(RT)^{-1} = R, \tag{32}$$

since R is rotation matrix (adj(R^T) is an adjoint matrix of R^T).

From Eq.(29),

$$RL' = AB^T \text{, where } L' = BB^T + L + \frac{1}{2}gI . \tag{33}$$

By transposing the both sides of Eq.(33), we obtain the following equation (note that L' is symmetric),

$$L'R^T = BA^T . \tag{34}$$

If we multiply each side of Eq.(33) with each side of Eq.(34), respectively, Eq.(35) is obtained since $R^T R = I$,

$$L'^2 = BA^T AB^T = VD^2V^T . \tag{35}$$

Obviously L' and L'^2 are commutative ($L'L'^2 = L'^2L'$), hence both can be reduced to diagonal forms by the same orthogonal matrix.

Thus we can write

$$L' = VDSV^T, \tag{36}$$

where $S = \mathrm{diag}(s_i)$, $s_i = 1$, or -1.

Now, from Eq.(36),

$$\det(L') = \det(VDSV^T) = \det(V)\det(D)\det(S)\det(V^T) = \det(D)\det(S). \tag{37}$$

On the other hand, from Eq.(33)

$$\det(L') = \det(R^T AB^T) = \det(R^T)\det(AB^T) = \det(AB^T). \tag{38}$$

Thus,

$$\det(D)\det(S) = \det(AB^T). \tag{39}$$

Since singular values are nonnegative, $\det(D) = d_1 d_2 \cdots d_m \geq 0$. Hence $\det(S)$ must be equal to 1 when $\det(AB^T) > 0$, and -1 when $\det(AB^T) < 0$.

Next, extremum values of $\|A - RB\|^2$ is derived as follows: from Eq.(33),

$$\begin{aligned} \|A - RB\|^2 &= \|A\|^2 + \|B\|^2 - 2\mathrm{tr}(AB^T R^T) \\ &= \|A\|^2 + \|B\|^2 - 2\mathrm{tr}(R^T AB^T) \\ &= \|A\|^2 + \|B\|^2 - 2\mathrm{tr}(L'). \end{aligned} \tag{40}$$

Substituting Eq.(36) into Eq.(40), we have

$$\begin{aligned} \|A - RB\|^2 &= \|A\|^2 + \|B\|^2 - 2\mathrm{tr}(VDSV^T) \\ &= \|A\|^2 + \|B\|^2 - 2\mathrm{tr}(DS) \\ &= \|A\|^2 + \|B\|^2 - 2(d_1 s_1 + d_2 s_2 + \cdots + d_m s_m). \end{aligned} \tag{41}$$

Thus, the minimum value of $\|A - RB\|^2$ is obviously achieved when $s_1 = s_2 = \cdots = s_m = 1$, if $\det(AB^T) \geq 0$, and $s_1 = s_2 = \cdots = s_{m-1} = 1$, $s_m = -1$, if $\det(AB^T) < 0$. This concludes the first half of the lemma.

Next, we determine a rotation matrix R achieving the above minimum value. When $\mathrm{rank}(AB^T) = m$, L' is nonsingular, thus it has its inverse $L'^{-1} = (VDSV^T)^{-1} = VS^{-1}D^{-1}V^T = VD^{-1}SV^T$ (note that $S^{-1} = S$, $SD^{-1} = D^{-1}S$). Therefore, from Eq.(33) we have

$$R = AB^T L'^{-1} = UDV^T VD^{-1}SV^T = USV^T. \tag{42}$$

Finally, when $\mathrm{rank}(AB^T) = m-1$, from Eq.(33) and Eq.(36)

$$RVDSV^T = UDV^T. \tag{43}$$

Multiplying V by both sides of Eq.(43) from the right and using $DS = D$ (since $d_m = 0$ and $s_1 = s_2 = \cdots s_{m-1} = 1$),

$$RVD = UD \tag{44}$$

is obtained. If we define an orthogonal matrix Q as follows,

$$Q = U^T RV , \tag{45}$$

we have

$$QD = D . \tag{46}$$

Let the column vectors of Q be $q_1, q_2, \cdots, q_m$ $(Q = [q_1, q_2, \cdots, q_m])$. The following equations are obtained by comparing both sides of Eq.(46),

$$d_i q_i = d_i e_i , \quad 1 \le i \le m-1 . \tag{47}$$

Hence,

$$q_i = e_i , \quad 1 \le i \le m-1 , \tag{48}$$

where e_i is a unit vector which has 1 as an ith element,

$$e_i = (0,0,\cdots,\overset{i}{1},\cdots,0)^T . \tag{49}$$

The last column vector q_m of Q is orthogonal to all other vectors q_i $(1 \le i \le m-1)$ since Q is an orthogonal matrix. Thus we have

$$q_m = e_m \quad \text{or} \quad q_m = -e_m . \tag{50}$$

On the other hand,

$$\det(Q) = \det(U^T)\det(R)\det(V) = \det(U)\det(V) . \tag{51}$$

Thus, $\det(Q) = 1$, if $\det(V) = 1$ and $\det(Q) = -1$, if $\det(U) \det(V) = -1$. Therefore we have

$$R = UQV^T = USV^T , \tag{52}$$

where

$$S = \begin{cases} I, & \text{if } \det(U)\det(V) = 1 ; \\ \operatorname{diag}(1,1,\cdots,1,-1), & \text{if } \det(U)\det(V) = -1 . \end{cases} \tag{53}$$

Q.E.D.

A Personalized Biological Data Management System Based on BSML

Kwang Su Jung, Sunshin Kim, and Keun Ho Ryu*

Database/Bioinformatics Laboratory, School of Electrical & Computer Engineering, Chungbuk National University, Cheongju, 361-763, Korea
{ksjung, sskim04, rkryu}@Chungbuk.ac.kr
http://dblab.chungbuk.ac.kr

Abstract. A mass of biological sequence and structural information have been produced in biological laboratories since the techniques to get the sequences of genomes or proteins have been improved in HGP (Human Genome Project). Unfortunately, there are scarcely software packages available to deal with the biological data in most of biological laboratories and they are just stored in file formats. An integrated management system of biological data is required to manage the sequence and annotation data taken from other open databases to improve the analysis of sequence data in biological laboratories. We therefore suggest a personalized system to edit, store and retrieve biological information, and convert the formats of sequence data, as well as to integrate and manage data.

1 Introduction

Biologists connect the public biological databases and retrieve sequences which are similar to sequences they have, then the results of retrieval are used in homology research, functional analysis, and prediction. Unfortunately, there are scarcely software packages available to deal with the sequence data in most biological laboratories and they are just stored in file formats.

The integration and management technique of heterogeneous sequence data from public sequence databases is widely used to manage diverse information and prediction. Thus a database management technique that is suitable for biological data is required. In particular, an integrated data model which handles the modification of program and data is needed for analysis on the various programs.

In this work, our system supports editing and converting among heterogeneous public biological database flat files as well as sequences users produce in laboratories. BSML (Bioinformatic Sequence Markup Language) based on XML is used for representing the complex and hierarchical biological data. Biologists easily analyze their accumulated biological data using our system and apply the results of biological data analysis to medical science and pharmacy at the level of biological laboratories.

---

* Corresponding author.

D.-S. Huang, K. Li, and G.W. Irwin (Eds.): ICIC 2006, LNBI 4115, pp. 362 – 371, 2006.

2 Related Work

There are some program packages that handle biological data. We will briefly introduce main function, its features, and formats which these analysis tools support.

2.1 BSML

BSML [1] (http://www.bsml.org/) is one of open standards of XML data in bioinformatics research. BSML represents good bio-physical features as well as visualization of biological sequence data comparing with other languages such as abstract annotations. It also supports converting formats among biological flat files base on flat-form independence. In this point, BSML is more complete and specifiable than other XML formats in bioinformatics. BSML which our system supports is used in the area of representing gene research results and biological molecules, and exchanging them, hence many laboratories and company choose BSML as a new standard.

2.2 Genomic Workspace

Recent tools which convert formats among standards are implemented by JAVA or Perl module. One of these worthy converting softwares is the Genomic Workspace [2]. It converts Genbank [3] flat file into BSML format and includes Genomic Viewer for a good graphic visualization of biological information. The Genomic Workspace doesn't have an own local database so that biological information which users modified is not stored into database.

2.3 Staden Package

Staden Package [4] is developed for managing and analyzing the sequence from sequencing machine by Medical Research Council Laboratory in U.K. It is constituted of several tools. *Trev* represents raw experimental file from sequencing machine. *Trace_diff* describes the mutation information of reference and trace data. *GAP4* edits contigs and assembly of sequences. *Pregap4* is used for pre-processing the assembly data. Spin analyzes the result sequence such as retrieving similar sequences and some other operations.

Staden Package analyzes and simply manages sequences in sequence experimental files but doesn't store sequences into database and retrieve. It can't process a mass of sequences.

2.4 Sequence Data Formats

Each flat file from public biological database has different format. Genbank, Swiss-Prot [5], PIR-Codata [6], and PDB [7] flat files are generally used. Also, each sequence analyzing software uses different format. One of them, FASTA [8] established by Pearson is the most common format.

The problems of biological formats we mentioned above are following. The formats can be modified when they are released. To understand the range of field and its value is difficult and data types in the same field in each format can be different. The conversion among different formats needs different parsers to extract the user-interesting field.

```
> SQ2002060300001
GGTACCTTCTGAGGCGGAAAGAACCAG
CCGGATCCCTCGAGGGATCCAGACATG
CTTACCGGATACCTG
```

Fig. 1. An example of FASTA format

Thus applying XML is essential to visualize and exchange the sequence data and bio-molecules because XML has high level format conversion technology and re-usability. There are a lot of markup languages to represent the genome data such as BSML, BioML (Biopolymer Markup Language) [9]. The public biological databases [3], [5], [6], [7] are also trying to produce the XML based formats to represent biological information.

3 System Architecture

Fig. 2 shows the system architecture we suggest. We briefly explain each component of our system which consists of File Transformation, Biological Information Editor, Query/Retrieval Processor, and Biological Data Storage Manager. In particular, Biological Information Editor constitutes of Mypage Editor, Annotation Editor, and Sequence Editor.

3.1 File Transformation

File Transformation Module extracts meaningful data fields from each flat file in order to integrate heterogeneous formats from various public biological databases. The Flat File Parser which is suitable for each flat file is developed to extract the interesting fields from divergent biological flat files. The integrated BSML model is created with extracted data. We developed each Flat File Parser such as Genbank, Swiss-Prot, PIR-Codata, PDB, FASTA, BSML, and so on. The mapping information between the parsed fields and BSML elements (attributes) is considered in order to load on the system.

3.2 Biological Information Editor

Biological Information Editor edits annotation and sequence data. Our Editor consists of MyPage Editor, Annotation Editor, and Sequence Editor.

MyPage Editor supports to make user defined formats that consist of interesting fields extracted from various flat files such as Genbank, Swiss-Prot, PDB, PIR-Codata, BSML, FASTA, and data on the local database. Once MyPage format is created by user, and then modification, insertion, and deletion of the field in the format are also performed by MyPage Editor. The user defined format through MyPage Editor is stored as an XML format on the disk or local database.

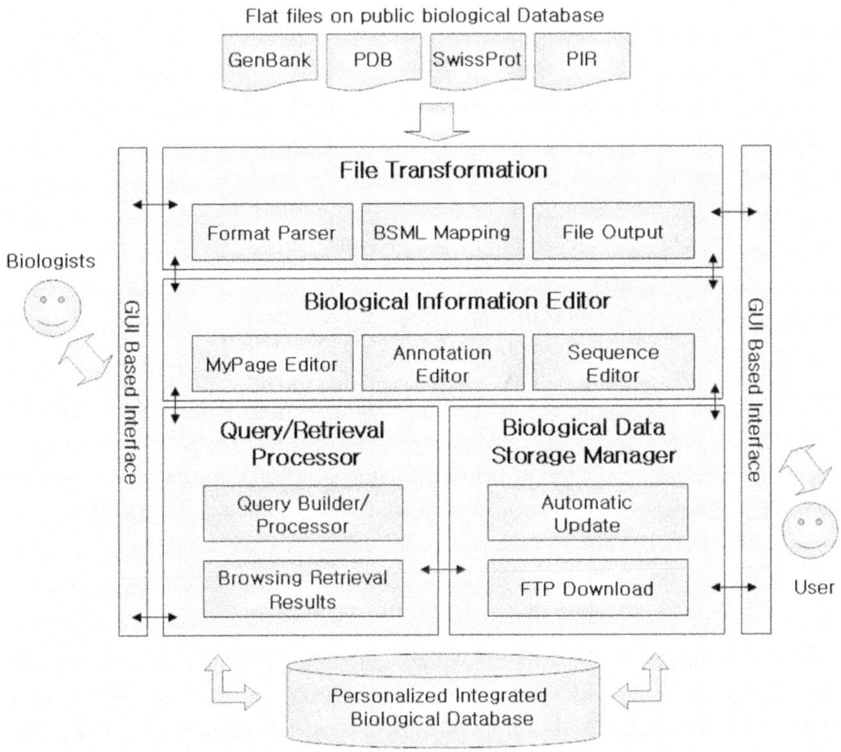

Fig. 2. System Architecture

While MyPage Editor creates user-defined formats and performs insertion, modification, and deletion of the interesting field extracted from each format, Annotation Editor performs the same operations for annotation data in each format. Therefore original source (flat file) is maintained their own frame. The modified original source stored as BSML format on the disk or local database if the user wants. Annotation Editor mainly handles annotation data of interesting sequence such as types, taxonomic information, citations, reference data, authors, and specific sites of sequences and so on.

Sequence Editor supports the useful operations to handle sequence, so it produces a new sequence from original sequence. We will briefly introduce several operations in

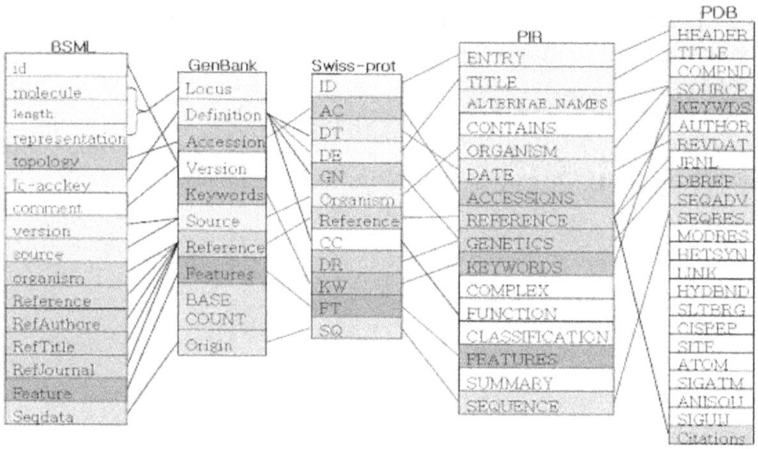

Fig. 3. Mapping Information among biological Formats

Fig. 4. *Base Composition* operation calculates the composition of DNA (Deoxyribo-Nucleic Acid) and shows us the percentage and the numbers of each Nucleotide (Adenine, Guanine, Cytosine, and Thymine). *Set-Subsequence* operation makes a new sequence from original sequence with starting and ending index which the user defines. *Complementary* operation produces complementary sequence of the original sequence. *Transcription* operation transforms the original sequence which is proven as gene, into mRNA (Messenger Ribo-Nucleic Acid) sequence. *Translation* operation transforms mRNA sequence into Protein sequence.

3.3 Query/Retrieval Processor

Through GUI (Graphic User Interface) environments, biologists input keywords to retrieve and Query/Retrieval Processor retrieves the local database and returns the suitable results which are compatible with the user-selected options such as AND and OR. Query results are displayed on the MyPage Editor when retrieving user-defined XML format through MyPage Editor. Also the results are showed on Annotation Editor when retrieving annotation data.

Query results on the MyPage and Annotation Editor are also modified by adequate Editors whenever user wants.

3.4 Biological Data Storage Manager

Sequence and its annotation data are separately stored because of the fast accession to biological data and retrieval. When biologists want to make other format from the database, for example FASTA format, the necessary data from the database are retrieved in order to make a new format.

The original Sequence:

```
01234567890123456789012345678901234567890123456789
atgttctacgtggttatatgcatactggaagcctacgacgtcatctaa
          10                20                30                40
```

(a) Base composition

A: 13 bp T: 15 bp G: 10 bp C: 10 bp

(b) Set-Subsequence

```
01234567890123456789012345678901234567890123456789
tctacgtggttatatgcatactggaagcctacgacgtcatc
5                15                25                35                45
```

(c) Complementary

```
01234567890123456789012345678901234567890123456789
ttagatgacgtcgtaggcttccagtatgcatataaccacgtagaacat
          10                20                30                40
```

(d) Transcription

```
01234567890123456789012345678901234567890123456789
auguucuacgugguuauaugcauacuggaagccuacgacgucaucuaa
          10                20                30                40
```

(e) Translation

```
01234567890123456789012345678901234567890123456789
Mfyvvicileaydvi
          10                20
```

Fig. 4. Sequence Operations

Our database maintains up-to-date biological information through automatic FTP (File Transfer Protocol) download and update. Users can monitor new release information of public biological database on the web using our system. Users store the absolute FTP path in our FTP lists. When biological databases release a new version, what users have to do is just choosing appropriate FTP path in FTP lists.

A new flat file from FTP downloading is compared with existing entry in the database to be updated. In this procedure, the system asks whether the database is updated as a new flat file or not.

4 Implementation

Implementation environments are as the following. MySQL 4.0.17 is used to store genomic sequence and annotation data. JAVA SWING by Sun-Mirco System is used to implement the major interface and JAXP (JAVA API for XML Processing) v1.2 for parsing XML documents. Our system is distributed on the web (http://dblab.chungbuk.ac.kr/~SeqMan) and an archive contains install-files such as MySQL 4.0.17 and our implementations. We have two examples to show our implementation.

Example 1. User wants to make Complementary, Transcription, Translation, and Set-Subsequence operations of original sequence (locus name 'DRONCX') in Genbank. The upper window in Fig. 5 shows the original sequence and the lower table shows the derived sequences from operations.

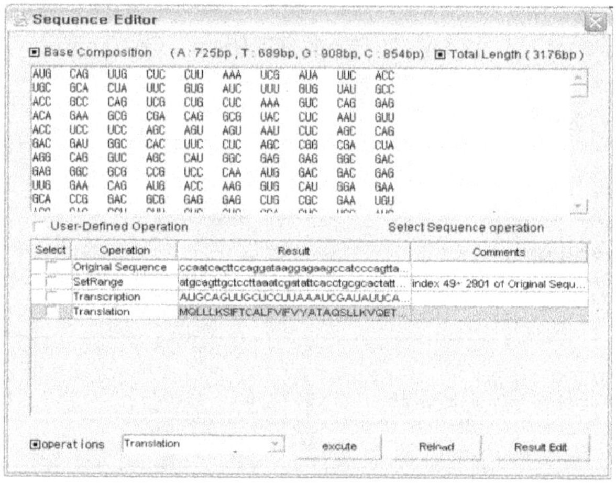

Fig. 5. Sequence Editor

Example 2. User wants to create a user defined format that consists of interesting fields from various file formats on MyPage Editor. Fig. 6 illustrates how a user defined format is created from Swiss-Prot (id 100K_RAT), Genbank (locus name DRONCX), and PIR-Codata (entry CCHU).

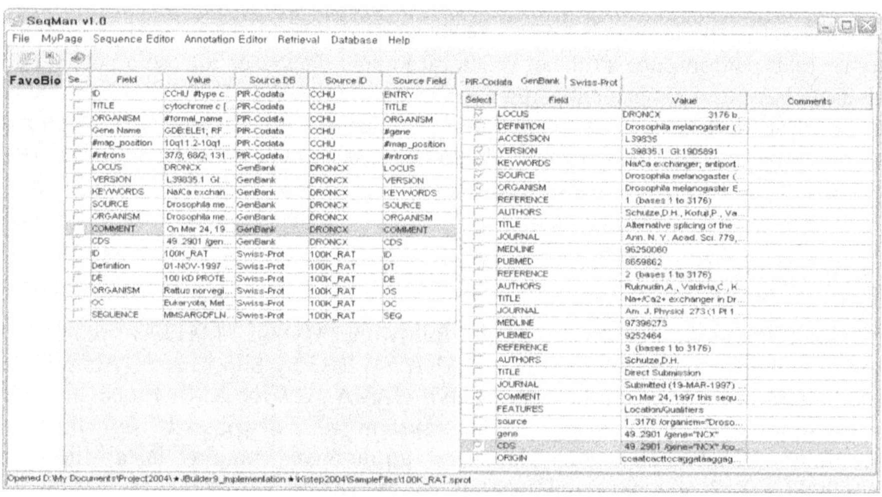

Fig. 6. MyPage Editor

5 Evaluation

In this paper, we implement new issues that we mentioned in introduction with deep consideration. The evaluation of the proposed system is performed according to comparing with Staden Package, Genbank, Genomic Workspace, and so on. Table 1 shows the result of comparison.

Table 1. Comparison with existing Sequence Management Systems

Item	Staden Package	Genbank	Genomic Workspace	Proposed System
Sequence Manipulation	Support	None	None	Support
File Format	EMBL format	Genbank /ASN.1 /BSML	BSML	XML Format based on BSML DTD
Storing Sequence	File	DBMS	BSML Documents/ Text	DBMS
Sequence Versions	None	Support	None	Support
File Transformation	None	None	Support	Support

Storing the historical data of released sequence (version) is useful to analyze the change of protein sequence. We expect that new useful protein and drug can be discovered based on analysis of version sequences.

Creating and designing new proteins from experiment and analyzing them using Sequence Editor are possible. The purpose of a PCR (Polymerase Chain Reaction) is to make huge copies of a gene. This is necessary to have enough starting template for sequencing. In this work, a complementary sequence of DNA strand is needed and a known Primer is required to synthesize. Therefore sequence operations we suggest are useful enough to design Primer.

Sequence information in our system is represented as XML format and converted into other formats to support efficient sequence analysis and easy format exchanging. That is, because of XML representation based on BSML DTD (Document Type Definition), sequence information is easily converted into other formats which are used in various analysis programs. FASTA is the most common format that the most analyzing software uses. One of advantages of our system is converting sequence from database or various sources into FASTA format.

6 Conclusion

According to improving techniques to get the biological information, a mass of biological data have been produced. Characteristics of these biological data including genome sequences are heterogeneous and various. Although the need of management systems which should reflect biological characteristic has been raised, the most current biological databases provide restricted function and act as repositories for biological data.

Therefore, this paper describes the management system of genome sequences and annotation data in order to manipulate the formats in bioinformatics. It includes format transforming, editing, storing and retrieving among collected nucleotide sequences from public databases, and handles sequence produced by experiments. It uses BSML format which is based on XML as a common format in order to extract data fields and exchange heterogeneous sequence formats. To manage sequences and their changes, that is, version management system for originated DNA is also useful to analyze and synthesize a new protein. This system automatically updates information in local database using FTP. This research is widely used for a long term study of biology, medical science, pharmacy and helps to accumulate the own sequence information at the level of biological laboratories.

Now, we are trying to extend our system in order to improve the analyzing ability. For instance, inserting sequence similarity module (such as Blast [10] and Psi-Blast [11]) and one-touch retrieval module which queries to public biological databases with one interface. After we implement sequence similarity checking modules, users don't need to connect sequence analyzing site on the web. In this process, we prevent our valuable information in the laboratory from revealing to exterior. One-touch querying interface using web agent and synonym semantic search make users retrieve results from more than one public database at same time, with one interface.

Acknowledgments

This work was supported by the Regional Research Centers Program of Ministry of Education & Human Resources Development in Korea.

References

1. Spitzner, J.: BSML(Bioinformatics Sequence Markup Language Manual) 3.1 Tutorials. LabBook Inc. (2002)
2. LabBook Genomic Workspace, LabBook Inc. (2005)
3. Benson, D.A., Karsch-Mizrachi, I., Lipman, D.J., Ostell, J., Rapp, B.A., Wheeler, D.L.: GenBank. Nucleic Acids Res., Vol. 30 (2002) 17-20
4. Bonfiled, J., Beal, K.F., Jordan, M., Cheng, Y., Staden, R.: The Staden Package Manual. Medical Research Council Labortory of Molecular Biology (2001)
5. Schneider, M., Tognolli, M., Bairoch, A.: The Swiss-Prot Protein Knowledgebase and ExPASy: Providing the Plant Ccommunity with High Quality Proteomic Data and Tools. Plant Physiol Biochem., Vol. 42 (2004) 1013-1021

6. Wu, C.H., Yeh, L.L., Huang, H., Arminski, L., Castro-Alvear, J., Chen, Y., Hu, Z. Kourtesis, P., Ledley, R.S., Suzek, B.E., Vinayaka, C.R., Zhang, J., Barker, W.C.: The Protein Information Resource. Nucleic Acids Res., Vol. 31 (2003) 345-347
7. Deshpande, N., Addess, K.J., Bluhm , W.F., Merino-Ott, J.C., Townsed-Merino, W., Zhang, Q., Knezevich, C., Chen, L., Feng, Z., Green, R.K., Flippen-Anderson, J.L., Westbrook, J.: The RCSB Protein Data Bank: a Redesigned Query System and Relational Database Based on the mmCIF Schema. Nucleic Acids Res., Vol. 33 (2005) 233-237
8. Pearson, W.R., Lipman, D.J.: Improved Tools for Biological Sequence Comparison. Proc. Narl. Acad. Sci, Vol. 85 (1988) 2444–2448.
9. Fenyo D.: The Biopolymer Markup Language. Oxford University Press (1999)
10. Altschul, S.F., Gish, W., Miller, W., Myers, E.W., Lipman, D.J.: Basic Local Alignment Search Tool. J. Mol. Biol., Vol. 215 (1990) 403-410
11. Altschul, S.F., Madden, T.L., Schaffer, A.A., Zhang, J., Zhang, Z., Miller, W., Lipman, D.J.: Gapped BLAST and PSI-BLAST:a New Generation of Protein Database Search Programs. Nucleic Acids Res., Vol. 25 (1997) 3389-3402
12. Gu, M.S., Hwang, J.H., Ryu, K.H.: Designing the Ontology of XML Documents Semi-Automatically. ICIC (2005) 818-827
13. Park, S.H., Ryu, K.H., Gilber, D.: Fast Smilarity Search for Protein 3d Structures Using Topological Pattern Matching Based on Spatial Relations. Int. J. Neural System, Vol. 15 (2005) 287-296
14. Park, S.H., Ryu, K.H.: Effective Filtering for Structural Similarity Search in Protein 3D Structure Databases. DEXA (2004) 761-770
15. Park, S.H., Ryu, K.H.: Fast Similarity Search for Protein 3D Strucure Databases Using Spatial Topological Patterns. DEXA (2004) 771-780
16. Park, S.H., Ryu, K.H.: Fast Filtering of Structural Similarity Search Using Discovery of Topological Patterns. IDEAL (2004) 396-401
17. Park, S.H., Ryu, K.H., Son, H.S.: A Protein Structural Information Management Based on Spatial Database and Active Trigger Rule. DEXA (2003) 413-422
18. Park, S.H., Ryu, K.H., Son, H.S.: Protein Structure Modeling Using a Spatial Model for Structure Comparison. IDEAL (2003) 490-497

An Automatic Nematode Identification Method Based on Locomotion Patterns

Bai-Tao Zhou[1] and Joong-Hwan Baek[2]

[1] Multimedia Retrieval Lab, School of Electronics and Communication Engineering, Hankuk Aviation University, South Korea
zhou@mail.hangkong.ac.kr
[2] School of Electronics and Communication Engineering, Hankuk Aviation University, South Korea
jhbaek@mail.hangkong.ac.kr

Abstract. Nematodes are primitive creatures that are endangering and devouring many of the essential resources beneficial to human beings. For effective inspection and quarantine, we have devised an image based system for quantitatively characterizing and identifying nematodes, and achieved average successful identification rate of 71.2% for the Uncoordinated (Unc) mutant types and 91.2% for other types. To enhance system performance, here we introduce the worm-body Trunk Coordinate System for defining and characterizing the locomotion patterns of representative mutants. At least 60,000 frames for each species, totally 480,000 frames, representing wild type and 7 other mutant types, were analyzed. The average correct classification rate, measured by Classification and Regression Tree (CART) algorithm, was 79.3% for Unc types. The scheme devised and the features extracted are good supplements for the previous automated nematode identification system.

1 Introduction

Nematodes are the most numerous multicellar animals on the earth, and there are nearly 20,000 described species classified in the phylum *Nemata*. Many of them are parasites of insects, plants and animals. Among them, some species are threatening the development of farming and forestry all over the world. Lesion nematode, *Pratylenchus Goodeyi*, for example, claims as high as 47% U.S. commercial pineapple reduction centered in Hawaii in 2002, according to the NCFAP report [6]. So detection and identification of nematodes are the prerequisite for controlling and checking them from spread. However the subtle differences between some species make it difficult to distinguish them simply by naked-eye observation. Therefore there exists the challenge of precisely defining and the quantitatively measuring nematode behavior patterns.

One way to address these problems is using compute-driven systems for recording and analyzing nematode behaviors. Some scholars have been doing relevant jobs with different methods and different features. Geng et al. [3] solved behavior phenotype classification of some *C. elegans*. Their work focused on the movement of centroid, the location of head and tail, and the position relationship between head to tail, head

D.-S. Huang, K. Li, and G.W. Irwin (Eds.): ICIC 2006, LNBI 4115, pp. 372 – 380, 2006.

to centroid, and tail to centroid. Silva et al. [7] proposed an intelligent nematode detection system. But the system only detected whether worms exist or not. We built an automatic identification system based on the head and tail locomotion patterns in the worm-body Trunk Coordinate System.

Based on our previous works [1] [9], we built an advanced system that can automatically identify different nematodes using Trunk Polar Coordinate System. The system can get approximately 80% for Unc types, and 90% or higher success rates for other types after we added some new locomotion features to head and tail upon the existing features. And the processing procedure in this system can be divided into several stages, seen in Fig. 1.

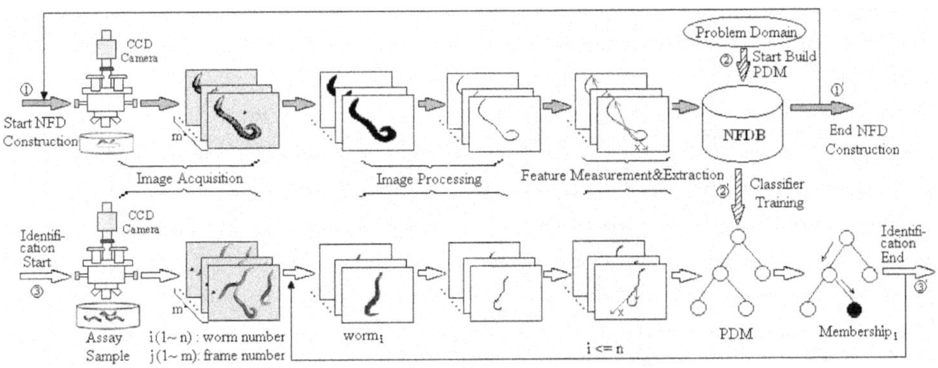

Fig. 1. The system architecture and processing stages

The system starts with constructing a general Nematode Feature Database (NFDB). By feeding Image Acquisition Block (IAB) each time with a new type-known assay sample, after Image Processing Block (IPB) and Feature Extraction Block (FEB), NFDB can record as much species information as possible so as to let it cover most of the problem domains. In the second phase, the system builds the Problem Domain Model (PDM) by querying NFDB. The querying terms come from the domain requirements, and the queried result is the domain-scope nematode features. These features, as training data, are used to build the classifier model for that domain nematode identification. In the third phase, the classifier decides the assay nematode memberships by feeding the assay sample features to the PDM.

The paper is organized according to the processing stages. The image acquisition and processing, feature extraction, and feature selection are discussed in section 2, section 3, and section 4 respectively. Section 5 gives the *C. elegans* worm experiment results. Section 6 concludes this article.

2 Image Acquisition and Processing

The original images are acquired from a stereomicroscope mounted with a device video camera [1], an automatic track system motorized microscope that could record an individual animal's behaviors at high magnification. In our system, the *C. elegans* locomotion was tracked with a stereomicroscope mounted with a high performance CCD video camera. For each species, in order to get one time-continuous footage, around 600 frames of that species were snapped with the frequency of 2Hz for time span of about 5 minutes, and this process was implemented 100 times for that species. So, at least 60,000 frames for each species, totally 480,000 frames, representing 8 types of sample worms, were analyzed. These captured image frames were trimmed by the smallest axis-aligned rectangles that contain the worm body, and then saved as eight-bit grayscale data with time-index.

To facilitate feature extraction, the grayscale data were subjected to image processing to reduce irrelevant information or noise and to enhance the image properties. Besides, considering the features to be extracted from skeleton images, thinning and pruning are necessary operations. The image processing includes several procedures shown in Fig. 2.

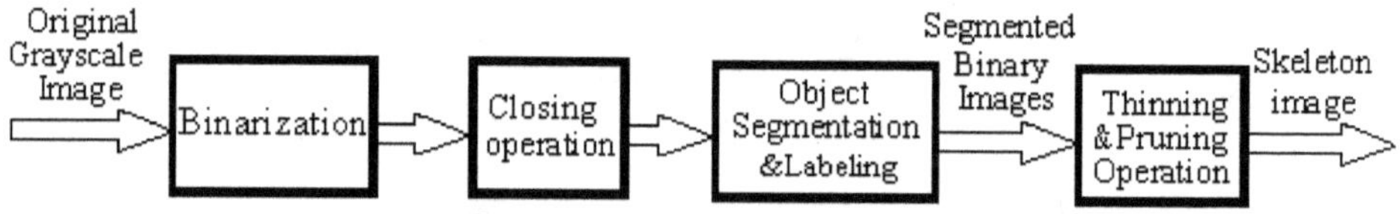

Fig. 2. Image processing procedure

The first step is to convert gray level image into binary image with a local threshold using a 5*5 scanning window (Fig. 3b). But we need decide the threshold value to discern the worm body and background by using statistic data such as histogram. Based on the histogram of the gray images, two peak values representing worm body and background can be found respectively according the distribution of the worm body gray value and the background gray value, and the relationship between them can be found. While considering the influence of factors as light and point of view, a moving window was scanned over the trimmed image, and the grayscale images can be converted smoothly into binary images according to the computation of the mean and standard deviation of the pixel inside the moving window.

The second step is to clean the spots inside the worm body with a morphological closing operator [4]. Because of the influence of light and view, some spots can be seen inside the worm body. Such morphological operations as dilation and erosion can be used to clean these spots (Fig. 3c). Yet these operations can also remove the looped holes at the same time. To remove the spots inside the worm body and to remain the holes simultaneously, we measure thickness of these two regions with 16 vectors (Fig. 3g). The number of pixels, or thickness, from the centroid to the background in each vector direction is counted for every spot region. If the total thickness is less than a threshold, in our experiment it is 25, the region is considered as noise. Otherwise, we preserve the region as a hole.

The third step is to remove isolated small objects with the sequential algorithm for component labeling [5]. The connected components were labeled by scanning the image in x and y directions sequentially, and the different components were segmented into different areas. The areas containing isolated objects such as eggs are removed by setting off pixel (Fig. 3d).

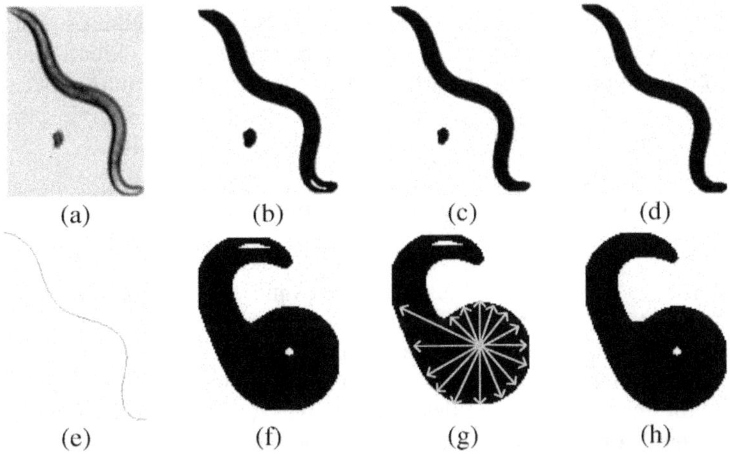

Fig. 3. (a) Original gray level image, (b)(f) Binary image after thresholding operation, (c) Binary image following closing operation, (d) Final clean binary image after removal of isolated object, (e) Skeleton obtained through thinning and pruning, (g) Hole detection using 16 measuring vectors, (h) The result of hole detection and noise removal

In addition, the morphological skeleton was obtained by applying a skeleton algorithm [8]. Finally, by iterating dilation operation along the binary image, a clean skeleton can be obtained. And by tracing the skeleton, we got the pixel sequence on the skeleton, which will be used to sample points.

3 Feature Extraction

Based on the segmented binary image and skeleton image, we can extract the features to describe the body shape and locomotion pattern of the Nematode. In order to avoid the extreme values introduced by noise, top 10% minimum value and top maximum values are discarded. At the same time, we use max, min and average to analyze some features. The measured features in our system include the minimum, maximum, and average values of the following: distance moved in 0.5, 5, 10,15, 20, 25, and 30s; and 5 min, number of reversals in 40, 80, and 120s; and 5 min, worm area, worm length, width at center and head/tail, ratio of thickness to length, fatness, eccentricity and lengths of major/minor axes of best-fit ellipse, height and width of minimum enclosing rectangle (MER), ratio of MER width and height, ratio of worm area to MER area, angle change rate. These features can be seen as [9]. Here we just discuss the new feature details.

We kept the features to describe the worm shape and movement in our previous system [9]. The features to describe the body shapes and locomotion patterns can be extracted based on the segmented binary images and skeleton images. The measured features in our system include the centroid moving distance, number of reversals, worm area, worm length, worm width, fatness, eccentricity and lengths of major/minor axes of best-fit ellipse, height and width of minimum enclosing rectangle (MER), angle change rate etc.

But as we mentioned in our previous work [9], using these features cannot classify the Unc classes well. In order to improve the performance of the identification results, we added the locomotion trend information of the nematodes into the feature space using the moving distance and absolute angle change of the sample points. In the following part, we will explain the details about these new features.

3.1 Head and Tail Recognition for Each Frame

Our work is based on two clues that come from human observers: 1) the movement of the worm starts from the head; 2) the head moves more frequently than the tail. Based on the domain knowledge, first, we can recognize the head and tail, and then select some sample points from the skeleton and contour to describe the locomotion current. We just used the movement degree to discern the head and tail in our experiments.

Based on the image processing described above, we can get binary image, skeleton image and contour image. We can get two end points and the skeleton pixel sequence. These two end points will be divided into two groups (group1 and group2) according to the position similarity between two consecutive frames. Let *Endptm*(*t*)=(*endptmx*(*t*), *endptmy*(*t*)) denotes the *m*th endpoint in frame *t*, and *Endptn*(*t*+1)=(*endptnx*(*t*+1), *endptny*(*t*+1)) as the *n*th endpoint in frame *t*+1. The distance between endpoint *n* in frame *t*+1 and the endpoint *m* in frame *t* is d(*m*,*n*). We initialize the *group1* as (*endpt1x*, *endpt1y*) and *group2* as (*endpt2x*, *endpt2y*).

For each following frame, if the head and tail are not being accidentally flipped, the *k*th endpoint in frame (*t*+1) *Endptk*(*t*+1) will be assigned as group *g* like the equations (1) provided that another point $\overline{Endptv(t+1)}$ which was assumed into the other group $\overline{g}$, and the distance between $\overline{Endptv(t+1)}$ and the endpoint that belongs to group $\overline{g}$ in frame *t* is not the maximum at the same time. Here g(*v*) is the group to which the *v*th endpoint in frame t belongs. This group rule actually is implemented according to the coordinate of the end points in adjacent two frames.

$$
\begin{aligned}
&(k,g)=\arg(k,g(v))\{\mathrm{d}(Endptk(t+1),Endptv(t))=\min\{\mathrm{d}(m,n)\}\} \\
&\text{where } \mathrm{d}(\overline{Endptk(t+1)},\overline{Endptv(t)}) \neq \max\{\mathrm{d}(m,n)\} \\
&m=\{Endpt1(t),\ Endpt2(t)\},\ n=\{Endpt1(t+1),\ Endpt2(t+1)\}
\end{aligned}
\tag{1}
$$

For these groups, we can mark them head group (*Headg*) with bigger offset or tail group (*Tailg*) with smaller offset.

3.2 Feature Extraction from Head and Tail Segments

We described the worm movement using sample point positions from head and tail segments in the Trunk Polar Coordinate System (Fig 4). From the figure, we can see the head and tail move trend can be reflected well under this new coordinate system. The head and tail segments are 1/3 skeleton-length. 10 points are sampled from two segments respectively. The center position on the skeleton is set as original point and the best-fit line from a 9-pixel long segment on the skeleton list near this position is set as the major axis, and the positive direction (p_x, p_y) is assigned as the same side

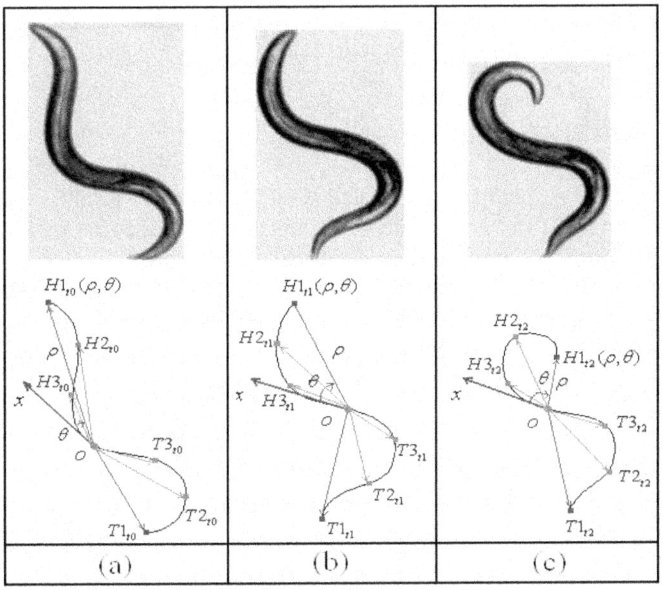

Fig. 4. Feature extracting based on Trunk Polar Coordinate System among three consecutive frames; the overstriking blue narrow line is the polar axis; the other narrow lines are the sample point vectors

with the head side as equation 2. The distance and angle of the sample point (ρ_i, θ_i) in this new system are the measurements by using equation 3 and 4.

$$(p_x, p_y)=\arg(l_{xk}, l_{yk}) \left\{ \cos^{-1} \frac{(x_i\text{-}x_c).l_{xk} + (y_i\text{-}y_c).l_{yk}}{\sqrt{(x_i\text{-}x_c)^2 + (y_i\text{-}y_c)^2} + \sqrt{l_{xk}^{\ 2} + l_{yk}^{\ 2}}} > 0 \right\}, \tag{2}$$

where $k \in \{1,2\}$, (l_{x1}, l_{y1}) is the vector that x increases on the fitting line, (l_{x2}, l_{y2}) is the opposite vector of (l_{x1}, l_{y1})

$$\rho_i = \sqrt{(x_i\text{-}x_c)^2 + (y_i\text{-}y_c)^2}\ , \tag{3}$$

where (x_c, y_c) is the center position on the skeleton

$$\theta_i = \cos^{-1} \frac{(x_i\text{-}x_c).p_x + (y_i\text{-}y_c).p_y}{\sqrt{(x_i\text{-}x_c)^2 + (y_i\text{-}y_c)^2} + \sqrt{p_x^{\ 2} + p_y^{\ 2}}}, \tag{4}$$

where (p_x, p_y) is the direction vector of the positive polar axis.

The difference of these two measurements $(\Delta\rho_i, \Delta\theta_i)$ is used to describe the moving trend. $\Delta\rho_i$ is the distance offset of the *i*th sample point to the original point in this new coordinator system from the *t*th frame to the (*t*+1)th frame. And the minimum, average and maximum statistic of these measurements are used in our system as new features to improve the identification performance.

4 Feature Selection

After adding the new features, we used the classification and regression tree (CART) algorithm [2] to evaluate the importance of the feature variables and 10-fold cross validation to determine the performance and the right size of the optimal tree. The CART approach involves recording a set of examples of each worm type according to the problem domain modeling. In our system, we set Unc type as the problem domain. And CART is used to evaluate the feature importance to classify or to identify Unc types. From the measured featured obtained in section 3, a training vector is generated, each vector element recording the values for each feature measurement. Using this learning sample, CART produced a binary classification tree in which each binary split of the data involves a splitting question of the form 'Is $x_{m\leq}$c?', where x_m is the measurements, and c is a threshold. The root node of the tree contains all the training cases; the worm types are equally mixed together in this node.

In our experiment, the original measurement vector is consisted of 94 features used in our previous system [9] and 12 new features as discussed in section 3. According to the report from CART, we built an optimal classification tree using only 14 parameters which are listed in Table 1. Here the point worthy mentioning is that the average distance offset of head, the average offset distance of tail, and the maximum head angle offset are added as important new features. And the cross validation classification probability for each type is given in Table 2. The success rates are listed along the diagonal and the misclassification error rates are listed in the off-diagonal entries. Compared with the results in [9], we can see the distinctly improved performance of *unc-36* and *unc-38*.

Table 1. Important features identified by CART

Category	Description
LNMFRMAX	Max ratio of worm length to MER fill
MVHLFAVG	Average distance moved in 0.5 seconds
ANCHRMAX	Max angle change rate
TOTRV	Total number of reversals in 5 minutes
CNTLRMIN	Min ratio of center thickness to length
RV80MAX	Max number of reversals in 40 sec/ 80 frames
LPSUM	Total number of frames the worm is looped
AREAMAX	Max area of the worm
CNTMVAVG	Average thickness at the center position
MVHLFMAX	Max distance moved in 0.5 seconds
LNMFRAVG	Average ratio of worm length to MER fill
HDISOFFAVG	Average offset distance of head
HANGOFFMAX	Max head angle offset
TDISOFFAVG	Average offset distance of tail

5 Experiment Results

We programmed in Visual C++ to process the image to get the filtered binary image and skeleton image. Based on these two images, some shape features such as length, fatness, best-fit ellipse height and width, worm area and so on and some locomotion features like reversals, angle change rate, the number of holes are computed. And the two endpoints in every frame are divided into two groups, and the sample point positions at the Trunk Polar Coordinate System are computed using the Visual C++ program. The offsets of the head position and tail position are obtained by using MATLAB codes based on the coordinate data in Visual C++. And we also built the optimized classification tree using CART, which is used to evaluate the feature importance and to select features.

Table 2. Classification results from 10-fold cross validation

Actual	Predicted							
	Wild	Goa-1	Nic-1	Unc-36	Unc-38	Egl-19	Unc-2	Unc-29
Wild	0.951	0.000	0.000	0.000	0.000	0.000	0.020	0.030
Goa-1	0.010	0.933	0.000	0.020	0.010	0.000	0.000	0.020
Nic-1	0.000	0.000	0.890	0.000	0.060	0.000	0.000	0.000
Unc-36	0.000	0.000	0.000	0.800	0.050	0.020	0.120	0.050
Unc-38	0.000	0.000	0.000	0.040	0.847	0.000	0.050	0.170
Egl-19	0.000	0.000	0.000	0.020	0.000	0.933	0.000	0.020
Unc-2	0.033	0.067	0.017	0.150	0.100	0.000	0.767	0.000
Unc-29	0.000	0.000	0.000	0.120	0.080	0.010	0.110	0.767

Table 3. The comparison of the proposed method with the previous method in [9]

Methods	Predicted (Success Ratio)							
	Wild	Goa-1	Nic-1	Unc-36	Unc-38	Egl-19	Unc-2	Unc-29
Previous	0.950	0.930	0.880	0.760	0.730	0.930	0.717	0.760
Proposed	0.951	0.933	0.891	0.800	0.847	0.933	0.767	0.767

In our experiment, the learning samples are from 800 data files. And we identified these 8 different uncoordinated mutants with distinct locomotion patterns: wild-type, *unc-36*, *unc-38*, *egl-19*, *nic-1*, *unc-29* and *unc-2*. Each strain has 100 5-min recordings, with images captured every 0.5s. Every stain has 100 data file, and every data file has about 600 frames. From the experiment results of CART analysis in Table 3, we can see better identification performance for Unc type worms compared with the result using method in [9]. However, the successful classification ratios of *unc-2* and *unc-29* are still less than 80%. We analyzed the data generated by this proposed system; lots of frames are regarded as invalid, which leads to the incomplete feature sets. And after observing the recording of *unc-2* and *unc-29*, we found that their locomotion patterns tend to coil which is not supported by our feature extraction block. The holes and loops in the worm body make it hard to describe the worm body using a sequence of data point. Because at the cross points, we have not found out a suitable method to get a correct sequence of data points.

6 Conclusion

This image based automatic identification of nematode system can avoid subjectivity from observation, and be much more reliable at detecting representative abnormalities and behavior patterns that are subtle during a short time span or manifested in long time span. And also we can find that better identification results can be made by combining both shape and locomotion features. Especially for the Unc types, which have special locomotion pattern, the identification results can be improved greatly by using the locomotion pattern with the offset of the head and tail position.

Yet the identification performances of some worm types are still not excellent. After analyzing the data generated by this proposed system, we found some of frames are marked as invalid by the proposed system due to the worms postures are not suitable for the further processing and feature extraction, which eventually leads to the incompleteness of the feature sets. We are still trying to find better ways to match the loops or the holes locomotion pattern for the Unc type.

Acknowledgements

This work was supported by the Korea Research Foundation Grant No. 04-2-031. This research was also supported partially by the Internet Information Retrieval Research Center (IRC) in Hankuk Aviation University. IRC is a Regional Research Center of Gyeonggi Province, designated by ITEP and Ministry of Commerce, Industry and Energy.

References

1. Baek, J., Cosman, P., Feng, Z., Silver, J., Schafer, W.R.: Using machine vision to analyze and classify *C. elegans* behavioral phenotypes quantitatively. J. Neurosci. Methods, Vol. 118 (2002) 9-21
2. Breiman, L., Fried J.H., Olshen R.A., Stone C.J.: Classification and regression trees. Belmont, CA, Wadsworth (1984)
3. Geng,W., Cosman, P., Berry, C.C., Feng, Z., Schafer, W.R.: Automatic Tracking, Feature Extraction and Classification of *C. elegans* Phenotypes. IEEE Trans. Biomedical Engineering. Vol. 51 (2004) 1181-1820
4. Gonzalez, R. C., Woods R. E.: Digital image processing (2nd Edition). PrenticeHall. (2002)
5. Jain, R., Rangachar, R., Schunck, B.: Machine Vision. McGraw-Hill New York (1995)
6. Leonard P. Gianessi, Cressida S. Silvers, Sujatha Sankula: Current and Potential Impact For Improving Pest Management In U.S. Agriculture An Analysis of 40 Case Study. National Center for Food & Agricultural Policy (2002)
7. Silva, C.A., Magalhaes, K.M.C., Doria Neto, A.D.: An intelligent system for detection of nematodes in digital images. Proceedings of the International Joint Conference on Neural Networks, Vol. 1 (2003) 20-24
8. Zhang, T.Y., Suen, C.Y.: A fast parallel algorithm for thinning digital patterns. Comm. ACM, Vol. 27, No.3. (1984) 236-239
9. Zhou B., Nah W., Lee K., Baek J.: A General Image Based Nematode Identification System Design – CIS 2005. Lecture Notes in Artificial Intelligence, Vol. 3802, Springer-Verlag Berlin Heidelberg (2005) 899-904.

An Efficient Attribute Ordering Optimization in Bayesian Networks for Prognostic Modeling of the Metabolic Syndrome

Han-Saem Park and Sung-Bae Cho

Department of Computer Science, Yonsei University
134 Shinchon-dong, Sudaemoon-ku, Seoul 120-749, Korea
sammy@sclab.yonsei.ac.kr, sbcho@cs.yonsei.ac.kr

Abstract. The metabolic syndrome has become a significant problem in Asian countries recently due to the change in dietary habit and life style. Bayesian networks provide a robust formalism for probabilistic modeling, so they have been used as a method for prognostic model in medical domain. Since K2 algorithm is influenced by an input order of the attributes, optimization of BN attribute ordering is studied. This paper proposes an evolutionary optimization of attribute ordering in BN to solve this problem using a genetic algorithm with medical knowledge. Experiments have been conducted with the dataset obtained in Yonchon County of Korea, and the proposed model provides better performance than the comparable models.

1 Introduction

The metabolic syndrome is composed of a cluster of metabolic disorders including abdominal obesity, insulin resistance, dyslipedemia and hypertension, and the correlation between metabolic syndrome and coronary heart disease was reported in previous studies [1]. It affects around 25% of adults over the age of 20 and up to 45% over age 50 in the United States [2]. In Asian countries, it has become a significant problem lately due to the change in dietary habit and life style. In situations like this, many groups have been studying the metabolic syndrome from all over the world [1-2].

The Bayesian network has emerged in recent years as a powerful technique for handling uncertainty in complex domains [3]. It is a model of a joint probability distribution over a set of random variables. The Bayesian network is represented as a directed acyclic graph where nodes correspond to variables and arcs correspond to probabilistic dependencies between connected nodes [3]. Bayesian networks have been used for prediction or classification problem in the medical domain and shown high performance. In particular, they have been applied successfully to the modeling of diagnosis and prognosis for diverse diseases [4-6]. There have been many black box tools that classify or predict several diseases, and neural networks are the representative example. Bayesian networks have strengths that they can use the domain knowledge easily and analyze the results compared to them [5]. Even though

D.-S. Huang, K. Li, and G.W. Irwin (Eds.): ICIC 2006, LNBI 4115, pp. 381 – 391, 2006.

sometimes they are not better than neural networks in terms of accuracies, Bayesian networks are appropriate methods in the medical domain that require analyzing the results with medical knowledge.

This paper deals with a problem that predicts the metabolic syndrome with the dataset obtained in Yonchon County of Korea. This paper makes a prognostic model using Bayesian network, and has used the K2 algorithm by Cooper and Herskovits in order to learn its structure [7]. Since the result of the K2 algorithm is influenced by an input ordering of the attributes, an optimization of this ordering has been also studied [7]. This paper proposes an efficient optimization method using a genetic algorithm with medical domain knowledge in order to solve this problem. Contrary to the conventional methods, after clustering similar attributes into each group, an ordering of the groups and an ordering of the attributes in each group have been performed in turns. By applying the medical domain knowledge, an efficient and reliable evolution process can be conducted. Subsequently, the experiments using the proposed prognostic model have been done after the structure and parameter learning processes in order to show its usefulness.

2 The Problem: The Metabolic Syndrome

The definition of the metabolic syndrome was provided by the National Cholesterol and Education Program, Adult Treatment Panel III (ATP III). It requires the presence of three or more of the following components [1, 2]:

1) Abdominal obesity
(waist circumference >102 cm in men and >88 cm in women),
2) Hypertriglyceridemia ($\geq$ 150 mg/dL),
3) Low high density lipoprotein (HDL) cholesterol
<40 mg/dL in men and <50 mg/dL in women),
4) High blood pressure (systolic $\geq$ 130 mmHg or diastolic $\geq$ 80 mmHg) and
5) High fasting glucose (>110 mg/dL).

Since this original standard is not appropriate for Asian, we have used modified the definition for Asian of the abdominal obesity (waist circumference >90 cm in men and >80 cm in women) in this paper [8].

3 Prognostic Modeling of the Metabolic Syndrome

Fig. 1 illustrates the flowchart that makes the proposed prognostic model. This process can be divided into four main parts: pre-processing, attribute ordering, BN learning, and prediction processes. For the pre-processing part, medical domain knowledge has been applied, so the prediction model would be more reliable. For the attribute ordering part, a genetic algorithm has been also used in order to optimize the ordering of the model efficiently. The structure and parameter learning processes have been conducted after the attribute ordering, and the whole process is finished with prediction of input samples.

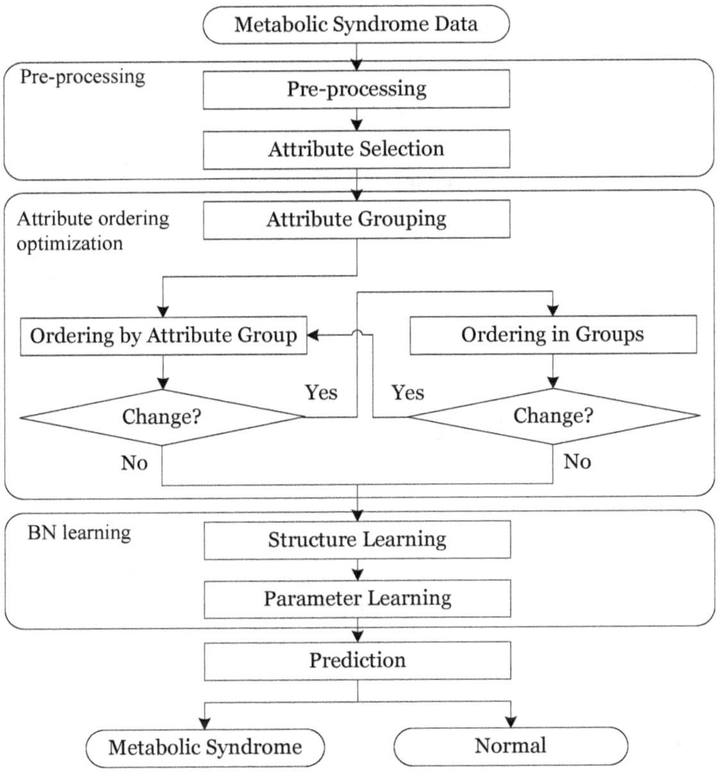

Fig. 1. The overall flowchart of the proposed method

3.1 Data Pre-processing and Attribute Selection

The pre-processing part subdivides into the pre-processing and an attribute selection. Here, we have decided the discrete states of the attributes since most attributes have continuous values and BN requires discrete states. The medical literatures [1, 2, 8] and experts have given a help for this process.

Table 1 demonstrates the attributes and their possible states. Explaining a few important attributes here, fasting glucose is divided into normal (<110), impaired glucose metabolism (110~125) and diabetes (>125). Triglyceride is increased if its value is larger than 150, or normal. In case of HDL cholesterol, it is decreased (abnormal) when it is smaller than standard value, 40 in men and 50 in women. The standards for state discrimination used here are significant since they have been used by medical experts [8].

We have chosen the informative attributes for the prognostic model after the pre-processing. As described in section 2, basic attributes required for defining metabolic syndrome are eight. However, there are more attributes that influence the metabolic syndrome such as age, 2-hour postprandial glucose, hypertension and body mass index [9]. Among these four attributes, the information of hypertension is redundant with the information of basic attributes. Therefore, we have proposed the Bayesian network model using 11 attributes, which include eight basic attributes and additional three informative attributes of age, 2-hour postprandial glucose and body mass index for the prognostic model of the metabolic syndrome. It is well-known that age is closely related to the metabolic syndrome and 2-hour postprandial glucose influence fasting glucose of an important attribute [9, 10]. Body mass index is selected because it is related to obesity, which is related to the metabolic syndrome.

Table 1. Attributes and the possible states

Attribute	Possible states
Age	young, middle aged, old
Sex	male, female
Fasting glucose	normal, impaired glucose metabolism, diabetes
2-hour postprandial glucose	normal, impaired glucose metabolism, diabetes
Waist circumference	normal, abdominal obesity
Triglyceride	normal, increased
HDL cholesterol	decreased, normal
Body mass index	low weight, normal, over weight, obesity
Ratio of waist-hip circumference	normal, abdominal obesity
Systolic blood pressure	normal, hypertension
Diastolic blood pressure	normal, hypertension

3.2 Evolutionary Optimization of Attribute Ordering

For the structure learning process of the Bayesian network, we have used the K2 algorithm, which is subject to an ordering of attributes when they are input. Therefore, an attribute ordering optimization has been studied to make Bayesian network more accurate. Larranaga *et al.* regarded this ordering problem as an ordering of cities in the TSP (Traveling Salesman Problem), and applied several genetic algorithms that had been used for the TSP [7]. We have also used the genetic algorithm to optimize attribute ordering and applied medical domain knowledge to make a model more efficiently.

1) Optimizing Attribute Group Ordering

As illustrated in attribute ordering part of Fig. 1, grouping of similar attributes is performed first. It is so complicated that we cannot simplify the causal relationship among attributes, but it is known that the attributes related to obesity are expressed first and ones related to metabolic disorder follow generally [10]. We have grouped twelve attributes, 11 attributes and a label, based on this information, so three attributes (waist circumference, body mass index, and ratio of waist-hip circumference) related to obesity and four attributes (fasting glucose, 2-hour postprandial glucose, triglyceride, and HDL cholesterol) related to metabolic disorder are grouped, respectively. The remaining five attributes are treated as five groups with a single attribute, respectively.

Subsequently, an ordering of these seven attribute groups is optimized. Ordering of attribute groups is optimized first, and then ordering of attributes in each group is optimized. These two processes are repeated until there is no change of ordering after each optimization step.

2) Application of the Genetic Algorithm

Details of chromosome representation, selection, and genetic operations are as follows.

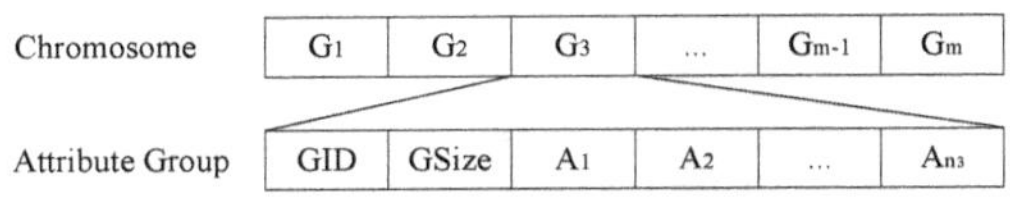

Fig. 2. Chromosome representation

As shown in Fig. 2, supposing each chromosome consists of *m* attribute groups, each group has group ID, the number of attributes in each group, and attributes themselves in each group, general genetic operations can be conducted. Attribute ordering of initial chromosome is decided randomly, and the fitness of each individual is evaluated in terms of the predictive accuracy because this paper focuses on the prognostic modeling.

After evaluating the fitness of individuals, individuals for the next generation are selected using rank-based selection method. In (1), $I_{(g,j)}$ means the *j*th individual in the *g*th generation, and Rank($f(I_{(g,j)})$) means the rank of each individual based on the fitness. *n* represents the number of individuals, and $p_{(g,j)}$, the probability that each individual $I_{(g,j)}$ is selected, can be provided as shown in (1).

$$p_{(g,j)} = \frac{n - Rank(f(I_{(g,j)})) + 1}{n(n+1)/2} \quad (1)$$

Crossover and mutation operations follow the selection process. Larranaga *et al.* compared several crossover operators on the TSP, and cycle crossover (CX) operator was the best [9]. This operator provides high performance even though the number of individuals is small. It attempts to create an offspring from the parents where every position is occupied by a corresponding element from one of the parents [7]. Fig. 3 illustrates an example of CX operation. There are two cycles in each individual, and they are crossed over each other.

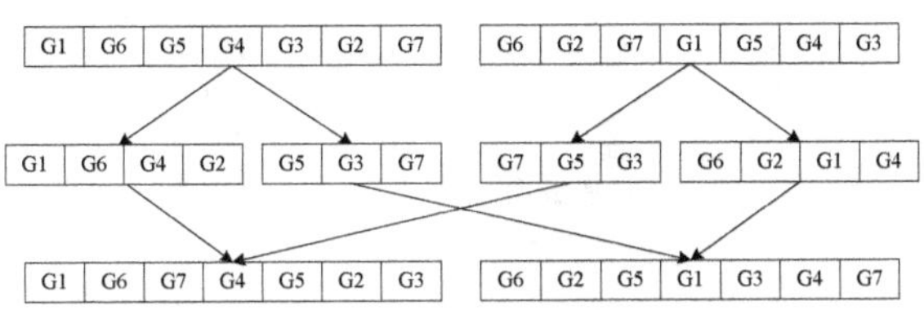

Fig. 3. An example of CX operation

Displacement mutation (DM) operator has been used for mutation operation. This operator provided the best performance when it was used with CX operator in Larranaga's work [7]. First, it selects a random substring and removes it, and it inserts this substring into random position. Fig. 4 shows an example of DM operation.

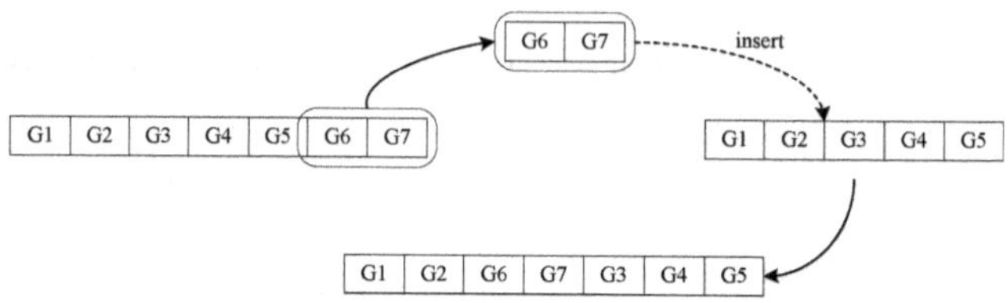

Fig. 4. An example of DM operation

```
The K2 Algorithm
Input:
    A set of n nodes,
    An ordering on the nodes,
    An upper bound u on the number of parents a node can
have,
    A dataset D including m cases.
Output:
    A printout of each node and its parents.
for ( i=1; i<n; i++ ) {
Π_i = 0;
P_old = g(i, Π_i );
OK2Proceed = TRUE;
    while ( OK2Proceed &&|Π_i|<u ) {
        Let Z be the node in Pred( X_i ) - Π_i that maximizes
g(i, Π_i∪{Z});
        P_new = g(i, Π_i∪{Z});
         if ( P_new > P_old ) {
            P_old = P_new ;
            Π_i = Π_i∪{Z} ;
        }
        OK2Proceed = FALSE;
    }
    print("node: ", X_i , "parents: ", Π_i );
}
```

Fig. 5. The K2 algorithm

3) Structure and Parameter Learning

With the evolved ordering decided before, we train the Bayesian network using the K2 algorithm [13]. The K2 algorithm narrows down the search space by fixing an order of the attributes. The probabilities of nodes have been calculated from the frequency counts of the learning data. Fig. 5 provides the pseudo-code of the K2 algorithm.

Given dataset *D*, the K2 algorithm searches the set of parent nodes that maximizes P(B_s , *D*) for every node. K2 starts by assuming that nodes do not have parents. For every step, it incrementally adds specific parent whose addition increases the probability of the result structure most. These processes continue until the addition of a single parent cannot increase the probability. K2 is a greedy heuristic, and it does not guarantee the structure with the highest probability. The details are provided in [13].

Once the structure of Bayesian network is decided, parameter values of each node are calculated from the frequency of learning data, and then the prognostic model of the metabolic syndrome is completed.

4 Experimental Results

4.1 Experimental Data

The dataset used in this paper was examined for epidemiological research of the local community. The surveys were conducted twice in 1993 and 1995 in Yonchon County of Korea. 2,293 subjects were participated in the first survey, and 1,193 of them were participated in the second survey [12]. We have used the data of 1,135 subjects who did not include missing values and the 18 attributes that could influence the prediction of the metabolic syndrome. After that we have decided whether each sample would belong to the metabolic syndrome or not. 18 attributes and the distribution by states are shown in Table 2.

4.2 Parameters and Settings

In order to compare the result of Bayesian network model with the other models, neural network (NN) and *k*-nearest neighbors (kNN) prediction models have been used. 11 input nodes, 20 hidden nodes, and 2 output nodes are used for NN, and *k* of kNN is fixed as 3 having shown the best performance in a preliminary experiment.

For the genetic algorithm, population size is set as 20, and the process has evolved by the generation of 100. The population size is small because the total search space is not so large. The number of possible combinations is 7!, but the time cost is expensive using an exhaustive method since the learning and inference processes are required for every individual of every generation. The selection rate of 0.8 and mutation rate of 0.02 are used. Crossover rate is set as 1.0, but it is less than that because the result of the CX operation sometimes is not changed.

For experiments before evolution, 10-fold cross validation is repeated 30 times, and the average is used for the result. For evolution process, we have divided the data into three parts with the ratio of 3:1:1. The first part is used for the learning, the second one is used for the validation, and the last one is used for the test.

Table 2. The attributes of tfhe dataset and the distribution by states

Attributes	n	Percentage
Age		
Young	111	9.78
Middle aged	512	45.11
Old	512	45.11
Sex		
Female	646	56.92
Male	489	43.08
Fasting glucose		
Normal	942	82.99
Impaired glucose metabolism	166	14.63
Diabetes	27	2.38
2-hour postprandial glucose		
Normal	985	86.78
Impaired glucose metabolism	150	13.22
Diabetes	0	0.00
Waist circumference		
Normal	456	40.18
Abdominal obesity	679	59.82
Triglyceride		
Normal	731	64.41
Increased	404	35.59
HDL cholesterol		
Decreased	873	76.92
Normal	262	23.08
Body mass index		
Low weight	36	3.17
Normal	405	35.68
Over weight	267	23.52
Obesity	427	37.62
Ratio of waist-hip circumference		
Normal	771	67.93
Abdominal obesity	364	32.07
Systolic blood pressure		
Normal	838	73.83
Hypertension	297	26.17
Diastolic blood pressure		
Normal	773	68.11
Hypertension	362	31.89

4.3 Experimental Results

First, we have conducted comparison experiments among Bayesian network models that have different numbers of attributes. The first model is BN with eight basic attributes, the second one is the proposed model with 11 selected attributes, and the last one is the BN model with all attributes. Table 3 indicates that the result of 11 attributes is the best.

Table 3. The comparison of accuracies by the number of attributes

Attribute size	8	**11**	18
Accuracy (%)	70.74 (± 0.0017)	**72.15 (± 0.0082)**	70.82 (± 0.0079)

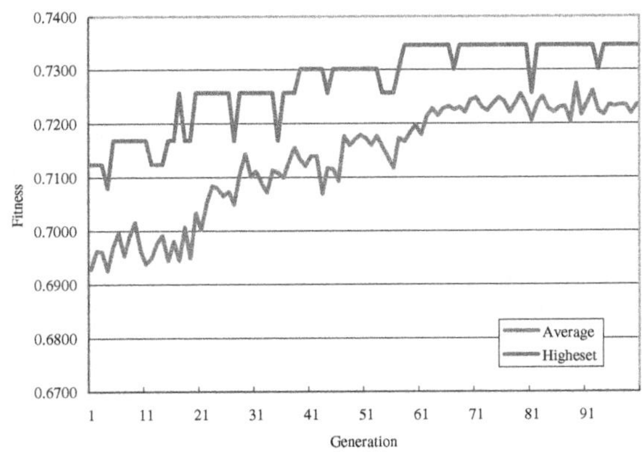

Fig. 6. The fitness transition graph during an evolution process

Table 4. The comparison of accuracies before and after optimization

Ordering	Accuracy (%)
Optimized	**72.12 (± 0)**
Random	70.09 (± 0.0513)

Table 5. The comparison of accuracies by the number of attributes

Prediction model	Accuracy (%)
BN	**72.12 (± 0)**
NN	63.19 (± 0.0152)
3NN	62.56 (± 0)

We have applied the proposed method to total 1,135 data, and Fig. 6 illustrates the evolution process. It evolves well since average and the highest fitness values get a little higher accordingly as the generation grows. The highest fitness converges after the 60th generation, and the average fitness converges after a few generations.

After that, we have compared the model of optimized ordering with the model of random ordering. Table 4 provides the comparison result in terms of accuracy. Here, the difference between two models is statistically significant ($p<0.001$).

Finally, we have compared the BN model after attribute optimization with two other models of neural networks and k nearest neighbors that have been frequently used in pattern recognition field in order to solve the prediction problem. Table 5 shows the comparison result with accuracy, and the differences of the proposed model and other models (BN vs. NN, BN vs. 3NN) are statistically significant ($p<0.001$). Generally, NN has strength in accuracy though it cannot be interpreted easily. However, BN model has provided better performance, and we can guess that the pre-processing and attribute selection processes using medical domain knowledge were effective.

5 Conclusions

This paper proposed a Bayesian network model with evolutionary algorithm in order to predict the metabolic syndrome. In processes of building the prognostic model, we applied the medical domain knowledge in order to make the model more reliable. We also adopted the genetic algorithm to optimize attribute ordering, and completed the model efficiently using the medical domain knowledge in the optimization process.

We verified that the proposed method provided better performance compared with the model before ordering optimization as well as other models such as neural networks and *k*-nearest neighbor.

Acknowledgments. This research was supported by the Ministry of Information and Communication, Korea under the Information Technology Research Center support program supervised by the Institute of Information Technology Assessment, IITA-2005-(C1090-0501-0019).

References

1. Mykkanen, L., Kuusisto, J., Pyorala, K., Laakso, M.: Cardiovascular Disease Risk Factors as Predictors of Type 2 (Non-Insulin-Dependent) Diabetes Mellitus in Elderly Subjects. Diabetologia, 50 (2004) 453-469
2. Mehta, N. N., Reilly, M. P.: Mechanisms of the Metabolic Syndrome. Drug Discov Today, 1(2) (2004) 187-194
3. Lee, S. M., Abbott, P. A.: Bayesian Networks for Knowledge Discovery in Large Datasets: Basics for Nurse Researchers. J Biomed Inform, 36 (2003) 389-399
4. Antal, P., Fannes, G., Timmerman, D., Moreau, Y., Moor, B. D.: Using Literature and Data to Learn Bayesian Networks as Clinical Models of Ovarian Tumors. Artif Intell Med, 30 (2004) 257-281
5. Wang, X.-H., Zheng, B., Good, W. F., King, J. L., Chang, Y.-H.: Computer-Assisted Diagnosis of Breast Cancer Using a Data-driven Bayesian Belief Network. Int J Med Inform, 54 (1999) 115-126
6. Sierra, B.: Using Bayesian Networks in the Construction of A Bi-level Multi-Classifier. A Case Study Using Intensive Care Unit Patients Data. Artif Intell Med, 22 (2001) 233-248
7. Larranaga, P., Kuijpers, C. M. H., Murga, R. H., Yurramendi, Y.: Learning Bayesian Network Structures by Searching for the Best Ordering with Genetic Algorithms. IEEE T Syst Man Cy A, 26 (4) (1996)
8. Moon, M. K., Cho, Y. M., Lim, K. S., Park S., Lee, H. K.: Metabolic Syndrome. The Korean Society of Endocrinology, 18 (2003) 105-116
9. Lindblad, U., Langer, R. D., Wingard, D. L., Thomas, R. G., Barrett-Connor, E. L.: Metabolic Syndrome and Ischemic Heart Disease in Elderly Men and Women. Am J Epidemiol, 153 (2001) 481-489

10. Girod, J. P., Brotman, D. J.: The Metabolic Syndrome as a Vicious Cycle: Does Obesity Beget Obesity. Med Hypotheses, 60(4) (2003) 584-589
11. Cooper, G. F., Herskovits, E. A.: A Bayesian Method for Induction of Probabilistic Networks from Data. Mach Learn, 9(4) (1992) 309-347
12. Park, Y.: Prevalence of Diabetes and IGT in Yonchon County, South Korea Diabetes Care, 18 (1995) 545-548

Analysis and Simulation of Synchronization for Large Scale Networks

Xinkai Chen[1] and Guisheng Zhai[2]

[1] Department of Electronic & Information Systems, Shibaura Institute of Technology, 307 Fukasaku, Minuma-ku, Saitama-city, Saitama 337-8570, Japan
chen@sic.shibaura-it.ac.jp

[2] Department of Mechanical Engineering, Osaka Prefecture University, 1-1 Gakuen-Cho, Sakai, Osaka 599-8531, Japan
zhai@mecha.osakafu-u.ac.jp

Abstract. Networks of coupled large scale oscillators have been studied in biology for a number of years. It has been recognized that transient in the nearest neighbor connected networks may take far too long to die out. In the model of mammalian rhythm, it is considered that a few long distance interconnections exist. Typically, these long distance interconnections are considered to occur in a random way. In this study, we discuss the synchronization problem for coupled oscillator networks which can model the mammalian rhythm. Then, the distribution model for the random long distance connections is proposed and is demonstrated by simulation. Furthermore, simulation also shows that synchronization still holds even a large part of the network is destroyed.

1 Introduction

Recently, trying to understand the behavior of symmetrically coupled oscillators in biological networks has been a well studied subject [1][9][10]. Traditionally, it is assumed that such networks only have nearest neighbor interconnections. Under suitable assumptions on the connectivity pattern and strength, it is shown that such networks begin to display synchronized oscillatory behavior. However, it has been recognized that this type of coupling has the drawback that transient time is too long. It has been suggested by Watts and Strogatz [8] that a few long distance interconnections may have a dramatic effect on speeding up transient response. The so-called "small world networks" have intermediate connectivity properties but exhibit a high degree of clustering as in the regular networks and a small average distance between vertices as in the random network. They also found that the small world networks of coupled phase oscillators can synchronize almost as readily as the globally coupled networks, despite the fact that they have much fewer edges [9].

One problem is how to quantify the effects of long distance connections on the transient response times of a small world network [4][7]. Many simulation results have shown that a couple of generic long distance connections have the same effect in reducing transients compared to a dozen or so nearest neighbor connections. Until now, no model has been reported that can support this comparison. It is considered

D.-S. Huang, K. Li, and G.W. Irwin (Eds.): ICIC 2006, LNBI 4115, pp. 392 – 401, 2006.

that transient dynamics of a typical oscillator network can be related to a dominant eigenvalue of the coupled matrix. Thus, the distribution of the random long distance interconnection must be firstly clarified. In this paper, firstly, we try to analyze the relation between the interconnection and the dominant eigenvalue of the coupled matrix of the coupled oscillator networks which can model the mammalian rhythm. Then, we try to give a distribution model of the long distance connections between the corresponding neurons. Finally, the proposed model is demonstrated by simulation.

2 Preliminaries

Circadian rhythms are observed in the physiology of mammals and other higher organisms. In mammals, they are generated in a pacemaker located in the suprachiasmatic neucleus (SCN) of the hypothalamus [2][3]. SCN consists of 16000 neurons arranged in a symmetric bilateral structure, and it is generally believed that each isolated SCN neuron behaves as an oscillator by itself. Kronauer [3] proposed a model to describe the oscillations in a single SCN cell in 1990. A van der Pol oscillator is used to model stable oscillations in SCN neurons, with an additive term to model response due to light,

$$\left(\frac{12}{\pi}\right)\dot{x} = y + \varepsilon\left(x - \frac{4}{3}x^3\right) + B, \tag{1}$$

$$\left(\frac{12}{\pi}\right)\dot{y} = -\left(\frac{24}{\tau}\right)^2 x + By, \tag{2}$$

where ε is a constant which determines the size of the periodic orbit of the van der Pol oscillator (usually chosen as 0.13) [5][6], τ is a constant which represents the period of the oscillator (roughly equals to 24), B is defined by $B = (1 - px)CL^{\frac{1}{3}}$, L is the light intensity in lux, and p and C are constants ($p = \frac{1}{3}$, $C = 0.018$).

We consider a network consisting of N nonlinear oscillators arranged in the form of a ring. Each of the oscillators represents an SCN neuron, and internal dynamics is represented by a van der Pol oscillator, where the term B in (1)-(2) will not be considered.

$$\left(\frac{12}{\pi}\right)\dot{x}_i = y_i + \varepsilon\left(x_i - \frac{4}{3}x_i^3\right) + \sum_{j=1}^{N} a_{ij} x_j, \tag{3}$$

$$\left(\frac{12}{\pi}\right)\dot{y}_i = -\left(\frac{24}{\tau}\right)^2 x_i + \sum_{j=1}^{N} a_{ij} y_j, \tag{4}$$

x_i and y_i are the state of the i-th neuron. The matrix $A = (a_{ij}) \in R^{N\times N}$ is called the coupling matrix which represents the coupling configuration of the network. For

$i \neq j$, $a_{ij} = a_{ji} \geq 0$ is the strength of the connectivity between the i-th neuron and the j-th neuron. If $a_{ij} = a_{ji} = 0$, then there is no connection between the i-th neuron and the j-th neuron. For $i = j$, $a_{ii} = -\sum_{\substack{j=1 \\ j \neq i}}^{N} a_{ij} = -\sum_{\substack{j=1 \\ j \neq i}}^{N} a_{ji}$. x_i and y_i satisfy $x_{i+N} = x_i$, $y_{i+N} = y_i$.

Suppose the network is connected in the sense that there are no isolated neurons. Then the coupling matrix A is irreducible. Thus, it can be shown that zero is an eigenvalue of A with multiplicity 1 and all the other eigenvalues are negative.

3 Synchronization Stability Analysis

The dynamical network (3)-(4) is said to achieve synchronization if

$$\begin{bmatrix} x_1 \\ y_1 \end{bmatrix} = \cdots = \begin{bmatrix} x_N \\ y_N \end{bmatrix} = S(t) \tag{5}$$

as $t \to \infty$, where $S(t)$ is the limit cycle generated by

$$\left(\frac{12}{\pi}\right)\dot{x} = y + \varepsilon\left(x - \frac{4}{3}x^3\right), \tag{6}$$

$$\left(\frac{12}{\pi}\right)\dot{y} = -\left(\frac{24}{\tau}\right)^2 x. \tag{7}$$

Clearly, the stability of the synchronized states of the network (3)-(4) is determined by the limit cycle $S(t)$ and the coupling matrix A.

Lemma 1: Let $0 = \lambda_1 > \lambda_2 \geq \cdots \geq \lambda_N$ be the eigenvalues of the coupling matrix A. If the following $N-1$ of 2-dimensional linear time-varying systems are exponentially stable

$$\left(\frac{12}{\pi}\right)\dot{w}_1 = \left(\varepsilon - 4\varepsilon x^2\Big|_{x \in S(t)} + \lambda_k\right)w_1 + w_2, \tag{8}$$

$$\left(\frac{12}{\pi}\right)\dot{w}_2 = -\left(\frac{24}{\tau}\right)^2 w_1, \tag{9}$$

for $k = 2, \cdots, N$, then the synchronized states (5) are exponentially stable.

Proof: Let $\begin{pmatrix} x_i \\ y_i \end{pmatrix} = S(t) + \begin{pmatrix} \rho_{1i} \\ \rho_{2i} \end{pmatrix}$. Now linearize (3)-(4) about *S*(*t*) yields

$$\left(\frac{\pi}{12}\right)\dot{\Theta} = \Theta\begin{pmatrix} \left(\varepsilon - 4\varepsilon x^2\big|_{x\in S(t)}\right) \\ -\left(\frac{24}{\tau}\right)^2 \end{pmatrix} + A\Theta, \tag{10}$$

where $\Theta = \begin{pmatrix} \rho_{11} & \rho_{21} & \cdots & \rho_{N1} \\ \rho_{11} & \rho_{22} & \cdots & \rho_{N2} \end{pmatrix}^T$. Since *A* is a real symmetric matrix, there exists an unitary matrix $\Phi = [\phi_1 \quad \phi_2 \quad \cdots \quad \phi_N]$ such that

$$A\phi_k = \lambda_k \phi_k, \quad k = 1, 2, \cdots, N. \tag{11}$$

By expanding the two columns of Θ on the basis of Φ, we have

$$\Theta = \Phi \upsilon. \tag{12}$$

where the matrix $\upsilon \in R^{N\times 2}$ obeys the following equations

$$\left(\frac{\pi}{12}\right)\dot{\upsilon} = \upsilon\begin{pmatrix} \left(\varepsilon - 4\varepsilon x^2\big|_{x\in S(t)}\right) \\ -\left(\frac{24}{\tau}\right)^2 \end{pmatrix} + \Lambda\upsilon \tag{13}$$

with $\Lambda = diag(\lambda_1, \lambda_2, \cdots, \lambda_N)$. Let $\upsilon_k = (\upsilon_{k1}, \upsilon_{k2})$ be the k-th row of $\upsilon \in R^{N\times 2}$. It yields

$$\dot{\upsilon}_k^T = \begin{pmatrix} \left(\varepsilon - 4\varepsilon x^2\big|_{x\in S(t)} + \lambda_k\right)\upsilon_{k1} + \upsilon_{k2} \\ -\left(\frac{24}{\tau}\right)^2 \upsilon_{k1} \end{pmatrix}. \tag{14}$$

Note that $\lambda_1 = 0$ corresponds to the synchronization of the system states. If the following *N-1* pieces of 2-dimensional linear time varying systems described in (8)-(9) are exponentially stable, then Θ will exponentially tend to the origin, which implies that the synchronized states (5) are exponentially stable. The lemma is proved.

Based on Lemma 1, we have the next result.

Lemma 2: If $\lambda_2 \le -\varepsilon$, then the synchronized states (5) are exponentially stable.

Proof: The result is obvious by observing equation (8)-(9).

By Lemma 1 and Lemma 2, it can be concluded that transient time of the oscillator network is governed by the dominant eigenvalue λ_2 of the coupled matrix A.

4 Synchronization in Scale-Free Networks

4.1 Synchronization Influenced by Connections

In this section, the influence of the connections on synchronization will be studied. In the following, we are interested in discussing the synchronization of a scale free network.

To begin with, we introduce the following lemma [11].

Lemma 3: Let $C \in R^{n \times n}$ and $D \in R^{n \times n}$ be symmetric matrices. If D is a negative semi-definite matrix, then

$$\lambda_2(C+D) \le \lambda_2(C), \tag{15}$$

where the inequality holds when D is negative definite.

First, let us study the nearest neighbor coupling configuration. In this case, the corresponding coupling matrix is described by

$$A_{nc} = \begin{bmatrix} a_{11} & a_{12} & & & a_{1N} \\ a_{12} & a_{22} & a_{23} & & \\ & \ddots & \ddots & \ddots & \\ & & a_{N-1,N-2} & a_{N-1,N-1} & a_{N-1,N} \\ a_{1N} & & & a_{N-1,N} & a_{NN} \end{bmatrix}. \tag{16}$$

Let $\min_{i \ne j} a_{ij} = q$. Since the eigenvalues of the matrix

$$A_{\min} = \begin{bmatrix} -2q & q & & & q \\ q & -2q & q & & \\ & \ddots & \ddots & \ddots & \\ & & q & -2q & q \\ q & & & q & -2q \end{bmatrix} \tag{17}$$

are $-4q\sin^2(\frac{k\pi}{N})$, $k=0, 1, \cdots, N-1$, by Lemma 3, it is obvious that the dominant eigenvalue λ_2 of the matrix A_{nc} described in (16) is smaller than or equal to $-4q\sin^2(\frac{\pi}{N})$. Thus, according to Lemma 2, the nearest neighbor coupled network will synchronize if

$$4q\sin(\frac{k\pi}{N}) \geq \varepsilon \tag{18}$$

Therefore, when the network is very large and the coupling strengths are not strong (which means that many a_{ij} $(i \neq j)$ are smaller than 1), the nearest neighbor coupled network may not synchronize. Furthermore, if the coupling strengths are very weak, then the nearest neighbor coupled network may not synchronize even when the network is not very large.

Secondly, let us study the global coupled network. In this case, the corresponding coupling matrix is described by

$$A_{nc} = \begin{bmatrix} a_{11} & a_{12} & a_{13} & \cdots & a_{1N} \\ a_{12} & a_{22} & a_{23} & \cdots & a_{2N} \\ \vdots & \vdots & \vdots & \ddots & \vdots \\ a_{1,\,N-1} & a_{2,\,N-1} & a_{3,\,N-1} & a_{N-1,\,N-1} & a_{N-1,\,N} \\ a_{1N} & a_{2N} & a_{3N} & a_{N-1,\,N} & a_{NN} \end{bmatrix}. \tag{19}$$

Let $\min\limits_{i \neq j} a_{ij} = q$. By Lemma 3, it can be easily seen that the dominant eigenvalue λ_2 of the matrix A_{nc} described in (19) is smaller than or equal to $-Nq$ (with *N-1* multiplicity). According to Lemma 2, the global coupled network will synchronize if $qN \geq \varepsilon$.

If many coupling strengths a_{ij} $(i \neq j)$ are very small, then the global coupled network may not synchronize.

Thirdly, let us consider the coupled network where each neuron is adjacent to its n neighboring neurons on each side. The corresponding coupling matrix is described by

$$A_{nc} = \begin{bmatrix} a_{11} & \cdots & a_{1n} & 0 & a_{1,N-n+1} & \cdots & a_{1N} \\ \vdots & \ddots & & \ddots & \ddots & \ddots & \vdots \\ a_{1n} & & & & \ddots & \ddots & a_{n,N-n+1} \\ 0 & \ddots & & \ddots & & \ddots & 0 \\ a_{1,N-n+1} & \ddots & \ddots & & & & a_{N-n+1,N} \\ \vdots & \ddots & \ddots & \ddots & & \ddots & \vdots \\ a_{1N} & \cdots & a_{n,N-n+1} & 0 & a_{N-n+1,N} & \cdots & a_{NN} \end{bmatrix}. \tag{20}$$

Let $\min_{i \neq j} a_{ij} = q$. Then, by some elementary computations, it can be calculated that the dominant eigenvalue λ_2 of the coupled matrix A_{nc} described in (20) is smaller than or equal to $-4q\sum_{j=1}^{n} \sin^2(\frac{j\pi}{N})$, where Lemma 3 is employed. Therefore, when the network is very large and the coupling strengths are not strong (which means that many a_{ij} $(i \neq j)$ are smaller than 1), this kind of coupled network may not synchronize. Furthermore, if the coupling strengths are very weak, then this kind of coupled network may not synchronize even when the network is not very large.

From the above analysis, it can be seen that long distance connections with "strong" coupling strength are necessary to synchronize a large scale network. Furthermore, the transient time of the network also depends on these long distance connections.

In order to describe the behavior of the coupled network, the concept "small world networks" are introduced by Watts and Strogatz [8]. It is considered that most of the connections are of the nearest neighbor type and a few are of a randomly formed long distance connections. On the other hand, Lemma 3 implies that adding long distance connections will move the dominant eigenvalue of the coupling matrix A to left. This can be supported by analyzing the behavior of the matrix A. In fact, adding a connection with strength b_{ij} $(j > i)$ between the i-th and j-th neuron implies adding the matrix

$$B = \begin{bmatrix} \ddots & & & & \\ & -b_{ij} & \cdots & b_{ij} & \\ & \vdots & \ddots & \vdots & \\ & b_{ij} & \cdots & -b_{ij} & \\ & & & & \ddots \end{bmatrix} \begin{matrix} \\ i \\ \\ j \\ \\ \end{matrix} \qquad (21)$$
$$\qquad\qquad i \qquad\qquad j$$

to the original coupling matrix. By Lemma 3, it yields that $\lambda_2(A+B) \leq \lambda_2(A)$ by considering the special structure of the coupling matrix.

4.2 Distribution Model of Random Long Connections

Now, let us consider the distribution of the random connection. Here the multiplicity of the connections is permitted. The multiple connection between a particular pair of neurons means the connection strength is enhanced by adding the connection strengths.

Distribution Model: For a coupled network, let us suppose that each neuron is adjacent to its n neighboring neurons on each side. This assumption guarantees that there is no isolated neuron. For the i-th neuron, suppose the "long distance" neurons means the neurons numbered from $i+\left[\frac{N}{4}\right]$ to $i+\left[\frac{3N}{4}\right]$. Furthermore, suppose there are m long distance connections in the network, and the distribution of each long distance connection obeys the uniform distribution. For simplicity, suppose these long connection strengths are same, say α which meets $\alpha \geq 1$.

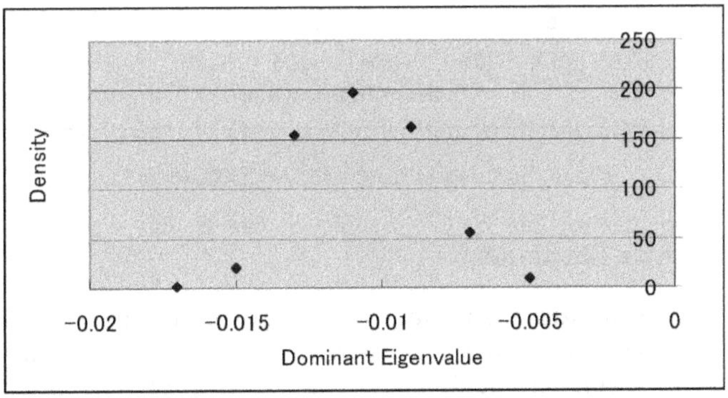

Fig. 1. Simulation results for N=2000, m=250

Now, let us give the simulation of the proposed distribution model. For simplicity, the connection strength is set to 1. Suppose the neurons are arranged in a ring, i.e. n=1. The simulations are done 600 times. The densities of the dominant value are shown in Figures 1-3. Figure 1 shows the simulation results for the distribution model when N=2000, m=250. Figure 2 shows the simulation results when N=2000, m=400. Figure 3 shows the simulation results when N=2000, m=600.

It can be seen that, when m is increased, the dominant value moves to left. This is also consistent to the analysis in the above section.

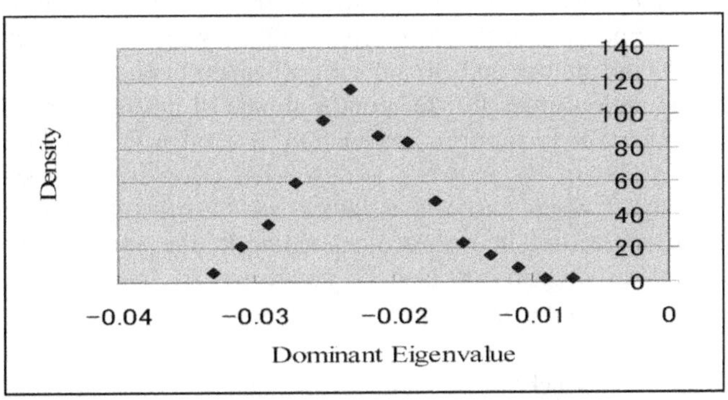

Fig. 2. Simulation results for N=2000, m=400

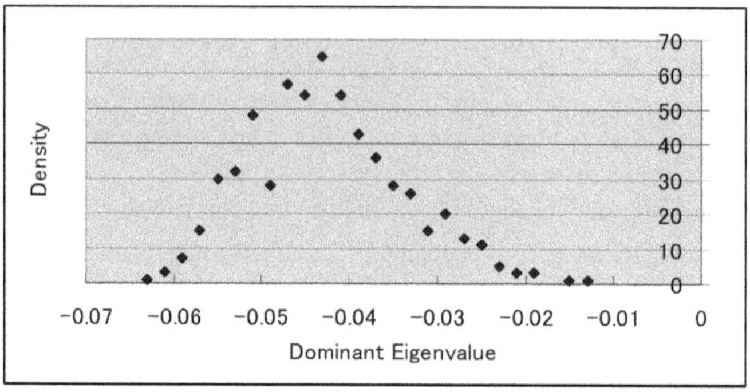

Fig. 3. Simulation results for N=2000, m=600

4.3 Synchronization Simulations

In this section, the synchronization time will be investigated by computer simulation based on the proposed long distance connection model for N=2000. In the presented simulation results, "synchronization" means that the norm of state difference between one oscillator and another oscillator becomes smaller than 5×10^{-4} after an instant t_s. The instant t_s is called "synchronization time". Table 1 shows the mean synchronization times for different values of m, where 600 times of the simulations are done for each m. It can be seen that the synchronization becomes faster if the number of long distance connections are larger. This is also consistent to the analysis in the above sections.

Table 1. The mean synchronization time for different m

Value of m	2×10^3	4×10^3	10^4	3×10^4
Synchronization Time (s)	24.57	8.65	2.54	0.50

Now, based on the above setting, let us discuss the case when a part of the network is destroyed. It is supposed that the destroyed oscillators are neighbored ones. For the network with 2000 oscillators and 30000 long distance connections, Table 2 shows the mean synchronization times for different numbers of destroyed oscillators, where 600 times of the simulations are done for each m. It is clear that synchronization still exists even a large part of the network is destroyed (However, the synchronization time becomes longer when more oscillators are destroyed). By observing the existence of synchronization in a destroyed network, the validity of the proposed model for mammalian rhythm can also be supported by the phenomenon that the instinct of the animals still works even the brain is seriously injured.

Table 2. The mean synchronization time via the number of destroyed oscillators

Number of destroyed oscillators	0	100	200	300	400	500
Synchronization Time (s)	0.50	0.55	0.63	0.73	0.87	1.00

5 Conclusion

This paper discusses the synchronization problem for the coupled large scale oscillator networks which can model the mammalian rhythm. Long distance connections with "strong" coupling strength are necessary to synchronize a large scale network. Furthermore, the transient time of the network also depends on these long distance connections. The long distance interconnections are considered to occur in a random way. The distribution model for the random long distance connections in a network is proposed. Simulations results are carried out to illustrate the synchronization theory and to show the validity of the proposed distribution model.

Acknowledgments

The first author is supported in part by the Japan Ministry of Education, Sciences and Culture under Grant-in-Aid for Scientific Research (B) 16760354.

References

1. Kopell, N.: We Got Rhythm: Dynamical Systems of The Nervous System. Notices American Math. Soc. 47 (2000) 6-16
2. Kronauer, R.E.: Mathematical Model of The Human Circadian System with Two Interacting Oscillators. American Journal of Physiology 242 (1982) R3-R7
3. Kronauer, R.E.: A Qualitative Model for The Rffects of Light on The Amplitude and Phase of The Deep Circadian Pacemaker. In: Horne, J. (ed.): Proceedings of the Tenth European Congress on Sleep Research.. Pontenagel Press, Dusseldorf (1990) 306
4. Li, X., *et al*: Synchronization and Desynchronization of Complex Dynamical Networks: An Engineering Viewpoint. IEEE Trans. On Circuits and Systems – I: Fundamental Theory and Applications 50 (2003) 1381-1390
5. Miller, J.D. *et al*: New Insights into The Mammalian Circadian Clock. Sleep 19 (1996) 641-667
6. Miller, J. *et al*: Phase Locking in The Mammalian Circadian Clock. In: Proceeding of the IEEE Conference on Decision and Control, Sydney (2002)
7. Wang, X.F., Chen, G.: Synchronization in Small World Dynamical Networks. International Journal of Bifurcation and Chaos 12 (2002) 187-192
8. Watts, D.J., Strogatz, S.H.: Collective Dynamics of Small World Networks. Nature 393(1998) 440-442
9. Watts, D.J.: Small Worlds: The Dynamics of Networks Between Order and Randomness. Princeton University Press, RI (1999)
10. Wu, C.W., Chua, L.O.: Synchronization in An Array of Linearly Coupled Dynamical Systems. IEEE Trans. Circuits and Systems – I: Fundamental Theory and Applications 42(1995) 430-447
11. Bellman, R.: Introduction to Matrix Analysis. McGraw-Hill, New York (1960)

Detection of Basal Cell Carcinoma by Automatic Classification of Confocal Raman Spectra

Seong-Joon Baek[1], Aaron Park[1], Jin-Young Kim[1], Seung Yu Na, Yonggwan Won[1], and Jaebum Choo[2]

[1] The School of Electronics and Computer Engineering, Chonnam National University, Gwangju, South Korea, 500-757
[2] Dept. of Applied Chemistry, Hanyang University, Ansan, South Korea, 426-791

Abstract. Raman spectroscopy has strong potential for providing non-invasive dermatological diagnosis of skin cancer. In this study, we investigated various classification methods with confocal Raman spectra for the detection of basal cell carcinoma (BCC), which is one of the most common skin cancer. The methods include maximum a posteriori (MAP) probability, probabilistic neural networks (PNN), *k*-nearest neighbor (KNN), multilayer perceptron networks (MLP), and support vector machine (SVM). The classification framework consists of preprocessing of Raman spectra, feature extraction, and classification. In the preprocessing step, a simple half Hanning method is adopted to obtain robust features. Classification results involving 216 spectra gave about 97% true classification rate in case of MLP and SVM, which is an evident proof of the effectiveness of confocal Raman spectra for BCC detection. In addition to it, spectral regions, which are important for classification, are examined by sensitivity analysis.

Keywords: Raman spectroscopy, pattern recognition, sensitivity analysis, basal cell carcinoma detection.

1 Introduction

Skin cancer is one of the most common cancers in the world. Recently, the incidence of skin cancer has dramatically increased due to the excessive exposure of skin to UV radiation caused by ozone layer depletion, environmental contamination, and so on. If detected early, skin cancer has a cure rate of 100%. Unfortunately, early detection is difficult because diagnosis is still based on morphological inspection by a pathologist. There are two common skin cancers: BCC and squamous cell carcinoma (SCC). Both BCC and SCC are nonmelanoma skin cancers, and BCC is the most common skin neoplasm[1].

The accurate detection of BCC has attracted much attention from clinical dermatologists since it is difficult to distinguish BCC tissue from surrounding noncancerous tissue. The routine diagnostic technique used for the detection of BCC is pathological examination of biopsy samples. This involves removal of tissue from suspected areas, which is then sliced and stained to enable the

D.-S. Huang, K. Li, and G.W. Irwin (Eds.): ICIC 2006, LNBI 4115, pp. 402–411, 2006.

pathologist to identify morphological abnormalities. The method relies upon a subjective judgment, which is dependent on the level of experience of the individual pathologist and can lead to the excessive biopsy of tissues. Thus, a fast and accurate diagnostic technique for the initial screening and selection of lesions for further biopsy is needed [2].

Raman spectroscopy has the potential to resolve this problem. It can be applied to provide an accurate medical diagnosis to distinguish BCC tissue from surrounding normal (NOR) tissue. Recently, some of the researchers carried out BCC detection using Raman spectroscopy. By far, the most widely used method is Fourier transform (FT) Raman spectroscopy [3] [4]. In the previous studies, a long wavelength excitation laser was used to minimize autofluorescence from skin tissue. However, the longer wavelength laser gives poor Raman scattering intensities compared to the shorter one. As a result, FT Raman spectra in poor signal-to-noise ratio suffer from so called background noise [2]. The background noise introduces both variance into the spectra and correlation between frequency components which makes robust feature extraction difficult. Thus FT Raman spectra require complicated statistical treatments to eliminate background noise as [3].

More recently, direct observation method based on the confocal Raman technique was presented for the dermatological diagnosis of BCC using the shorter wavelength argon ion laser [2]. According to the study, confocal Raman spectra provides promising results for detection of precancerous and noncancerous lesions without special treatment. Hence, with the confocal Raman spectra, we could expect that an automatic classifier having robust detection results would be designed. In this paper, we investigate various classification methods and compare the classification results. The methods include MAP, PNN, KNN, MLP, and SVM. Experimental results will be given to confirm that the confocal Raman spectra provide a promising results for automatic detection of BCC.

2 Confocal Raman Measurements and Preprocessing of Data

The preparation of data includes confocal Raman measurement and preprocessing. Raman spectra were obtained using a commercial Renishow 2000 Raman microscope system. After the measurements, normalization and windowing were carried out for the spectra. Dimension reduction was followed using principal component analysis (PCA) for feature extraction. We will describe the procedures in turn.

2.1 Sample Preparation and Raman Measurements

The tissue samples were prepared with the conventional treatment, which is exactly the same as [2]. BCC tissues were sampled from 10 patients with a routine biopsy. Cross sections of $20\mu m$ were cut with a microtome at -20°C and stored in liquid nitrogen. Two thin sections of every patients were used for experiments. One section was used for classification and the other section was stained with H&E and used as a reference after locating the boundaries between BCC and NOR by an expert pathologist with a routine cancer diagnosis.

A Renishaw 2000 confocal Raman microscope system was set up in such a way that the light from the region in focus be accepted, while the light from the adjacent regions is rejected. In this way, the spectral interference or contamination from the surrounding regions could be minimized compared to nonconfocal Raman spectra. An argon ion laser operating at $\lambda = 514.5nm$ was used as an excitation source with about $20mW$ power.

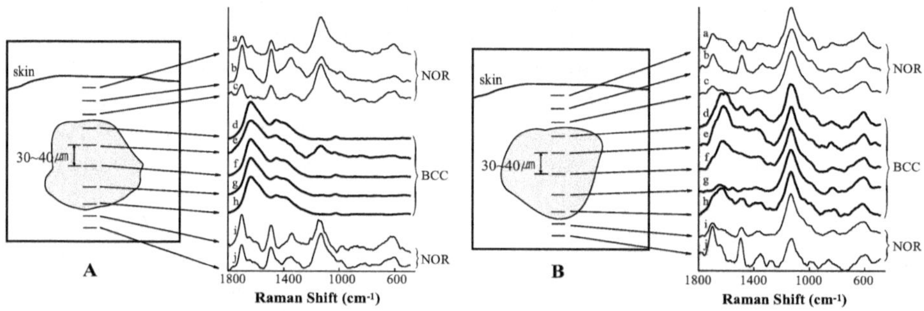

Fig. 1. Confocal Raman profiles of skin tissue with an interval of 30-40 μm

The confocal Raman spectra for the skin samples are shown in Figure 1, where no strong background noise is not observed as expected. In the Figure, most of the spectra show a clear distinction between BCC and NOR tissues while it is not so evident for the case of Fig. 1 B(g) and Fig. 1 B(i). Figure 1 B(g) is considered to be an outlier because it is very different from the surrounding spectra. On the other hand, Fig. 1 B(i) looks similar to a BCC spectrum and can not be considered as an outlier. It is probably caused by the fact that the spectrum was obtained from the vicinity of the boundary between BCC and NOR. There might be a false marking.

A skin biopsy was performed in the perpendicular direction from the skin surface, and it is the same for the spectral measurements. That is the direction from the epidermis to the dermis in Fig. 1. Raman spectra of BCC tissues were measured at different spots with an interval of 30-40μm. In this way, 216 Raman spectra were collected from 10 patients.

2.2 Preprocessing of Data and Dimension Reduction

It is known that the near infrared FT Raman spectra suffer from background noise that originates from skin fluorescence [3]. On the other hand, the confocal Raman spectra do not show any strong background noise as you see in the Fig. 1. That is the reason why we should use confocal Raman spectroscopy. We normalized the spectra so that they fall in the interval [-1,1], which is often called *minmax* method. Needless to say, there are many normalization methods. For example, one can normalize a given data set so that the inputs have means of zero and standard deviations of 1 or have the same area. According to our preliminary experiments, however, the *minmax* method gave the best results.

Next to the data normalization, a clipping window was applied so that unnecessary data should be discarded. We adopted the same clipping window in [5] so as to discard unnecessary data and lower the false positive ratio. For dimension reduction, well known principal component analysis (PCA) was applied. Since PCA identifies orthogonal bases on which projections are uncorrelated, it is the most preferred method for data reduction. Principal components can be obtained via eigenvalue decomposition of the following scatter matrix $\boldsymbol{S}$.

$$\boldsymbol{S} = \sum_k (\boldsymbol{d}_k - \boldsymbol{\mu})(\boldsymbol{d}_k - \boldsymbol{\mu})^T \tag{1}$$

where $\boldsymbol{d}_k$ is a k-th input pattern and $\boldsymbol{\mu}$ is the mean of $\boldsymbol{d}_k$. If we let $\boldsymbol{D}$ be a diagonal matrix of eigenvalues in descending order and $\boldsymbol{E}$ be an orthogonal matrix whose columns are the corresponding eigenvectors, we can obtain principal components $\boldsymbol{x}_k$ as follows.

$$\boldsymbol{S} = \boldsymbol{E}\boldsymbol{D}\boldsymbol{E}^T \tag{2}$$

$$\boldsymbol{x}_k = \boldsymbol{E}^T \boldsymbol{d}_k \tag{3}$$

Data reduction is accomplished by discarding the unimportant elements of $\boldsymbol{d}_k$. The number of retained principal components was determined experimentally according to classification methods.

3 Classification Methods and Experimental Results

To see how well the feature vectors of two different classes, BCC and NOR, are separated, we plotted them in the Fig. 2, where x, y, z correspond to the first, the second, and the third component of the transformed feature vector. The figure clearly shows that the input feature vectors are reasonably well separated as expected. Thus we could expect that a classifier operate well, which will be confirmed at the latter of this section.

3.1 Classification Methods

Five types of classifiers including MAP, PNN, KNN, MLP, and SVM were examined. In the MAP classification, we select the class, w_i, that maximizes the posterior probability $P(w_i|\boldsymbol{x})$. Given the same prior probability, it is equivalent to the selection of the class that maximizes the class conditional probability density. Let w_1, w_2 be BCC class and NOR class respectively. MAP classification rule is expressed as follows.

$$\text{Decide } w_1 \text{ if } P(\boldsymbol{x}|w_1) \geq P(\boldsymbol{x}|w_2) \tag{4}$$

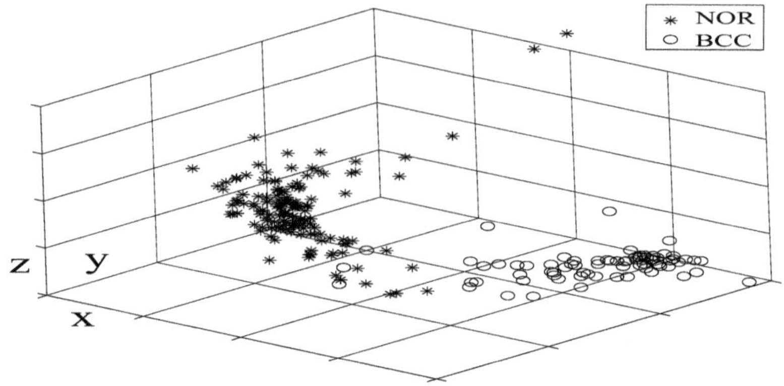

Fig. 2. Distribution of the first three components of PCA transformed spectra

Since input vectors are well separated and could be thought to be distributed around the class mean, we can model the class conditional probability with the multivariate Gaussian probability density function. The parameters, mean vector $\boldsymbol{\mu}$ and covariance matrix $\boldsymbol{\Sigma}$, were estimated in the maximum likelihood sense. Let n_i be the number of data in w_i. Then the decision rule is expressed with a discriminant function, $g_i(\boldsymbol{x})$, as follows.
Decide w_1 if $P(\boldsymbol{x}|w_1) \geq P(\boldsymbol{x}|w_2)$:

$$g_i(\boldsymbol{x}) = -(1/2)\boldsymbol{x}^T \boldsymbol{\Sigma}_i^{-1}\boldsymbol{x} + \boldsymbol{\Sigma}_i^{-1}\boldsymbol{\mu}_i + r_i \tag{5}$$

$$r_i = -(1/2)\boldsymbol{\mu}_i^T \boldsymbol{\Sigma}_i^{-1}\boldsymbol{\mu}_i - (1/2)ln|\boldsymbol{\Sigma}_i^{-1}| \tag{6}$$

$$\boldsymbol{\mu}_i = (1/n_i)\sum_{k=1}^{n_i} \boldsymbol{x}_k \tag{7}$$

$$\boldsymbol{\Sigma}_i = (1/n_i)\sum_{k=1}^{n_i}(\boldsymbol{x}_k - \boldsymbol{\mu})(\boldsymbol{x}_k - \boldsymbol{\mu})^T \tag{8}$$

In the PNN classification, each feature vectors in the training set is scaled to have unit length. An input vector is also to be normalized before classification [6]. The activation function, $\varphi_i(\boldsymbol{x})$, is evaluated for every data in the training set. Let $\boldsymbol{p}_i$ be the normalized feature vector in the training set. The discriminant function of PNN is given by

$$\varphi_i(\boldsymbol{x}) = exp((\boldsymbol{p}_i^T \boldsymbol{x} - 1)/\sigma^2) \tag{9}$$

$$g_i(\boldsymbol{x}) = (1/n_i)\sum_{\boldsymbol{p}_i \in w_i} \varphi_i(\boldsymbol{x}), \tag{10}$$

where parameter σ was determined to be 0.01 experimentally.

We used the following Mahalanobis distance for the KNN classification.

$$d(\boldsymbol{x}, \boldsymbol{y}) = (\boldsymbol{x} - \boldsymbol{y})^T \boldsymbol{\Sigma}(\boldsymbol{x} - \boldsymbol{y}) \tag{11}$$

where $\boldsymbol{\Sigma}$ is a sample covariance matrix computed from all the data in the training set. The discriminant function of the KNN classifier, $g_i(\boldsymbol{x})$, is the number of training data belonging to i class among the k nearest neighbors of $\boldsymbol{x}$. The number of nearest neighbors k was set to be 5. The KNN algorithm requires a large number computation in proportion to the number of training data. Fortunately, there are many fast algorithms. In the experiments, we used the algorithm in [7].

MLP is the most powerful and flexible classifier since they can adapt to arbitrarily complex posterior probability functions [8]. Such extreme flexibility calls for careful control of overfit and detection of outliers. But for the well separated data, overly careful adjustment of the parameters of the networks is not necessary. That is the benefit of confocal Raman spectra. The network operation is characterized by the equation, $\boldsymbol{o}_k = f(net_k)$. The input to the unit, net_k, and the bipolar sigmoidal function, $f()$, is given by the following equations.

$$net_k = \sum_i w_{ik}\boldsymbol{o}_i + bias_k \tag{12}$$

$$f(net_k) = \frac{2}{1 + exp(-2net_k)} - 1 \tag{13}$$

In the experiment, the number of hidden unit was set to be 9. Since there are only two classes, we used one output unit. MLP models were trained to ouput -1 for the NOR class and +1 for the BCC class using back propagation algorithm. At the classification, output value is hard limited to give a classification result. The performance of MLP undergoes a change according to the initial condition. Thus the experiments were carried out 20 times and the results were averaged.

SVM is also a powerful methodology for solving problems in nonlinear classification. In the simplest pattern recognition tasks, SVM uses a linear separating hyperplane to create a classifier with a maximal margin. In cases when given classes cannot be linearly separated in the original input space, SVM first nonlinearly transforms original features into a higher dimensional feature space. This transformation can be achieved by using various nonlinear mappings: polynomial, sigmoid, and radial basis function (RBF). After the nonlinear transformation, linear optimal separating hyperplane can easily be found. The resulting hyperplane will be optimal in the sense of being a maximal margin classifier with respect to training data [9]. In the experiments, we used the following Gaussian RBF kernel.

$$K(\boldsymbol{x}_i, \boldsymbol{x}_j) = e^{-\frac{||\boldsymbol{x}_i - \boldsymbol{x}_j||^2}{\sigma^2}} \tag{14}$$

where σ^2 is set to 20. Least squares SVM were used as an optimization method, where the regularization parameter is set to 1 [10].

3.2 Experimental Results

Overall 216 data were divided into two groups. One is a training set and the other is a test set. Actually, the data from 9 patients were used as a training set and the data from the remaining patient were used as a test set. Once the classification completes, the data from one patient are eliminated from the training set and used as new test data. The previous test data are now inserted into the training set. In this way, the data from every patients were used as a test set. That is the process of 'leaving one out'. The average number of BCC and NOR spectra in the test set is 8 and 14 and that in the training set is 68 and 126 respectively.

The classification results without the data windows are summarized in the Table 1. The number of principal components was 5 for all the methods. In the table, we can see that the the sensitivity of every methods is over 91.5%. Among them, MAP, MLP, and SVM show the sensitivity over 93% and outperform the other methods. Since there aren't enough BCC data, nonparametric methods such as PNN and KNN might be inferior to the others for BCC detection. But the situation is somewhat different in the case of NOR data. The specificity of PNN and KNN is now nearly equal to the others.

Table 1. Classification results with original spectra. Stars indicate the decision of an expert pathologist.

	MAP		PNN		KNN		MLP		SVM	
	BCC	NOR	BCC	NOR	BCC	NOR	BCC	NOR	BCC	NOR
BCC*	93.0	7.0	91.6	8.4	91.8	8.2	93.2	6.8	93.4	6.6
NOR*	4.2	95.8	3.6	96.4	6.9	97.1	5.6	96.4	4.3	95.7

To show the effectiveness of the clipping window, another experiments were carried out with a half Hanning window. The results are shown in the Table 2. With the clipping window, the sensitivity of every methods is improved over 94%. But the averaged specificity remains nearly the same when comparing the results with the previous ones. Considering that the overall performance of each methods gets better, this indicates that the clipping window contributes to lowering the false positive ratio more than the false negative ratio. In case of MLP and SVM, overall true classification rate is over 96.7%.

Table 2. Classification results processed with the half Hanning window. Stars indicate the decision of an expert pathologist.

	MAP		PNN		KNN		MLP		SVM	
	BCC	NOR	BCC	NOR	BCC	NOR	BCC	NOR	BCC	NOR
BCC*	94.6	5.4	94.6	5.4	95.9	4.1	97.3	2.7	96.1	3.9
NOR*	2.3	97.7	5.9	94.1	2.9	97.1	3.5	96.5	2.9	97.1

To see an example of decision boundary, we plotted them in the Fig. 3 in case of SVM, where x, y axis corresponds to the first and the second principal component. The figure clearly shows that some misclassified data lie in the vicinity of the decision boundary.

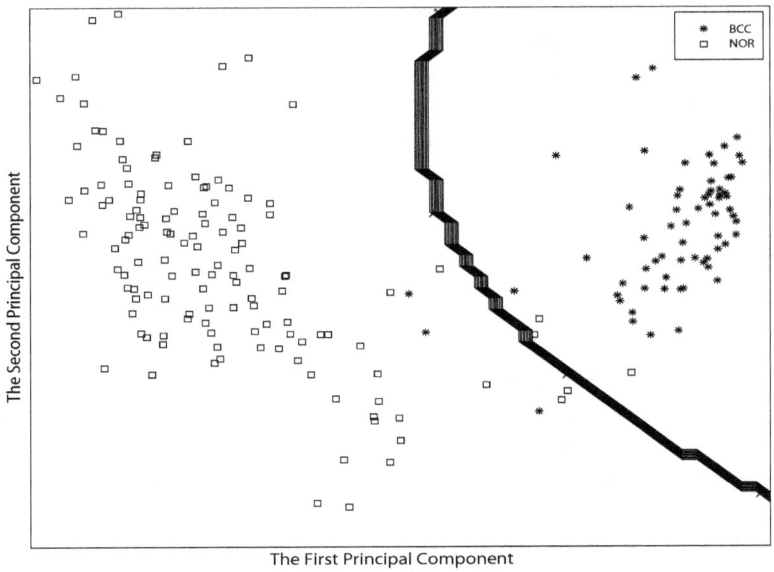

Fig. 3. The decision region of SVM with two principal components

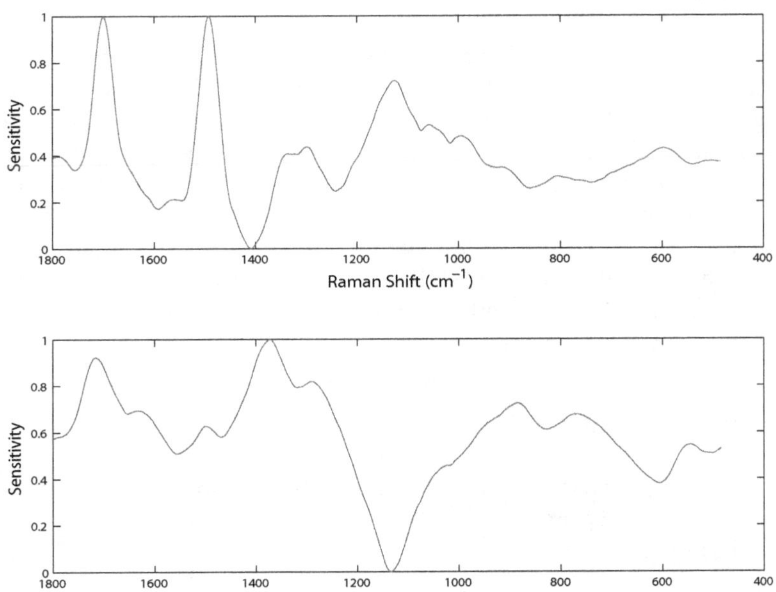

Fig. 4. Examples of sensitivity map for BCC (upper) and NOR (lower)

The sensitivity analysis, which is analogous to the factor analysis in statistics, can be used to examine the impact of each input variable. Generally it provides a gross indicator of key factors via measuring the effect of altering an input variable on the output value by integrating over all input patterns. We performed a sensitivity analysis of MAP for simplicity. Let the absolute value average sensitivity $\boldsymbol{s}_i$ given for class ω_i. Given the same prior probability, it can be derived as follows.

$$\boldsymbol{s}_i = \frac{1}{n_i}\sum_{k=1}^{N_i} \left|\frac{\partial p(\boldsymbol{x}^k|\omega_i)}{\partial \boldsymbol{d^k}}\right| \tag{15}$$

$$\frac{\partial p(\boldsymbol{x}^k|\omega_i)}{\partial \boldsymbol{d^k}} = p(\boldsymbol{x}^k|\omega_i)\widetilde{\boldsymbol{E}}\boldsymbol{\Sigma}_i^{-1}(\boldsymbol{\mu}_i - \boldsymbol{x}^k)^T \tag{16}$$

$$p(\boldsymbol{x}^k|\omega_i) = \frac{1}{(2\pi)^{d/2}|\Sigma|^{1/2}} exp[-\frac{1}{2}(\boldsymbol{x}^k - \boldsymbol{\mu}_i)^T \boldsymbol{\Sigma}_i^{-1}(\boldsymbol{x}^k - \boldsymbol{\mu}_i) \tag{17}$$

where $\boldsymbol{d}^k$ is the k-th data before PCA, n_i is the number of pattern in class ω_i, and $\widetilde{\boldsymbol{E}}$ is a reduced PCA transformation matrix with 5 column vectors. Then we normalize the sensitivity factor for each class to have unit vector length. Figure 4 shows examples of sensitivity map for BCC and NOR. Large sensitivity peaks are observed at 1658, 1457, 1340, 1317 cm^{-1} which correspond to amide I, lipid and protein, amide III vibrational mode. The results are consistent with those in [2] and confirm that the adopted clipping is appropriate.

4 Conclusion

In this paper, we investigated various classification methods with confocal Raman spectra to detect BCC. At the preprocessing stage, data was processed with a clipping window to lower the false positive classification rates. The experimental results with and without the clipping window revealed that the application of the clipping window could lower the false positive ratio. The MLP and SVM classification performance involving 216 Raman spectra was over 96.7%. The results indicate that the confocal Raman spectra provides discriminating spectra for the classification of BCC and NOR tissue. In addition to it, we showed that there is a consistency between automatic classification and direct observation by human with sensitivity analysis.

Acknowledgement

This work was supported by grant No. RTI-04-03-03 from the Regional Technology Innovation Program of the Ministry of Commerce, Industry and Energy(MOCIE) of Korea.

References

1. Jijssen, A., Schut, T. C. B., Heule, F., Caspers, P. J., Hayes, D. P., Neumann, M. H., Puppels, G. J.: Discriminating Basal Cell Carcinoma from its Surrounding Tissue by Raman Spectroscopy. Journal of Investigative Dermatology, 119 (2002) 64–69

2. Choi, J., Choo, J., Chung, H., Gweon, D. G., Park, J., Kim, H. J., Park, S., Oh, C.H.: Direct Observation of Spectral Differences Between normal and Basal cell Carcinoma (BCC) Tissues Using Confocal Raman Microscopy. Biopolymers, 77 (2005) 264–272
3. Sigurdsson, S., Philipsen, P. A., Hansen, L. K., Larsen, J., Gniadecka, M., Wulf, H. C.: Detection of Skin Cancer by Classification of Raman Spectra. IEEE Trans. on Biomedical Engineering, 51 (2004) 1784–1793
4. Nunes, L. O., Martin, A. A., Silveira J. L., Zampieri, M., Munin, E.: Biochemical Changes between Normal and BCC Tissue: a FT-Raman Study. Proceedings of the SPIE, 4955 (2003) 546–553
5. Baek, S. J., Park, A.: Basal Cell Carcinoma Detection by Classification of Confocal Raman Spectra. (to appear in ICIC 2006 Proceedings)
6. Duda, R. O., Hart, P. E., Stork, D. G.: Pattern Classification. Jone Wiley & Son, Inc, New Work (2001)
7. Baek, S. J., Sung, K. M.: Fast KNN Search Algorithm for Nonparametric Classification. IEE Electronics Letters, 35 (2000) 2104–2105
8. Gniadecka, M., Wulf, H., Mortensen, N., Nielsen, O., Christensen, D.: Diagnosis of Basal Cell Carcinoma by Raman Spectra. Journal of Raman Spectroscopy, 28 (1997) 125–129
9. Kecman, V.: Learning and Soft Computing. The MIT Press, London (2001)
10. Suykens, J. A. K., Vandewalle, J.: Recurrent Least Squares Support Vector Machines. IEEE Transactions on Circuits and Systems, 47 (7) (2000) 1109-1114

Clustering Gene Expression Data for Periodic Genes Based on INMF

Nini Rao[1] and Simon J. Shepherd[2]

[1] School of Life Sciences & Technology, University of Electronic Science & Technology of China, Chengdu 610054, P.R. China
Raonn@126.com
[2] Advanced Signals Laboratory, School of Engineering, Design & Technology, University of Bradford, BD7 1DP, UK

Abstract. In this paper, we have explored the use of improved non – negative matrix factorization (INMF) to analyze gene expression data. Firstly, the mathematical principle of INMF algorithm is analyzed; Secondly, we proposed an INMF - based method for clustering periodic genes, which can provide valuable information for gene network research. Using simulated data, our approach is able to extract periodic genes subsets even when the signal-to-noise ratio is low. Subsequently, our approach is tested by real gene expression datasets from Yeast and is compared with the related other approaches. Our results showed that our scheme is feasible and effective.

1 Introduction

One particular aim when analyzing microarray data is to find statistical evidence of periodicity, and then to identify this subset of genes that is responsible during the cell cycle. A comprehensive catalog of cell cycle regulation genes can serve as a starting point for functional discovery [1-4]. Usually only a small fraction of the genes under investigation exhibits some evidence of periodically varying expression during the cycle so that the overall signal in the data is dominated by non – periodic components.

There are a number of methods discussed in the literature for obtaining this subset of genes with periodic variation. These methods include: 1) cluster analysis, which has been demonstrated to be of significant value for exploration of gene expression profiles [5-8]. However, such cluster analyses provide little insight into the relationships between groups of co-regulated genes or the behavior of biological networks as a whole. 2) parametric methods, such as Principal Component Analysis (PCA) and the singular value composition (SVD) [9-13], independent component analysis (ICA) [14-15] and separation correlation metric (SCM), alternatively known as the Fisher's discrimination criterion (FDC) [16]; and 3) nonparametric methods, such as support vector machines[17], neural networks, expectation maximization[18], statistical analysis[19 - 20], etc. It is well known that the accuracy and reliability of the results are often dependent on the measurement approaches applied, and no single measurement so far is guaranteed to generate a satisfactory result [21].

The recent controversial argument with regard to the suitability of both available methods and current microarray data showed that it is difficult to identify correctly

D.-S. Huang, K. Li, and G.W. Irwin (Eds.): ICIC 2006, LNBI 4115, pp. 412 – 423, 2006.

the subset of genes with clear periodic signature. For instance, Shedden and Cooper [22] questioned the presence of generic cell – cycle specific signal in the datasets concerning yeast cells that were previously analyzed by Spellman *et al.* Their results were inconsistent with those obtained by Spellman *et al*[2]. In another foundational work, Wichert *et al.* presented a statistical method based the average periodogram and Fisher's g – statistic for identifying genes relevant for the cell cycle, which has some potential drawbacks. First, the g – statistic assumes as null – model a purely Gaussian process, while this assumption has not been verified so far. Second, it is sensitive to the size of the sample number and the size of the subset of cyclic genes in the data[20].

The above analysis shows that a further development of methodology for identifying periodic genes is necessary. In this paper, we propose a method based on improved non – negative matrix factorization (INMF) to extract the subset of the genes periodically expressed in gene expression data. Pascual-Montano, A. D.*et al.* have applied NMF to gene expression data to identify highly correlated genes and experiments that behave in a similar manners in only a sub portion of the data, and also applied this methodology to freeform texts in MEDLINE abstracts for a semantic analysis of scientific articles[23]. In their work, NMF algorithm is classical one, while we improved the classical NMF algorithm in this work. Otherwise, the focusing problems are different in two works. Previously, we have also used NMF to extract the characteristic patterns from gene expression profiles [24]. In our previous work, NMF algorithm was still traditional one. This work, which improves the classical NMF algorithm and further applies INMF to extract the periodic genes subsets, extends our previous work.

2 INMF Algorithms

Firstly, we analyzed the traditional NMF algorithm. A positive $n \times m$ matrix $\mathbf{V}$ is factorized into two low-rank factors $\mathbf{W}$ and $\mathbf{H}$ of sizes $n \times r$ and $r \times m$ respectively, where r is the desired rank of the factorization.

$$\mathbf{V}^{(n\times m)} = \mathbf{W}^{(n\times r)}\,\mathbf{H}^{(r\times m)} \tag{1}$$

In general, $(n+m)r < nm$. At each step of the algorithm there will exist an error between the product $\mathbf{WH}$ and the target matrix $\mathbf{V}$. This error $\mathbf{E}$ is defined as

$$\mathbf{E} = \mathbf{WH} - \mathbf{V} \tag{2}$$

The natural way to quantify the magnitude of $\mathbf{E}$ is in terms of the 2-norm $N_{(2)}$, which is the sum of the squares of the element of $\mathbf{E}$. It can also be expressed (and computed) more conveniently in terms of the *trace* function $tr(\mathbf{X})$, which is defined as the *sum of the elements* on the lead diagonal of $\mathbf{X}$. In these terms, $N_{(2)}$ is:

$$N_{(2)}(\mathbf{X}) = tr(\mathbf{X}^T\mathbf{X}) = tr(\mathbf{X}\mathbf{X}^T)$$

(To minimize computation in the case of non-square $\mathbf{X}$, we choose the order of multiplication to minimize the dimension of the product.) Thus, the error norm s of $\mathbf{E}$ is:

$$s = tr(\mathbf{E}^T\mathbf{E}) = tr((\mathbf{WH} - \mathbf{V})^T\,(\mathbf{WH} - \mathbf{V})) \tag{3}$$

We can infer that the change δs in s at each step is:

$$\delta s = -2\mu \, \{tr(\mathbf{Y}) + tr(\mathbf{Z})\} \tag{4}$$

where $\mathbf{Y} = \mathbf{HE}^T\mathbf{EH}^T$ and $\mathbf{Z} = \mathbf{W}^T\mathbf{EE}^T\mathbf{W}$. In this form, it is clear that the matrices $\mathbf{Y}$ and $\mathbf{Z}$ are of the form $\mathbf{U}^T\mathbf{U}$ that is always at least a semi-positive definite matrix, and hence its trace must be *positive*. Therefore, δs is always *negative* and the error is always *reduced* at each step. Although we obviously cannot achieve it, our ambition would be to minimize the error in one step, that is, we would like $\delta s = -s$. Using this to solve for the critical value of μ. we have:

$$\mu = s \,/\, \{2\, [N_{(2)}(\mathbf{HE}^T) + N_{(2)}(\mathbf{W}^T\mathbf{E})]\} \tag{5}$$

The classical NMF algorithm is as follows:

- Input $\mathbf{V}$
- Set $\mu \leftarrow \mu 0$ (say 0.001, or anything small)
- Set $\mathbf{W} \leftarrow$ random positive $n \times r$ matrix
- Set $\mathbf{H} \leftarrow$ random positive $r \times m$ matrix
- Loop until $s <$ *tolerance*
  - $\mathbf{W}_{temp} \leftarrow \mathbf{W} - \mu(\mathbf{WH} - \mathbf{V})\mathbf{H}^T$
  - $\mathbf{H} \leftarrow \mathbf{H} - \mu\mathbf{W}^T(\mathbf{WH} - \mathbf{V})$
  - $\mathbf{W} \leftarrow \mathbf{W}_{temp}$
  - $\mathbf{E} \leftarrow \mathbf{WH} - \mathbf{V}$
  - $\mu \leftarrow s \,/\, \{4\, [N_{(2)}(\mathbf{HE}^T) + N_{(2)}(\mathbf{W}^T\mathbf{E})]\}$
- Return $\mathbf{W}$ and $\mathbf{H}$

Von Neumanm first described this algorithm in 1950 in terms of finding the projection onto the intersection (or direct sum) of two closed sub-spaces of a Hilbert space [25]. He showed that projecting a point into each sub-space in turn would ultimately yield the projection onto the intersection. His classic paper includes proof that such an algorithm (if properly formed) will always converge.

Note that we have divided s by 4 times the trace and not 2 times the trace, as we derived. This is because the value of μ derived is the critical value – it is the *absolutely maximum* value μ can take for stability. However, remember that our analysis omitted to take account of the second order terms and occasionally these can become sufficiently significant that the μ generated goes past the point of criticality and the algorithm goes into positive feedback and blows up. Therefore, for safety we "throttle back" on μ by halving its calculated value. Actually, this is very over-cautious. In practice, scaling μ to around 0.9 of its calculated value is usually fine.

This algorithm will work very well for many matrices with strongly positive elements. Note, however, that there is nothing *inherent* in the algorithm to stop it generating negative elements in the factors. In cases where $\mathbf{V}$ has a low-rank approximation with *sparse* factors, the above algorithm can occasionally generate negative elements in those places where zeros should appear in the factors. To cure this, we make improvement for the classical algorithm.

For the improved NMF (INMF) algorithm, we define $\mathbf{W} = \mathbf{A} \bullet \mathbf{A}$ and $\mathbf{H} = \mathbf{B} \bullet \mathbf{B}$ where [•] denotes the element-wise (Hadamard) product of matrices. If we now recast the algorithm in terms of updates to **A** and **B**, when we compute **W** and **H**, their elements cannot be negative as each is a square of an element in **A** and **B** respectively. Once an element is zero, it stays zero. Thus

$$\mathbf{W} = \mathbf{A} \bullet \mathbf{A} \quad \text{and} \quad \mathbf{H} = \mathbf{B} \bullet \mathbf{B}$$

$$\delta\mathbf{W} = 2\mathbf{A} \bullet \delta\mathbf{A} \quad \text{and} \quad \delta\mathbf{H} = 2\mathbf{B} \bullet \delta\mathbf{B}$$

So

$$\delta\mathbf{A} = \delta\mathbf{W} \bullet/\, 2\mathbf{A} = -\,\mu\mathbf{E}\mathbf{H}^{T} \bullet/\, 2\mathbf{A}$$

$$\delta\mathbf{B} = -\,\delta\mathbf{H} \bullet/\, 2\mathbf{B} = -\,\mu\mathbf{W}^{T}\mathbf{E} \bullet/\, 2\mathbf{B}$$

where $\delta\mathbf{W} = -\,\mu\mathbf{E}\mathbf{H}^{T}$, $\delta\mathbf{H} = -\,\mu\mathbf{W}^{T}\mathbf{E}$ and [•/] indicates element-wise (Hadamard) division of matrices. Thus, we have the improved NMF algorithm:

- Input **V**
- *Set* $\mu \leftarrow \mu 0$ *(say 0.001, or anything small)*
- Set $\mathbf{A} \leftarrow$ random positive $n \times r$ matrix
- Set $\mathbf{B} \leftarrow$ random positive $r \times m$ matrix
- Loop until $s < tolerance$
    - $\mathbf{W} \leftarrow \mathbf{A} \bullet \mathbf{A}$
    - $\mathbf{H} \leftarrow \mathbf{B} \bullet \mathbf{B}$
    - $\mathbf{E} \leftarrow \mathbf{WH} - \mathbf{V}$
    - $\mu \leftarrow s / \{4\, [N_{(2)}(\mathbf{H}\mathbf{E}^{T}) + N_{(2)}(\mathbf{W}^{T}\mathbf{E})]\}$
    - $\mathbf{A} \leftarrow \mathbf{A} - \mu(\mathbf{E}\mathbf{H}^{T} \bullet/\, 2\mathbf{A})$
    - $\mathbf{B} \leftarrow \mathbf{B} - \mu(\mathbf{W}^{T}\mathbf{E} \bullet/\, 2\mathbf{B})$
- Return **W** and **H**

Our simulation results show that this algorithm is *stable* and converges *very fast.* The improvement for NMF algorithm is a little, but very efficient.

Compared with other matrix factorization methods, such as Principal Component Analysis (PCA), Singular Value Decomposition (SVD) and Independent Component Analysis (ICA), Improved Non – negative Matrix Factorization (INMF) has three characteristics:

(1) There is no the distribution requirements for the latent variables in the dataset;
(2) The localized information hidden in the dataset can be extracted by INMF;
(3) The original dataset can be directly processed, that is, need not be normalized.

3 INMF – Based Cluster Method for Periodic Genes

The method can be divided into two steps. In the first step, various characteristic expression patterns hidden in gene expression data are extracted by INMF algorithm. Usually, it is easy to select periodic expression patterns from the characteristic pattern sets. According to our previous analysis [24], the method for extracting characteristic patterns can be stated as follows. Suppose that **V** is gene expression dataset, v_{ij} in **V** is the expression level of the ith gene in the jth sample. The elements of the ith row of **V** form an n–dimensional vector g_i, which we refer to as the *transcription response* of the ith gene. In practical terms, this factorization finds a small set of "basis vectors" **W** and a set of "hidden" describing factors **H**. Since the columns of **H** are in one – to – one correspondence with the columns of **V**, the results can be interpreted as each column of **V** being described as weighted of a few basis vectors, the weights being the corresponding column of **H**. Our analysis shows that the vector in the lth row of the matrix **H**, e_l lists the expression of the lth eigengene across the different samples. So, we infer that e_l represent a characteristic expression pattern across all samples, while this pattern is biological interpretable. From **H,** we select out whole periodic patterns and establish a new transformation matrices $\mathbf{H}^t$ with dimension $k \times m$, where k is the number of periodic expression patterns. In step two, the periodic expression genes in **V** are identified. For this purpose, the Pearson correlation method is used to identify interesting genes that have similar periodic expression patterns to those eigengenes in $\mathbf{H}^t$, that is, we calculated the correlation coefficients R between the lth eigengene in $\mathbf{H}^t$ ($l = 1, 2, \ldots, k$) and all genes g_i ($i = 1, 2, \ldots, n$) in **V**. If the correlation coefficient R between the ith gene and the lth eigengene exceeds a specified *threshold of similarity* θ_R (empirically or analytically), the ith gene is deemed to have a similar expression pattern to that of the lth eigengene and so they are in a cluster. The larger θ_R is, the higher the accuracy of the identification is.

4 Experiment Results

In order to verify the feasibility and the effectiveness of our methodology, we designed simulated datasets and select Yeast gene expression datasets, which have been used as benchmark dataset in previous studies, to do test experiments.

4.1 Simulated Data Analysis

Firstly, genes in the simulated dataset have one of three kinds of transcriptional response, inspired by experimentally observed patterns in the Cho *et al.* cell cycle data [1]:

- noise (1600 genes)
- noisy sine pattern (200 genes)
- noisy exponential pattern (200 genes).

Noise for all three groups of genes was modeled by sampling from a uniform distribution within the interval (0, 1). The sine pattern has the functional form $a\sin(2\pi t/140)$ and the exponential pattern the form $be^{-t/100}$ where a is sampled uniformly over the interval (1.5 - 3), b is sampled uniformly over (4 - 8), t is the time (in minutes) associated with each sample, and time points are sampled every ten minutes beginning at $t = 0$.

Four characteristic patterns (or eigengenes) of the transcriptional response are extracted by INMF when taking $r = 4$. The results are showed in Fig. 1. Obviously, only eigengene 1 in Fig. 1(a) has periodic structure, the eigengene 2 shows a clear exponential pattern, while eigengene 3 and 4 denote two noisy patterns because they do not exhibit any regulatory. These patterns are same with the known patterns.

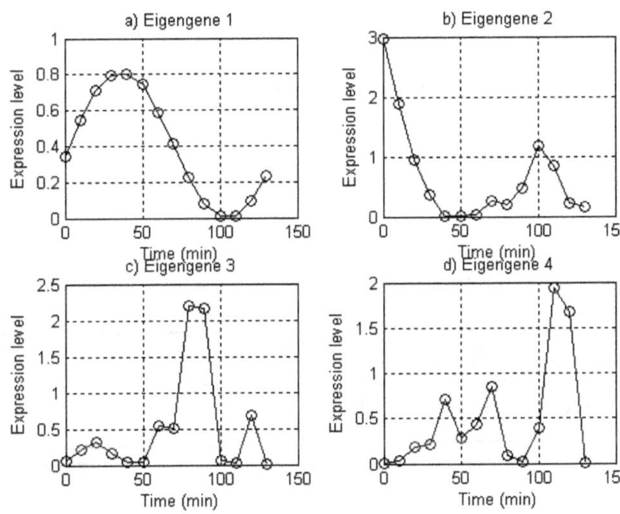

Fig. 1. The expression patterns extracted by INMF from the synthetic data set

To illustrate the ability of our method to extract *weak* expression patterns, we increased the noise intensity into about 67% of the maximum sine amplitude and about 50% of the maximum exponent initial value. Fig. 2 shows the characteristic patterns extracted by our method. We know from Fig. 2, even under these difficult signal-to-noise ratio conditions, eigengene 1 still reveals the sine pattern very clearly, eigengene 2 still displays an exponential pattern and eigengene 3 and 4 again lack any obvious structure. So, our method has good ability to resist the noise.

Further, we studied the sensitivity of our method with respect to the number of cyclic genes in the dataset by decreasing 200 sine genes into 125, 100, 75, 60, 50, 40, 20 genes respectively. The results showed that the above changes have no effect on the pattern extraction ability of INMF method and the desired sine expression patterns can still be extracted (data not show).

In order to obtain the periodic gene clusters, we identify respectively those genes with periodic structure in the simulated data using the extracted sine pattern. The

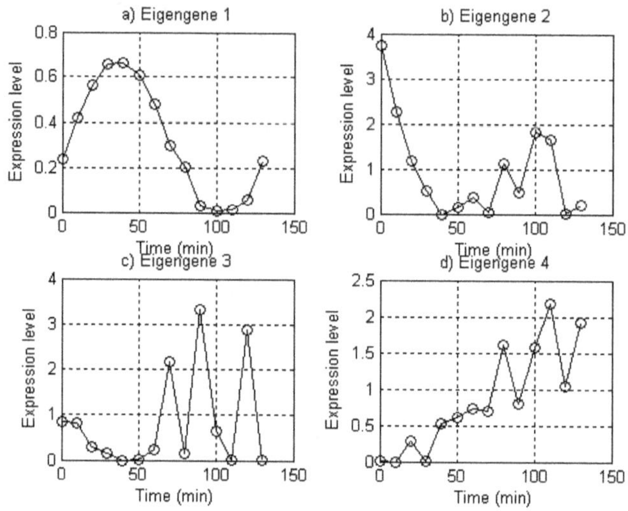

Fig. 2. The Extracted weak patterns by INMF under these difficult signal-to-noise ratio conditions

Table 1. The clustering results of periodic genes by three different methods ($\theta_R = 0.992$)

Case	Signal – to – noise radio 1			Signal – to – noise radio 2		
Methods	INMF	K – means	SOM	INMF	K – means	SOM
Number of clustered periodic genes	200	136	200	199	40	194

experimental results under above two different signal - to - noise radio cases are showed in table 1 and are compared with those of K – means and Self – Organizing Maps (SOM) cluster methods when noisy sine genes are 200 in the dataset.

From a large number of experiments, we found that 200 sine genes can be precisely identified by our method when the specified threshold θ_R of similarity was taken as 0.992 in the case of signal - to - noise radio 1. We checked these sine genes and found that they are exactly 200 known sine genes. In the case, SOM method can also cluster 200 sine genes. But K – means method only clustered 136 sine genes and 64 sine genes were missed. In the case of signal – to - noise radio 2, the noise intensity was increased. 199 sine genes are clustered by our method when $\theta_R = 0.922$ and they are among 200 known sine genes. Only one gene was missed. When we decrease θ_R into 0.919, 205 genes were identified. We checked the results and found that 5 false genes from noise were detected. Thus, the reasonable θ_R value is 0.922 in this case. SOM clustered 194 sine genes, while K – means clustered only 40 sine genes.

The above analysis showed that our method is effective for clustering periodic genes when the appropriate θ_R is taken.

4.2 Biological Data Analysis

Spellman *et. al.* [2] used DNA microarrays and samples from yeast cultures synchronized by three independent methods: α factor arrest, elutriation, and arrest of a cdc15 temperature-sensitive mutant, and identified the approximately 800 genes that meet an objective minimum criterion for cell cycle regulation using periodicity and correlation algorithms. The related information on three biological datasets from Spellman *et al.* is showed in table 2. In addition, some other authors also selected these biological datasets to test their methods [9, 11-12, 20]. Hence, alpha dataset (DS), elutriation DS and cdc15 DS (the results not show) are used to validate our methodology in this work.

Table 2. Summary of the biological datasets

Dataset	Method of arrest	start	end	sample	sample number
alpha DS	Alpha mating factor	0 m	119 m	every 7 m	18
elutriation DS	elutriation	0 m	390 m	every 30 m	14
cdc15 DS	Temp. sensitive cdc muant	0 m	290 m	ev. 20 m for 1 hr. ev. 10 m for 3 hr. ev. 20 min for final hr	24

4.2.1 Alpha DS

For alpha dataset, genes with missing values and constant expression levels for all sample points were removed from the study. Two characteristic patterns, which show a periodic behaviour, are extracted by our method when taking $r = 5$. The results are showed in Fig. 3. Two cell – cycle behaviour is mainly of the form $A \sin(\omega t + \phi)$ with the same ω but differing ϕ value and possibly due to desynchronization, where $\omega = 1 / T$, $T \cong 67$ minutes. Two periodic characteristic patterns can be explained to represent cell cycle expression oscillations.

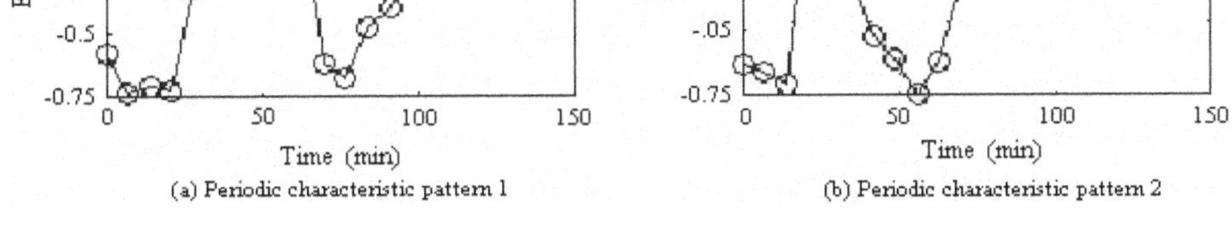

(a) Periodic characteristic pattern 1 (b) Periodic characteristic pattern 2

Fig. 3. The extracted periodic patterns from alpha dataset

Further, we take threshold criterion $\theta_R = 0.922$ according to the previous analysis for the simulated data and clustered the periodic genes in alpha DS. The results and the comparisons with those results of Spellman, *et al* [2]., Alter, *et al*[12] and Wichert[20], *et al* are showed in table 3.

Table 3. The cluster results and the comparisons for alpha DS

Method	Our method	Spellman's method	Alter's method	Wichert's method
Results	636/4579	638/4579	544/4579	468/4415

For alpha DS, Spellman *et al.* identified 638 cell cycle-regulated genes from 4579 genes and Alter *et al.* identified 544 genes from 4579 genes. 468 periodic expression genes were identified by the method of Wichert, *et al.* Using our method when taking $\theta_R = 0.922$, the number of identified genes (636) with periodic structure is almost same as that of Spellman *et al.* Two genes were missed by our method.

4.2.2 Elutriation DS

For elutriation dataset, genes with missing values and constant expression levels for all sample points were removed from the study. Firstly, two periodic characteristic patterns called pattern 1 and pattern 2 were extracted when taking $r = 5$. The results are showed in Fig. 4. The time variation of periodic characteristic pattern 1 fits a normalized sine function of period T (about 390 min), A sin $(2\pi t/T)$. The time variation of periodic characteristic pattern 2 fits a cosine function of period T (about 390 min) with the amplitude B of a normalized cosine with this period, $B\cos(2\pi t/T)$. We infer that the periodic characteristic pattern 1 and 2 are consistent with gene expression oscillations during a cell cycle.

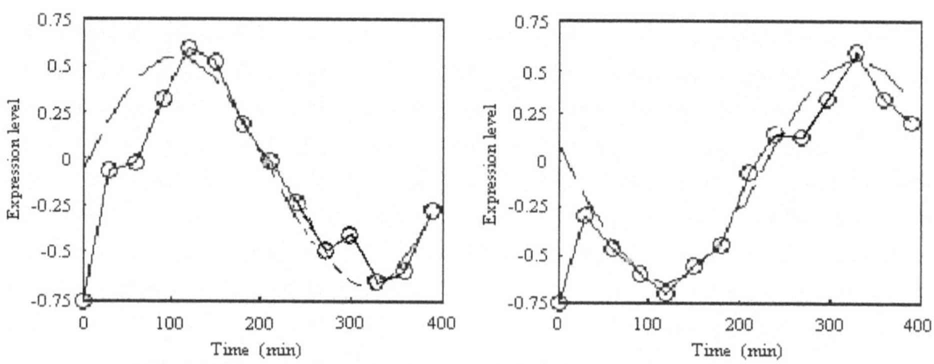

(a) Periodic characteristic pattern 1 (b) Periodic characteristic pattern 2

Fig. 4. The extracted periodic patterns from elutriation dataset

Secondly, we clustered the periodic genes in elutriation DS. The results and the comparisons with those results of Spellman, *et al.*, Alter, *et al* and Wichert, *et al* are showed in table 4.

Table 4. The cluster results and the comparisons for elutriation DS

Method	Our method	Spellman's method	Alter's method	Wichert's method
Results	785/5981	784/5981	641/5981	193/5695

For the elutriation dataset, Spellman *et al.* identified 784 cell cycle-regulated genes from 5981 genes and Alter *et al.* identified 641 periodic genes from 5981 genes. Only 193 periodically expressed genes were identified by the method of Wichert, *et al.* They explained that the elutriation data provided only little statistically significant information with regard to cell cycle regulation. The number of periodic genes clustered by our method is 785 when taking $\theta_R = 0.922$, which is closed to that of Spellman *et al.* When we compared these periodic genes with the results of Spellman *et al.*, we found that only one periodic gene is not belong to Spellman's periodic genes cluster and is possibly from noise. Our results seem to support again Spellman *et al* and better than the results of Alter, *et al.* and Wichert, *et al.*

5 Conclusions

In this paper, we have analyzed the weak of the traditional non – negative matrix factorization algorithm and proposed an improved non – negative matrix factorization algorithm for gene expression data analysis. Based on INMF algorithm, we presented an approach to cluster gene expression data for periodic gene subsets. Our approach includes two steps. Firstly, the periodic characteristic expression patterns are extracted by improved non – negative matrix factorization algorithm. Secondly, Pearson correction was suggested to identify individual periodic genes for the periodic gene clusters. Compared with other related approaches, our approach can be applied to gene expression data with any size and distribution and has strong ability to extract the characteristic expression patterns hidden in gene expression data, even under the condition of high noise levels. Our approach performed well in clustering genes periodically expressed for the simulated dataset and Yeast gene expression datasets, which deliver valuable information for gene network research and gene function discovery.

Acknowledgment

This research is supported by grants from the National Natural Science Foundation of China (Grant No. 60571047), the Academic Leader Training Foundation of Sichuan (Grant No. 901008) and the Youth Talent Training Program of UESTC (Grant No. 601016).

References

1. Cho, R.J., Campbell, M.J., et al.: A Genome–Wide Transcriptional Analysis of the Mitotic Cell Cycle. Mol. Cell. 2 (1998) 65–73
2. Spellman, P.T., Sherlock, G., Zhang, M.Q., et al.: Comprehensive Identification of Cell Cycle–Regulated Gene of the Yeast Saccharomyces Cerevisiae by Micro-Array Hybridation. Mol. Biol. Cell. 9 (1998) 3273–3297
3. Cho, R.J., Huang, M.X., Dong, H., et al.: Transcriptional Regulation and Function During the Human Cell Cycle. Nature Genetics. 27 (2001) 48–54
4. Lockhart, D.J., et al.: Expression Monitoring by Hybridization to High – Density Oligonucleotide Arrays. Nature Biotechnology. 14 (1996) 1675–1680
5. Eisen, M.B., Spellman, P.T., Brown, P.O., Botstein, D.: Cluster Analysis and Display of Genome–wide Expression Patterns. Proc. Nat. Acad. Science USA. 95 (1998) 14863–14868
6. Lukashin, A.V., Fuchs, R.: Analysis of Temporal Gene Expression Profiles: Clustering by Simulated Annealing and Determining the Optimal Number of Clusters. Bioinformatics, 17 (2001) 405–414
7. Mclachlan, G.J., Bean, R.W.: A Mixture Model–Based Approach to the Clustering of Micro-Array Expression Data. Bioinformatics. 18 (2002) 413–422
8. Deutsch J.M.: Evolutionary Algorithms for Finding Optimal Gene Sets in Microarray Prediction. Bioinformatics. 19 (2003) 45–52
9. Raychaudhuri, S., Stuart, J.M., et al.: Principal Components Analysis to Summarize Micro-array Experiments: Application to Sporulation Time Series. Pacific Symposium on Biocomputing. (2000) 455–466
10. Wall, M.E., Rechtsteiner, A. et al.: Singular Value Decomposition and Principal Component Analysis. In A Practical Approach to Micro-array Data Analysis, ed. Berrar, D. P., Kluwer, Chapter 5 (2003) 91–109
11. Holter, N.S., Mitra, M., Maritan, A., et al.: Fundamental Patterns Underlying Gene Expression Profiles: Simplicity From Complexity. Proc. Nat Acad. Science USA. 97 (2000) 8409–8414
12. Alter, O., Brown, P.O., Botstein, D.: Singular Value Decomposition for Genome–Wide Expression Data Processing and Modelling. Proc. Nat. Acad. Science USA. 97 (2000) 10101–10106
13. Wall, M.E., Dyck, P.A., Brettin, T.S.: SVDMAN - Singular Value Decomposition Analysis of Micro-array Data. Bioinformatics. 17 (2001) 566–568
14. Liebermeister, W.: Linear Modes of Gene Expression Determined by Independent Component Analysis. Bioinformatics. 18(1) (2002) 51 – 60
15. Yamanishi, Y., Itoh, M., Kanehisa, M.: Extraction of Organism Groups from Phylogenetic Profiled using Independent Component Analysis. Genome Informatics. 13 (2002) 60 -70
16. Fisher, R.A.: The Use of Multiple Measurements in Taxonomic Problems. Annu. Eugenics, pt. II, 7 (1936) 179 – 188
17. Brown, M., Grundy, W.N., Campbell, M.J., et al: Knowledge - based Analysis of Microarray Data by Using Support Vector Machines. Proc. Nat. Acad. Science USA. 97 (2000) 262 – 267
18. Duda, R.O., Hart, P.E., Stork, D.G.: Patten Classification, 2nded. New York: Wiley (2001)
19. Whitfield, M.L., Sherlock, G., Saldanha, A.J., et al.: Identification of Genes Periodically Expressed in the Human Cell Cycle and Their Expression in Tumor. Mol. Biol. Cell. 13 (2002) 1977 – 2000

20. Wichert, S., Fokianos, K., Strimmer, K.: Identifying Periodically Expressed Transcripts in Microarray Time Data. Bioinformatics. 20(1) (2004) 5 – 20
21. Zhu, Q., Cui, H., Cao, K., Chan, W.C.: Algorithm Fusion of Gene Expression Profiling for Diffuse Large B – Cell Lymphoma Outcome Prediction. IEEE Trans. On Information Tech. In Biomedical. 8(2) (2004) 79 - 88
22. Shedden, K., Cooper, S.: Analysis of Cell – Cycle Gene Expression in Saccharonization Cerevisiae using Microarrays and Multiple Synchronization Methods. Nucleic Acids Res. 30 (2002) 2920 – 2929
23. Pascual, A.D., et al.: Non – negative Matrix Factorization for Gene Expression and Scientific Text Analysis. Proceedings of 11th International Conference on Intelligent Systems for Molecular Biology, Australia (2003)
24. Rao, N., Shepherd, S.J.: Extracting Characteristic Patterns from Genome – Wide Expression by NMF. Proceedings of IEEE Internaltional Conference on Computational Systems Bioinformatics, Stanford, California (2004) 570 - 571
25. Neumann, V.J.: Functional Operators, Vol. II. Annals of Mathematics No. 22, Princeton University Press (1950)

Feature Selection for Microarray Data Analysis Using Mutual Information and Rough Set Theory

Wengang Zhou, Chunguang Zhou, Hong Zhu, Guixia Liu, and Xiaoyu Chang

College of Computer Science and Technology, Jilin University,
Changchun 130012, P.R. China
wgzhou@email.jlu.edu.cn

Abstract. Cancer classification is one major application of microarray data analysis. Due to the ultra high dimension of gene expression data, efficient feature selection methods are in great needs for selecting a small number of informative genes. In this paper, we propose a novel feature selection method MIRS based on mutual information and rough set. First, we select some top-ranked features which have higher mutual information with the target class to predict. Then rough set theory is applied to remove the redundancy among these selected genes. Binary particle swarm optimization (BPSO) is first proposed for attribute reduction in rough set. Finally, the effectiveness of the proposed method is evaluated by the classification accuracy of SVM classifier. Experiment results show that MIRS is superior to some other classical feature selection methods and can get higher prediction accuracy with small number of features. Generally, the results are highly promising.

1 Introduction

The development of microarray technology has made it easy to monitor the expression pattern of thousands of genes simultaneously and a huge amount of gene expression data has been produced during microarray experiments. These data has widely been applied to accurate prediction and diagnosis of cancer. Especially cancer classification [1] is an important issue because it can identify many genes relevant to cancer. The results reported in the literature have confirmed the effectiveness of mining cancer information from gene expression data. But microarray data often consists of small number of samples and large number of genes. The ultra high dimension of gene expression data makes it necessary to develop effective feature selection methods in order to reduce the computation cost and improve the classification accuracy.

There are two general approaches to feature selection: filters [2] and wrappers [3]. In a filter method, features are selected based on the intrinsic characteristics which determine their relevance with the target classes. In wrapper type methods, the usefulness of a feature is directly judged by the estimated accuracy of a learning method and typically requires huge computational effort. Thus, it is difficult for wrappers to deal with large feature sets such as gene expression data. We mainly focus on the selection of a few tens features among several thousands by developing an efficient filter methods for cancer classification. When a small number of informative genes are selected, their biological relationship with the target disease can easily be identified.

D.-S. Huang, K. Li, and G.W. Irwin (Eds.): ICIC 2006, LNBI 4115, pp. 424 – 432, 2006.

Mutual information [4] has recently been proposed for feature selection. One common practice to use this method is to simply select the top-ranked genes with higher mutual information. But a deficiency of this simple ranking approach is that the features can be correlated among themselves. If gene g_i is ranked high for classification task, other genes highly correlated with gene g_i are also likely to be selected. This raises the issue of redundancy in feature set. Rough sets theory provides a feasible way to deal with redundancy [5]. An important concept is reduct in rough sets theory. Reduct is those minimal attribute sets of information system, which keep the same classification capability with original attribute set. The aim of reduction is to find out a minimum set of relevant attributes (features) that describe the dataset as well as all the original attributes do. Thus finding reduct can select the most relevant genes with the target class to predict and remove the redundancy among the selected features.

In this paper, we propose a novel feature selection method MIRS based on mutual information and rough set theory. First, mutual information is used to select some top-ranked genes which have higher mutual information from each data set. Then rough set theory is applied to remove the redundancy among these selected genes. Binary particle swarm optimization (BPSO) is first suggested as an attribute reduction algorithm for rough sets. Finally, the effectiveness of MIRS is evaluated by the classification accuracy of SVM classifiers. The entire system is shown in Figure 1. Experiment results show that the proposed method is superior to some other classical feature selection methods and can always get higher classification accuracy with fewer features.

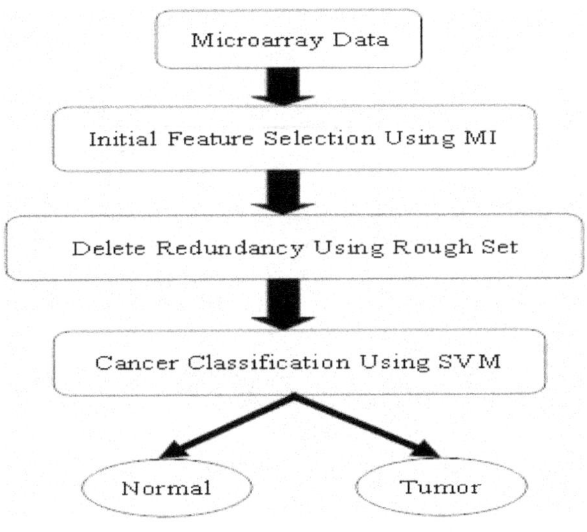

Fig. 1. Cancer classification system

2 Mutual Information for Feature Selection

In accordance with Shannon's information theory [6], the uncertainty of a random variable Y can be measured by the entropy $H(Y)$. For two variables X and Y, the conditional entropy $H(Y \mid X)$ measures the remaining uncertainty of Y when X is known. The mutual information (MI) $I(X;Y)$ measures the certainty about Y that is resolved by X. Apparently, the relation of $H(Y)$, $H(Y \mid X)$ and $I(X;Y)$ is as follows:

$$I(X;Y) = H(Y) - H(Y \mid X) \tag{1}$$

The objective of training a classification model is to minimize the uncertainty about predictions on class labels Y for the known observations X. Thus, it is equivalent to increase the MI $I(X;Y)$ as much as possible for training a classifier. The goal of feature selection process for classification is naturally to achieve the smallest subset of possible features which have higher values of $I(X;Y)$. With the entropy defined by Shannon, the prior entropy of Y is expressed as follows:

$$H(Y) = -\sum_{y \in Y} P(y) \log P(y) \tag{2}$$

Where $P(y)$ represents the probability of Y. The conditional entropy $H(Y \mid X)$ is computed according to the following formula:

$$H(Y \mid X) = -\int P(x)\left(\sum_{y \in Y} P(y \mid x) \log P(y \mid x)\right) dx \tag{3}$$

The mutual information MI between X and Y is presented formally as follows:

$$I(X;Y) = \sum_{y \in Y} \int P(y \mid x) \log \frac{P(y,x)}{P(y)P(x)} dx \tag{4}$$

The probability density function is required in order to estimate the mutual information. One of the most popular methods is to use a histogram. In a two dimensional space, histogram can be constructed feasibly. But there would be many problems in higher dimensional data space. The increase of data dimension may decrease the estimation accuracy. In addition, the required memory will increase exponentially.

In this paper, the estimation of probability density is based on the distribution of mutual information [7] in a Bayesian framework by a second-order Dirichlet prior distribution. Beta approximation of the distribution is adopted in this paper. The top-ranked 700 genes and 500 genes with higher mutual information for the leukemia data set and the colon data set discussed in section 5 are selected respectively.

If too many genes are selected, there must be much noise retained in the data. On the other hand, if we select very few genes, some information contained in the data set for classification may be lost. The number of genes we select for the two data sets is determined with respect to the observation from classification experiments. Subsequently the redundancy among these selected genes will be removed in section 4.

3 Background on Rough Set Theory

Rough set theory is first introduced by Pawlak [8] in the 1980s as a mathematical tool to deal with uncertainty. In this section, we will introduce the principal concepts of rough sets theory related to our attribute reduction approach.

Information System: In rough sets theory, an information system S is denoted as $S = \{U, A, V, f\}$, where U is a finite set of instances $U = \{x_1, x_2, \cdots, x_n\}$. A is a finite set of attributes (features) and consists of condition attribute set C and decision attribute set D. $f : U \times A \rightarrow V$ is a function that $f(x_i, q) \in V_p$ for every $q \in A, x_i \in U$.

Indiscernibility Relation: Let $P \subseteq A, x_i, x_j \in U$, a binary relation IND called indiscernibility relation is defined as follows:

$$IND(P) = \{(x_i, x_j) \mid (x_i, x_j) \in U \times U, a \in P, f(x_i, a) = f(x_j, a)\} \tag{5}$$

Let $U / IND(P)$ denote the family of all equivalence classes of the relation $IND(P)$. For simplicity notation $U / IND(P)$ will be written as U / P.

Lower Approximation: Let $R \subseteq C$ and $X \subseteq U$, the R-lower approximation set of X is the set of all elements of U which can be certainly classified as elements of X according to knowledge R. It can be presented formally as follows:

$$\underline{R}X = \bigcup\{Y \in U / R : Y \subseteq X\} \tag{6}$$

Positive Region: The positive region of decision attribute set D with respect to R is the set of all objects from universe U that can be classified with certainty to classes of U / D employing attributes from R. It can be defined as follows:

$$POS_R(D) = \bigcup_{X \in U / D} \underline{R}X \tag{7}$$

Dispensable and Indispensable Features: Let $c \in C$. A feature c is dispensable if $POS_{(C-|c|)}(D) = POS_C(D)$; Otherwise feature c is indispensable.

Reduct: A set of features $R \subseteq C$ is called a reduct of C if all $c \in R$ are indispensable and $POS_R(D) = POS_C(D)$. In other words, a reduct is the minimal feature subset preserving the above condition.

4 Support Vector Machine

The Support Vector Machine (SVM) [9] is a statistic learning method first proposed by Vapnik in 1995 and its structure is shown in Figure 2. It is based on the theories of VC dimension and structure risk minimization. For two-class classification problem, when two types of examples are nonlinear separable, SVM uses a nonlinear mapping known as kernel function to map the training data into a higher dimensional feature space, and then constructs an optimal separating hyperplane in the higher dimensional space corresponding to a nonlinear classifier in the input space. The kernel function can be linear, polynomial, sigmoid, or Gaussians (RBF). With the kernel functions and the high dimensional space, the hyperplane computation requires a quadratic programming, which is computationally intensive.

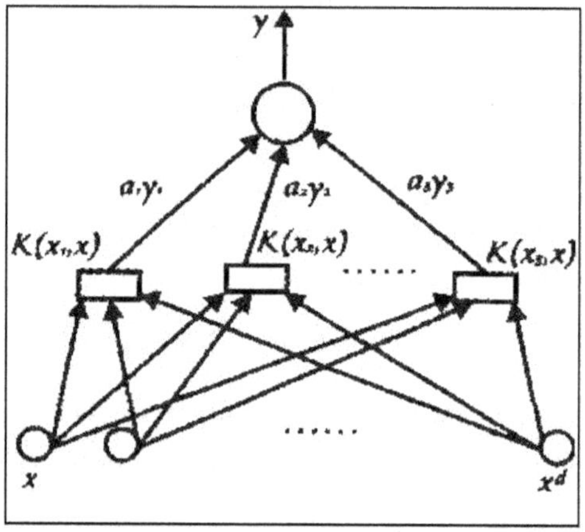

Fig. 2. Structure of support vector machine

5 BPSO for Attribute Reduction in Rough Set

Particle swarm optimization (PSO) is an evolutionary computation technique first introduced for use in real number space by Kennedy and Eberhart in 1995. It has been shown to be a powerful optimization method in many practical applications. In 1997, a binary version of particle swarm optimization (BPSO) is proposed and its performance has been tested on five benchmark functions [10]. But it has not been widely used and still need much further research.

Rough set can be used to find out all possible feature subsets. However, examining exhaustively all subsets of features for selecting the optimal one has been proved to be NP-hard [11]. Heuristic algorithms provide a new way to solve this NP-hard optimization problem. In this section, we suggest binary particle swarm optimization as an attribute reduction algorithm in rough set and apply it to find minimal reduct by removing the redundancy among the genes selected by mutual information.

5.1 Data Preprocessing

The values of gene expression level are continuous. But rough set can only handle discrete attribute value. Hence in order to use the attribute reduction algorithm, all the express level values of selected genes must be discretized firstly. The Entropy/MDL discretization algorithm of Rosetta [12] is used in our experiments. During the entire procedure of attribute discretization and attribute reduction, we combine the training samples with the testing samples together for each data set.

5.2 Population Initialization

Let n be the number of selected features (genes) by mutual information from the original data set. The velocity of i_{th} particle is initialized as a n-dimensional vector with the following form: $V_i = (v_{i1}, v_{i2}, \cdots, v_{in})$. Then the initial position of the i_{th} particle $X_i = (x_{i1}, x_{i2}, \cdots, x_{in})$ can be computed according to Eq. (11). Where $v_{ij} \in [-6,6]$, $x_{ij} \in \{0,1\}$, $j \in \{1,2,\cdots,n\}$. x_{ij} is equal to 1 or 0 which indicates the corresponding feature is selected or not. We put all the selected features of the i_{th} particle into the attribute set P_i.

5.3 Fitness Evaluation

The goal of reduction is to use fewer features to achieve the same or better performance compared with that obtained using the complete feature set. Hence, individual evaluation contains the following two objectives: (1) Minimization of the feature numbers; (2) Maximization of the classification capability. We have to make some tradeoffs between the two objectives. In this paper, classification capability is always have higher priority. If two individuals have the same classification capability, the individual with fewer features will have higher fitness. A simple weighting method is adopted to define the fitness of particles as follows:

$$f(i) = w_1 \cdot fc(i) + w_2 \cdot (1 - fn(i)) \tag{8}$$

$$fc(i) = card(POS_{P_i}(D)) / card(U) \tag{9}$$

Where w_1 and w_2 are weight coefficients, $fc(i)$ is the classification capability we can get by using the feature set P_i, $fn(i)$ is the number of features contained in P_i, $card(U)$ represents the cardinality of the set U.

5.4 Update Velocity and Position

Each particle represents a candidate solution with four state variables: v_i, x_i, p_i, p_g. These variables present the current velocity, current position, previous best position and current global best position of the i_{th} particle respectively. The velocity and position vector are updated according to the following equations:

$$V_{ij} = w \cdot V_{ij} + c_1 \cdot rand() \cdot (P_{ij} - X_{ij}) + c_2 \cdot rand() \cdot (P_{gj} - X_{ij}) \tag{10}$$

$$X_{ij} = \begin{cases} 0, if\ \rho \geq sig(V_{ij}) \\ 1, if\ \rho < sig(V_{ij}) \end{cases} \tag{11}$$

Where c_1 and c_2 are known as acceleration coefficients, X_{ij} represents the j_{th} element of the n-dimensional vector X_i. *Rand* () produces a random number between 0 and 1. ρ is a random number selected from the uniform distribution in [0, 1]. The function $sig(V_{ij})$ is a sigmoid limiting transformation.

6 Gene Expression Data Sets

There are several microarray data sets published from cancer gene expression studies. Two data sets of them are used to test the effectiveness of our proposed method. Because the benchmark data sets have been studied in many papers, we can compare the results of our method with others conveniently.

Leukemia data set [13] consists of 72 samples: 25 samples of acute myeloid leukemia (AML) and 47 samples of acute lymphoblastic leukemia (ALL). Each sample contains the expression levels of 7129 genes measured using high density oligonucleotide microarrays. In our experiments, 38 samples are used as training data and the remaining 34 samples are used as test data.

Colon data set consists of 62 samples of colon epithelial cells taken from cancer patients [14]. Each sample contains 2000 gene expression levels. 20 out of 62 samples are normal samples and the remaining are cancer samples. In our different experiments, 31 or 44 samples are used as training data and the remaining 31 or 18 samples are used as test data.

7 Experiment Results

In all the experiments, BPSO is run with a population size of 30 and it is terminated when the maximum generation of 600 is arrived. The five parameters in the Eq. (8) and Eq. (10) are set to as follows: $w = 1, c_1 = c_2 = 2, w_1 = 100, w_2 = 0.06$. All the algorithms are implemented in matlab 6.5. The features selected by BPSO are used for testing the classification accuracy by support vector machine. The classification accuracy is measured by the number of correct predictions made by the classifier over the test set.

In table 1, experiment results are displayed by using different feature selection methods (No indicates no feature selection). The classification accuracy is obtained by using linear SVM. For the colon data set, 44 samples are used as training data and 18 samples as testing data in this experiment. We can observe that our proposed feature selection method MIRS can always obtain higher classification accuracy with fewer features compared with using MI only. The effectiveness of MIRS is also verified by the remarkable improvement of classification accuracy compared with not using feature selection.

Table 1. The comparison results on feature number and classification accuracy

Dataset	Original No.	FS method	Selected No.	Accuracy
Leukemia	7129	No	7129	58.8%
	7129	MI	700	94.1%
	7129	MIRS	48	97.1%
Colon	2000	No	2000	50.0%
	2000	MI	500	66.7%
	2000	MIRS	32	94.4%

We have also compared the performance of MIRS with some other classical feature selection techniques on the two real gene expression data sets. The comparison results are shown in table 2. The results of other techniques are extracted from a survey reported by Sung [15]. These feature selection techniques being compared include principal components (PC) and correlational coefficient (CC).

In this experiment, we use 31 samples as training data and the other 31 samples as testing data for the colon data set so that we can compare with the results of Sung directly. We try the following two kinds of support vector machines: (1) Linear SVM (no kernel); (2) Radial basis function SVM (RBF kernel). It is obvious that our proposed method is consistently better than the above methods in all the two data sets.

Table 2. Comparison of classification accuracy between MIRS and other methods

Data set	Feature selection	Linear SVM	Rbf SVM
Leukemia	MIRS	97.1%	97.1%
	PC	79.4%	79.4%
	CC	85.3%	85.3%
Colon	MIRS	80.7%	83.9%
	PC	64.5%	64.5%
	CC	64.5%	64.5%

8 Conclusions

Mutual information has recently been proposed for feature selection. But it often contains redundancy in the feature set selected by this method. Attribute reduction in rough set theory provides a feasible way to deal with redundancy and does not reduce the contained information. In this paper, we propose a novel feature selection method MIRS based on mutual information and rough set. First, we select some top-ranked features which have higher mutual information with the target class to predict from two public available real gene expression data sets. Then rough set theory is applied to remove the redundancy among these selected genes. Binary particle swarm optimization (BPSO) is first proposed for attribute reduction in rough set. Finally, the effectiveness of the proposed method is evaluated by the classification accuracy of SVM classifier. Experiment results show that MIRS can always get higher prediction accuracy with small number of features compared with using MI only and is superior to some other classical feature selection methods. Generally, the results are highly promising.

Acknowledgments

This work is supported by the National Natural Science Foundation of China under Grant No. 60433020 and the Key Laboratory of Symbol Computation and Knowledge Engineering of the Ministry of Education of China.

References

1. Furey, T., Cristianini, N., Duffy, N.: Support Vector Machine Classification and Validation of Cancer Tissue Samples Using Microarray Expression Data. Bioinformatics, 16 (2000) 909-914
2. Model, F., Adorjan, P., Olek, A., Piepenbrock, C.: Feature Selection for DNA Methylation Based Cancer Classification. Bioinformatics, 17 (2001) 157-164
3. Kohavi, R., John, G.: Wrapper for Feature Subset Selection. Artificial Intelligence, 97 (1997) 273-324
4. Chow, T., Huang, D.: Estimating Optimal Feature Subsets Using Efficient Estimation of High-Dimensional Mutual Information. IEEE Transactions on Neural Networks, 16 (2005) 213-224
5. Zhong, N., Dong, J. Z.: Using Rough Sets with Heuristics for Feature Selection. Journal of Intelligent Information System, 16 (2001) 199-214
6. Cover, T., Thomas, J.: Elements of Information Theory. Wiley Series in Telecommunications, New York (1991)
7. Zaffalon, M., Hutter, M.: Robust Feature Selection by Mutual Information Distributions. Proceedings of the 14th International Conference on Uncertainty in Artificial Intelligence, (2002) 577-584
8. Pawlak, Z.: Rough Sets. International Journal of Computer Information Science, 11 (1982) 341-356
9. Vapnik, V.: The Nature of Statistical Learning Theory. Springer-Verlag, Berlin Heidelberg New York (1995)
10. Kennedy, J., Eberhart, R. C.: A Discrete Binary Version of the Particle Swarm Algorithm. Proceedings of the 1997 Conference on Systems, Man, and Cybernetics. Piscataway NJ, IEEE Press, (1997) 4104-4109
11. Skowron, A., Rauszer, C.: The Discernibility Matrices and Functions in Information Systems. Intelligent decision support: Handbook of applications and advances of rough set theory, 11 (1992) 331-362
12. Aleksander: Institute of Mathematics, University of Warsaw, Poland. http://rosetta.lcb.uu.se/
13. Golub, T. R., Slonim, K. D., Tamayo, P. et al.: Molecular Classification of Cancer: Class Discovery and Class Prediction by Gene Expression Monitoring. Science, 286 (1999) 531-537
14. Alon, U., Barkai, N. et al.: Broad Patterns of Gene Expression Revealed by Clustering Analysis of Tumor and Normal Colon Cancer Tissues Probed by Oligonucleotide Arrays. PNAS, 96 (1999) 6745-6750
15. Cho, S., Won, H.: Machine Learning in DNA Microarray Analysis for Cancer Classification. Proceedings of the First Asia-Pacific bioinformatics conference on Bioinformatics, 19 (2003) 189-198

Feature Subset Selection for Protein Subcellular Localization Prediction

Qing-Bin Gao and Zheng-Zhi Wang

Institute of Automation, National University of Defense Technology, Changsha 410073, Hunan, People's Republic of China
qbgao@nudt.edu.cn

Abstract. Most of the existing methods for protein subcellular localization prediction are based on a large number of features that are considered to be potentially useful for determining protein subcellular localizations. However, predictors with large numbers of input variables usually suffer from the curse of dimensionality as well as the risk of overfitting. Using only those features that are relevant for protein subcellular localization might improve the prediction performance and might also provide us with some biologically useful knowledge. In this paper, we present a feature ranking based feature subset selection approach for subcellular localization prediction of proteins in the context of support vector machines (SVMs). Experimental results show that this method improves the prediction performance with selected subsets of features. It is anticipated that the proposed method will be a powerful tool for large-scale annotation of biological data.

1 Introduction

The availability of massive amounts of genomic data from large-scale sequencing projects has given impetus in recent years to a large effort in developing statistical and machine learning methods to infer biological models from data. Predicting protein subcellular localization is a hot topic in bioinformatics community, which attempts to classify proteins by their amino acid sequence into sets of cellular compartments using a reliable computational method. The subcellular localization resource is helpful to understand protein function because each cell compartment has maintained a characteristic physicochemical environment [1].

Current computational methods for protein subcellular localization prediction are mostly based on multiple information resources that could be extracted from amino acids and amino acid properties, such as sorting signal sequence [2,3], amino acid composition or dipeptide composition [4-10], n-peptide composition [11], amino acid property [12,13], quasi-sequence-order [14], pseudo-amino acid composition [15], functional domain composition [16], Zp curve [17], gene ontology (GO) [18], homology [19,20], and cellular automata image [21]. Though protein N-terminal sorting signal contains all the information used to localize a cellular protein, the assignments of gene 5'-regions or protein N-terminal sequences are often unreliable in genome sequencing projects. Thus, the existing methods on this basis might be

D.-S. Huang, K. Li, and G.W. Irwin (Eds.): ICIC 2006, LNBI 4115, pp. 433 – 443, 2006.

problematic in cases that these leader sequences are missing or only partially included. Moreover, prediction methods based on amino acid composition alone may lose some other kinds of information relevant to the prediction of protein subcellular localizations. It has been shown that incorporating such information resource may improve the prediction performance to some extent [22-28].

Since it remains unknown how many and which features are relevant to protein subcellular localization, the dimension of feature set used in some methods is usually very high. However, predictors with large numbers of variables may suffer from the curse of dimensionality as well as the risk of overfitting. On the other hand, apart from using machine learning methods merely for prediction, biologists are also interested in the discovery of domain-specific knowledge behind the complex process. A well-known method to get more insight into original data is the application of feature selection approaches [29-31]. By eliminating irrelevant or redundant features, a subset of relevant features can be discovered. One potential advantage of feature selection is that it would decrease the cost of computation as well as the risk of overfitting. Another is that it might improve the prediction performance and might also provide us with some biologically useful knowledge.

In this paper, we present a feature selection approach for the prediction of protein subcellular localizations. A classifier based on support vector machines (SVMs) was used to estimate the prediction performance of subsets of features by 5-fold cross-validation. The proposed method was applied to a freely available protein dataset to demonstrate its efficiency and the experimental results show that our approach selects features that improve the prediction performance against the use of all features.

2 Materials and Methods

2.1 Protein Dataset

Dataset used in this paper was constructed by Park and Kanehisa [10]. All proteins in the dataset were extracted from SWISS-PROT release 39.0 and only those eukaryotic proteins with unique and experimentally determined subcellular localization were included. After a sequence similarity check operation, no sequence has more than 80% similarity to any other sequences in the dataset. The dataset contains 7579 proteins for 12 subcellular locations. The number of proteins in each location is shown in Table 1.

2.2 Protein Representation

An important issue in the prediction of protein subcellular localization is to represent the protein with certain encoding scheme, i.e. to quantify the protein as the input of a machine learning algorithm. However, most of the existing methods for the prediction of protein subcellular localization are based on a large number of input features with comparably few samples, such as the SVM-based method by Park and Kanehisa [10]. They characterize each protein in the training dataset by a 400D (dimensional) feature vector, whereas the size of samples in several compartments is very small, such as the cytoskeleton (40), golgi apparatus (47) and vacuolar (54). Therefore, feature selection is necessary to alleviate the curse of dimensionality.

Table 1. The number of proteins in each location of the dataset used in this paper

Subcellular location	Number of proteins
Chloroplast	671
Cytoplasmic	1241
Cytoskeleton	40
Endoplasmic reticulum	114
Extracellular	861
Golgi apparatus	47
Lysosomal	93
Mitochondrial	727
Nuclear	1932
Peroxisomal	125
Plasma membrane	1674
Vacuolar	54
Total	7579

2.2.1 Feature Extraction

There are 20 natural amino acids that may occur in a protein sequence and are denoted as {A, C, D, E, F, G, H, I, K, L, M, N, P, Q, R, S, T, W, V, Y}. In this paper, we represent proteins with features of amino acid composition and amino acid pair composition extracted directly from their primary sequence. Amino acid composition comprises the percentages of the 20 amino acids in a protein sequence. Although this is a simple representation of proteins, its correlation with protein function, secondary structure and subcellular localization are well-known. Amino acid pair composition can be considered as a 2-gram method as stated by Wu *et al.* [32]. The 2-gram encoding scheme extracts various patterns of 2 consecutive amino acid residues from a protein sequence in a sliding window fashion and counts the number of occurrences of the extracted residue pairs. So there are 400 possible amino acid pairs in total. The occurrence frequency of each amino acid pair is then calculated and used as the element of a 400D feature vector for characterizing the protein. Amino acid pair composition has been used for predicting protein secondary structure and classifying protein function in the past [33,34]. It is proved to be a better representation of proteins than amino acid composition because it encapsulates some information about the order of residues.

We also employ the 9-letter exchange group $\{e_1, e_2, e_3, e_4, e_5, e_6, e_7, e_8, e_9\}$ to represent a protein, which is a re-substitution of the 20-letter amino acids by a 9-letter amino acids according to their physicochemical properties, where $e_1 \in \{C\}$, $e_2 \in \{M\}$, $e_3 \in \{N,Q\}$, $e_4 \in \{D,E\}$, $e_5 \in \{S,T\}$, $e_6 \in \{P,A,G\}$, $e_7 \in \{I,V,L\}$, $e_8 \in \{F,Y,W\}$ and $e_9 \in \{H,K,R\}$, see Table 2 [33]. Thus, we get 90 additional features with the similar encoding scheme for the 20 amino acids. Finally, we extract an original set of 510 features in total. Due to the problem of curse of dimensionality, we attempt to use feature selection to reduce the dimension of feature space.

Table 2. The 9-letter exchange group for the 20 amino acids based on their physicalchemical pproperties

Group	Residues	Description
1	C	Cysteine, remains strongly during evolution
2	M	Hydrophobic
3	N, Q	Amides, polar
4	D, E	Acids, positive, polar
5	S, T	Alcohols
6	P, A, G	Aliphatic, small
7	I, V, L	Aliphatic
8	F, Y, W	Aromatic
9	H, K, R	Bases, charged

2.2.2 Feature Selection

In this paper, we introduce the feature subset selection approach by Wang, *et al.* [34], and apply it to the problem of protein subcellular localization prediction. We define a distance function to measure the relevance score of each feature and use it afterwards to rank and select features. Let $\mathbf{S}_i$ and $\mathbf{S}_j$ be two classes of proteins. Let X be a feature and x be its value. Let $D_{i,j}(X)$ denote the distance function between $\mathbf{S}_i$ and $\mathbf{S}_j$, which measures the discriminatory power of feature X to separate the two classes, defined as

$$D_{i,j}(X) = \frac{(m_i - m_j)^2}{d_i^2 + d_j^2}. \tag{1}$$

where m_i and d_i (m_j and d_j) are the mean value and standard deviation of feature X in class $\mathbf{S}_i$ ($\mathbf{S}_j$), respectively. The mean value m and the standard deviation d of feature X in class $\mathbf{S}$ are defined by

$$m = \frac{1}{N}\sum_{i=1}^{N} x_i. \tag{2}$$

$$d = \sqrt{\frac{1}{N-1}\sum_{i=1}^{N}(x_i - m)^2}. \tag{3}$$

where x_i is the value of feature X with respect to protein sequence $S_i \in \mathbf{S}$, and N is the total number of protein sequences in $\mathbf{S}$. We define average distance function $D(X)$ by

$$D(X) = \frac{\sum_{i,j(i<j)}^{k} D_{i,j}(X)}{k}. \tag{4}$$

where k is the number of classes, here $k = 12$. The distance measure prefers feature X to feature Y if $D(X) > D(Y)$. Intuitively, this means it is easier to distinguish between two classes by using feature X than using feature Y. The values of $D(X)$ are ranked and features with the smallest values are removed first from the original set. Let $X_1, X_2, \cdots, X_{N_g}$ be the top N_g features with the largest $D(X)$ values, they are used in the present work to characterize proteins. Instead of returning an optimal subset of features, this feature selection approach ranks the features from most relevant to least relevant and removes the least relevant features.

2.3 Support Vector Machine

Support vector machine (SVM) [35] is a popular machine learning technique based on structural risk minimization for pattern classification. The basic idea of SVM can be described briefly as follows.

Given a series of training vectors $\mathbf{x}_i \in \mathbf{R}^d$ $(i=1,\ 2,\ \cdots,\ n)$ in two classes with corresponding labels $y_i \in \{+1,\ -1\}$ $(i=1,\ 2,\ \cdots,\ n)$, here +1 and -1 are used to stand for the two classes, respectively. The goal is to construct a binary classifier or derive a decision function from the available training samples which has a small probability of misclassifying a future sample. SVM performs a mapping of the training vectors $\mathbf{x}_i$ $(i=1,\ 2,\ \cdots,\ n)$ from the training space $\mathbf{R}^d$ into a higher dimensional space $\mathbf{H}$ by a kernel function $K(\mathbf{x}_i,\ \mathbf{x}_j)$ and finds an optimal separating hyperplane (OSH), which maximizes the margin between the hyperplane and the nearest data points of each class in the space $\mathbf{H}$. Training a SVM is equivalent to resolving the following convex quadratic optimization problem:

$$\textbf{Min}: \quad \frac{1}{2}\sum_{i=1}^{n}\sum_{j=1}^{n}\alpha_i\alpha_j y_i y_j K(\mathbf{x}_i,\mathbf{x}_j)-\sum_{i=1}^{n}\alpha_i. \tag{5}$$

$$\text{subject to } \sum_{i=1}^{n} y_i\alpha_i = 0, \tag{6}$$

$$\text{and } 0 \le \alpha_i \le C, \quad i=1,\ 2,\cdots,\ n. \tag{7}$$

where n is the number of training samples, C is the regularization parameter used to decide the trade-off between training error and the margin, and α_i $(\mathrm{i}=1,\ 2,\ \cdots,\ n)$ are the coefficients. The decision function is

$$\operatorname{sgn}(\sum_{i=1}^{n} y_i\alpha_i K(\mathbf{x}_i,\ \mathbf{x})+b). \tag{8}$$

Two popular kernel functions are defined by

$$K(\mathbf{x}_i,\ \mathbf{x}_j)=(\mathbf{x}_i \bullet \mathbf{x}_j+1)^d. \tag{9}$$

$$K(\mathbf{x}_i,\ \mathbf{x}_j)=\exp(-\gamma \| \mathbf{x}_i\text{-}\mathbf{x}_j \|^2). \tag{10}$$

Equation (9) is the polynomial kernel function with degree d. Equation (10) is the Radial Basic Function (RBF), where $\gamma = 1/\sigma$ and σ is called the width of the kernel.

For a given dataset, only the kernel function and the regularization parameter C are selected to specify one SVM.

Protein subcellular localization prediction is a multiclass classification problem. Here the number of classes is 12. One simple strategy to handle the multiclass classification problem is to convert the multi-classification into a series of binary classifications. For the 12-class classification problem, we constructed 12 SVM classifiers. The *i*th SVM is trained with the positive samples in *i*th class and negative samples in the other classes. We refer to SVMs trained in this way as 1-*v*-*r* (one-versus-rest) SVMs, which classify an unknown sample into the class that corresponds to the SVM with the highest output value. This strategy has been sufficiently justified in some literatures [36,37].

We used the LibSVM library by Chang [38] to develop the classifier of SVMs, which can be downloaded freely from http://www.csie.ntu.edu.tw/~cjlin/libsvm/ for scientific use.

2.4 Performance Measurement

The prediction performance was evaluated by 5-fold cross-validation test. In this test, the training data set of proteins was separated into five balanced sets. Each of these sets contained approximately the same number of proteins. In each round of cross-validation, we used four sets to train the SVMs while one set was set aside for evaluating the method. This procedure was repeated five times, once for each set. The total accuracy (TA), location accuracy (LA), and Matthew's correlation coefficient (MCC) were used to summarize the prediction performance of the present method. They are defined by [9,10]

$$\mathrm{TA} = \frac{\sum_{i=1}^{k} p(i)}{N}. \tag{11}$$

$$\mathrm{LA}(i) = \frac{\sum_{i=1}^{k} \frac{p(i)}{N_i}}{k}. \tag{12}$$

$$\mathrm{MCC}(i) = \frac{p(i)n(i) - u(i)o(i)}{\sqrt{(p(i)+u(i))(p(i)+o(i))(n(i)+u(i))(n(i)+o(i))}}. \tag{13}$$

where N is the total number of proteins in the dataset, k is the number of subcellular locations, N_i is the number of proteins in location i, $p(i)$ is the number of correctly predicted proteins in location i, $n(i)$ is the number of correctly predicted proteins not in location i, $u(i)$ is the number of under-predicted proteins and $o(i)$ is the number of over-predicted proteins. To assess the prediction reliability, a Reliability Index (RI) is usually assigned according to the difference between the highest and the second-highest output value of the SVMs in multiclass classification [9], which is a useful indication of the level of confidence in the prediction of a protein. The assignment of RI is not performed in this study.

3 Results and Discussion

3.1 Results

The prediction performance based on different subsets of features was evaluated by SVMs. In this paper, RBF kernel with parameters $C = 100$ and $\gamma = 60$ was used to train the SVMs. The RBF kernel has been investigated in other domains and achieved good performance. Feature selection was done according to a backward elimination procedure. In this work, backward elimination started with the complete set of 510 ranked features. For the limitation of computational power, we removed 50 features each time from the current set of features. Features with the least relevance score were removed first. The total accuracy (TA) and location accuracy (LA) were calculated by 5-fold cross-validation performed on the dataset. The dependency relationship of TA and LA with the number of features before and after feature selection is shown in Figure 1 and Figure 2, respectively. Figure 1 demonstrates that our approach select features with the highest scores that achieve better performance when compared with the use of the same number of features before feature selection. Figure 2 indicates LA changes smoothly after feature selection, whereas it changes sharply before feature selection.

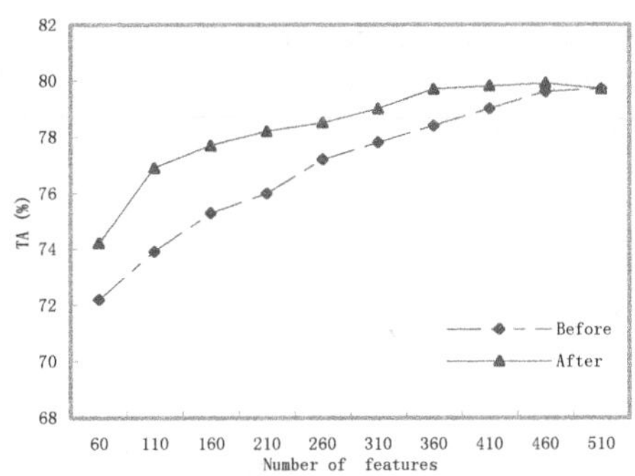

Fig. 1. The dependency relationship of TA with the number of features before and after feature subset selection

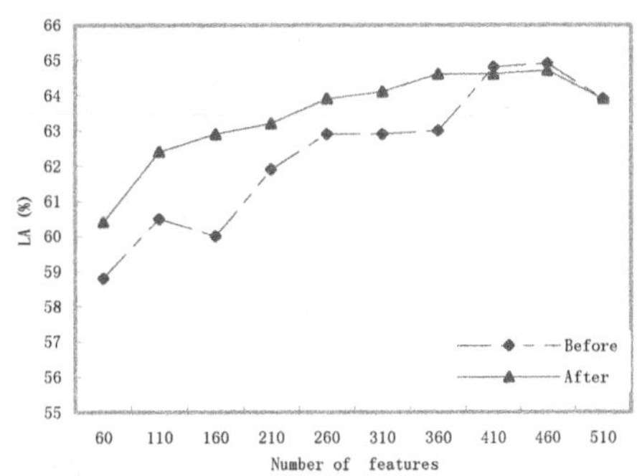

Fig. 2. The dependency relationship of LA with the number of features before and after feature subset selection

The prediction accuracy for each subcellular location based on the 510 features and a subset of 460 features are shown in Table 3. The TA and LA were calculated to measure the prediction performance of the present method. The prediction accuracy and MCC for each location were also calculated to measure the performance of each classifier. From Table 3 we know that the TA and LA based on the selected 460 features were 79.9% and 64.7%, respectively, higher than 79.7% and 63.9% based on all 510 features. The results indicate that our method improves the prediction performance by selecting relevant features.

Table 3. Prediction performance for the 12 subcellular locations based on all 510 feature and the selected 460 features

Location	510 features		460 features	
	Accuracy (%)	MCC	Accuracy (%)	MCC
Chloroplast	76.8	0.75	77.9	0.75
Cytoplasmic	73.8	0.66	73.5	0.66
Cytoskeleton	70.0	0.81	70.0	0.81
Endoplasmic reticulum	57.9	0.70	58.8	0.68
Extracellular	82.5	0.80	82.0	0.81
Golgi apparatus	31.9	0.50	34.0	0.46
Lysosomal	67.7	0.73	68.8	0.72
Mitochondrial	55.4	0.58	56.4	0.58
Nuclear	89.5	0.75	89.5	0.76
Peroxisomal	31.2	0.44	34.4	0.44
Plasma membrane	91.5	0.91	91.9	0.91
Vacuolar	38.9	0.54	38.9	0.56
TA	79.7	—	79.9	—
LA	63.9	—	64.7	—

3.2 Comparison with Previous Methods

In order to examine the performance of our method, we made comparison with the method by Park and Kanehisa [10]. They also used SVMs to evaluate the prediction performance of their method on the same dataset. The comparison results are shown in Table 4. Although the size of the dataset was small (7579 versus 7589 entries), our method improved the LA dramatically, from 57.9% to 64.7%, and improved the TA

Table 4. Performance comparison of our method with Park and Kanehisa [10] by 5-fold cross-validation test

Location	Park and Kanehisa	Our work
	Accuracy (%)	Accuracy (%)
Chloroplast	72.3	77.9
Cytoplasmic	72.2	73.5
Cytoskeleton	58.5	70.0
Endoplasmic reticulum	46.5	58.8
Extracellular	78.0	82.0
Golgi apparatus	14.6	34.0
Lysosomal	61.8	68.8
Mitochondrial	57.4	56.4
Nuclear	89.6	89.5
Peroxisomal	25.2	34.4
Plasma membrane	92.2	91.9

Vacuolar	25.0	38.9
TA	78.2	79.9
LA	57.9	64.7

to some extent, from 78.2% to 79.9%. The results in Table 4 indicate that out method outperforms their method particularly for some small groups, such as cytoskeleton (58.5% versus 70.0%), endoplasmic reticulum (46.5% versus 58.8%), golgi apparatus (14.6% versus 34.0%), peroxisomal (25.2% versus 34.4%) and vacuolar (25.0% versus 38.9%). In contrast, their method excels our method for some large groups, the examples are nuclear (89.6% versus 89.5%) and plasma membrane (92.2% versus 91.9%).

4 Conclusion

In this paper, a feature ranking based feature subset selection approach is proposed for the prediction of protein subcellular localization in the context of SVMs. High prediction performance has been achieved in a 5-fold cross-validation test performed on a freely available protein dataset. This demonstrates the effectiveness of the present method. One potential benefit of feature selection is that it might improve the prediction performance of a predictor. Another is that it might also provide us with some biologically useful knowledge. It is especially important when the prediction models are not transparent to biologists, such as the SVMs. It is anticipated that the proposed method will be a powerful tool for large-scale annotation of biological data in post-genome era.

References

1. Andrade, M.A., O'Donoghue, S.I., Rost, B.: Adaptation of Protein Surfaces to Subcellular Location. J. Mol. Biol. 276 (1998) 517-525
2. Nakai, K., Horton, P.: PSORT: a Program for Detecting Sorting Signals in Proteins and Predicting their Subcellular Localization. Trends Biochem. Sci. 24 (1999) 34-36
3. Emanuelsson, O., Nielsen, H., Brunk, S., Von Heijne, G.: Predicting Subcellular Localization of Proteins Based on their N-terminal Amino Acids Sequences. J. Mol. Biol. 300 (2000) 1005-1016
4. Nakashima, H., Nishikawa, K.: Discrimination of Intracellular and Extracellular Proteins using Amino Acid Composition and Residues-pair Frequencies. J. Mol. Biol. 238 (1994) 54-61
5. Cedano, J., Aloy, P., Perez-Pons, J.A., Querol, E.: Relation between Amino Acid Composition and Cellular Location of Proteins. J. Mol. Biol. 266 (1997) 594-600
6. Reinhardt, A., Hubbard, T.: Using Neural Networks for Prediction of the Subcellular Location of Proteins. Nucleic Acids Res. 26 (1998) 2230-2236
7. Chou, K.C., Elrod, D.W.: Protein Subcellular Location Prediction. Protein Eng. 12 (1999) 107-118
8. Yuan, Z.: Prediction of Protein Subcellular Location using Markov Chain Models. FEBS Lett. 451 (1999) 23-26
9. Hua, S., Sun, Z.: Support Vector Machine Approach for Protein Subcellular Location Prediction. Bioinformatics, 17 (2001) 721-728.

10. Park, K.J., Kanehisa M.: Prediction of Protein Subcellular Locations by Support Vector Machines using Compositions of Amino Acids and Amino Acid Pairs. Bioinformatics. 19 (2003) 1656-1663
11. Yu, C.S., Lin, C.J., Hwang, J.K.: Predicting Subcellular Localization of Proteins for Gram-negative Bacteria by Support Vector Machines based on N-peptide Compositions. Protein Sci. 13 (2004) 1402-1406
12. Feng, Z.P., Zhang, C.T.: Prediction of the Subcellular Location of Prokaryotic Proteins Based on the Hydrophobic Index of the Amino Acids. Int. J. Biol. Macromol. 14 (2001) 255-261
13. Sarda, D., Chua, G.H., Li, K.B., Krishnan, A.: pSLIP: SVM based Protein Subcellular Localization Prediction using Multiple Physicochemical Properties. BMC Bioinformatics. 6 (2005) 152
14. Chou, K.C.: Prediction of Protein Subcellular Locations by Incorporating Quasi-sequence-order Effect. Biochem. Biophys.Res. Commun. 278 (2000) 477-483
15. Chou, K.C.: Prediction of Protein Cellular Attributes using Pseudo-amino Acid Composition. Proteins Struct. Funct. Genet. 43 (2001) 246-255
16. Chou, K.C., Cai, Y.D.: Using Functional Domain Composition and Support Vector Machines for Prediction of Protein Subcellular Location. J. Biol. Chem. 277 (2002) 45765-45769
17. Feng, Z.P., Zhang, C.T.: A Graphic Representation of Protein Primary Structure and its Application in Predicting Subcellular Locations of Prokaryotic Proteins. Int. J. Biochem. Cell Biol. 34 (2002) 298-307
18. Chou, K.C., Cai, Y.D.: A New Hybrid Approach to Predict Subcellular Localization of Proteins by Incorporating Gene Ontology. Biochem. Biophys. Res. Commun. 311 (2003) 743-747
19. Bhasin, M., Raghava, G.P.: ESLpred: SVM-based Method for Subcellular Localization of Eukaryotic Proteins using Dipeptide Composition and PSIBLAST. Nucleic Acids Res. 32 (2004) 414-419
20. Xie, D., Li, A., Wang, M., Fan, Z., Feng, H.: LOCSVMPSI: a Web Server for Subcellular Localization of Eukaryotic Proteins using SVM and Profile of PSI-BLAST. Nucleic Acids Res. 33 (2005) 105-110
21. Xiao, X., Shao, S., Ding, Y., Huang, Z., Chen, X., Chou, K.C.: Using Cellular Automata to Generate Image Representation for Biological Sequences. Amino Acids. 28 (2005) 29-35
22. Cai, Y.D., Chou, K.C.: Predicting Subcellular Localization of Proteins in a Hybridization Space. Bioinformatics. 20 (2004) 1151-1156
23. Bhasin, M., Garg, A., Raghava, G.-P.S.: PSLpred: Prediction of Subcellular Localization of Bacterial Proteins. Bioinformatics. 21 (2005) 2522-2524
24. Gao, Q.B., Wang, Z.Z., Yan, C., Du, Y.H.: Prediction of Protein Subcellular Location using a Combined Feature of Sequence. FEBS Lett. 579 (2005) 3444-3448
25. Matsuda, S., Vert, J.P., Saigo, H., Ueda, N., Toh, H., Akutsu, T.: A Novel Representation of Protein Sequences for Prediction of Subcellular Location using Support Vector Machines. Protein Sci. 14 (2005) 2804-2813
26. Xiao, X., Shao, S., Ding, Y., Huang, Z., Huang, Y., Chou, K.C.: Using Complexity Measure Factor to Predict Protein Subcellular Location. Amino Acids. 28 (2005) 57-61
27. Pan, Y.X., Li, D.W., Duan, Y., Zhang, Z.Z., Xu, M.Q., Feng, G.Y., He, L.: Predicting Protein Subcellular Location using Digital Signal Processing. Acta. Biochim. Biophys. Sin. 37 (2005) 88-96
28. Hoglund, A., Donnes, P., Blum, T., Adolph, H.W., Kohlbacher, O.: MultiLoc: Prediction of Protein Subcellular Localization using N-terminal Targeting Sequences, Sequence Motifs, and Amino Acid Composition. Bioinformatics. 22 (2006) 1158-1165

29. Chuzhanova, N.A., Jones, A.J., Margetts, S.: Feature Selection for Genetic Sequence Classification. Bioinformatics. 14 (1998) 139-143
30. Degroeve, S., Baets, B.D., de Peer, Y.V., Rouze, P.: Feature Subset Selection for Splice Site Prediction. Bioinformatics. 18 (2002) S75-S83
31. Wang, M., Yang, J., Xu, Z.J., Chou, K.C.: SLLE for Predicting Membrane Protein Types. J. Theor. Biol. 232 (2005) 7-15
32. Wu, C., Whitson, G., McLarty, J., Ermongkonchai, A., Chang, T.C.: Protein Classification Artificial Neural System. Protein Sci. 1 (1992) 667-677
33. Yang, M.Q., Yang, J.K., Zhang, Y.Z.: Extracting Features from Primary Structure to Enhance Structural and Functional Prediction. RECOMB (2005)
34. Wang, J.T.L., Ma, Q., Shasha, D., Wu, C.H.: New Techniques for Extracting Features from Protein Sequences. IBM Sys. J. 40 (2001) 426-441
35. Vapnik, V.: Statistical Learning Theory. Wiley, New York (1998)
36. ScholkÖpf, B., Burges,C., Vapnik,V.: Extracting Support Data for a Given Task. Proc. First Int. Conf. KDDM, AAAI Press (1995)
37. Hsu, C.W., Lin, C.J.: A Comparison of Methods for Multi-class Support Vector Machines. IEEE Trans. Neural Networks. 13 (2002) 415–25
38. Chang, C.C., Lin, C.J.: LIBSVM: a Library for Support Vector Machines. Software is available at http://www.csie.ntu.edu.tw/~cjlin/libsvm (2001)

Fuzzy k-Nearest Neighbor Method for Protein Secondary Structure Prediction and Its Parallel Implementation

Seung-Yeon Kim[1], Jaehyun Sim[2], and Julian Lee[3]

[1] Computer Aided Molecular Design Research Center, Soongsil University, Seoul 156-743, Korea
sykim@ssu.ac.kr
[2] School of Dentistry, Seoul National University, Seoul 110-749, Korea
jhsim@snu.ac.kr
[3] Department of Bioinformatics and Life Science, Soongsil University, Seoul 156-743, Korea
jul@ssu.ac.kr
http://bioinfo.ssu.ac.kr/~jul/jul_eng.htm

Abstract. Fuzzy k-nearest neighbor method is a generalization of nearest neighbor method, the simplest algorithm for pattern classification. One of the important areas for application of the pattern classification is the protein secondary structure prediction, an important topic in the field of bioinformatics. In this work, we develop a parallel algorithm for protein secondary structure prediction, based on the fuzzy k-nearest neighbor method, that uses evolutionary profile obtained from PSI-BLAST (Position Specific Iterative Basic Local Sequence Alignment Tool) as the feature vectors.

1 Introduction

Although the prediction of the three-dimensional structure of a protein from its amino acid sequence is one of the most important problems in bioinformatics [1,2,3,4], *ab initio* prediction of the tertiary structures based solely on sequence information has not been successful so far. For this reason, lots of research efforts have been made for the determination of the protein secondary structure [5,6,7,8,9,10,11,12,13,14,15,16], which can serve as an intermediate step toward determining its tertiary structure.

The most common definition of the secondary structure is based on Dictionary of Secondary Structure of Proteins (DSSP) [17] where the secondary structure is classified as eight states. By grouping these eights states into three classes Coil (C), Helix (H), and Extended (E), one obtains three state classification, which is more widely used. Therefore, the protein secondary structure prediction is a typical pattern classification problem, where one of the three possible states is assigned to each residue of the query protein.

D.-S. Huang, K. Li, and G.W. Irwin (Eds.): ICIC 2006, LNBI 4115, pp. 444–453, 2006.

The first step for solving such a problem is the feature extraction, where the important features of the data are extracted and expressed as a set of numbers, called feature vectors. The performance of the pattern classifier depends crucially on the judicious choice of the feature vectors. It has been shown that constructing feature vectors from the evolutionary profile obtained from PSI-BLAST (Position Specific Iterative Basic Local Alignment Search Tool) [18], a bioinformatics tool for the search of homologous protein sequences, gives better prediction results than other choices [6,16] (see Sect. 2.1).

Once an appropriate feature vector has been chosen, a classification algorithm is used to partition the feature space into disjoint regions with decision boundaries. The decision boundaries are determined using feature vectors of a reference sample with known classes, which are also called the reference dataset or training set. The class of a query data is then assigned depending on the region it belongs to. Various pattern classification algorithms such as artificial neural network or support vector machine have been used for the protein secondary structure prediction.

The k-nearest neighbor method is the simplest algorithm for the pattern classification. Moreover, it can be easily adapted for parallel computation. Although the k-nearest neighbor method has been used for the secondary structure prediction [11,12,14,15], the fuzzy variant of the algorithm [19] has never been used for the secondary structure prediction, although it has been used for the solvent accessibility prediction [20].

In this work, we develop a parallel algorithm for the protein secondary structure prediction, based on the fuzzy k-nearest neighbor method [19], where PSI-BLAST profiles are used as the feature vectors. As a test of our algorithm, we perform a benchmark test on EVA common set 1 consisting of 60 proteins [22].

2 Methods

2.1 The Feature Vectors

In order to construct the feature vector for a protein residue, we first perform database search with PSI-BLAST [18]. PSI-BLAST then calculates the rate of substitution of each residue of the query protein to another amino acids. By multiplying appropriate normalization factors, taking logarithms, and rounding off to integer values, these numbers are converted to what is called the position specific scoring matrix, also called profile, a matrix of the size (protein length) $\times$ 20. This PSI-BLAST profile contains evolutionary information that cannot be obtained from the raw sequence only. For a protein residue whose secondary structure is to be predicted, one takes a window of size N_w centered around this residue, and uses the matrix of size $N_w \times 20$ as the feature vector to be input into the pattern classification algorithm (see Fig. 1). We use $N_w = 15$ in this work. The resulting feature vector is a $15 \times 20 = 300$ dimensional matrix. This feature vector is the same as the one used in previous works [6,16] based on other pattern classification methods.

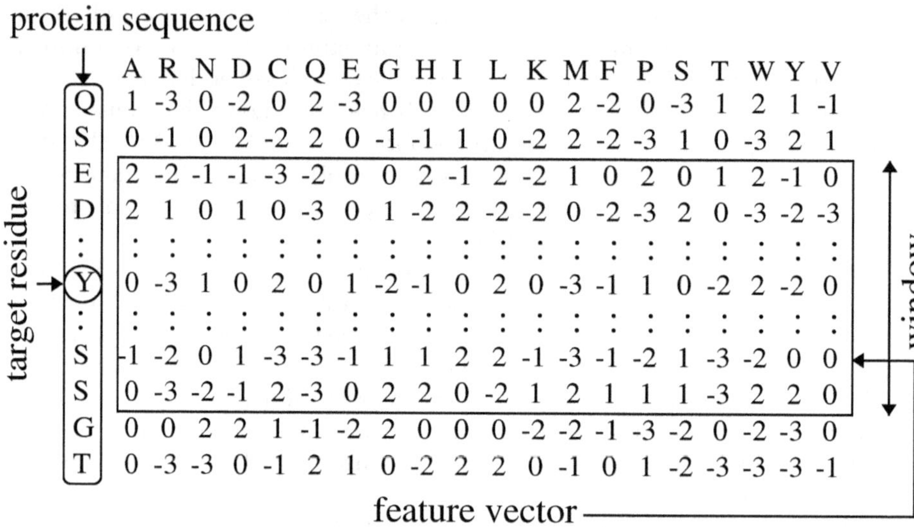

Fig. 1. The relation between PSI-BLAST profile and the feature vector of a residue. The feature vector corresponding to a target residue is constructed from the PSI-BLAST profile by considering a window of finite size (15 residues in this work) centered on the residue.

2.2 The Distance Measure

There are various ways of defining the distance between two feature vectors A and B, but in this work we use three methods, Euclidean, Cosine, and Correlation distances, defined as

$$D_{AB}(Euc) = \sum_{i=1}^{N_w} w_i \sum_j (P_{ij}(A) - P_{ij}(B))^2, \tag{1}$$

$$D_{AB}(Cos) = 1 - \sum_{i=1}^{N_w} w_i \frac{\sum_j P_{ij}(A) \cdot P_{ij}(B)}{\sqrt{\sum_p P_{ip}(A)^2 \sum_q P_{iq}(B)^2}}, \tag{2}$$

$$D_{AB}(Corr) = 1 - \sum_{i=1}^{N_w} w_i \frac{\sum_j (P_{ij}(A) - \bar{P}_i(A)) \cdot (P_{ij}(B) - \bar{P}_i(B))}{\sqrt{\sum_p (P_{ip}(A) - \bar{P}_i(A))^2 \sum_q (P_{iq}(B) - \bar{P}_i(B))^2}}, \tag{3}$$

respectively, where $P_{ij}(A)(i = 1, 2, \cdots, 15; j = 1, 2, \cdots, 20)$ is a component of the feature vector A, w_i a weight parameter, and

$$\bar{P}_i(A) \equiv \frac{1}{20} \sum_{j=1}^{20} P_{ij}(A).$$

Since we expect the profile elements for residues nearer to the target residue to be more important in determining the local environment of the target residue, we use weights $w_i = (8 - |8 - i|)^2$.

2.3 The Reference Dataset

In order to construct the reference dataset consists of representative protein chains without bias, we utilize the ASTRAL SCOP database, where the protein chains are hierarchically classified into structural families, and representative proteins are selected for each of them. In particular, we used ASTRAL SCOP (version 1.63) chain-select-95 subset and chain-select-90 subset [21]. We then clustered these sequences with BLASTCLUST (NCBI BLAST 2.2.5, http://www.ncbi.nlm.nih.gov/BLAST/) and selected the representative chain for each cluster, in order to remove additional homologies. The resulting reference dataset consists of 4362 non-redundant proteins (905684 feature vectors) that have less than 25% sequence identity with each other.

2.4 Fuzzy k-Nearest Neighbor Method

In the simplest version of the fuzzy k-nearest neighbor (FKNN) method [19], the fuzzy class membership $u_s(\mathbf{x})$ to the class s is assigned to the query data $\mathbf{x}$ according to the following equation:

$$u_s(\mathbf{x}) = \frac{\sum_{sec(j)=s} D_j^{-2/(m-1)}}{\sum_{j=1}^{k} D_j^{-2/(m-1)}} \tag{4}$$

where the summation of j in the numerator is restricted to those belonging to the class s, m is a fuzzy strength parameter, which determines how heavily the distance is weighted when calculating each neighbor's contribution to the membership value, k is the number of nearest neighbors, and c is the number of classes, and D_j is the distance between the feature vector of the query data $\mathbf{x}$ and the feature vector of its j-th nearest reference data $\mathbf{x}(j)$.

The advantage of the fuzzy k-nearest neighbor algorithm over the standard k-nearest neighbor method is quite clear. The fuzzy class membership $u_s(x)$ can be considered as the estimate of the probability that the query data belongs to class i, and provides us with more information than a definite prediction of the class for the query data. Moreover, the reference samples which are closer to the query data are given more weights, and an optimal value of m can be chosen along with that for k, in contrast to the standard k-nearest neighbor method with fixed value of $2/(m-1) = 0$. In fact, the optimal value of k and m are found from the leave-one-out cross-validation procedure (see below), and the resulting value for $2/(m-1)$ is indeed nonzero.

The optimal values of k and m were determined by leave-one-out cross validation test, where the prediction was performed for one of the chains in the reference dataset, using the remaining 4361 chains as the reference dataset, procedure being repeated for each of the 4362 chains. The optimal values of k and

m are determined as the ones yielding the maximum average value of Q_3 score, which is define as:

$$Q_3 \equiv 100\% \times \frac{N_{corr}}{N} \tag{5}$$

with N and N_{corr} being the total number of residues of the query protein, and the total number of correctly predicted residues, respectively.

The optimal value of m turns out to be 1.29, and that of k is 85 when using Euclidean and Correlation distances, and 70 when using Cosine.

2.5 The Parallel Implementation

The FKNN method can be easily adapted for parallel computation. In the parallel implementation, the computational load is shared between computational nodes, resulting in drastic increase in computational speed. The advantage of the parallel program in terms of computational time can also be seen from Fig. 2 (see Results and Discussions). To elaborate on the parallel algorithm, each of the nodes is assigned a distinct subset of the feature vectors in the reference dataset, and each member of this set is compared with query vector, and k_{nn} of them with the smallest distance from the query vector are chosen. The numbers of the feature vectors assigned to the nodes are all equal up to roundoff error, so that the loads are balanced. The 0-th node, which we call the master, performs the job of collecting k_{nn} candidates of nearest neighbors from each of the nodes. It then sorts these $N_{nodes} \times k_{nn}$ indices with respect to the distance D to select the final k_{nn} nearest neighbors. The master then produces the final output. The pseudo-code for the parallel algorithm is given in algorithm 1., along with the sub-algorithms 2., 3., and 4..

Algorithm 1. parallel FKNN algorithm for the protein secondary structure prediction

k_{nn} = Number of nearest neighbors (constant)
N_{nodes} = Total number of computing nodes
$Rank$ = The number of this node, a number between 0 and $N_{nodes} - 1$
Construct the feature vector for each residue of the query protein {algorithm2.}
Construct the feature vector for each residue in the database {algorithm3.}
$st = Rank * N_f/N_{nodes} + 1$
$ed = (Rank + 1) * N_f/N_{nodes}$ {st and ed is the starting and ending number of feature vectors the current node will look into. This is to divide computational load between nodes.}
for $j_q = 1$ to L_q **do**
  Calculate the probabilities $prob(j_q, s)$ of the residue q being in each of the conformational state s (=C,H,E) {algorithm4.}
  The predicted secondary structure $\mathrm{S}(j_q)$ = (s that maximizes $\mathrm{prob}(j_q, s)$)
  print out j_q, $\mathrm{S}(j_q)$, and $prob(j_q, s)$ (s = C,H,E)
end for

Algorithm 2. Constructing feature vectors for each residue of the query protein

```
Read in query profile
L_q = Length of the query sequence
for j_q = 1 to L_q do
   Construct matrix P_q(j_q) of size 15 × 20, centered around the residue j_q, from the
   query profile
end for
```

Algorithm 3. Constructing feature vectors for each residue in the database

```
N_f = 0 {N_f will be the total number of feature vectors in the reference dataset,
equal to the total number of residues in the dataset}
N_p = Number of protein chains in the reference dataset
Read in profiles in the reference dataset (database profiles)
for i = 1 to N_p do
   L(i) = Length of the i-th protein chain
   for j = 1 to L(i) do
      N_f ⇐ N_f + 1
      Construct matrix P_DB(N_f) of size 15 × 20, centered around the residue j of the
      i-th protein, from the database profile
   end for
end for
```

3 Results and Discussions

The benchmark test was performed on EVA common set 1 consisting of 60 proteins [22] and RS126 set consisting of 126 non-homologous protein [5], with the optimal values of m and k determined by the leave-one-out cross-validation on the reference dataset derived from ASTRAL SCOP (see Methods). The performance on EVA common set 1 was compared with three neural network based prediction methods, PSIPRED (v2.3) [6], PROFking (v1.0) [7], and SABLE (v2.0) [8], and the performance on RS126 set was compared with two methods based on support Vector Machine (SVM), SVM freq [9] and SVMpsi [10].

In addition to Q_3 score (see section 2.4), two additional performance scores, SOV score [23] and three state correlation coefficient (Corr(3)) [24], are used for the assessment of performance. The average values and the standard errors of these scores for the performance on EVA common set 1, of the fuzzy k-nearest method with various distance measures, and the other three methods, are displayed in Table 1. The results of the test on RS126 set are shown on Table 2.

We see that in both of these test, the performance is best when the Correlation distance measure is used. We see that in the first test, average performance scores are lower than those of PSIPRED and SABLE, but higher than PROFking. However, considering the magnitudes of the standard error, these differences are not drastic, and we may say that the performances are more or less comparable to other methods. Also, the actual performances of the prediction algorithms depend on their versions and the set of proteins used for the test, and it should

Algorithm 4. Calculating the fuzzy membership of a query residue to each of the secondary structural class

```
for s = C,H,E do
   membership(j_q, s) = 0
end for
for m_DB = st to ed do
   D(j_q, m_DB) = Distance between P_q(j_q) and P_DB(m_DB)
end for
Sort indices of the feature vectors the current node is examining, with respect to
D(j_q, m_DB), in descending order.
if Rank == 0 then {This node is the master, so collect the results and re-sorts
them, and print the final output}
   indx() ⇐ save indices of k_nn nearest neighbors among the feature vectors exam-
   ined by the master
   dscore() ⇐ save distances of k_nn nearest neighbors among the feature vectors
   examined by thc master
   for i = 1 to N_nodes − 1 do
      Receive indices and distances of k_nn nearest neighbors among the feature vectors
      examined by the i-th node
      indx() ⇐ add indices of k_nn nearest neighbors among the feature vectors ex-
      amined by the i-th node
      dscore() ⇐ add distances of k_nn nearest neighbors among the feature vectors
      examined by the i-th node
   end for
else
   Send indices and distances of k_nn nearest neighbors among the feature vectors
   examined by the i-th node to the master
end if
if Rank == 0 then
   Sort indices with respect to dscore() {The collection consists of N_nodes × k_nn
   results, so master must sort them again to select k_nn nearest neighbors}
   for j_DB = 1, k_nn do {Calculate the fuzzy membership from k_nn nearest neigh-
   bors}
      s(j_DB) = secondary structural class corresponding to the j_DB-th feature vector

      membership(j_q, s(j_DB)) ⇐ membership(j_q, s(j_DB))+ fuzzy membership calcu-
      lated from D(j_q, j_DB)
   end for
   for s = C,H,E do
      prob(j_q, s) = membership(j_q, s)/ Σ_{s'∈{C,H,E}} membership(j_q, s')
   end for
end if
```

be emphasized that the result is not to be considered as an extensive test of these methods.

Since the programs based on SVM are not available for public use, we quote the values from the literature [9,10]. The values of performance measures not

Table 1. Average scores of secondary structure prediction on EVA common set 1, using fuzzy k-nearest neighbor (FKNN) method with Euclidean (Euclid), Cosine (Cos), and Correlation (Corr) distance measures. The average scores are given also for three other methods for comparison. The numbers in the parentheses are the standard errors.

	Q3	SOV	Corr(3)
FKNN(Euclid)	70.9 (1.8)	64.5 (2.3)	0.495 (0.024)
FKNN(Cos)	70.9 (1.8)	64.5 (2.3)	0.499 (0.034)
FKNN(Corr)	71.8 (1.9)	67.9 (2.4)	0.527 (0.026)
PSIPRED	75.1 (1.8)	75.3 (2.4)	0.557 (0.024)
PROFking	67.2 (2.3)	64.3 (2.8)	0.463 (0.029)
SABLE	75.6 (1.5)	73.1 (2.5)	0.532 (0.029)

Table 2. Average scores of secondary structure prediction on RS126 set, using fuzzy k-nearest neighbor (FKNN) method with Euclidean (Euclid), Cosine (Cos), and Correlation (Corr) distance measures. The average scores are given also for two other methods based on SVM, for comparison.

	Q3	SOV	Corr(3)
FKNN(Euclid)	88.6	83.1	0.791
FKNN(Cos)	88.6	83.1	0.744
FKNN(Corr)	89.0	84.0	0.796
SVMfreq	75.1	-	-
SVMpsi	76.1	72.0	-

reported in the references are omitted. We see that the fuzzy k-nearest neighbor method also shows good performance when compared with SVM-based methods.

The parallel code was implemented in mpi C, and run on 32 Intel Xeon processors. For 60 proteins in the EVA set, for the Euclidean, cosine, and correlation distance measures, respectively, the calculation took 47, 58, and 60 minutes of wall clock time, defined as the time elapsed between the start and end of the program.

The advantage of the parallel algorithm we introduced in this work is that the communication between computational nodes are kept to a minimal level. In fact, the most of the computations are performed by each of the nodes independently, and the communication occurs only at the end of such computations, and only between the king and slaves, when the master collects the results from the slaves and sorts them again to predict the secondary structure. In order to examine the parallel efficiency, we repeated the computation for EVA common set 1 using the correlation distance measure for different number of CPUs in order to obtain the response curve in Fig. 2. In the figure, the inverse of the time is plotted against the number of CPUs involved in the computation, in order to show the dependence of the computational speed on the number of CPUs. The result shows that, although the dependence is not exactly linear, the scalability is reasonably good, demonstrating the advantage of parallel computation over serial version.

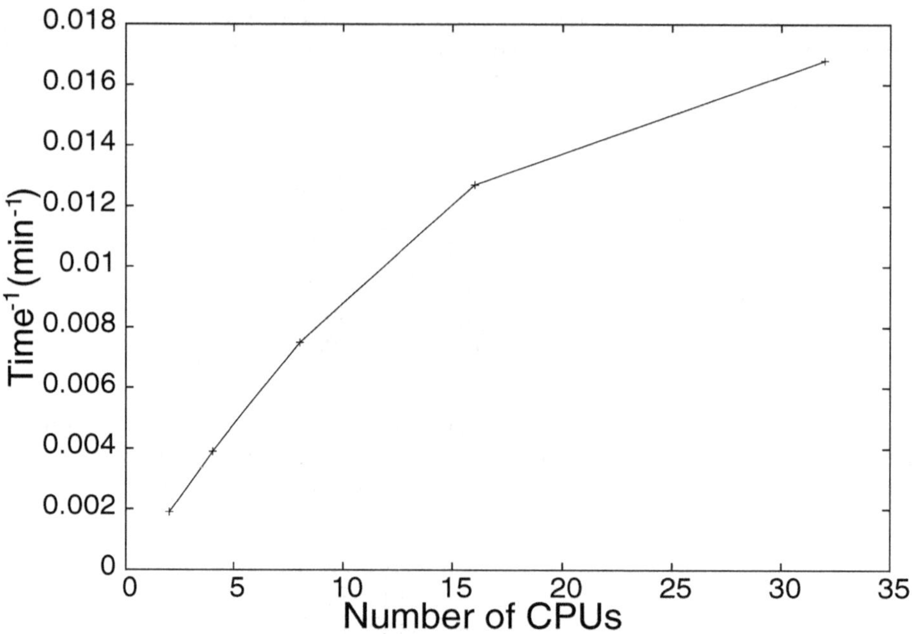

Fig. 2. The inverse of wall time in $\min^{-1}$ (*vertical axis*) plotted against the number of CPUs used for the computation (*horizontal axis*). The curve shows excellent scalability of the parallel FKNN algorithm, due to minimal amount of communication between CPUs.

Acknowledgement

This work was supported by the Korean Research Foundation Grant funded by the Korean Government (MOEHRD) (KRF-2005-005-J01101).

References

1. Kryshtafovych, A., Venclovas, C., Fidelis, K., Moult, J.: Progress over the First Decade of CASP Experiments. Proteins. vo. 61 (2005) 225–236
2. Lee, J., Kim, S.-Y., Joo, K., Kim, I., Lee, J.: Prediction of Protein Tertiary Structure using PROFESY, a Novel Method Based on Fragment Assembly and Conformational Space Annealing. Proteins. vol. 56 (2004) 704–714
3. Lee, J., Kim, S.-Y., Lee, J.: Protein Structure Prediction Based on Fragment Assembly and Parameter Optimization. Biophys. Chem. vol. 115 (2005) 209–214
4. Lee, J., Kim, S.-Y., Lee, J.: Protein Structure Prediction Based on Fragment Assembly and Beta-strand Pairing Energy Function. J. Korean Phys. Soc. vol. 46 (2005) 707–712
5. Rost, B., Sander, C.: Prediction of Secondary Structure at Better than 70% Accuracy. J. Mol. Biol. vol. 232 (1993) 584–599
6. Jones, D.: Protein Secondary Structure Prediction Based on Position-specific Scoring Matrices. J. Mol. Biol. vol. 292 (1999) 195–202

7. Ouali, M., King, R.: Cascaded Multiple Classifiers for Secondary Structure Prediction. Protein Science. vol. 9 (1999) 1162–1176
8. Adamczak, R., Porollo, A., Meller, J.: Combining Prediction of Secondary Structure and Solvent accessibility in proteins. Proteins. vol. 59 (2005) 467–475
9. Hua, S., Sun, Z.: A Novel Method of Protein Secondary Structure Prediction with High Segment Overlap Measure: Support Vector Machine Approach. J. Mol. Biol. vol. 308 (2001) 397–407
10. Kim, K., Park, H.: Protein Secondary Structure Prediction based on improved Support Vector Machines Approach. Protein Eng. vol. 16 (2003) 553–560
11. Joo, K., Lee, J., Kim, S.-Y., I., Kim, Lee, S.J., Lee, J.: Profile-based Nearest Neighbor Method for Pattern Recognition. J. Korean Phys. Soc. vol. 44 (2004) 599–604
12. Joo, K., Kim, I., Lee, J., Kim, S.-Y., Lee, S.J., Lee, J.: Prediction of the Secondary Structure of Proteins Using PREDICT, a Nearest Neighbor Method on Pattern Space. J. Korean Phys. Soc. vol. 45 (2004) 1441–1449
13. Pollastri, G., McLysaght, A. Porter: a new, Accurate Server for Protein Secondary Structure Prediction. Bioinformatics vol. 21 (2004) 1719–1720
14. Jiang, F.: Prediction of Protein Secondary Structure with a Reliability Score Estimated by Local Sequence Clustering. Protein Eng. vol. 16 (2003) 651–657
15. Salamov A. A., Solovyev V. V.: Protein Secondary Structure Prediction Using Local Alignments. J. Mol. Biol. vol. 268 (1997) 31–35
16. Kim, H., Park, H.: Prediction of Protein Relative Solvent Accessibility with Support Vector Machines and Long-range Interaction 3D Local Descriptor. Proteins. vol. 54 (2004) 557–562
17. Kabsch, W., Sander, C.: Dictionary of Protein Secondary Structure: Pattern Recognition of Hydrogen-bonded and Geometrical Features. Biopolymers vol. 22 (1983) 2577–2637
18. Altschul, S.F., Madden, T.L., Schaffer, A.A., Zhang, J., Zhang, Z., Miller, W., Lipman, D.J.: Gapped BLAST and PSI-BLAST: a New Generation of Protein Database Search Programs. Nucleic Acids Res. vol. 25 (1997) 3389–3402
19. Keller, J. M., Gray, R., Givens, J. A. : A Fuzzy k-nearest Neighbor Algorithm. IEEE Trans. Systems Man Cybernet. vol. 15 (1985) 580–585.
20. Sim, J. H., Kim, S.-Y., Lee, J.: Prediction of Protein Solvent Accessibility Using Fuzzy k-Nearest Neighbor Method. Bioinformatics vol. 21 (2005) 2844–2849.
21. Brenner, S.E., Koehl, P., Levitt, M.: The ASTRAL Compendium for Protein Structure and Sequence Analysis. Nucleic Acids Res. vol. 28 (2000) 254–256
22. Koh, I. Y., Eyrich, V., Marti-Renom, M. A., Przybylski, D., Madhusudhan, M. S., Eswar, N., Grana, O., Pazos, F., Valencia, A., Sali, A., Rost, B.: EVA: Evaluation of Protein Structure Prediction Servers. Nucleic Acids Res. vol. 31 (2003) 3311–3315
23. Zemla, A., Venclovas, C., Fidelis, K., Rost, B.: A Modified Definition of Sov, a Segment-Based Measurement for Protein Secondary Structure Prediction Assessment. Proteins. vol. 34 (1999) 220–223
24. Gorodkin, J.: Comparing two K-category Assignment by a K-category Correlation Coefficient. Comput. Biol. and Chem. vol. 28 (2004) 367–374

Gene Selection Based on Mutual Information for the Classification of Multi-class Cancer

Sheng-Bo Guo[1,2], Michael R. Lyu[3], and Tat-Ming Lok[4]

[1] Department of Automation, University of Science and Technology of China, Hefei, Anhui, 230026, China
sbguo@iim.ac.cn
[2] Intelligent Computation Lab, Hefei Institute of Intelligent Machines, Chinese Academy of Sciences, P.O. Box 1130, Hefei, Anhui, 230031, China
[3] Computer Science & Engineering Dept., The Chinese University of Hong Kong, Shatin, Hong Kong
[4] Information Engineering Dept., The Chinese University of Hong Kong, Shatin, Hong Kong

Abstract. With the development of mirocarray technology, microarray data are widely used in the diagnoses of cancer subtypes. However, people are still facing the complicated problem of accurate diagnosis of cancer subtypes. Building classifiers based on the selected key genes from microarray data is a promising approach for the development of microarray technology; yet the selection of non-redundant but relevant genes is complicated. The selected genes should be small enough to allow diagnosis even in regular laboratories and ideally identify genes involved in cancer-specific regulatory pathways. Instead of the traditional gene selection methods used for the classification of two categories of cancers, in the present paper, a novel gene selection algorithm based on mutual information is proposed for the classification of multi-class cancer using microarray data, and the selected key genes are fed into the classifier to classify the cancer subtypes. In our algorithm, mutual information is employed to select key genes related with class distinction. The application on the breast cancer data suggests that the present algorithm can identify the key genes to the BRCA1 mutations/BRCA2 mutations/the sporadic mutations class distinction since the result of our proposed algorithm is promising, because our method can perform the classification of the three types of breast cancer effectively and efficiently. And two more microarray datasets, leukemia and ovarian cancer data, are also employed to validate the performance of our method. The performances of these applications demonstrate the high quality of our method. Based on the present work, our method can be widely used to discriminate different cancer subtypes, which will contribute to the development of technology for the recovery of the cancer.

1 Introduction

Microarray technology, a recent development in experimental molecular biology, provides biomedical researchers the ability to measure expression levels of thousands of genes simultaneously. Such gene expression profiles are used to understand the

D.-S. Huang, K. Li, and G.W. Irwin (Eds.): ICIC 2006, LNBI 4115, pp. 454–463, 2006.

molecular variations among disease related cellular processed, and also to help the increasing development of diagnostic tools and classification platforms in the cancer research.

With the development of the microarray technology, the necessary processing and analysis methods grow increasingly critical. It becomes gradually urgent and challenging to explore the appropriate approaches because of the large scale of microarray data comprised of the large number of genes compared to the small number of samples in a specific experiment. For the data obtained in a typical experiment, only some of genes are useful to differentiate samples among different classes, but many other genes are irrelevant to the classification. Those irrelevant genes not only introduce some unnecessary noise to gene expression data analysis, but also increase the dimensionality of the gene expression matrix, which results in the increase of the computational complexity in various consequent researches such as classification and clustering. As a consequence, it is significant to eliminate those irrelevant genes and identify the informative genes, which is a feature selection problem crucial in gene expression data analysis [1, 2].

In the present paper, we propose a novel gene selection method based on the mutual information for the multi-class cancer classification using microarray data. Our method firstly calculates the mutual information (MI) between the discreted gene expression profiles and the cancer class label vector for all the samples. Then, the genes are ranked according to the calculated MI. These selected genes with high ranks are fed into the nearest neighbor method.

The rest of the present paper is organized as follows. In section 2, we first introduce the method to discretize the gene expression data, and then we in detail formulate the principle of GSMI. Section 3 describes the test statistics. And Section 4 describes the experiment. In Section 5, GSMI is applied to analyze the breast cancer dataset. Section 6 contains the conclusions.

2 Methods

Among the many thousands of genes simultaneously measured in a specific microarray experiment, it is impossible that all of their expressions are related to a particular partition of the samples. In the analysis of a biological system, the following ‘rules of thumb’ regarding gene functions are often assumed. 1) A gene can be in either the ‘on’ or ‘off’ state; 2) not all genes simultaneously respond to a single physiological event; 3) gene functions are highly redundant [3]. According to these assumptions, we consider the genes as random variables with two values, in which 1 denotes the ‘on’ state and 0 denotes the ‘off’ state. As a consequence, the gene expression data can be discretized into two states 0 and 1, respectively. The discretization of the gene expression data will be formulated later in Section 3.

Assume that a microarray dataset can be represented as a $G \times S$ matrix A with generic element a_{gs} representing the expression level of the gene, g in sample, s . All the samples are divided into n categories, and with the class label denoted by C with its element a_{gs} standing for the class of *ith* sample. From the biological point of view, those genes, having higher mutual agreement with class label of the cancer microarray data, contribute more significantly on the classification of the cancer subtypes. Consequently, these genes should be selected as the key genes and used to the sequent classification and clustering. According to the information theory, mutual information can be used to measure the mutual agreement between two object models. We then employ the mutual information to rank every gene according to mutual information between the gene and the class label of the cancer microarray data.

Based on the information-theoretic principle of mutual information, the mutual information of two random variables X and $h_i = w_i / \sum w_i$ with a joint probability mass function $p(x, y)$ and marginal probability mass functions $p(x)$ and $p(y)$ is defined as [4]:

$$I(X;Y)=\sum_{x_i,y_j} p(x_i,y_j)\log\frac{p(x_i,y_j)}{p(x_i)p(y_j)} . \tag{1}$$

Let us suppose that the domain of $G_i, i \in \{1,...,G\}$, is distcretized into two intervals. After discretization, the domains of all the genes can be represented by $dom(G_i) = \{v_{ik}\}, k = 1,2$ where $v_{i1=0}$ and $v_{i2=0}$. Denoted by σ the SELECT operation from relational algebra and $|S|$ denote the cardinality of set S [8]. The probability of a gene in microarray data having $G_i = v_{ik}, i \in \{1,...,G\}, k \in \{1,2\}$ is then given by:

$$P(G_i = v_{ik}) = \frac{|\sigma_{G_i = ik}(A)|}{|\sigma_{G_i \neq \Phi}(A)|} . \tag{2}$$

And the joint probability of the gene in the gene expression data has $G_i = v_{ik}$ and the class label $C = c_i, i \in \{1,...,n\}$ is calculated by:

$$P(G_i = v_{ik} \wedge C = c_i) = \frac{|\sigma_{G_i = v_{ik} \wedge C = c_i}(A)|}{|\sigma_{G_i \neq NULL}(A)|} . \tag{3}$$

Definition 1. The interdependence measure I between the gene and the class label, G_i and C, $i \in \{1,...,G\}$, is defined as:

$$I(G_i:C)=\sum_{k=1}^{2}\sum_{l=1}^{n} P(G_i=v_{ik} \wedge C=c_i)\log\frac{P(G_i=v_{ik} \wedge C=c_i)}{P(G_i=v_{ik})P(C=c_i)} . \tag{4}$$

$I(G_i : C)$ measures the average reduction in uncertainty about G_i that results from learning the value of C [9]. If $I(G_i : C) > I(G_j : C), i, j \in \{1,...,G\}, i \neq j$, the dependence of G_i and the class label C is greater than the dependence of G_j and C. Before ranking the genes according to the mutual information, the redundancy in the microarray should be decreased because of the fundamental principle of microarray technology. Due to the principle of micorarray technology, the gene expression matrix contains high redundancy since some genes are measured more than once.

Definition 2. The mutual information matrix of the microarray, named as M with its element m_{ij}, is given by:

$$m_{ij} = \sum_{k=1}^{2}\sum_{l=1}^{2} P(G_i = v_{ik} \wedge G_j = v_{jl}) \log \frac{P(G_i = v_{ik} \wedge G_j = v_{jl})}{P(G_i = v_{ik})P(G_j = v_{jl})} . \tag{5}$$

For simplicity, the mutual information matrix M is normalized to M^* with its element m^* given as follows,

$$m_{ij}^* = \frac{m_{ij}}{H(G_i, G_j)} \tag{6}$$

where the joint entropy of the gene G_i and G_j is denoted by $H(G_i, G_j)$, which is given by:

$$H(G_i, G_j) = -\sum_{k=1}^{2}\sum_{l=1}^{2} P(G_i = v_{ik} \wedge G_j = v_{jl}) \log P(G_i = v_{ik} \wedge G_j = v_{jl}) . \tag{7}$$

The redundancy in the microarray data is reduced by the following method. In the matrix M^*, the elements on the diagonal are all with the same value 1. These rows without containing the value less than 0.95 are labeled in the normalized mutual information matrix except these elements in the diagonal of the matrix. We then select these genes corresponding to these rows. Due to the error in the process of computing the mutual information, the cutoff value is set to 0.95 so that the redundancy can be reduced as much as possible. Otherwise, if the cutoff value is set to 1, the redundancy cannot be reduced to the expected extent.

After selecting the genes with little redundancy, if any, the selected genes (SGS) are ranked according to the interdependence measure I between the gene expression profiles and the class label. Then the SGS are used to train the RBF neural network to classify the cancer subtypes in the designed experiment formulated in the next section.

3 Test Statistics

A general statistical model for gene expression values will be firstly introduced followed by several test statistics in this section.

Assume that there are more than two kinds of distinct tumor tissue classes for the problem under consideration and there are p genes (variables) and n tumor mRNA samples (observations). After introducing the novel gene selection method, we now turn to some test statistics used for testing the equality of the class means for a fixed gene. The following five parametric test statistics will be considered [10].

3.1 ANOVA F Test Statistics

The definition of this test is given by:

$$F = \frac{(n-k)\sum_i n_i(\overline{Y}_{i\cdot} - \overline{Y}_{\cdot\cdot})^2}{(k-1)\sum_i (n_i - 1)s_i^2} . \tag{8}$$

where $\overline{Y}. = \sum_{j=1}^{n_i} Y_{ij} / n_i, \overline{Y}.. = \sum_{i=1}^{k} n_i \overline{Y}_{i\cdot} / n_i$, and $s_i^2 = \sum_{j=1}^{n_i} (Y_{ij} - \overline{Y}_{i\cdot})^2 / (n_i - 1)$.

3.2 Brown-Forsythe Test Statistic

Brown-Forsythe Test Statistic [11] is given by:

$$B = \frac{\sum_i n_i(\overline{Y}_{i\cdot} - \overline{Y}_{\cdot\cdot})^2}{\sum_i (1 - n_i / n)s_i^2} . \tag{9}$$

3.3 Welch Test Statistics

Welch test statistics [12] is defined as

$$W = \frac{\sum_i w_i(\overline{Y}_{i\cdot} - \sum_i h_i \overline{Y}_{i\cdot})^2}{(k-1) + 2(k-2)(k+1)^{-1}\sum_i (n_i - 1)^{-1}(1-h_i)^2} \tag{10}$$

with $w_i = n_i / s_i^2$ and $h_i = w_i / \sum w_i$.

4 Introduction to the Experiment

To evaluate the performance of GSMI, we applied it to the well-known gene expression data sets: the breast cancer data [5], in which RNA from samples of primary breast tumors from 7 carriers of the BRCA1 mutation, 8 carries of the BRCA2 mutation, and 7 patients with sporadic cases of breast caner have been hybridized to a cDNA microarray containing 6512 complementary DNA clones of 5361 genes [6]; Leukemia 72 with 6817 genes, 38 ALL-Bcell, 9 ALL-Tcell, and 25 AML, and Ovarian with 7129 genes, 27 epithelial ovarian cancer cases, 5 normal tissues, and 4 malignant epithelial ovarian cell lines [13].

Before calculated the mutual information, the microarray expression level should firstly be preprocessed according to an alternative idea of the Optimal Class-Dependence Discretization Algorithm (OCDD) [11]. OCDD is a new method to convert variables into discrete variables for inductive machine learning, which can thus be employed for pattern classification problems. The discretization process is formulated as an optimization problem, then the normalized mutual information that measures the interdependence between the class labels and the variable to discretized as the objective function, and then iterative dynamic programming is applied to find its optimum [14]. For each continuous gene expression profile in microarray expression matrix A , its domain is typically discretized into two intervals for gene selection, which are denoted by 0 and 1, respectively. We then use the normalized mutual information measure that reflects interdependence between the class label and the attribute to be discretized as the objective function to find a global optimal solution separating the domain of the gene expression data.

We employ the nearest neighbor method to classify the cancer cases with different cancer subtypes. The leave-one-out cross-validation (LOOCV) is used to evaluate the accuracy of classification.

5 Experiment Results and Discussion

By using our method, genes are ranked by the mutual information between the genes and the class label. Then, the nearest neighbor classifier is employed as the benchmark to classify the three cancer microarray expression datasets.

In the classification performance evaluation process, we employed LOOCV, which is a widely used method for evaluating the performance of the classification of gene expression data [7].

The results of our method on the three datasets are given in Figure 1, Figure 2 and Figure 3, respectively. From figure 1, the classification error rate minimized to 0% when 18 genes are selected according to our method, but the genes selected by all the test statistics used for classification are not as effective as ours since the classification accuracies maximize to 73% at 404 genes for the ANOVA test statistic, 78% at 119 genes for Brown-Forsythe test statistic, 73% at 116 genes for Welch test statistic. From figure 2, the classification error rate minimized to 1.39% when 19 genes are selected according to our method, but the genes selected by all the test statistics used for classification are not as effective as ours since the classification accuracies maximize to 85% at 70 genes for the Brown-Forsythe test statistic, 71% at 156 genes for ANOVA test statistic, 80.5% at 354 genes for Welch test statistic. From figure 3, the classification error rate minimized to 0 % when 21 genes are selected according to our method, but the genes selected by all the test statistics used for classification are not as effective as ours since the classification accuracies maximize to 97.3% at 38 genes for the Welch test statistic, 97.3% at 370 genes for ANOVA test statistic, 94.45% at 105 genes for Brown-Forsythe test statistic.

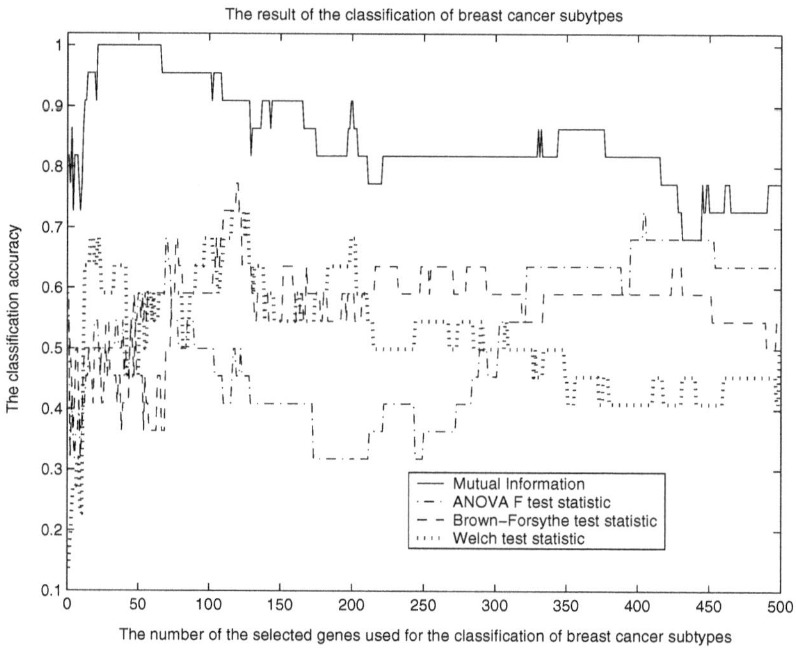

Fig. 1. Comparison on the breast cancer cases

Demonstrated by these results, the present method based on the mutual information obviously outperforms the three test statistics. The major reason for the superiority of our method is that mutual information between the genes and the class labels reflects the potential relation and correlation, and thus indicates the discriminability of genes. What's more, there is no assumption about the probability distribution of the microarray data. The three test statistics are based on the default probability distribution, but it is not clear whether microarray data are according to the default probability till now without adequate samples of the cancer cases.

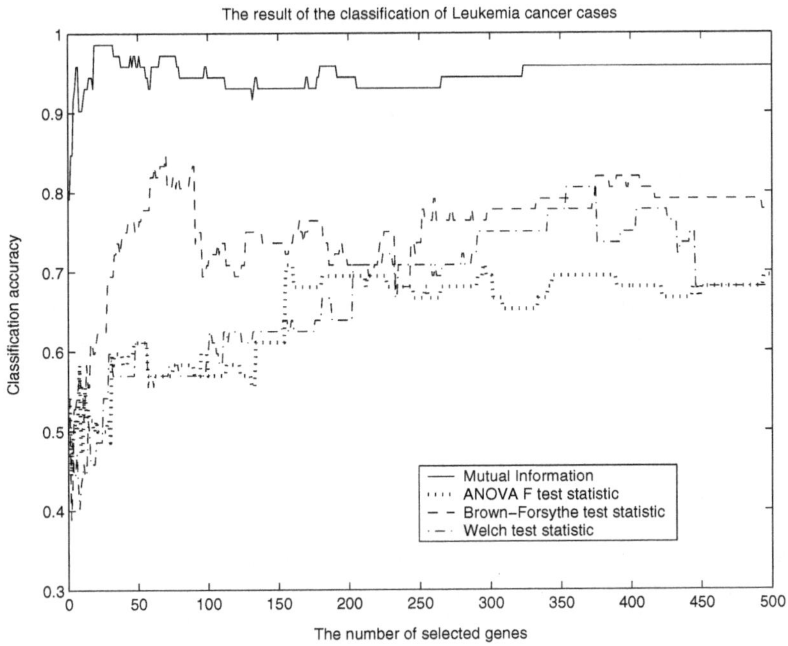

Fig. 2. Comparison on the leukemia cancer cases

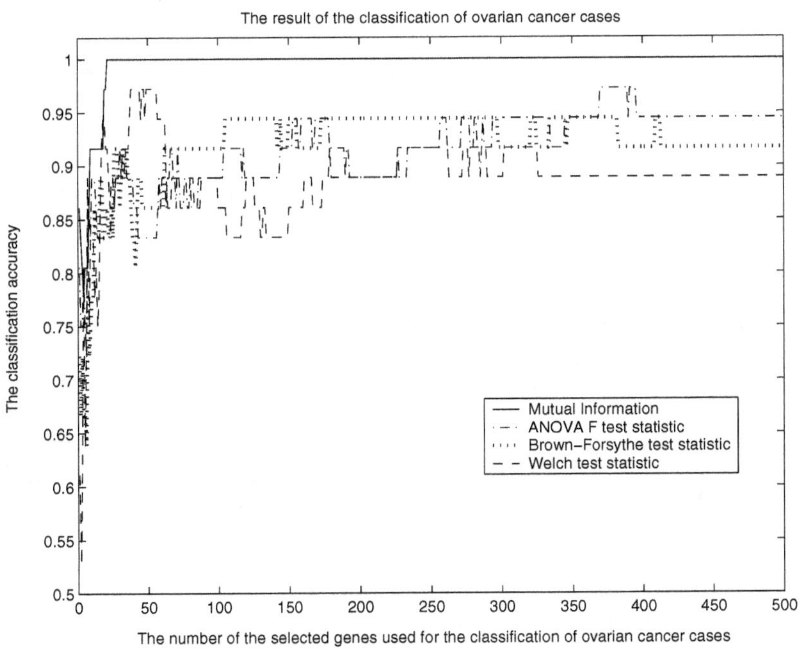

Fig. 3. Comparison on the ovarian cancer cases

6 Conclusions

In this paper, contrary to other work managing discriminating one cancer subtype from the other, we present the novel method for the classification of the multi-class cancer subtypes, which will contribute the development of the technology for the research and cure of cancer. In our work, an information theoretic approach is proposed for gene selection algorithm based on the mutual information, which proves promising in the classification of multi-class cancer microarray datasets. The mutual information between the gene expression data and the class label is calculated. And the genes are selected according to the calculated mutual information after removing the redundancy. The successful applications on the breast, as well as ovarian and leukemia cancer datasets prove that our algorithm is effective, robust and appropriate to the classification of the multi-class cancers since it can discovery the informative key genes.

Future work will concentrate on the further research on the method for extracting the key features from gene expression data. And we will try to fully evaluate the performance of the classification by employing SVM and neural network, such as RBF neural network. Furthermore, we will also concentrate on the research of biological significance of the found key genes and try to find the specific feature of the key genes and understand the function of these genes. By doing so, the microarray technology can be fully used.

Acknowledgement

The authors are grateful to Dechang Chen for sharing the micorarray data sets of breast, ovarian as well as leukemia with us.

References

1. Ben-Dor, A.: Tissue Classification with Gene Expression Profiles, Journal of Computational Biology, 7 (2000) 559-583
2. Weston, J., Mukherjee, S., Chapelle, O., Pontil, M., Poggio, T., Vapnik, V.: Feature Selection for SVMs. In Advances in Neural Information Processing Systems, MIT Press, 13 (2001)
3. Xing, E. P., Richard, M. K.: CLIFF: Clustering of High-Dimensional Microarray Data via Iterative Feature Filtering Using Normalized Cuts. Bioinformatics, 17 (1) (2001) 306-315
4. Cover, T., Thomas J.: Elements of Information Theory. John Wiley and Sons, Inc (1991)
5. Hedenfalk, I., Duggan, D., Chen, Y., Radmacher, M., Bittner, M., Simon, R., Meltzer, P., Gusterson, B., Esteller, M., Raffeld, M.: Gene-Expression Profiles in Hereditary Breast Cancer. New Eng. J. Med. 344 (2001) 539-548
6. Nathalie, P., Frank, D. S., Johan, A. K. S., Bart, L. R. D. M.: Systematic Benchmarking of Microarray Data Classification: Assessing the Role of Non-linearity and Dimensionality Reduction. Bioinformatics, 20 (17) (2004) 3185-3195
7. Simon, R.: Supervised Analysis when The Number of Candidate Features Greatly Exceeds the Number of the Cases. SIGKDD Explorations, 5 (2) (2003) 31-36

8. Au, W. H., Keith, C.C. C., Andrew, K.C. W., Wang, Y.: Attribute Clustering for Grouping, Selection and Classification of Gene Expression Data. IEEE/ACM Transactions on computational biology and bioinformatics, April-June, 2 (2) (2005) 83-101
9. MacKay, D. J. C.: Information Theory, Inference, and Learning Algorithm. Cambridge Univ. Press (2003)
10. Chen, D. C., Liu, Z. Q., Ma, X. B., Hua, D.: Selecting Genes by Test Statistics. Journal of Biomedicine and Biotechnology, 2 (2005) 132-138
11. Brown, M. B., Forsythe, A. B.: The Small Sample Behavior of Some Statistic which Test the Equality of Several Means. Technometrics (1974) 129-132
12. Welch, B. L.: On the Comparison of Several Mean Values: An Alternative Approach. Biometrika., 38 (1951) 330-336
13. Chen, D. C., Hua, D., Jaques, R., Cheng, X. Z.: Gene Selection for Multi-class Prediction of Microarray Data. Bioinformatics Conference, 2003, CSB'03, Proceedings of the 2003 IEEE, (2003) 492-495
14. Liu, L., Andrew, K.C. W., Wang, Y.: A Global Optimal Algorithm for Class-Dependent Discretization of Continuous Data, 8 (2) (2004) 151-170

Gene Selection by Cooperative Competition Clustering

Shun Pei[1,2], De-Shuang Huang[1], Kang Li[3], and George W. Irwin[3]

[1] Intelligent Computing Lab, Institute of Intelligent Machines, Chinese Academy of Sciences, P.O. Box 1130, Hefei, Anhui 230031, China

[2] Department of Automation, University of Science and Technology of China

dshuang@iim.ac.cn

[3] School of Electrical & Electronic Engineering Queen's University Belfast

Abstract. Clustering analysis of data from DNA microarray hybridization studies is an essential task for identifying biologically relevant groups of genes. Attribute cluster algorithm (ACA) has provided an attractive way to group and select meaningful genes. However, ACA needs much prior knowledge about the genes to set the number of clusters. In practical applications, if the number of clusters is misspecified, the performance of the ACA will deteriorate rapidly. In fact, it is a very demanding to do that because of our little knowledge.We propose the Cooperative Competition Cluster Algorithm (CCCA) in this paper. In the algorithm, we assume that both cooperation and competition exist simultaneously between clusters in the process of clustering. By using this principle of Cooperative Competition, the number of clusters can be found in the process of clustering. Experimental results on a synthetic and gene expression data are demonstrated. The results show that CCCA can choose the number of clusters automatically and get excellent performance with respect to other competing methods.

1 Introduction

In a typical biological system, it is often not clearly known how many genes are sufficient to fully characterize a macroscopic phenotype. However, in practice a working mechanistic hypothesis that is testable and largely captures the biological truth seldom involves more than a few dozens of genes. And knowing the identity of those relevant genes is just as important as finding the grouping of samples they induced [1]. Thus, the selection of the relevant features is key to analyze gene expression microarray data.

So far, several useful approaches have been taken to select the most meaningful genes. One approach is to use a clustering algorithm to group genes into coherent sets and then select representative genes to represent these clusters [3]. This approach only deals with the gene redundancy problem, while it fails to detect non-discriminating genes. Principal component analysis (PCA) may remove non-discriminating genes by restricting attention to so-called eigenfeatures corresponding to the large eigenvalues, but each basis element of the new genes subspace for this method is the linear combination of all the original genes, which make it difficult to identify the important genes[4]. Furthermore, the t -statistics is widely used to select discriminating genes [5]. It is important to note that the t-statistics can only be used when the samples are

D.-S. Huang, K. Li, and G.W. Irwin (Eds.): ICIC 2006, LNBI 4115, pp. 464–474, 2006.

preclassified and cannot solve the multiclass problem directly. In this paper, we use a two-steps method to select discriminating and meaningful genes, i.e., first using the mutual information between the class label vector and the sample vector to remove most of the non-discriminating genes, then the meaningful genes are selected by using the Cooperative Competition Clustering technique.

Clustering analysis of gene expression microarray data is a key step to understand how the activity of genes varies during biological processes and is affected by disease states and cellular environment. Clustering can be used to group genes according to their expression in a set of samples. There are rich literatures on clustering analysis and various techniques have been developed. Some well-known examples are: k-means algorithm, Attribute Cluster Algorithm [1], Kohonene's self-organizing maps (SOM) [6], and various hierarchical clustering algorithms [7], etc. However, these clustering algorithms need prior knowledge, which is but unknown in practical applications. K-means algorithm, ACA as well as SOM need to in advance know the number of clusters, and hierarchical clustering algorithms need to set a suitable stopping condition. Without prior knowledge, it is hard to find rational results. Recently, some researchers proposed competition clustering algorithm [15] to select cluster model automatically, those algorithms overemphasize the competition and overlook the cooperation between clusters so that some interesting but small cluster will not been found. In this paper, we propose a new clustering algorithm referred to as Cooperative Competition cluster algorithm (CCCA), which can detect the number of clusters automatically also. We suppose that the competition and the cooperation exist at the same time. In the process of competition, winners will move to the centers of clusters and represent clusters gradually while the losers are slowly died out. By this way, true clusters can be found effectively. In some situations, there are some small but interesting clusters. In the competition with other bigger and nearer clusters, those small clusters are in the inferior position and may die out. In order to avoid such situation, we suppose that cooperation exist simultaneously and can help the clusters compete with other clusters. In our experiment, the algorithm can find the suitably interesting and meaningful clusters effectively.

So far, Euclidean distance and Pearson's correlation coefficient have been widely used as the distance measure for clustering analysis. However, when Euclidean distance is applied to measure the similarity between genes, it is not effective to reflect functional similarity such as positive and negative correlation, interdependence as well as closeness in values. In fact, Euclidean distance actually accounts only for the last. In other words, the primary interest of the overall shapes of genes [8] is not well considered. Consequently, some researchers proposed Pearson's correlation coefficient [9]. An empirical study [10] has also shown that Pearson's correlation coefficient is not robust to outliers and it may assign a high similarity score to a pair of dissimilar genes. Recently, some researches [11] proposed using mutual information to measure the similarity between genes. Mutual information has several merits: Firstly, mutual information makes full use of the high-order statistical information contained in the gene expression data. Secondly, there are many missing data in gene expression data, both Euclidean distance and Pearson's correlation coefficient need to estimate those missing data; this operating may induce some extra errors. However, mutual information doesn't need to estimate these missing data. Thirdly, mutual information not only can measure the similarity between genes, but also the importance of gene by computing the mutual information between gene vector and the class label vector. In our paper, we shall use mutual information to measure the importance of genes and the similarity between genes.

To demonstrate CCCA's usefulness for mining and analyzing gene expression data and to evaluate its performance, one gene expression data set, colon cancer, are used. We first applied CCCA to each of them, for selecting the most promising genes, then fed the selected genes into several well-known classification algorithms and compared their classification accuracies with those yielded by other gene selection methods. These classification algorithms, including a decision-tree-based algorithm, as well as Multilayer Feedforward Neural Networks (MFNN) and the nearest neighbor approach, are used in this paper. The experimental results demonstrate that CCCA outperforms other algorithms significantly.

2 Methods

Gene clustering is the process which finds c disjoint clusters, $C_1, C_2 \ldots C_c$, of correlated genes by assigning each genes in $\{A_1, A_2 \ldots A_p\}$ to one of these clusters. Formally, we define gene clustering as the process that $\forall A_i, i \in \{1, \ldots p\}$, A_i is assigned to a Cr, $r \in \{1, \ldots, c\}$, where $C_r \cap C_s = \Phi$, for all $s \in \{1, \ldots, c\} - \{r\}$.

To find meaningful clusters, gene clustering is conducted so that genes within a cluster should have high correlation or high interdependence to each other whereas genes in different clusters are less correlated or more independent. Most of the conventional clustering methods use some distance metric to measure the dissimilarity or distance between two objects. In this paper, we use mutual information to measure similarity between two genes, which we believe is more meaningful if interdependent patterns are the most significant characteristics of a cluster reflecting the interrelationship among genes.

2.1 The Gene Interdependence Measure

For each continuous fingerprint in gene expression data, its domain is typically discretized into a finite number of intervals for data mining. In this paper, we use an Optimal Class-Dependence Discretization Algorithm (OCDD) [12] to discretize the continuous data. This method uses the normalized mutual information measure, which reflects interdependence between the class label vector and the gene, as the objective function and iterative dynamic programming to find a global optimal solution.

Let us suppose that the domain of $A_i, i \in \{1,2 \ldots p\}$, is discretized by OCDD into mi intervals. After discretization, the domains of all the attributes in R can be represented by $dom(A_i) \in \{v_{i1}, v_{i2} \ldots v_{ip}\}$, $i \in \{1,2 \ldots p\}$, where $v_{ik} = a_{ik}$; $k = 1,2 \ldots m_i$, if A_i is discrete and $v_{ik} = [b_{ik}, b_{i(k+1)}] \subseteq [l_i, u_i]$, $k = 1,2 \ldots m_i$, $b_{i1} = l_i, b_{i(m_{i+1})} = u_i$ if A_i is a discretized continuous attribute.

Let σ denote the SELECT operation from relational algebra and $|S|$ denote the cardinality of set S. The probability of a record in R having $A_i = v_{ik}$; $i \in \{1,2\ldots p\}$; $k = 1,2\ldots m_i$, is then given by:

$$\Pr(A_i = v_{ik}) = \frac{\left|\sigma_{A_i = v_{ik}}(R)\right|}{\left|\sigma_{A_i \neq NULL}(R)\right|} \tag{1}$$

and the joint probability of a record in R having $A_i = v_{ik}$ and $A_j = v_{jl}$, $j \in \{1,\ldots,p\}, i \neq j, k \in \{1,\ldots,m_i\}, l \in \{1,\ldots,m_j\}$ is calculated by:

$$\Pr(A_i = v_{ik} \wedge A_j = v_{jl}) = \frac{\left|\sigma_{A_i = v_{ik} \wedge A_j = v_{jl}}(R)\right|}{\left|\sigma_{A_i \neq NULL \wedge A_j \neq NULL}(R)\right|} \tag{2}$$

For two fingerprints, $x = \{x_1,\ldots,x_p\}$ and $y = \{y_1,\ldots,y_p\}$, similarity can be measured by the interdependence redundancy measure[1]:

$$R(A_i : A_j) = \frac{I(A_i : A_j)}{H(A_i : A_j)} \tag{3}$$

where $I(A_i : A_j)$ is the mutual information between A_i and A_j,which is given by :

$$I(A_i : A_j) = \sum_{k=1}^{m_i}\sum_{l=1}^{m_j} \Pr(A_i = v_{ik} \wedge A_j = v_{jl}) \log \frac{\Pr(A_i = v_{ik} \wedge A_j = v_{jl})}{\Pr(A_i = v_{ik})\Pr(A_j = v_{jl})} \tag{4}$$

$H(A_i : A_j)$ is the joint entropy of A_i and A_j, which is given by :

$$H(A_i : A_j) = -\sum_{k=1}^{m_i}\sum_{l=1}^{m_j} \Pr(A_i = v_{ik} \wedge A_j = v_{jl}) \log \Pr(A_i = v_{ik} \wedge A_j = v_{jl}) \tag{5}$$

In order to investigate the interdependency of a gene with all the other within a group, we use the multiple interdependency redundancy measure [12]. The multiple interdependence redundancy measure of a gene A_i within a gene group or cluster, $C = \{A_j \mid j = 1,2,\ldots,p\}$, is defined as:

$$MR(A_i) = \sum_{j=1}^{p} R(A_i, A_j) \tag{6}$$

where $R(A_i : A_j)$ is the interdependence redundancy measure between A_i and A_j.

We also use the concept of 'mode', denoted by M_i, that is the gene with highest multiple interdependence redundancy in a gene cluster $C_i = \{A_j \mid j = 1,2,\ldots,p\}$. So it can be defined as:

$$M(i) = A(j), j = \max_{A_n \in C_i}(MR(A_n)) \tag{7}$$

2.2 The Cooperative Competition Clustering Algorithm

To group genes, $A_1, A_2 \ldots A_p$, into different clusters, we propose a new Cooperative Competition Clustering algorithm. We suppose that cooperation and competition exist simultaneously among clusters in the process of clustering, at the same time, the cooperation, which comes from the nearest cluster, can help the cluster to compete with other clusters. In the competition, winners will move to centers of clusters and represent clusters gradually, while the losers will die out. By this way, true clusters can be found in the end. In some situation, there will be some small but interesting clusters. In the competing with other bigger and nearer clusters, those small clusters are in inferior position and maybe die out. In order to avoid occurring this situation, we suppose that cooperation exist simultaneously in competing, which help the cluster to compete with other clusters.

In many existing algorithms, only the distance information is used to determine which cluster a sample belongs to. But for many practical problems, such as interaction between electric charges, only distance information is not enough to represent the interaction strength, while the masses for two electric charges have to be considered. So, in our proposed Cooperative Competition Clustering algorithm, we introduce a concept of cluster quality to measure the cluster competitive force.

Definition 1: The cluster density of a cluster, $C_i = \{A_j \mid j = 1,2,\ldots,p\}$, is defined as:

$$D(i) = \frac{\sum_{A_n, A_m \in C_i, n \neq m} R(A_n, A_m)}{N_i(N_i - 1)} \tag{8}$$

where $R(A_i : A_j)$ is the interdependence redundancy measure between A_i and A_j; N_i is the number of genes in the cluster, C_i

Nevertheless, in practical applications, only cluster density is not enough to measure the cluster competitive force, the number of genes in the cluster will also impact on the cluster competitive force. So we give the following definition:

Definition 2: The cluster quality of a cluster, $C_i = \{A_j \mid j = 1,2,\ldots,p\}$ is defined as:

$$Q(i) = D(i) * N_i^{\,k} \tag{9}$$

where $D(i)$ is the cluster density of cluster C_i; N_i is the number of genes in the cluster C_i; k is the coefficient which measures the importance of the number of genes in cluster. If $k = 0$, the cluster quality is only determined by cluster density, and not related to the number of genes in a cluster; If $k = 1$, the cluster density and the number of genes in a cluster have the same impact on the cluster quality; If $k > 1$, the number of genes in a cluster is more important to the cluster density; if $0 < k < 1$, the number of genes in a cluster is less important to the cluster density. We suppose that the number of genes in a cluster is less important than the cluster density, so we set $k = 0.5$ in this paper.

In the competition with other bigger clusters, sometimes those small clusters are in inferior position and may die out at beginning of the algorithm. In order to avoid occurring such situation, we suppose that cooperation exist simultaneously in competing, which can help the clusters to compete with other clusters. So we introduce a concept referred to as Inter-cluster aid to measure the strength of cooperation.

Definition 3: The Inter-cluster aid of a cluster, $C_i = \{A_j \mid j = 1,2,\ldots,p\}$, is defined as:

$$IA(i) = I(M_i, M_j)^2 * Q(j), j = \arg(\max_{n \neq i}(I(M_i, M_n))) \tag{10}$$

where $Q(j)$ is the cluster quality of cluster, c_i, M_i is the center mode of cluster, c_i, M_i is the center mode of cluster c_j; $I(M_i, M_j)$ is the interdependence redundancy measure between M_i and M_j.

The total competitive force of a cluster is thereby composed of the cluster quality of the cluster and the Inter-cluster aid of the cluster. Thus, we introduce the definition of the total competitive force of a cluster.

Definition 4: The total competitive force of a cluster, $C_i = \{A_j \mid j = 1,2,\ldots,p\}$, is defined as:

$$F(i) = Q(i) + IA(i) \tag{11}$$

where $Q(i)$ is the cluster quality of cluster C_i, $IA(i)$ is the inter-cluster aid of cluster C_i.

As a result, the Cooperative Competition Clustering algorithm can be summarized as follows:

Step 1. Initialization. Let us select the number of clusters, k, where k is greater than the true number of clusters k^*. Assume k genes to be randomly selected as cluster modes $M_i, i \in \{1,2,\ldots,k\}$.

Step2. Assigning every gene to one of the clusters. Take a sample $A_j, j \in \{1,2\ldots p\}$ from the data set. For cluster i, $i \in \{1,2,\ldots,k\}$, let

$$f(C_i \mid A_j) = \begin{cases} 1 & j = \arg(\max(F(i)) * R(A_j, M_i)) \\ 0 & else \end{cases} \quad (12)$$

where $F(i)$ is the total competitive force of cluster C_i , $R(A_j, M_i)$ is the interdependence redundancy measure between gene, A_j ,and cluster mode, M_i .

Step 3. Update cluster modes and the total competitive force for every clusters. Firstly, determine these clusters, in which the number of samples is few, then these clusters should be removed. And then, for every cluster, $C_i, i \in \{1,2,\ldots,k\}$, we set the $M_i = A_j, j = \arg(\max_{A_n \in C_i}(MR(A_n)))$; and compute the competitive force for every clusters.

Step 4. Termination. Steps 2 and 3 are repeated until cluster modes $M_i, i \in \{1,2,\ldots,k\}$ do not change.

In practical applications, how does CCCA work well? There are two hypotheses done: 1). Samples in the same cluster are uniformly distributed. 2). Different clusters are well separated. We can explain the working principle of CCCA by Fig-1. When two (or more) selected cluster modes, say, mode A and B, are in the same true cluster as in Fig-1 (a). At beginning stage, the cluster quality of mode A almost equals to that of mode B and the determination plane, i.e., the plane p, is in the middle of modes A and B. We suppose that mode A is much near to the mode of true cluster, i.e., mode C. Consequently more samples are determined to belong to cluster A so that the cluster quality of cluster A gets much bigger. On the other hand, few samples are determined to belong to cluster B and the cluster quality of cluster B gets much smaller. After that, mode A moves forward to mode C, while mode B moves to the margin of the true cluster and the determination plane move forward to mode B. Finally, mode A moves to mode C and represents the true cluster; while mode B will be removed. When two true clusters are represented by mode A and mode B, respectively, as shown in Fig-1 (b). The cluster quality of cluster A equals to that of cluster B, and the determination plane is in the middle of modes A and B at the beginning stage. We suppose that more samples are near to cluster A and few samples are near to cluster B. Consequently more samples are determined to belong to cluster

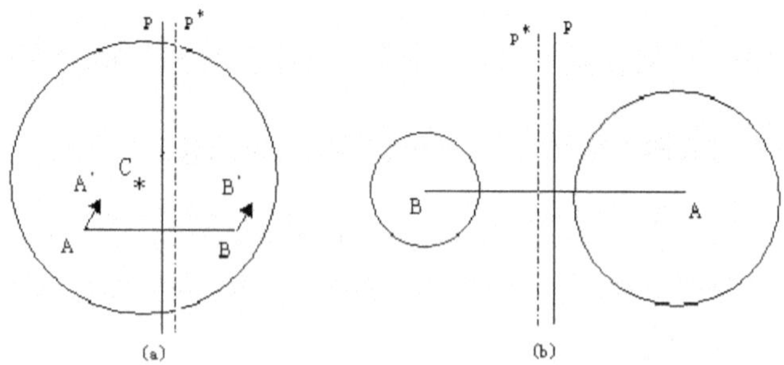

Fig. 1. A and B are initial mode of cluster A and B, respective. The plane p is the initial determination plane. A^* and B^* are new modes of cluster A and B after update, respective. The plane p^* is the new determination plane after update.

A and the cluster quality of cluster A gets much bigger, while few samples are determined to belong to cluster B and cluster quality of cluster B gets much smaller. In the following update, the determination plane moves forward to mode B for bigger cluster quality of cluster A. However, because the two true clusters are well separated and very few samples in the interspaces between cluster A and cluster B, the moving of the determination plane cannot make cluster A to become much bigger and bigger and the determination plane will be stable finally.

3 Experimental Results on Gene Expression Data Sets

We applied CCCA to two widely used gene expression data sets: the colon cancer data set. The colon cancer data set consists of 62 samples and genes. The samples are composed of tumor biopsies collected from tumors and normal biopsies collected from the healthy part of the colons of the same patient. Each sample has been preclassified into one of the two classes: normal and cancer.

It can be evidently that there are so much irrelevant and redundant gene expression data. The data structure of relevant and determinative genes is most important and interesting, while the data structure of irrelevant and non-determinative genes is not. So we removed the most irrelevant genes before using the CCCA by using the interdependence redundancy measure ($R(F, A_i)$) between the class label vector (F) and the sample vector A_i. We selected genes with $R(F, A_i) > T$, where T is a cutoff parameter that controls the strength of selection. To show the efficiency of this selection method, we used the selected genes to do classification tests.

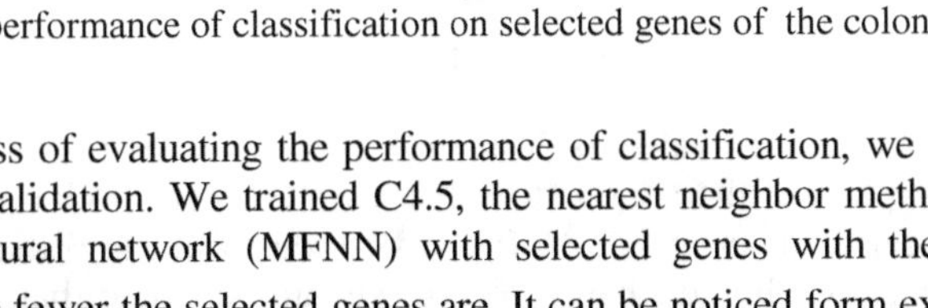

Fig. 2. The performance of classification on selected genes of the colon cancer data set

In the process of evaluating the performance of classification, we employ the leave-one-out cross-validation. We trained C4.5, the nearest neighbor method and Multilayer feedforward neural network (MFNN) with selected genes with the $R(F, A_i) > T$. The bigger T is, the fewer the selected genes are. It can be noticed form experimental results that increasing the features to be considered by a classifier may degrade its performance, if the number of training examples that are used to design the classifier is small relative to the number of features. This paradoxical behavior is termed as the peaking phenomena [13,14]. The explanation stands in the following: If T is small, non-determinative genes are removed and the classification accuracies increase; on the contrary, if T is larger, some determinative genes are removed and the classification accuracies decrease. To avoid removing the determinative genes, it is better to select a smaller T. However, there will be too many non-determinative genes if T is too small. In the following experiments, we set T to 0.1 in colon cancer data set.

We used CCCA to test the colon cancer data set. Initially, we set the number of clusters to be 20. Then, CCCA divides the colon cancer data set into 4 clusters. After that, we selected top k genes to form every genes cluster, so a total of $4 \times k$ genes are selected for $k = 1,2,3$ in the colon cancer data set. We then used C4.5, MFNN and nearest neighbor method to build classifiers for testing on the selected genes.

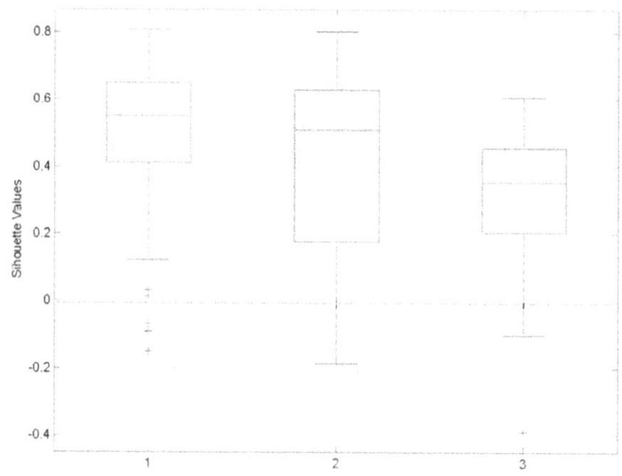

Fig. 3. Boxplot of silhouette value of genes in cluster on the breast cancer data set

To evaluate the quality of the clusters formed by CCCA, we used the silhouette measure proposed by Rousseeuw (1987). For computing the silhouette value of a gene A_i, we first estimated two a_i and b_i. Let us denote C_r as the cluster to which gene belong. The scalar a_i is average similarity between A_i and all other genes of C_r. For any other cluster $C_s \neq C_r$, let $d(A_i, C_s)$ denote the average similarity of gene to all genes of C_s. The scalar b_i is the biggest of these $d(A_i, C_s), r \neq s = 1 \ldots K$, the silhouette value $S(x_i)$ of gene A_i is the ratio $\frac{a_i - b_i}{\max(a_i, b_i)}$. The silhouette value lies between –1 and 1.when its value is less than zeros, the corresponding is poorly classified. The Boxplots of the silhouette value are shown for both colon cancer data set and breast cancer data set in Fig 3. It can be seen from Fig 3 that the silhouette values of most genes are bigger than zero. So we can conclude that most genes are well classified and the quality of the clusters found by CCCA is very well.

To evaluate the gene selection performance of CCCA, the selected genes are fed to the same classification algorithms. In literature (Wai et al., 2005), the performance of ACA has been proved better than that of k-means, Biclustering and SOM. Here, we do not repeat this problem. So in this paper, we only compared the performance of CCCA with that of ACA. The experimental results are showed in Tables 1.

The performance of CCCA almost corresponds to that of ACA, but ACA doesn't have the ability to set K automatically. ACA needs user to set the K in advance, If K

Table 1. The performance of classification on selected genes of the colon cancer data set

	Number of selected genes	Classification accuracy	
		CCCA	ACA
C4.5	4	83.9%	80.6%
	8	85.5%	83.9%
	12	85.5%	83.9%
Nearest neighbor	4	90.3%	83.9%
	8	85.5%	82.3%
	12	85.5%	82.3%
MFNN	4	91.9%	90.3%
	8	91.9%	90.3%
	12	91.9%	90.3%

is not set correctly, the performance of ACA deteriorates rapidly. The performance of classification algorithms on the top genes selected by ACA with different number of clusters is given in Tables 2. It is interesting to find that ACA yields the promising performance with the numbers of cluster determined by CCCA.

Table 2. The performance of classification on selected genes by ACA of the colon cancer data set

	Number of cluster							
	2	3	4	5	6	7	8	9
C4.5	77.4%	77.4%	80.7%	82.3%	82.3%	83.9%	82.3%	80.7%
NN	80.7%	80.7%	83.9%	83.9%	85.5%	83.9%	83.9%	82.3%
MFNN	80.7%	80.7%	90.3%	85.5%	85.5%	83.9%	83.9%	87.1%

4 Conclusions

Even though the Attribute cluster algorithm has been proposed for clustering on gene expression data, the use of this method has been hindered so far by problems associated with the choice of K. Recently many researchers proposes the competition clustering method to solve this problem, but the most of existing competition algorithms overemphasize the competition and overlook the cooperation between clusters so that many small but interesting clusters have been omitted.

In this paper, we proposed a new clustering algorithm known as Cooperative Competition Clustering (CCCA). In this algorithm, we took into account the cooperation between clusters to devise the Cooperative Competition Clustering algorithm. In other words, CCCA uses the idea of cooperative competition: Cooperation and competition exist simultaneously between clusters in the process of clustering and the cooperating, which comes from the nearest cluster, can help the cluster to compete with other clusters. By using this principle of Cooperative Competition, the number of clusters can be found in the process of clustering efficiently. We devised the experiments on two gene data sets, the colon cancer data set and the breast cancer data set. IT is can be seen that CCCA can find the k automatically and select interesting and meaningful genes. The clusters found by the CCCA are well separated and those selected genes can yield very high classification performance on both data sets.

References

1. Au, W-H., Keith, C.C. C., Andrew, K.C. W., Wang, Y.: Attribute Clustering for Grouping, Selection, and Classification of Gene Expression Data. IEEE Trans. Computation Biology and Bioinformatics, Vol. 2, No. 2 (2005) 83-101
2. Xing, EP., Karp, RM.: CLIFF: Clustering of High-Dimensional Microarray Data via Iterative Feature Filtering Using Normalized Cuts. Bioinformatics, Vol. 17 Suppl. 1: S306-15 (2001)
3. Hastie, T., Tibshirani, R., Eisen, M., Brown, P., Scherf, U., Weinstein, J., Alizadeh, A., Staudt, L., Botstein, D.: Gene Shaving: a New Class of Clustering Methods for Expression Arrays. In Tech. Report, Stanford University (2000)
4. Alter, O., Brown, PO., Botstein, D.: Singular Value Decomposition for Genome-Wide Expression Data Processing and Modeling. Proc Natl Acad Sci USA (2000) 10101-10106
5. Piatetsky-Shapiro, G., Khabaza, T., Ramaswamy, S.: Capturing Best Practice for Microarray Gene Expression Data Analysis. Proc. Ninth ACM SIGKDD Int'l Conf. Knowledge Discovery and Data Mining (2003) 407-415
6. Tamayo, P., Solni, D., Mesirov, J., Zhu, Q., Kitareewan, S., Dmitrovsky, E., Lander, E.S., Golub, T.R.: Interpreting Patterns of Gene Expression with Self-Organizing Maps: Methods and Application to Hematopoietic Differentiation, Proc. Nat'l cademy of Sciences of the United States of Am., Vol. 96, No. 6 (1997) 2907-2912
7. Alon, U., Barkai, N., Notterman, D.A., Gish, K., Ybarra, S., Mack, D., Levine, A.J.: Broad Patterns of Gene Expression Revealed by Clustering Analysis of Tumor and Normal Colon Tissues Probed by Oligonucleotide Arrays. Proc. Nat'l Academy of Sciences of the United States of Am., Vol. 96, No. 12 (1999) 6745-6750
8. Jiang, D., Tang, C., Zhang, A.: Cluster Analysis for Gene Expression Data: A Survey. IEEE Trans. Knowledge and Data Eng., Vol. 16, No. 11 (2004) 1370-1386
9. Eisen, MB., Spellman, PT., Brown, PO., Botstein, D.: Cluster Analysis and Display of Genome-Wide Expression Patterns. Proc. Natl. Acad. Sci. USA (1998) 14863--14868
10. Heyer, L.J., Kruglyak, S., Yooseph, S.: Exploring Expression Data: Identification and Analysis of Coexpressed Genes. Genome Research, Vol. 9 (1999) 1106-1115
11. Wong, A.K.C., Liu, T.S.: Typicality, Diversity and Feature Patterns of an Ensemble. IEEE Trans. Computers, Vol. 24, No. 2 (1975) 158-181
12. Liu, L., Wong, A.K.C., Wang, Y.: A Global Optimal Algorithm for Class-Dependent Discretization of Continuous Data. Intelligent Data Analysis, Vol. 8, No. 2 (2004) 151-170
13. Jain, A.K., Chandrasekaran, B.: Dimensionality and Sample Size Considerations in Pattern Recognition Practice. In P.P. Krishnaiah and L.N. Kanal, Editors, Handbook of Statistics, North Holland (1982) 835–855
14. Raudys, S.J., Jain, A.K.: Small Sample Size Effects in Statistical Pattern Recognition: Recommendations for Practitioners. IEEE Transactions on Pattern Analysis and Machine Intelligence, 13(3) (1991) 252–264
15. Xu, L.: Rival Penalized Competitive Learning, Finite Mixture, and Multisets Clustering. Proc.1998 IEEE Int. Joint Conf. Neural Networks 3 (1998) 2525-2530

Genetic Algorithm and Neural Network Based Classification in Microarray Data Analysis with Biological Validity Assessment

Vitoantonio Bevilacqua, Giuseppe Mastronardi, and Filippo Menolascina

Dipartimento di Elettrotecnica ed Elettronica, Polytechnic of Bari
Via E. Orabona, 4, 70125, Bari, Italy
bevilacqua@poliba.it

Abstract. Microarrays allow biologists to better understand the interactions between diverse pathologic states at the gene level. However, the amount of data generated by these tools becomes problematic. New techniques are then needed in order to extract valuable information about gene activity in sensitive processes like tumor cells proliferation and metastasis activity. Recent tools that analyze microarray expression data have exploited correlation-based approach such as clustering analysis. Here we describe a novel GA/ANN based method for assessing the importance of genes for sample classification based on expression data. Several different approaches have been exploited and a comparison has been given. The developed system has been employed in the classification of ER+/- metastasis recurrence of breast cancer tumours and results were validated using a real life database. Further validation has been carried out using Gene Ontology based tools. Results proved the valuable potentialities and robustness of similar systems.

1 Introduction

Microarrays have gained in this time a great fame . This technology enables the measurement of the levels of mRNA molecules inside a cell and, consequently, the proteins being produced. In this context, the comparison between gene expression patterns through the measurement of the levels of mRNA in healthy versus unhealthy cells can supply important information about pathological states, as well as information that can lead to earlier diagnosis and more efficient treatment.

The real challenge, then, is to find a set of genes, out of the thousands mapped, which can be used to develop a classifier with the highest accuracy [5]. Identification of a set of differentially expressed genes could serve to identify disease subtypes that may benefit from distinct clinical approaches to treatment. This was the primary objective of Foekens et al in [3] and [4]; to predict accurately patient's risk of recurrence is an important aspect of lymph node negative cases treatment planning. Gene signature extracted by Foekens et al consists of 76 genes (60 for ER+ and 16 for ER- cases). In [3] Foekens et al. have employed statistical methods and supervised/ unsupervised clustering techniques in order to extract knowledge from a 286 x 22482

D.-S. Huang, K. Li, and G.W. Irwin (Eds.): ICIC 2006, LNBI 4115, pp. 475 – 484, 2006.

array of gene expression values. Although correlation-based approaches have been widely applied in analyzing the patterns of gene expression [1][2], it's commonly believed they may not fully extract the information from data corrupted by high-dimensional noise. Therefore, these ranking based techniques select the genes which individually provide better classification, but they may not result in meaningful gene combinations for an overall classification task. Hence approaches capable of performing an efficient search in high dimensional spaces, such as evolutionary algorithms (EAs), should prove to be ideal candidates. What is more, while high-throughput technology has significantly accelerated the rate at which biological information is acquired, tools that can successfully mine the resulting large data sets are needed. Some research groups have exploited the potentialities of soft computing techniques applied to bioinformatics and some these works were carried out in the field of microarray data analysis [6] [7].

With this work we have tried to address the problem of *gene selection* using a Genetic Algorithm that evolves populations of possible solution and uses an Artificial Neural Network in order to test gene signatures' ability to correctly classify cases belonging to the test set. Therefore, results returned by the GA/ANN based system are then validated using Gene Ontology, a biological validity assessment tool that can show interesting cues of research for biologist and physicians.

This paper firstly gives some details of the problem of classification problem in "post genomic" era. Then a description of the GA and of genetic operators is given. An outlook on the ANN classifier follows. In final paragraphs system results and brief investigations of biological plausibility are exposed.

2 Methods

2.1 Data Acquisition and Preprocessing

The dataset used to evaluate performances of the system proposed is publicly available and can be downloaded from the Gene Expression Omnibus (GEO) web site. GEO is a data repository of high-throughput gene expressions and hybridization arrays maintained by the National Center for Biotechnology Information. GEO databases have been used in the recent years by researchers of all over the world in order to give publicity to results of specific researches. Each dataset submitted to GEO receives an ID code (a unique identifier) called "Accession Number".Our focus is on the dataset *GSE2034*, submitted by Tim Jatkoe on the February, 23rd 2005 and provided to us by the I.R.C.C.S. of Bari, Italy. This dataset collects the results of a multi-center research carried out by Veridex LLC in collaboration with the Department of Medical Oncology, Erasmus MC-Daniel den Hoed of Rotterdam. The research, which involved 286 patients, aimed at discovering gene signatures able to identify patients at high risk of distant recurrence [3]. The ability to identify patients who have a favorable prognosis could, after independent confirmation, allow clinicians to avoid adjuvant systemic therapy or to choose less aggressive therapeutic options.

The data matrix obtained contains quasi-raw data. Gene expression levels, in fact, are characterized by statistical properties that force researchers to apply preprocessing algorithms. In the common practice of microarray data analysis the "normalization" is a key step. Data normalization for microarray experiments is an open field of research and many alternative algorithms have been proposed in order to accomplish this delicate task [9]. Together with normalization, filtering techniques met more and more often the consensus of researcher thanks to their ability to exclude in early stage of the research less meaningful variables that add computational costs to the data mining of similar datasets [10]. Even in this field many different approaches have been proposed. For the peculiarities of the dataset *GSE2034* an "entropy information" based gene filtering has been employed in order to remove genes with low entropy expression values [10]. At the end of these processes a normalised matrix 286 x 20055 have been obtained.

2.2 Gene Selection Using Genetic Algorithms

In recent years some research groups have focused their attention on the exploitation of GAs' potentialities in information extraction from biomedical database. In [11] a multi-objective algorithm has been employed in order to build a reliable classification tool small in size and, at the same time, able to produce as accurate a classification as possible. In [12] the problem of gene assessment and sample classification for gene expression data have been addressed using a Genetic Algorithm and a K-Nearest Neighbor classifier.

The proposed approach is based on a hybrid system that uses a GA to select subsets of genes and an ANN that classifies cases and returns a metric of the error which is used as a fitness function for the selected subset of genes. Given the high variety of approaches reported in literature we have tried herein to carry out a comparative study of the different systems and their results on the chosen dataset.
As known GA, are basically inspired by natural evolution and selection. In biological systems, genetic information is stored in chromosomes. Chromosomes are replicated and passed onto the next generation with selection depending on fitness. Genetic information can, however, also be altered through genetic operations such as mutation and crossover. In GAs, each "chromosome" is a set of genes, which constitutes a candidate solution to the problem. There are four major components of GA:

- Chromosome;
- Fitness;
- Selection;
- Crossover / Mutation.

The modular approach followed in the design phase of the system has allowed to simply evaluating the performances of the other compared classifier systems described below, acting on the fitness function module.

Brief descriptions of particular implementations of the described systems are given below.

2.2.1 Chromosome Representation

In this work a 20 genes long chromosome has been used in order to codify a 20 genes long gene signature. A binary codification of the chromosome has been selected.

2.2.2 Fitness Function

For the calculation of the fitness function the Sum of the Squared Errors (SSE) error returned by an ANN based classifier has been used. This is a metric of the error made by the ANN in the classification task, other evaluation systems could be simply implemented.

2.2.3 Selection Criterion

The selection criterion was the elitistic one. Best perform-ing gene signatures from each population are allowed to pass to the other generation producing offspring.

2.2.4 Crossover /Mutation

In the crossover and mutation operators some constraints have been implemented in order to maintain acceptability of the solution (e.g. in order to avoid the repetition of a gene in the same chromosome). Crossover probability has been set to 40% and mutation probability was time dependent.

2.3 Artificial Neural Network Based Classification

Given the characteristics of the problem and the dimensionality of the dataset, an ANN-Feed-Forward (ANN-FF) has been chosen. Then, for each fitness function evaluation an ANN-FF is trained on 200 cases each of which is defined by 20 parameters (genes selected by the GA and passed to the fitness function for evaluation) and validated. On the basis of results of research described in [14], and having observed high similarity of dimensionality of the two datasets, a three layer ANN has been set up. The selected topology provided for 25 neurons in the first layer, 12 in the hidden layer, and 1 output neuron. Activation functions were: "*tansig*" for the first two layers and "*pure linear*" for the last layer. Stop criterion for the training phase were: 50000 epochs or SSE less than 0.004. Initial learning rate has been set to 0.3 and modified by the descent gradient momentum algorithm. These choices provided a solution able reach a good equilibrium between learning and generalization capabilities of the system.

3 Gene Selection

After the preprocessing stage and the system development, the experimentation phase has been carried out. The GA/ANN hybrid system has been set up and executed on the *GSE2034* dataset. The GA/ANN as well as the other compared GA/X hybrid systems were executed 100 times and each GA run accounted for 500 generations. Good convergence ability has been reached with described parameters as can be seen in figure 1.

From a computational standpoint, it's worth noting that statistical classifiers and K-NN based system are characterized by low CPU times when compared to SVM and ANN hybrid solutions. Even though the GA/ANN system showed a computationally intensive behaviour results returned by this system are characterized by an important factor. The amount of variance in the genes extracted, registered along the 100 runs of

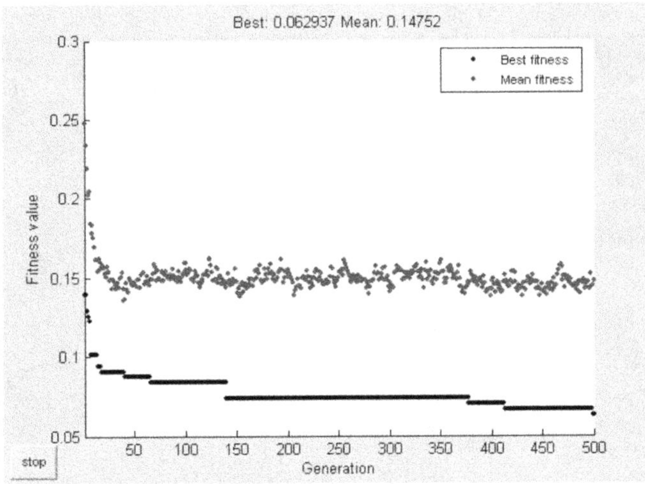

Fig. 1. Best fitness, in black, and mean fitness values of generations

the GA, is quite low. This particular aspect distinguishes the GA/ANN approach from the others reported as comparison in this work. As it can be seen in the "Results and Comparisons" section, the GA/ANN returns results coherent with the problem; furthermore it has been observed that this system shows a particular ability in extracting relevant genes with a considerably higher probability than other genes.

4 Results and Comparisons

Comparative results of the hybrid systems previously described are provided in this section. After 100 executions of the GA a ranking of selected genes has been compiled. This ranking considered selection frequencies for each gene extracted and returned and interesting overview of systems' performances. It's worth noting that all of the solutions analyzed selected sets of genes that demonstrated to be overlapping. In table 1, the rankings of the 20 most selected genes for each hybrid system are reported. As it can be seen, a considerable part of the most selected genes are reported by all of the systems analyzed. However GA/ANN system showed lower variability in results returned; this aspect is peculiar of this particular hybrid system only. Genes extracted by the GA/ANN system remained quite the same over all the 100 runs of the algorithm. However, as confirmed by competitive results returned by the other systems, the choice of using a GA in order to extract relevant genes has a considerably positive impact on the overall system accuracy; evidently the sensitivity of GA employment on system's performances is slightly higher than that of the classifier related choice.

However gene signatures returned by similar techniques are useless until an interpretation is formulated regarding their activities, interaction, and possible involvements in critical contexts (e.g. estrogen synthesis).

Table 1. Genes extracted by the compared systems, rank positions for each solution are reported

Gene ID	GA/ANN Rank	GA/SVM Rank	GA/KNN Rank	GA/Stat Rank
219340_s_at	1	1	1	1
217771_at	2	2	2	2
202418_at	3	3	6	4
206295_at	4	4	4	7
200726_at	5	5	5	3
210314_x_at	6	7	3	6
219588_s_at	7	6	8	8
212567_s_at	8	8	7	10
55081_at	9	10	10	18
218430_s_at	10	9	9	9
217404_s_at	11	11	12	13
205848_at	12	12	13	12
214915_at	13	13	16	-
202687_s_at	14	15	14	-
221241_s_at	15	16	11	11
210593_at	16	14	-	-
204028_s_at	17	17	15	-
201112_s_at	18	19	17	-
209825_s_at	19	18	-	16
209602_s_at	20	-	19	14
209604_s_at	-	20	-	5
201579_at	-	-	18	15
210347_s_at	-	-	-	17
209603_at	-	-	20	19
200827_at	-	-	-	20

In this work we provide a two-fold validation of the sets of genes. In current research in the bioinformatics field there are two main trends as for the validation of results returned by any kind of computational approach. The first is referred to as "Data-driven" approach which mainly includes statistical tests or validity indices (e.g. Dunn's index or Silhouette method) applied to the data clustered. The second is referred to as "Knowledge-driven" method [15]. In the Molecular Biology and Bioinformatics field the "Knowledge-base" model has been translated in a set of "genetic ontology" maintained by a specific consortium: the "Gene Ontology Consortium"[16]. Biological validity assessment of gene signatures obtained is given in the next section. In this section "Statistical-Driven" validation is carried out. The focus is obviously on the gene signature extracted by the GA/ANN, however, given the similarities with other gene subsets, similar analyses could be done for the other results. The first gene, "219340_s_at", has been selected 85 times over 100 executions. This gene's product is a transmembrane protein involved in signal transmission between cells. The putative CLN8 protein gene has been selected as the most statistically meaningful gene even in the work of Foekens[3]. In the following positions in the ranking there is a set of genes "201112_s_at", "202687_s_at","

204028_s_at", "205848_at", "206295_at", "210314_x_at" and "221241_s_at" that show frequencies between 62 and 83 and that belong to the same class of genes. In fact all of these genes are involved in the "cell cycle" regulation, included apoptosis that is, the way cells die in an ordered or programmed way. Apoptosis is one of the main types of Programmed Cell Death that allow organism to handle cell proliferation. These mechanisms have been discovered to be strictly correlated to tumor proliferation [18]; in particular down-regulation of apoptosis-related genes can lead to uncontrolled cell proliferation which is a key aspect of cancer.

In order to give a comprehensive outlook on all the techniques involved in this comparative study, table 2 collects computational times required by each system. Measurements refer to an Intel P4 EE 2.8 GHz with 2 GB of DDR RAM memory. An average value of all the measurements made has been calculated, approximated to next integer number and reported.

Table 2. Computational times (in seconds) required by each system

GA/Stat	GA/KNN	GA/SVM	GA/ANN
1385	9910	44500	144200

5 Biological Validity Assessment

In this section we provide a biological validity assessment of the results obtained in previous steps. Obviously the focus is primarily on the GA/ANN solution, however, given the similarities among the extracted genes subsets, following considerations could be considered applicable even to the other systems.

The Gene Ontology (GO) is composed of three related ontologies covering basic areas of biological research: the molecular function of gene products, their role in multi-step biological processes, and their physical structure as cellular components. GO Consortium provides to constantly control and update databases. The GO defines a shared, structured and controlled vocabulary to annotate molecular attributes across models organisms. As shown in figures 2 and 3 each ontology is constructed as a directed acyclic graph. GO is composed by GO terms and each GO term consists of a unique alphanumerical identifier, a common name, and a definition. Terms are classified into only one of the three ontologies. In this way sets of genes, extracted by statistical techniques as well as computational intelligence based systems, could be analyzed in order to observe if there is any "interaction" or involvement in biological path considered critical for the disease or phenomenon examined.

One of the most interesting GO based software is the GO Surfer of the Harvard School of Public Health [15].

Fed with the list of the 20 most selected genes extracted by the GA/ANN hybrid system, GO Surfer puts in evidence quite interesting results. In figure 2 the GO trees of the "biological processes" category are shown. Each node represents an individual GO term and all GO terms at display are associated with at least one out of 20 specific genes. It is remarkable that a great part of these genes belong to GO terms in some way correlated to the regulation of cell life (cellular proliferation, apoptosis and necrosis).

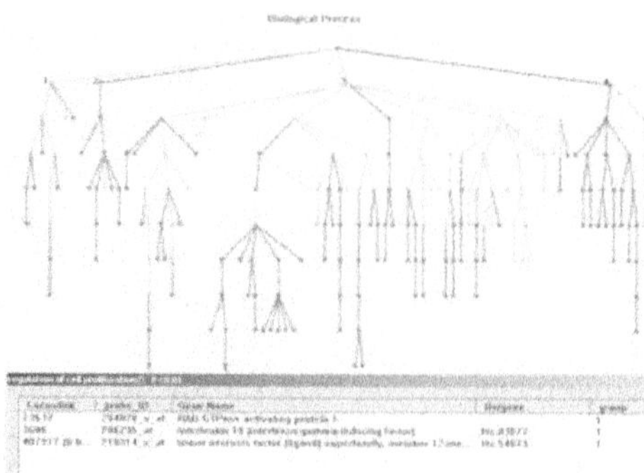

Fig. 2. "Biological Process" GO Tree. "Regulation of Cell proliferation" GO Term in evidence.

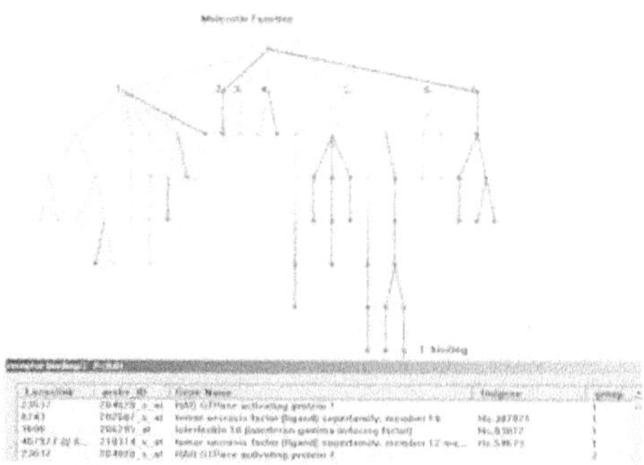

Fig. 3. "Molecular Process" GO Tree. "Receptor binding" GO Term in evidence.

Down-regulation of these genes could results in uncontrolled cell proliferation and then oncogenesis. In figure 3 GO tree of the molecular function is reported. From this standpoint it is interesting to observe that in one of the most meaningful term ($p < 0.01$), the "receptor binding" one, two "tumor necrosis factor" genes are present. Even in this case irregular activity of these genes can bring to dysfunction in cell cycle regulation. In the most interesting of the "cellular component" tree branch ($p < 0.01$) "Cytoskeleton" the "205848_at" gene is included. This is a growth arrest specific gene, that is to say a gene that could regulate the growth of cells (furthermore in this process the well known P53 onco-suppressor gene is involved). On the basis of all these considerations and observations it can be argued that the research in the field of Breast Cancer should focus on the activity and regulation of "221241_s_at"," 202687_s_at", "210314_x_at", "205848_at" and "55081_at" genes. As we have previously seen, these genes are not only able to build an accurate classifier, but they are even involved in biological and molecular processes that could be individuated as strictly correlated to typical tumour pattern, just like dysfunction in cell life cycle and receptor activity.

6 Concluding Remarks

The development of gene expression data analysis methods is one of the most important challenges of the post-genomic era. With the advent of microarray technology, the scientific community has assisted at the growth of datasets with a peculiar aspect: high disproportion between the two dimensions. This is a critical challenge for both the data miners and the tools they employed.. Machine learning techniques, in this context, have demonstrated to be a powerful tool especially for what concerns clustering and combinatorial optimization problems. These are two of the most active research branch in the bioinformatics field. Soft computing techniques, still, are gaining more and more attention due to their abilities and potentialities in the two fields described. System's performances were tested using a real-world publicly available dataset (*GSE2034*). It is worth noting that the proposed solution shows large applicability: other datasets could be explored with similar results (further experiments were successfully carried out on [19]).

A comparative study of performances of other systems has been given in this paper as additional evaluation tool. As analyzed in previous sections, the results returned by the GA/ANN are quite competitive: the proposed algorithm demonstrated to be robust and be affected by very low variability of results. This is the most interesting characteristics of the system described. In the various experiments carried out the GA/ANN approach selected small subsets of genes with a high frequency; this can be interpreted as the ability of the proposed system, to focus its research on subsets of features and to avoid local minima entrapment. According to the "Knowledge-driven" principles, on the other hand, a Gene Ontology based validity assessment has been carried out. As demonstrated in previous sections validity assessment can be considered to be satisfied both from a statistical and from a biological standpoint. Furthermore we observed that some of the most relevant genes extracted were included in the gene signature proposed by Foekens [3]. Further works will be mainly oriented on the optimization of the classifier. Artificial Immune System (AIS) based approaches are being studied. Given the AIS-ANN theory similarity [20], AIS based systems in the limited resources approach [21] could reveal interesting potentialities, optimizing both classification abilities and resource employment. Further researches have to be carried out but the results in the primary stage are encouraging.

Acknowledgements

The authors would like to thank M.D. Angelo Paradiso of the I.R.C.C.S. of Bari for his valuable support.

References

1. Golub, T.R., Slonim, D.R., Tamayo, P., et al.: Molecular Classification of Cancer: Class Discovery and Prediction by Gene Expression Monitoring. Science, 286 (1999)
2. Alizadeh, A.A, Eisen, M.B, et al.: Distinct Types of Diffuse Large B-Cell Lymphoma Identified by Gene Expression Profiling. Nature 403, (2000) 503-11

3. Wang, Y., Klijn, J.G.M., Zhang, Y., et al.: Gene-expression Profiles to Predict Distant Metastasis of Lymph-node-negative Primary Breast Cancer. Lancet (2005)
4. Foekens, J. A., David Atkins, et al.: Multi-center Validation of a Gene Expression Based Prognostic Signature in Lymph Node-Negative Primary Breast Cancer, to appear
5. Zhang, M.: Extracting Functional Information from Microarrays: A Challenge for Functional Genomics. PNAS, 99(20), (2002) 12509-12511
6. Chakraborty, A., Maka, H.: Biclustering of Gene Expression Data Using Genetic Algorithm, CIBCB (2005)
7. Juliusdottir, T, Corne, D., Keedwell, E., Narayanan, A.: Two-Phase EA/k-NN for Feature Selection and Classification in Cancer Microarray Datasets, CIBCB (2005)
8. Bolstad, B.M., Irizarry, R.A, Åstrand, M., Speed, T.P.: A Comparison of Normalization Methods for High Density Oligonucleotide Array Data Based on Variance and Bias, Bioinformatics 19(2), (2003)
9. Virginie, M. A., Cody, M. J., Cheng, J., Dermody, J.J., Soteropoulos, P., Recce, M., Tolias., P.P.: Noise Filtering and Nonparametric Analysis of Microarray Data Underscores Discriminating Markers of Oral, Prostate, Lung, Ovarian and Breast Cancer. BMC Bioinformatics (2004)
10. Kohane, I.S., Kho, A.T., Butte, A.J.: Microarrays for an Integrative Genomics, MIT Press, (2003)
11. Kalyanmoy, D., Raji, R.,: Classification of Two and Multi-Class Cancer Data Reliably Using Multi-objective Evolutionary Algorithms. KanGAL Report No. 2003006
12. Li, L., et al.: Gene Assessment and Sample Classification for Gene Expression Data Using a Genetic Algorithm/k-nearest Neighbor Method. Combinatorial Chemistry and High Throughput Screening, (2001) 727-739
13. Cristianini, N., Shawe-Taylor, J.: An Introduction to Support Vector Machines (And Other Kernel-Based Learning Methods). Cambridge University Press, Cambridge, UK. (2000)
14. Bevilacqua, V., Mastronardi, G., Menolascina, F.: Hybrid Data Analysis Methods and Artificial Neural Network Desing in Breast Cancer Diagnosis: IDEST experience, CIMCA 2005, in press (2005)
15. Zhong, S., Storch, F., Lipan, O., Kao, M.J., Weitz, C., Wong W.H.: GoSurfer: A Graphical Interactive Tool for Comparative Analysis of Large Gene Sets in Gene Ontology Space. Applied Bioinformatics, An Introduction to Support Vector Machines (And Other Kernel-Based Learning Methods), An Introduction to Support Vector Machines (And Other Kernel-Based Learning Methods), (2004)
16. http://www.geneontology.org
17. Bonnotte, B., Favre, N., Moutet, M., Fromentin, A., Solary, E., Martin, M., Martin, F.: Role of Tumor Cell Apoptosis in Tumor Antigen Migration to the Draining Lymph Nodes. Journal of Immunoly (2000), 1995-2000
18. Tang, K., Ponnuthurai N.S., Xin Y.: Feature Selection for Microarray Data Using Least Squares SVM and Particle Swarm Optimization, CIBCB (2005)
19. Huang, H., Cheng S.E.C.,et al: Gene Expression Predictors of Breast Cancer Outcome, Lancet (2003) 361, 1590-1596
20. Dasgupta, D.: Artificial Neural Networks and Artificial Immune Systems: Similarities and Differences, Proc. of the IEEE SMC, 1, (1997) 873-878
21. Watkins, A., Timmis, J., Boggess, L.: Artificial Immune Recognition System (AIRS): An Immune Inspired Supervised Machine Learning Algorithm. Genetic Programming and Evolvable Machines, (2004)

Inferring Species Phylogenies: A Microarray Approach

Xiaoxu Han

Department of Mathematics and Bioinformatics Program
Eastern Michigan University, Ypsilanti, MI 48197 USA
xiaoxu.han@emich.edu

Abstract. The incongruence between the gene trees and species remains a challenge in molecular phylogenetics. In this work, we propose a novel microarray approach to resolve this problem based on our previously proposed phylogenomic mining method. In our microarray approach, we first selected 28 genes from a set of statistically significant housekeeping genes from the *S. cerevisiae* cell cycle time series microarray data. Then we employed the BLAST and synteny criteria to identify homologs and orthologys of the selected genes among the genomes of other species. Finally, we applied the phylogenomic mining method for the aligned genes to infer the species phylogeny. The phylogenetic mining method used the self-organizing map mining, hierarchical clustering and entropy measure to concatenate the phylogenomically informative genes to infer species phylogenies. Compared with the original gene concatenation approach, our method not only overcome the ad-hoc mechanism and prohibitive phylogenetic computing problem of the species inference for the large number of taxa but also first integrated the microarray techniques in the species phylogeny inference.

1 Introduction

The gene tree / species tree problem is a fundamental problem in molecular phylogenetics [1,2]. It refers that the gene trees, the phylogenies obtained from individual genes, are incongruent with the species tree. The species phylogeny reflects the true evolution history of species where all lineages are results of speciation and divergence. A gene phylogeny represents the evolutionary history of the gene for a set of organisms. The incongruence between gene trees and species trees can be observed from all levels of taxa. However, the probability of incongruence between the gene trees and species tree is high for a set of closely related species [1,2,3]. Such incongruence may originate from reasons including gene duplication/loss, lineage sorting, horizontal gene transfer and the technical factors including the artifacts of phylogenetic analysis and possible insufficient data sampling [3,4]. Resolving the incongruence among the gene trees and species trees has fundamental impacts on molecular evolution and phylogenetics.

Besides the previous phylogentic reconciliation models to resolve the gene tree and species tree problem [3], there are currently two recently proposed resolutions to deal with this problem. The first is called the *genome- tree* approach, which uses the complete genome data to build a phylogenetic tree and the final evolutionary tree is a

D.-S. Huang, K. Li, and G.W. Irwin (Eds.): ICIC 2006, LNBI 4115, pp. 485 – 493, 2006.

species tree [5,6]. However, this promising method suffers from naïve phylogenetic computing models and faces the computational hurdle in the genome tree reconstruction. Furthermore, small numbers of relevant genomes are currently available although thousands of genomes are being sequenced [5]. The other method is the genome based gene concatenation method proposed by Rokas *et al.* [7]. This method is widely accepted by biologists for the sake of its simplicity and convenience. The basic idea of their method is to select and align a set of widely distributed orthologous genes from the genomes of multiple species. Then they empirically determine the required number of genes to robustly infer phylogenetic tree by conducting phylogenetic analysis for the randomly selected genes until the final species tree is a phylogeny with the maximum support. In their experiment, they screened 106 orthologous genes from eight yeast genomes and showed the phylogenetic analysis (ML and MP) with at least 20 genes concatenation led to a species tree with the maximum support on each inferred branch [7].

Although the genome based gene concatenation method is easy to use, there are possible problems of this ad hoc approach. First, the gene concatenation number in the method has to be empirically determined for each dataset. Under the situation where large numbers of taxa are available in the species phylogeny inference, the computational burden for the phylogenetic analysis in this method will be prohibitive; Secondly, the random gene concatenation does not consider that some genes may have better phylogenetic information than others in the species phylogeny inference. On the other hand, it assumes all genes in the concatenation are equally phylogenetically informative. Thirdly, if the orthology criterion is not robust in the data collection, then genes used in phylogenetic reconstruction may have different evolutionary histories. The random gene concatenation under such a situation will bring noise signals in the phylogenetic tree reconstruction and the final tree inferred may not be robust or represent the true evolutionary relationships between organisms.

To support our concern about the genome based random gene concatenation method, we create a "worst" case of the random gene concatenation method. The 106 genes in the original dataset are classified into 45 congruent genes ("good" genes) and 61 incongruent genes ("bad" genes) under the Bayesian analysis and maximum likelihood analysis with the Shimodaira-Hasegawa (SH) test [8,9,10]. A congruent (good) / incongruent (bad) gene is a gene whose gene tree is/not the species tree after the Bayesian analysis and maximum likelihood analysis with the SH test. We only concatenate incongruent genes (bad genes) randomly and conduct the Bayesian analysis for the randomly combined bad genes. There are 10 testing trials for each combination case. We found it needed at least 28 genes under this case to reach the satisfactory result; that is, the final tree is a species tree where each inferred branch has the maximum support. On the other hand, the random concatenation experiments are also done for the congruent/good genes and random genes. We found it needed at least 4 congruent genes or at least 15 random genes in the combination to infer a species phylogeny where each inferred branch has the maximum support [11]. From this example, it is clear that different genes have unequal phylogenetic information and the implicitly equal-phylogenetic information assumption for each gene in the random gene concatenation method is improper.

In our previous work, we proposed a phylogenetic mining method to infer the species phylogeny based on the selected 106 orthologous genes to conquer the ad-hoc mechanism [12]. In this work, we propose a novel microarray based approach to infer the species phylogeny for the same problem under the framework of the phylogenomic mining. This is the first work to resolve the incongruence between gene tree and species tree problem from the point view of the microarray data. In the following sections, section 2 analyzes the characteristics of the gene expression profiles of the 106 selected genes in the *S. cerevisiae* cell cycle time series microarray data. Section 3 presents the microarray based approach to reconstruct the species phylogeny and section 4 describes the details about the phylogenomic mining method. Finally, we discuss the ongoing work and future work in the last section.

2 Gene Expressions of the Selected 106 Genes in the *S. cerevisiae* Microarray Data

According to the data collection criteria in [7], the retained 106 genes are distributed throughout the *S. cerevisiae* genome on all 16 chromosomes of the yeast genome and the synteny method is employed to build orthology before the gene alignments. We believe that the data collection approach is effective but a little bit ad-hoc since many genes can be found widely distributed on all the yeast chromosomes. Before we describe the details of our approach, we check the gene expressions of the selected 106 genes in the *S .cerevisiae* cell cycle time series microarray data.

Figure 1 shows the box plots of the gene expressions of good genes, bad genes, selected 106 genes and whole yeast genome including 6330 genes in *S. cerevisiae* cell cycle microarray data. Figure 2 shows the gene expressions (colored in white) of the 106 selected genes among the gene expressions of the whole yeast genome visualized by the *GeneSpring* 7.2. From the two figures, we can see the selected 106 gene set has the relatively "conservative" expression patterns among the whole yeast genome, that is, the variances of their gene expression data are relatively small compared with those of the whole expression data. We also visualize the gene expression patterns of the good genes and bad genes of the 106 genes. Figure 3 compares the variances of the expression data of the yeast genome, good genes and bad genes of the selected 106

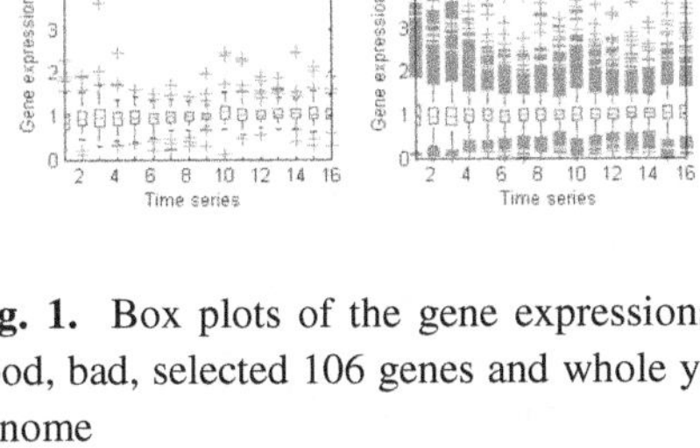

Fig. 1. Box plots of the gene expressions of good, bad, selected 106 genes and whole yeast genome

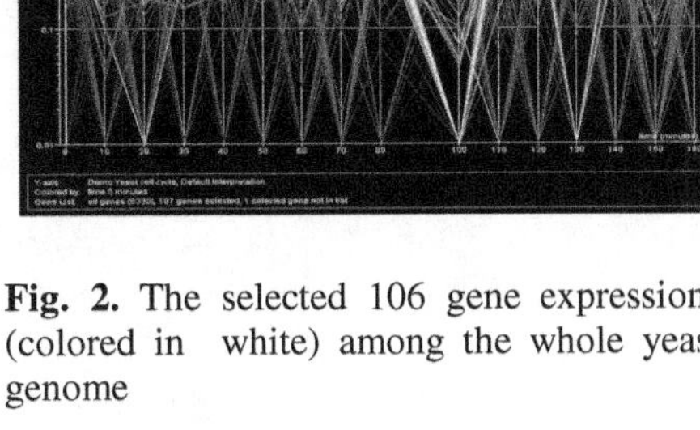

Fig. 2. The selected 106 gene expressions (colored in white) among the whole yeast genome

genes, selected 106 genes and two randomly selected 100 genes from the yeast genome. The gene expression data of the selected 106 genes have variance between 0 and 1. This variance is much lower than the variances of expression data of the yeast genome and two randomly selected gene sets! On the other hand, the expression data variances of the good genes are also lower than the expression data variances of the bad genes in the selected 106 genes. Figure 4 compares the expression data means of the yeast genome, good genes, bad genes and the selected 106 genes. It is easy to see that the selected 106 genes have means concentrated around the 1.0 expression level. The good gene expression means are "nearer" the gene expression level 1.0 than the bad gene expression means. From the two sets of results, we can tell that the gene expression levels of the selected genes are more conservative than the average gene expression level of the whole yeast genome. The good gene expression levels are more conservative than the bad gene expression levels among the selected 106 genes. The same conclusion can also be obtained by observing the box plots of the good genes, bad genes, selected 106 genes and whole yeast genome expressions in Figure 1.

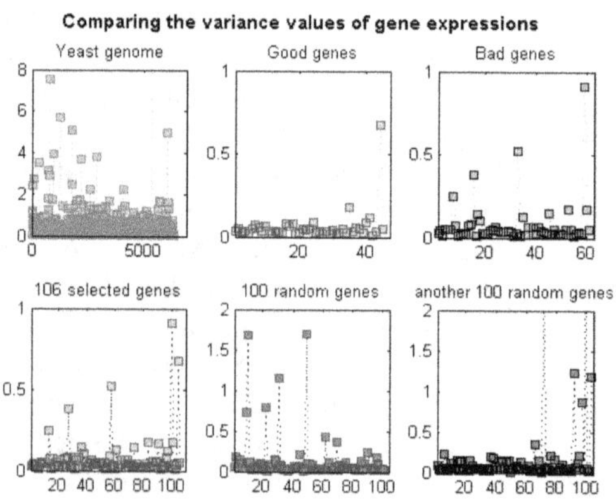

Fig. 3. Compare the gene expression variances of the different gene sets

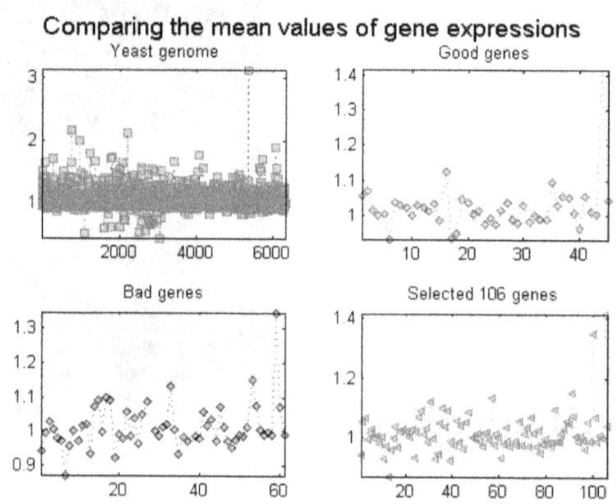

Fig. 4. Compare the gene expression means of the different gene sets

3 A Microarray Based Approach to Resolve Incongruence Between the Gene Trees and Species Tress

According to the gene expression analysis results of the congruent, incongruent, 106 selected genes and the whole yeast genome, it's natural to start from the housekeeping genes in the microarray data to collect a new dataset because their gene expressions have the least variations. We start from the *S .cerevisiae* cell cycle time series microarray data to collect housekeeping genes. The housekeeping genes can be viewed as potentially conservative genes in the yeast genome. This is similar to the widely-distributed gene finding in the original data collection [7]. Our microarray based species phylogeny inference method has the following steps.

1. Identify the statistically important house-keeping genes in the *Saccharomyces cerevisiae* cell cycle time series microarray data and select a set of genes from the statistically important house-keeping genes.
2. Use the BLAST and synteny criteria to identify homologs and orthologys for the selected genes.
3. Construct multiple alignments for the selected orthologous genes.
4. Conduct the phylogenomic mining for the aligned genes to infer the species phylogeny.

We first selected 114 statistically significant housekeeping genes from 1014 potential housekeeping genes with the lowest 20 percent variance from the *S. cerevisiae* microarray data by running the *GeneSpring* 7.2 software. We then randomly sampled 28 housekeeping genes from the 114 statistically significant housekeeping genes and blasted them to other seven genomes to build an orthologous gene set. In our sampling process, those genes which have gene loss in one species were dropped automatically. The following multiple alignments for each gene are conducted by the *ClustalW* and *MacClade* under a machine running Mac OS X system. After the alignments, the total length of the

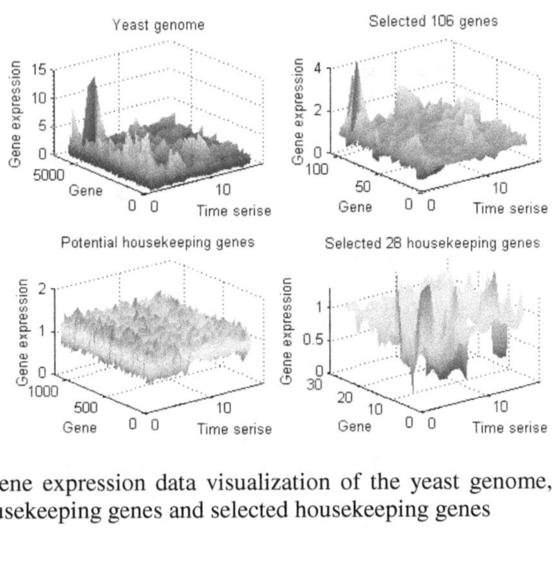

Fig. 5. The 3D gene expression data visualization of the yeast genome, selected 106 genes, 1014 potential housekeeping genes and selected housekeeping genes

28 aligned genes has 26876 nucleotide sites. In the selected 28 gene set, there are 18 congruent and 10 incongruent genes in this gene sample. The 10 incongruent genes are phylognetically informative genes according to the SH test; that is, their gene trees have closer distances to the species tree statistically. The SH-test is conducted under the GTR+Γ model with reestimated log likelihoods (RELL) approximation for 21 tree topologies for each gene with 1000 bootstrap replicates [10], Figure 5 shows the gene expression 3D visualization of the yeast genome, 106 selected genes, 1014 potential housekeeping gene sets and the sampled 28 genes from the 114 housekeeping genes. In each sub-figure, the x-axis represents the cell cycle time series, the y-axis represents genes and z-axis represents the gene expression levels.

4 Using Phylogenetic Mining to Infer Species Phylogenies

The basic idea of the phylogenomic mining is to discover the phylogenetic knowledge from a gene set G by the self-organizing map (SOM) mining, hierarchical clustering and entropy based phylogenetically informative gene finding approaches [11,12]. The gene set G is a set of aligned orthologous genes obtained from the species genome data. The knowledge includes the prototypes of the each species, clustering of genes and the predicted phylogenetically informative genes. After the phylogenetic knowledge discovery, the predicted phylogenetically informative genes are concatenated to be a "super-gene" to conduct the phylogenetic analysis to infer the species phylogenies. A super-gene is the concatenation of phylogenetically informative genes. Figure 6 shows the flowchart of the phylogenetic mining method. It can be sketched as following four steps.

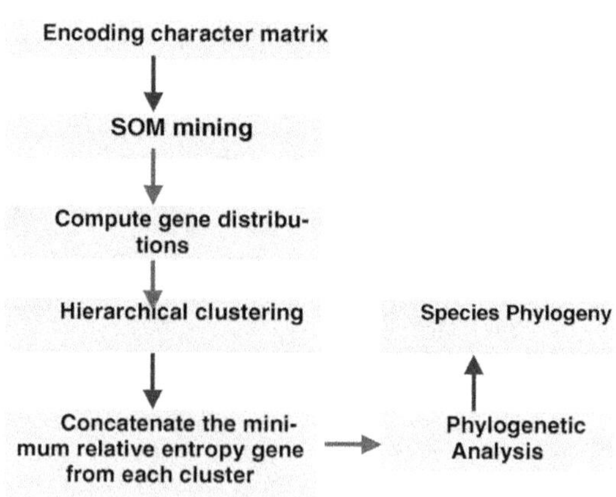

Fig. 6. The phylogenetic mining method to infer species phylogenies

The first step of the phylogeneic mining method is to encode the all the aligned gene data to the numeric data for the convenience of the knowledge discovery. In the data encoding, four nucleotides are encoded as the following orthogonal bases respectively and the missing nucleotides are encoded as a zero vector.

$$A = (1,0,0,0)^t,\ T = (0,1,0,0)^t$$
$$C = (0,0,1,0)^t,\ G = (0,0,0,1)^t \quad (1)$$

After the encoding, the 28 selected statistically important housekeeping genes are converted to a 26876×32 numeric matrix where each gene takes certain number of rows in the matrix according to its length.

The second step is to cluster gene sequences by the SOM mining. The SOM based clustering consists of two stages. The first stage is use the SOM mining to obtain the prototypes of the whole gene set and the gene distributions of all genes on the SOM plane [11,12]. The gene distribution of a gene on the SOM plane is an approximation of its probability mass function in the sequence space [12]. In our SOM mining, we implement the *conscience* learning mechanism [13,14] to guarantee that there is a faithful representation of the underlying distribution of the input gene data set. That is, the gene distribution on the SOM plane is a faithful approximation of the probability mass function of a gene in the sequence space. The second stage is to cluster all the genes by hierarchically clustering their gene distributions on the SOM plane. The gene distribution on the SOM plane for each gene is computed by retracing the projection of each gene site on the SOM plane [12].

It is proved that the complexity of the SOM learning scales linearly with the number of data sample m if the number of map units k is chosen to be proportional to the $\sqrt{m}$ [15]. According to this result, in our SOM mining, the SOM takes an input/output plane with $k = 169\,(13\times13)$ neurons for the 26876×32 input data matrix. Our SOM mining employs the *conscience* learning, which is a sequence learning algorithm. The sequence learning is not prone to trap in the local minimum in the learning than the batch map learning, although the latter is generally faster than the former [13,14]. The reference vectors associative with neurons on the SOM plane are initialized by the principal component analysis (PCA) instead of the random initialization for the sake of fast convergence. The neighborhood kernel function used in the SOM training is the Gaussian function. In the following hierarchical clustering, we decide the cutoff of the clustering by computing the U-matrix on the SOM plane [12,14]. For this 28 housekeeping dataset, we have 7 identified clusters in the hierarchical clustering (Figure 7).

The third step in the phylogenomic mining is to identify the phylogenetically informative genes in the gene set. It also consists of two steps. The first step is to identify the most conservative gene by finding the minimum entropy gene in the gene set. The second step is to find the genes with mostly similar sequence patterns with respect to the most conservative gene. This is equivalent to computing the relative entropy, also called the K-L distance, for all the genes with respect to the most conservative gene. The minimum entropy gene among the 28 housekeeping genes is the gene *YPR165W*. The five genes with the nearest K-L distances with the minimum entropy gene are {*YIL142, YOR259C, YPL117C, YLR243W, YIR002C*}. Among the five minimum relative entropy genes with respect to the minimum entropy gene, the gene *YLR243W* is a phylogenetically informative gene according to the SH-test, although it is not convergent under the Bayesian analysis and other four genes are convergent genes.

The fourth step is to build a "super-gene" by concatenating the phylogenetically informative genes and use the "super-gene" to infer the species phylogeny. The "super-gene" can be constructed by concatenating genes with the nearest K-L distances in the gene set or by picking the minimum relative entropy gene from each cluster and combining them together. We already proved that the genes with nearer K-L distance to the minimum entropy gene were more likely to be phylogenetically informative by the SH-test. We concatenated the seven minimum relative entropy genes including the gene *YER165W* itself. The final tree from the Bayesian and ML analysis was a species tree with the maximum support on each branch (Figure 8). We also built the super-gene by concatenating the seven minimum relative entropy genes from each cluster. Similarly, the phylogenetic tree reconstructed from the seven minimum relative entropy gene selected from each cluster was also the species tree with the maximum support on each inferred branch. We also changed the cutoff in the hierarchical clustering for the number of clusters to 6,5,4 and built super-genes by concatenating the minimum relative entropy from each cluster. We obtained the same species tree with the maximum supports in the phylogenetic reconstruction! Figure 8 shows the species tree with maximum support in the phylogenomic mining method.

Compared with the original random gene concatenation approach [7], our method achieves even better results than the just good gene combination case! The microarray based approach not only overcomes the ad-hoc mechanism in the original method but also provides a systematic resolution to the species tree reconstruction problem from the initial data collection to phylogenetically informative gene selection before the phylogenetic analysis. Our method is easy to extend to other dataset and can avoid the prohibitive phylogenetic computing problem of the species phylogeny inference of the large number of taxa because of the phylogenetically informative gene selection mechanism.

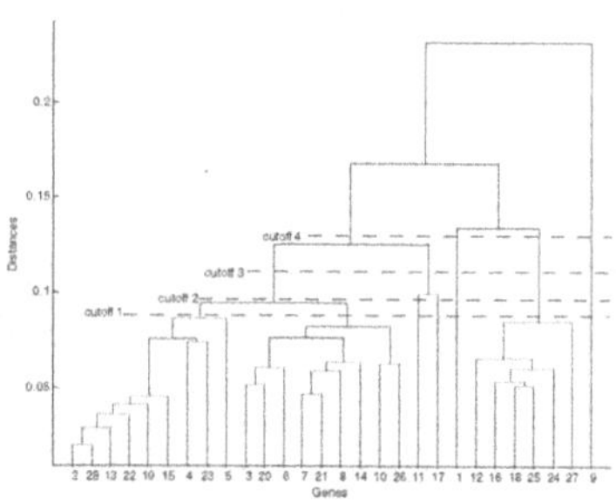

Fig. 7. Hierarchical clustering of 28 housekeeping genes

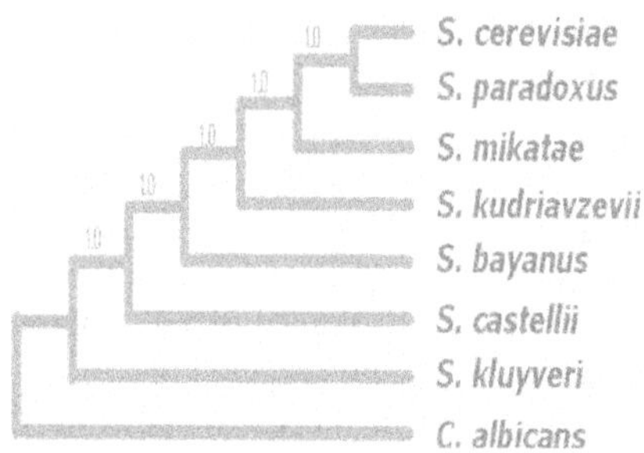

Fig. 8. The species tree with the maximum support

5 Conclusions

In this work, we showed our microarray based species tree inference approach was a robust and systematic approach to infer species phylogenies than the random gene concatenation method under a same problem [7]. In the on-going work, we are integrating few more microarray datasets other than the current microarray data to select statistically important housekeeping genes for the phylogenomic mining. In the future work, we are going to compare the clustering results between the SOM based clustering in the phylogenomic mining and the SOM clustering of microarray data themselves for the same set of genes. The SOM based clustering is a sequence based clustering and the latter SOM clustering is expression data based clustering. The comparison of two clustering results can give make the gene clustering analysis more representative and robust.

References

1. Nei, M., Kumar, S.: Molecular Evolution and Phylogenetics 2nd Edition. Oxford University Press (2000)
2. Page, R., Charleston, M.: From Gene to Organismal Phylogeny: Reconciled Trees and the GeneTree/Species Tree Problem. Molecular Phylogenetics and Evolution, Vol. 7 (1997) 231-240
3. Ma, B., Li, M., Zhang, L.: From Gene Trees to Species Trees. SIAM Journal on Computing, 30. (2000) 729-752
4. Huelsenbeck, J. P.: Performance of Phylogenetic Methods in Simulation. Systematic Biology, 44 (1995) 17–48
5. Delsuc F, Brinkmann H, Philippe H.: Phylogenomics and the Reconstruction of the Tree of Life. Nature Review Genetics, 6 (2005) 361-75.
6. Moret, B., Tang, J., Warnow., T.: Reconstructing Phylogenies from Gene-Content and Gene-order Data. In Mathematics of Evolution and Phylogeny, Oxford Univ. Press (2005) 321-352
7. Rokas, A., Williams, B., King, N., Carroll, S.: Genome-scale Approaches to Resolving Incongruence in Molecular Phylogenies, Nature. 425 (2003) 798-804
8. Huelsenbeck, J., Ronquist, F.: MRBAYES: Bayesian Inference of Phylogenetic Trees. Bioinformatics. 17 (2001) 754-755
9. Felsentein, J.: Inferring Phylogenies, Sinauer Associates, Inc. (2004)
10. Shimodaira, H., Hasegawa, M.: Multiple Comparisons of Log-likelihoods with Applications to Phylogenetic Inference. Molecular Biology and Evolution. 16 (1999) 1114-1116
11. Han, X.: Resolve the Gene Tree and Species Tree Problem by Phylogenetic Mining. Proceedings of 4th Asia-Pacific Bioinformatics Conference (APBC), Imperial College Press, London (2006) 287-296
12. Han, X.: A Machine Learning Approach to Resolving Incongruence in Molecular Phylogenies and Visualization Analysis. Proceedings of the 2005 IEEE Symposium on Computational Intelligence in Bioinformatics and Computational Biology (2005) 346-353
13. Haykin, S.: Neural Networks: A Comprehensive Foundation, 2nd Edition. Prentice-Hall. (1999)
14. Kohonen, T.: Self-Organizing Maps, 3rd edition. Berlin: Springer-Verlag. (2001)
15. Nikkila, J., Toronen,, P., Kaski, S., Venna, J., Castren, E., Wong, G.: Analysis and Visualization of Gene Expression Data using Self-organizing Maps. Neural Networks, 15. Special issue on New Developments on Self-Organizing Maps (2002) 9530-9636

Penalized Independent Component Discriminant Method for Tumor Classification

Chun-Hou Zheng[1], Li Shang[2], Yan Chen[1], and Zhi-Kai Huang[2]

[1] School of Information and Communication Technology, Qufu Normal University, Rizhao, Shandong, 276826, China
[2] Department of Automation, University of Science and Technology of China, Hefei, Anhui, China
zhch1972@eyou.com

Abstract. This paper proposes a new method for tumor classification using gene expression data. In this method, we first employ independent component analysis (ICA) to model the gene expression data, then apply optimal scoring algorithm to classify them. To show the validity of the proposed method, we apply it to classify two DNA microarray data sets involving various human normal and tumor tissue samples. The experimental results show that the method is efficient and feasible.

1 Introduction

A reliable and precise classification of tumors is essential for successful diagnosis and treatment of cancer. Current methods for classifying human malignancies are mostly rely on a variety of morphological, clinical, and molecular variables. Despite the progress recently made, there are still many uncertainties in diagnosis.

With the wealth of gene expression data from microarrays being been produced, more and more new prediction, classification and clustering techniques are being used for analysis of the data. Up to now, several studies have been reported on the application of microarray gene expression data analysis for molecular classification of cancer [1,2,4,6,18,19,20].

One feature of microarray data is that the number of tumor samples collected tends to be much smaller than the number of genes. The number for the former tends to be on the order of tens or hundreds, while microarray data typically contain thousands of genes on each chip. In statistical terms, it is called “large p, small n” problem [17], i.e., the number of predictor variables is much larger than the number of samples. To overcome this difficult, Ghosh [5] proposed a methodology using regularized regression models for the classification of tumors. In this literature, he focused on three types of regularized regression models, i.e., ridge regression, principal components regression and partial least squares regression. One drawback of these techniques is that only second-order statistical information of the gene data is used. However, in the task such as classification, much of the important information may be contained in the high-order relationships among samples. And thus, it is important to investigate whether or not

D.-S. Huang, K. Li, and G.W. Irwin (Eds.): ICIC 2006, LNBI 4115, pp. 494 – 503, 2006.

the generalizations of PCA, which are sensitive to high-order relationships (not just second-order relationships), are advantageous. Usually, independent component analysis (ICA) [2,15] is one of such generalizations. In this paper, we shall employ FastICA, which was proposed by Hyvärinen [10] and proven successful in many applications, to address the problems of tumor classification.

2 Methods

2.1 Independent Component Analysis

Considering an $n \times p$ data matrix $\mathbf{X}$, whose rows $\mathbf{r}_i$ ($i = 1, \cdots, n$) correspond to observational variables and whose columns $\mathbf{c}_j$ ($j = 1, \cdots, p$) are the individuals of the corresponding variables, the ICA model of $\mathbf{X}$ can be written as:

$$\mathbf{X} = \mathbf{AS} \tag{1}$$

Without loss of generality, $\mathbf{A}$ is an $n \times n$ mixing matrix, and $\mathbf{S}$ is an $n \times p$ source matrix subject to the condition that the rows of $\mathbf{S}$ are as statistically independent as possible. Those new variables contained in the rows of $\mathbf{S}$ are called as 'independent components', i.e., the observational variables are linear mixtures of independent components. The statistical independence between variables can be quantified by mutual information $I = \sum_k H(s_k) - H(S)$, where $H(s_k)$ is the marginal entropy of the variable s_k, and $H(S)$ is the joint entropy. Estimating the independent components can be accomplished by finding the right linear combinations of the observational variables, since we can invert the mixing matrix as:

$$\mathbf{U} = \mathbf{S} = \mathbf{A}^{-1}\mathbf{X} = \mathbf{WX} \tag{2}$$

There are a number of algorithms for performing ICA. In this paper, we shall employ the FastICA algorithm, which was proposed by Hyvärinen [10], to address the problems of tumor classification. In this algorithm, the mutual information is approximated by a 'contrast function':

$$J(s_k) = (E\{G(s_k)\} - E\{G(v)\})^2 \tag{3}$$

where G is an any nonquadratic function and v is a normally distributed variable. The interested readers can refer to literature [10] for further details.

Like PCA, ICA can remove all linear correlations. By introducing a non-orthogonal basis, it also takes into account higher-order dependencies in the data. Particularly, ICA is in a sense superior to PCA, which are just sensitive to second-order relationships of the data. And, the ICA model usually leaves some freedom of scaling and sorting by convention, the independent components are generally scaled to unit deviation, while their signs and orders can be chosen arbitrarily.

2.2 ICA Models of Gene Expression Data

Now let the $n \times p$ matrix $\mathbf{X}$ denote the gene expression data (generally speaking, n<<p), x_{ij} is the expression level of the *j*th gene in the *i*th assay. $\mathbf{r}_i$ (a *p*-dimensional vector), the *i*th row of $\mathbf{X}$, denotes the snapshot of the *i*th assay (cell sample)[1]. Alternatively, $\mathbf{c}_j$ (an *n*-dimensional vector), the *j*th column of $\mathbf{X}$, is the expression profile of the *j*th gene. We suppose that the data have already been preprocessed and normalized, i.e., every sample has mean zero and standard deviation one.

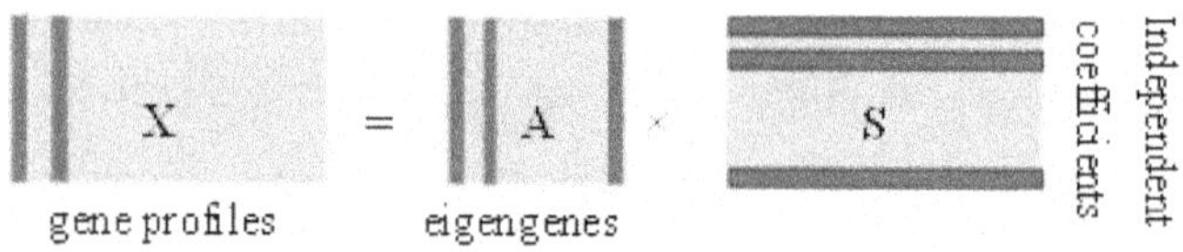

Fig. 1. The second gene expression data synthesis model. Each gene profile in the data matrix was considered to be a linear combination of underlying basis expression profiles (eigengenes) in the matrix $\mathbf{A}$ (the columns in $\mathbf{A}$). Each of the basis expression profiles was associated with a set of independent "causes (coefficient)", given by a vector of coefficients in S. The basis profiles were estimated by $\mathbf{A} = \mathbf{W}^{-1}$, where $\mathbf{W}$ is the learned ICA weight matrix.

Regardless of which algorithm is used to compute ICA, we can apply ICA to model gene expression data as shown in Fig 1. In this model, the gene expression profiles (columns of $\mathbf{X}$) can be regarded as points in a multidimensional space with dimensions corresponding to the number of samples. The linear ICA model $\mathbf{X} = \mathbf{AS}$ represents the gene expression profiles (the columns of $\mathbf{X}$) by a new set of basis vectors (the columns of $\mathbf{A}$, Fig. 1). This idea is based on the assumptions that, first, the gene expression profiles are determined by a combination of hidden regulatory variables, which were called 'expression modes'. Second, the genes' responses to these variables can be approximated by linear functions [9,11]. Expression mode *k* is characterized by its profile over the samples (*k*th column of $\mathbf{A}$) and by its linear influences on the genes (kth row of $\mathbf{S}$).

2.3 Penalized Regression Models

In this section, we briefly outline two types of regularized regression models, i.e., ridge regression and principal component regression.

2.3.1 Ridge Regression

Consider the standard regression model:

$$\mathbf{y} = \mathbf{X}\boldsymbol{\beta} + \boldsymbol{\varepsilon} \tag{4}$$

[1] In the gene data literature, the problem is usually formulated using the transposed matrix $\mathbf{X}^T$.

where $\mathbf{y}$ is an n-dimensional response vector; $\mathbf{X}$ is an $n \times p$ predictor matrix; $\boldsymbol{\beta}$ is a p-dimensional vector of unknown regression parameters; $\boldsymbol{\varepsilon}$ is a random vector with zero mean and one variance. In this paper, because n is smaller than p, the usual ordinary least squares estimator will not be well-defined. An alternative is to use the ridge regression estimator of β in equation (4):

$$\hat{\beta} = (\mathbf{X}^T\mathbf{X} + \lambda\mathbf{I})^{-1}\mathbf{X}^T\mathbf{y} \tag{5}$$

where $\mathbf{I}$ is a $p \times p$ identity matrix, and λ a constant. The parameter λ controls the amount of shrinkage in the data.

2.3.2 Principal Component Regression

The method of principal components regression can be traced back to the literature [12]. To use this method, we first perform a singular value decomposition (SVD) of the gene data $\mathbf{X}$:

$$\mathbf{X} = \mathbf{UDV}^T \tag{6}$$

where $\mathbf{U}$ is the $n\times n$ singular value decomposition matrix, and it has both orthonormal rows and columns. The diagonal matrix $\mathbf{D}$ contains the ordered eigenvalues of $\mathbf{XX}^T$ on the diagonal elements so that $\mathbf{D} = \text{diag}(d_1, d_{2,} \cdots, d_n)$, where $d_1 > d_2 > \cdots > d_n > 0$. $\mathbf{V}$ is a $p \times n$ matrix with orthonormal columns. Plugging this decomposition into equation (4), we have:

$$\mathbf{y} = \mathbf{X}\boldsymbol{\beta} + \boldsymbol{\varepsilon} = \mathbf{UDV}^T\boldsymbol{\beta} + \boldsymbol{\varepsilon} = \mathbf{H}\boldsymbol{\gamma} + \boldsymbol{\varepsilon} \tag{7}$$

where $\mathbf{H} = \mathbf{UD}$ and $\boldsymbol{\gamma} = \mathbf{V}^{\mathrm{T}}\boldsymbol{\beta}$. We can fit the model in equation (4) using ordinary least squares and get an estimate of $\boldsymbol{\beta}$ by multiplying $\mathbf{V}$ to the least squares estimator of $\boldsymbol{\gamma}$ in equation (7).

2.4 Optimal Ccoring

In the previous section, we have described two penalized regression models that have been used successfully in other applications such as in chemometrics. However, in this paper, the goal for our interest is classification. Thus, we should firstly re-express the classification problem as a regression problem. This is done using the optimal scoring algorithm. The point of optimal scoring is to turn categorical variables into quantitative ones by assigning scores to classes (categories).

Let g_i denote the tumor class for the ith sample ($i = 1, \cdots, n$), we assume that there are G tumor classes so that g_i takes values $\{1, \cdots G\}$. We first convert $\mathbf{g} = [g_1, \cdots g_n]^T$ into an $n \times G$ matrix $\mathbf{Y} = [Y_{ij}]$, where $Y_{ij} = 1$ if the ith sample falls into class j, and 0 otherwise. Let $\theta_k(\mathbf{g}) = [\theta_k(g_1), \cdots \theta_k(g_n)]^T$ ($k = 1, \cdots, G$) be the $n \times 1$ vector of quantitative scores assigned to $\mathbf{g}$ for the kth class. The optimal scoring problem involves finding the coefficients β_k and the scoring maps θ_k that minimize the following average squared residual:

$$ASR = \frac{1}{n}\sum_{k=1}^{G}\sum_{i=1}^{n}(\theta_k(g_i) - \mathbf{X}_i^T\beta_k)^2 \tag{8}$$

Let $\Theta_{G\times J}$ $(J \le G-1)$ be a matrix of J score vectors for the G classes, i.e., its kth row is the scores, $\theta_k(\bullet)$. Assume that the minimization of equation (12) is subject to the constraint $n^{-1}\|\mathbf{Y\Theta}\|^2 = 1$, then, as mentioned by [8], the minimization of this constrained optimization problem leads to the estimates of β_k that are proportional to the discriminant variables in linear discriminant analysis. The interested readers can refer to the literatures[7,8].

2.5 Penalized Optimal Scoring for Classification

So far, we have outlined the components necessary for the implementation of our procedure. In this section, we give out our algorithm for classifying the tumor samples.

We propose to use ICA for regularizing the gene expression data and then use a penalized optimal scoring procedure for classification. The outline of our method is shown as follows:

Step 1: Using the ICA model $\mathbf{X} = \mathbf{AS}$ to present the gene expression data, i.e., using ICA and consensus sources algorithm to calculate the eigengenes (columns of $\mathbf{A}$) and the independent coefficients (rows of $\mathbf{S}$).

Step 2: Choose an initial score matrix $\Theta_{G\times J}$ with $J \le G-1$ satisfying $\Theta^T D_p \Theta = I$, where $D_p = \mathbf{Y}^T\mathbf{Y}/n$. Let $\Theta_0 = \mathbf{Y}\Theta$.

Step 3: Fit a multivariate penalized regression model of Θ_0 on $\mathbf{A}$, yielding the fitted values $\hat{\Theta}_0$ and the fitted regression function $\hat{\eta}_0(\mathbf{A})$. Let $\hat{\eta}(\mathbf{X}) = S^{+}\hat{\eta}_0(\mathbf{A})$ be the vector of the fitted regression function on $\mathbf{X}$, where S^{+} is the pseudoinverse of $\mathbf{S}$.

Step 4: Obtain the eigenvector matrix $\mathbf{\Phi}$ of $\Theta_0^T\hat{\Theta}_0$, and hence the optimal scores $\Theta_1 = \Theta\mathbf{\Phi}$.

Step 5: Let $\eta(\mathbf{X}) = \mathbf{\Phi}^T\hat{\eta}(\mathbf{X})$.

What should be explained is that the objective function we are minimizing in *Step* 3 is the following expression:

$$ASR = \frac{1}{n}\sum_{k=1}^{G}\sum_{i=1}^{n}(\theta_k(g_i) - \mathbf{A}_i^T\gamma_k)^2 \tag{9}$$

where $\mathbf{A}_i$ is the ith column of $\mathbf{A}$, $\gamma_k = \mathbf{S}\beta_k$. Another problem is how to choose the initial values for Θ. Readers can refer to the discussions about this problem in literature [8]. In fact, our algorithm is somewhat similar to the algorithm proposed by Ghosh [5], except that we replace principal components with independent components.

Once the algorithm has been run, we now have a discriminant rule for classifying new samples. We use the nearest centroid rule to form the classifier, i.e., assign a new sample $\mathbf{X}_{new}$ to the class j that minimizes:

$$\delta(\mathbf{X}_{new}, j) = \left\| \mathbf{D}(\eta(\mathbf{X}_{new}) - \bar{\eta}^{j} \right\|^{2} \tag{10}$$

where $\bar{\eta}^{j} = \frac{\sum_{g_i=j} \eta(\mathbf{X}_i)}{n_j}$ denotes the fitted centroid of the jth class. $\mathbf{D}$ is a matrix with diagonal element:

$$D_{kk} = \left(\frac{1}{\lambda_k^2(1-\lambda_k^2)}\right)^{1/2} \tag{11}$$

where λ_k is the kth largest eigenvalue calculated in *Step* 4 of the algorithm.

3 Results

In this section, we shall demonstrate the efficiency and effectiveness of the proposed methodology described above by classifying two data sets with various human tumor samples.

3.1 Datasets

In this study, two publicly available microarray datasets are used to study the tumor classification problem. They are colon cancer data [1], and High-grade glioma data [13], respectively. In these datasets, all data samples have already been assigned to a training set or test set.

An overview of the characteristics of all the datasets can be found in Table 1. Preprocessing of this dataset was done by setting threshold and log-transforming on the original data, similar as in the original publication. Threshold technique is generally achieved by restricting gene expression levels to be larger than 20. In other words, the expression levels which are smaller than 20 will be set to 20. Regarding the log-transformation, the natural logarithm of the expression levels usually is taken. In addition, no further preprocessing is applied to the rest of the datasets.

Table 1. Summary of the datasets for thetwo binary cancer classification problems. (Explanation of the abbreviations used in Table 1: D, datasets; TR, training set; TE, test set; C1, class 1; C2, class 2; Levels, the number of genes; M, microarray technology; T1, oligonucleotide; 1. Colon cancer data; 2. High-grade glioma data).

D	TR		TE		Levels	M
	C1	C2	C1	C2		
1	14	26	8	14	2000	T1
2	21	14	14	15	12625	T1

3.2 Experiments Results

We now use the proposed methodology to classify the tumor data. Since all data samples in these four datasets have already been assigned to a training set or test set, we built the classification models using the training samples, and estimated the classification correct rates using the test set.

Table 2. The summary of the results of the numerical experiments on four datasets

Experiments		Colon data	
No.	Methods	Training set	Test set
1	LS-SVM linear kernel	99.64±0.87	82.03±7.49
2	LS-SVM RBF kernel	98.33±2.36	81.39±9.19
3	LS-SVM linear kernel (no regularization)	49.40±8.93	51.73±12.19
4	PCA + FDA (unsupervised PC selection)	90.95±5.32	80.30±9.65
5	PCA + FDA (supervised PC selection)	95.24±5.56	76.84±7.41
6	kPCA lin + FDA (unsupervised PC selection)	90.95±5.32	80.30±9.65
7	kPCA lin + FDA (supervised PC selection)	95.24±5.56	76.84±7.41
8	kPCA RBF + FDA (unsupervised PC selection)	87.86±11.24	75.11±15.02
9	kPCA RBF + FDA (supervised PC selection)	100.00±0.00	64.07±1.94
10	P-RR	97.75±2.36	83.88±6.53
11	P-PCR	91.25±2.02	85.54±4.45
12	P-ICR	93.63±2.05	85.95±5.16
13	PAM	91.50±4.29	83.63±5.82

Because the gene numbers of the last three data sets are so great that our computer (CPU: 2 GHz Pentium Ⅳ) can not process them using ridge regression model, we have not classified Glioma data using ridge regression model. On the contrary, ICA and PCA have the ability of compressing the gene expression data (as shown in Method Section), so we can use principal component regression or independent component regression to deal with large scale data.

To obtain reliable experimental results showing comparability and repeatability for different numerical experiments, this study not only uses the original division of each data set in training and test set, but also reshuffles all datasets randomly. In other words, all numerical experiments were performed with 20 random splitting of the two original datasets. And, they are also stratified, which means that each randomized training and test set contains the same amount of samples of each class compared to the original training and test set.

We used penalized independent component regression (P-ICR) proposed in this paper to analyze the four gene expression data sets. For comparison, we also used penalized ridge regression (P-RR), penalized principal component regression (P-PCR) proposed in[5], and PAM [16] to do the same tumor classification experiment.

Table 2. The summary of the results of the numerical experiments on four datasets (*continued*)

Experiments		Glioma data	
No.	Methods	Training set	Test set
1	LS-SVM linear kernel	90.02±14.16	61.25±11.75
2	LS-SVM RBF kernel	98.41±7.10	69.95±8.59
3	LS-SVM linear kernel (no regularization)	50.79±12.75	48.93±10.88
4	PCA + FDA (unsupervised PC selection)	92.29±7.12	68.72±7.24
5	PCA + FDA (supervised PC selection)	92.97±10.14	65.52±11.01
6	kPCA lin + FDA (unsupervised PC selection)	92.52±6.98	68.31±6.78
7	kPCA lin + FDA (supervised PC selection)	95.24±8.57	67.32±11.04
8	kPCA RBF + FDA (unsupervised PC selection)	94.78±9.05	64.20±11.19
9	kPCA RBF + FDA (supervised PC selection)	96.15±7.29	58.13±12.24
10	P-RR		
11	P-PCR	93.33±8.16	70.35±8.19
12	P-ICR	91.10±9.87	74.30±7.30
13	PAM	98.57±2.17	67.24±6.58

The classification results for tumor and normal tissues using our proposed penalized methods are displayed in Table 2. For each classification problem, the experimental results gave the statistical means and standard deviations of accuracy on the original data set and 20 randomizations as described above. Since the random splits for training and test set are disjoint, the results given in Table 2 are unbiased and can in general also be too optimistic.

To show the efficiency and feasibility of the method proposed in this paper, the results using other 9 methods (Methods 1-9) are also listed in Table 2 for comparison. These 9 methods can be subdivided in two steps: dimensionality reduction and classification. For dimensionality reduction, classical PCA as well as kernel PCA (with linear or RBF kernel) are used. Fisher discriminant analysis (FDA) and least squares support vector machine (LS-SVM) are then used for classification. Note that these methods and results were ever reported in literature[14], where the divisional method of each training and test data set is the same as ours. Readers can see the details about the first 9 methods from literature [14].

From Table 2 depicted above we can see that our proposed method is indeed efficient and feasible.

4 Conclusions

In this paper, we presented independent component analysis methods for the classification of tumors based on microarray gene expression data. The methodology involves regularizing gene expression data using ICA, followed by the classification applying penalized discriminant method. We have compared the experimental results of our method with other 12 methods, which show that our method is effective and efficient in predicting normal and tumor samples from four human tissues. Furthermore, these results hold under re-randomization of the samples.

In future works, we will in deep study the ICA model of gene expression data, how to apply the method proposed in this paper to solving multiclass problems of tumor classification, and also study how to make full use of the information contained in the gene data to restrict ICA models so that more exact prediction of tumor class can be achieved.

References

1. Alon,U., Barkai,N., Notterman,D.A., Gish,K., Ybarra,S., Mack,D., Levine,A.J.: Broad Patterns of Gene Expression Revealed by Clustering Analysis of Tumor and Normal Colon Tissues Probed by Oligonucleotide Arrays. Proc. Natl Acad. Sci. USA, 96 (1999) 6745–6750
2. Bartlett,M.S., Movellon,J.R., Sejnowski,T.J.: Face Recognition by Independent Component Analysis. IEEE Trans. Neural Netw, 13 (2002) 1450–1464
3. Bittner,M., Meltzer,P., Chen,Y., Jiang,Y., Seftor,E., Hendrix,M., Radmacher,M., Simon,R., Yakhini,Z., Ben-Dor,A. et al.: Molecular Classification of Cutaneous Malignant Melanoma by Gene Expression Profiling. Nature, 406 (2000) 536–540
4. Furey,T.S., Cristianini,N., Duffy,N., Bednarski,D.W., Schummer,M.: Haussler,D. Support Vector Machines Classification and Validation of Cancer Tissue Samples Using Microarray Expression Data. Bioinformatics, 16 (2000) 906–914
5. Ghosh,D.: Penalized Discriminant Methods for the Classification of Tumors from Microarray Experiments. Biometrics, 59 (2003) 992-1000
6. Golub,T.R., Slonim,D.K., Tamayo,P., Huard,C., Gaasenbeek,M., Mesirov,J.P., Coller,H., Loh,M.L., Downing,J.R., Caligiuri,M.A., Bloomfield,C.D., Lander,E.S.: Molecular Classification of Cancer: Class Discovery and Class Prediction by Gene Expression Monitoring. Science, 286 (1999) 531–537
7. Hastie,T., Buja,A., Tibshirani,R.: Penalized Discriminant Analysis by Optimal Scoring. Annals of Statistics, 23 (1995) 73-102
8. Hastie,T., Tibshirani,R., Buja,A.: Flexible Discriminant Analysis by Optimal Scoring. Journal of the American Statistical Association, 89 (1994) 1255-1270
9. Hori,G., Inoue,M., Nishimura,S., Nakahara,H.: Blind Gene Classification Based on ICA of Microarray Data. Proc. 3rd Int. Workshop on Independent Component Analysis and Blind Signal Separation (ICA2001), (2001) 332-336
10. Hyvärinen,A.: Fast and Robust Fixed-point Algorithms for Independent Component Analysis. IEEE Trans. Neural Netw, 10 (1999) 626–634
11. Liebermeister,W.: Linear modes of gene expression determined by independent component analysis. Bioinformatics, 18, (2002) 51-60
12. Massy,W.F.: Principal Components Regression in Exploratory Statistical Research. Journal of the American Statistical Association, 60 (1965) 234-246
13. Nutt,C.L., Mani,D.R., Betensky,R.A., Tamayo,P., Cairncross,J.G., Ladd,C., Pohl,U., Hartmann,C., McLaughlin,M.E. et al.: Gene Expression-based Classification of Malignant Gliomas Correlates Better with Survival Than Histological Classification. Cancer Res., 63 (2003)1602–1607
14. Pochet,N., De Smet,F., Suykens,J.A.K., De Moor,B.L.R.: Systematic Benchmarking of Microarray Data Classification: Assessing the Role of Non-linearity and Dimensionality Reduction. Bioinformatics, 20 (2004) 3185-3195
15. Teschendorff,A.E, Wang,Y.Z., Barbosa-Morais,N.L., Brenton,J.D., Caldas,C.: A Variational Bayesian Mixture Modelling Framework for Cluster Analysis of Gene-Expression Data. Bioinformatics, 21 (2005) 3025-3033

16. Tibshirani,R., Hastie,T., Narasimhan,B., Chu,G.: Diagnosis of Multiple Cancer Types by Shrunken Centroids of Gene Expression. Proc. Natl. Acad. Sci. USA, 99 (2002) 6567-6572
17. West,M.: Bayesian Factor Regression Models in the 'Large p, Small n' Paradigm. Bayesian Statistics, 7 (2003)723-732
18. Wang, H.Q., Huang, D.S.: Non-linear Cancer Classification Using a Modified Radial Basis Function Classification Algorithm. Journal of Biomedical Science, 12 (2005)819-826
19. Wang, H.Q., Huang, D.S.: Regulation Probability Method for Gene Selection. Pattern Recognition Letters, 27 (2006) 116-122
20. Wang, H.Q., Huang, D.S., Wang, B.: Optimization of Radial Basis Function Classifiers Using Simulated Annealing Algorithm for Cancer Classification. IEE Electronics Letters, 41 (2005) 630-632

Practical Linear Space Algorithms for Computing String-Edit Distances

Tony Y.T. Chan

The University of Akureyri, Solborg,
Akureyri 600, Iceland
tonychanyt@gmail.com

Abstract. String-edit operations consist of insertion of a symbol, deletion of a symbol, and substituting one symbol with another. String-edit distances have been applied in problems of error correction and pattern recognition. In this paper, two practical algorithms for computing the edit distance between two strings are presented. The space complexity for the first is $m+n+O(1)$, where m and n are the lengths of the input strings. The second requires only $\min(m,n)+O(1)$.

Keywords: String-edit distance, time and space complexities, algorithms.

1 Introduction

The string-editing problem that we are dealing with here is well-known in the fields of pattern recognition [1–4], file comparison [5], spelling correction and genome study [6]. A lower bound on time-complexity was established [7]. We are given a finite set of symbols, called the alphabet, and a finite set of edit operations. Each operation is the insertion, deletion, or substitution of one symbol and each operation has a cost which is a non-negative real number. Then, given two strings generated from the alphabet, we want to find the minimum total cost of a sequence of operations to transform one string into the other by these edit operations. This minimum total cost is known as the distance between the two input strings.

Using dynamic programming approach, Wagner and Fischer [8] described Algorithm A (see below) to find the edit distance in $O(mn)$ space as well as time, where m and n are the numbers of symbols in the first and the second input string respectively. In this paper, two practical linear-space algorithms are presented to find the distance between two strings. Algorithm B runs in $O(m+n)$ space and $O(mn)$ time but in actual CPU time units, it runs faster than Algorithm A. Algorithm C requires only $O(\min(m,n))$ space, not counting the storage locations for the input strings. It runs in $O(mn)$ time and in practice, it runs slightly faster than Algorithm A.

Edit distances has been applied to spelling correction software to identify a misspelled word's closest neighbors. For example, assuming all edit operations have the same cost of 1, the distance between the strings *December* and *ecema* is 4. It costs

D.-S. Huang, K. Li, and G.W. Irwin (Eds.): ICIC 2006, LNBI 4115, pp. 504–513, 2006.

1 to delete the *D*, 1 to change the *b* to *a*, and 2 to delete *er*. The total cost is 4 to change *December* to *ecema*.

In bioinformatics, a gene sequence is often coded simply as a string from the alphabet {A,T,G,C} of the 4 standard nucleotide bases. Human genes vary in size from a few hundred bases to more than 2 million bases. In this case, the saving in space from quadratic to linear would be tremendous. The algorithms presented here can be used to find homologous genes. Two sequences are homologous (related to each other through evolution) when they share a lot of similar subsequences. In pattern recognition, chromosomes are often represented by strings. Chan [9] used edit distance to distinguish between median and telocentric chromosomes.

2 Algorithm A

Let Σ be a finite set of symbols, i.e., the alphabet; $\text{del}(a), a \in \Sigma$, be a non-negative real number representing the cost of deleting a from a string; $\text{ins}(a), a \in \Sigma$, be a non-negative real number representing the cost of inserting a into a string; and $\text{sub}(a,b), a,b \in \Sigma$, be a non-negative real number representing the cost of substituting a for b. Now given two strings A and B, generated from the alphabet, find the minimum total cost to transform A into B by a sequence of these edit operations. Let $A[i] \in \Sigma, 1 \le i \le m$, be the ith symbol of A and similarly for $B[j]$. Wagner and Fischer described a quadratic time and space algorithm to find the distance as follows:

```
Algorithm A
Global inputs: del, ins, sub
Inputs: A, B
Output: D[m, n]
m := length(A)
n := length(B)
D [0, 0] := 0
for j := 1 to n do D[0, j] := D[0, j – 1]+ ins(B[j])
for i := 1 to m do D[i, 0] := D[i – 1, 0]+ del(A[i])
for i := 1 to m do
   for j := 1 to n do begin
      m1 := D[i – 1, j – 1]+ sub(A[i], B[j])
      m2 := D[i – 1, j]+ del(A[i])
      m3 := D[i, j – 1]+ ins(B [j])
      D[i, j] := min(m1, m2, m3)
   end
end
```

In practice, to satisfy the metric axioms, the insertion and deletion costs are the same for the same symbol, so we only need one function indel for both ins and del; also sub(a, b) = sub(b, a), and sub(a, a) = 0, i.e., the distance between a and b is the same as the distance between b and a, and the distance between a and a must be zero. From the implementation point of view, if we represent a symbol by an integer which is an index to the alphabet set, we can conveniently implement A and B as

1-dimensional integer arrays of length m and n respectively and we can implement indel as a real vector and sub as a real matrix. For example, if there are five symbols in the alphabet, then the integers 1, 2, 3, 4, and 5 can represent the five symbols and indel is a 1-dimensional array of length 5 containing non-negative real numbers, while sub is a 5 by 5 symmetric matrix with 0's on the main diagonal. An APL implementation can be found in [10] with its CPU time summary statistics.

Algorithm A takes O(mn) space because the matrix D has exactly (m + 1)(n + 1) cells. Line 3 initializes the cell D[0, 0]. Line 4 is a loop initializing the rest of the cells in the 0th row. Line 5 is a loop initializing the rest of the cells in the 0th column. Lines 6 to 12 contain the double lo op that fills out the remaining cells of D one by one in row-major order. Note that at any given point inside the double loop, the calculation of D[i, j] depends only on the cell directly above, on the cell directly to the north-west corner, and on the cell directly to the left. It is precisely because of this observation that linear space algorithms are possible. Fig. 1 shows this.

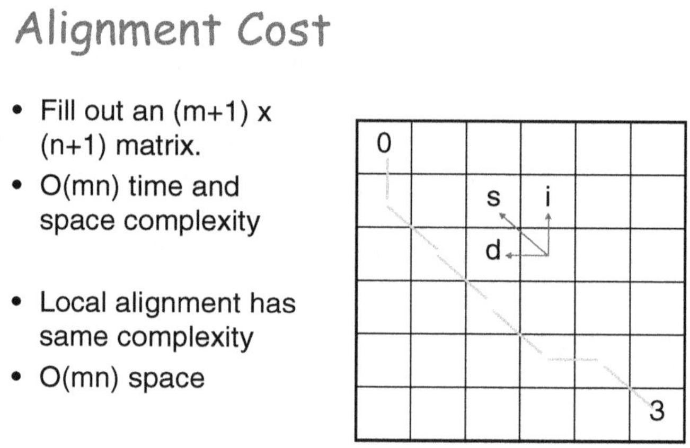

Fig. 1. Initial cost is 0. Final distance is 3; d is for deletion, s for substitution, and i for insertion.

3 Algorithm B

Algorithm B below takes advantage of the ideas that we do not need to store the whole matrix D to calculate the distance, and that insertion and deletion are symmetric operations.

```
Algorithm B
Global inputs: indel, sub
Inputs: A, B
Output: d[n]
m := length(A)
n := length(B)
q := m
d[q] := 0
for j := 1 to n do begin
```

```
6.    q := q + 1
7.    d[q] := d[q – 1]+ indel(B [j])
8. end
9. p := m
10. for i := 1 to m do begin
11.     p := p – 1
12.     q := p
13.     d[q] := d[q + 1]+ indel(A[i])
14.      for j := 1 to n do begin
15.         q := q + 1
16.         m1 := d[q]+ sub(A[i], B[j])
17.         m2 := d[q + 1]+ indel(A[i])
18.         m3 := d[q – 1]+ indel(B[j])
19.         d[q] := min(m1, m2, m3)
20.     end
21. end
```

4 Proof of Correctness of Algorithm B

Wagner and Fischer [8] proved that Algorithm A correctly computes the edit distance between strings A and B. The correctness of Algorithm B can be proved by showing that d[q] in statement 19 of Algorithm B has the same value as D[i, j] in statement 11 of Algorithm A.

Lemma. In Algorithm B, $q = m - i + j$.

Proof. There are four cases which need to be considered.

First, when $i = j = 0$, from statement 3, we have $q = m = m - 0 + 0$.

Second, when $i = 0$ and $j > 0$, we can seen from the single loop, i.e., statements 5 to 8, that every time j is incremented by 1, q also is incremented by 1 so that $q = m + j = m - 0 + j$.

Third, when $i > 0$ and $j = 0$, we can see from the outer loop, specifically statements 9 to 12, that every time i is incremented by 1, q is decremented by 1 so that $q = m - i = m - i + 0$.

Fourth, when $i > 0$ and $j > 0$, we can see from statements 14 and 15 that every time j is incremented by 1, q is also incremented by 1. From the third case, we know that in statement 12, q = m-i so that in statement 15, $q = m - i + j$. Hence, in Algorithm B, $q = m - i + j$ as stated in the lemma.

Theorem. For every assignment to D[i, j] in Algorithm A, there is a corresponding assignment to $d^i[q]$ in Algorithm B, where $q = m - i + j$ and $d^i[q]$ is the value of d[q] at the ith pass.

Proof. There are four cases.

First, in statement 3 of Algorithm A, when $i = j = 0$, D[0, 0] := 0. This is translated as statement 4 in Algorithm B as d[q] := 0 where $q = m - i + j$.

Second, in statement 4 of Algorithm A, when i = 0 and j > 0, we have

D[0, j] := D[0, j − 1]+ ins(B[j]).

This is translated in the single loop, as statement 7 of Algorithm B, when i = 0 and j > 0, which reads

d[q] := d[q − 1]+ indel(B[j]).

Since the two lines of codes here have the same i and j, by the lemma, D[0, j] corresponds to d[q]. We know that ins(B [j]) = indel(B [j]). It remains only to be shown that d^0[q − 1] = D[0, j − 1] in order to establish d^0[q] = D[0, j]. By the lemma, [0, j − 1] maps to m − 0 + j − 1 = m + j − 1 = q − 1 so that d^0[q − 1] = D[0, j − 1]. The correctness of the value in d^0[q −1] is inherited from the correctness of the value in D[0, j −1]. Alternatively, the correctness of the value in d[q − 1] can be see from analyzing the single loop in Algorithm B. Each time through the j loop, d[q − 1] is simply the value of the previous calculation.

Third, when i > 0 and j = 0, we compare statement 5 of Algorithm A with statement 13 of Algorithm B. The proof for this case is analogous to case two. We need only to show that [i − 1, 0] maps to m − (i − 1) + 0 = m − i + 1 = q + 1.

Lastly, we have the double loops, when i > 0 and j > 0. Comparing statement 8 of Algorithm A with statement 16 of Algorithm B, we can show that [i − 1, j − 1] maps to m − (i − 1) + j − 1 = m − i + 1 + j − 1 = q. Comparing statement 9 of Algorithm A with statement 17 of Algorithm B, we can show that [i − 1, j] maps to m−(i−1)+ j = m−i +1 +j = q +1. Comparing statement 10 of Algorithm A with statement 18 of Algorithm B, we can show that [i, j − 1] maps to m − i + j − 1 = m − i + j − 1 = q − 1. Now we have proved the correctness of Algorithm B by finding a mapping that relates every [i, j] to q and proving that for every D[i, j] there is a corresponding identical value d^i[q].

5 Space and Time Analysis of Algorithm

Apart from storage for the two input strings, the dominating data structure is the 1-d array d, which has exactly 1 + m + n words. In fact, the algorithm takes m + n + K words where K is a constant independent of m and n. So the space complexity is O(m+n) which is substantially better than that of O(mn) for Algorithm A.

The most time-consuming part of the algorithm is the inner loop which is executed exactly mn times. So the time complexity is O(mn) which is the same as that of Algorithm A. But a closer analysis reveals that Algorithm B has a lower multiplier in the quadratic term, so that, in practice, Algorithm B actually takes less CPU time than Algorithm A. For Algorithm A, each time through the inner loop, there are four 2-d array indexings. For Algorithm B, each time through the inner loop, there are four 1-d array indexings plus one extra addition and one more assignment operation (statement 15).

6 Algorithm C

Observe that in Algorithm B, the single loop basically initializes the 0th row of D and packs it to the right end of d starting at location d[m] ending at location d[m + n]; then at the ith pass through the double loop, it packs the ith row of D starting at location d[m − i] ending at location d[m − i + n]. At any point in time, we need only n+1 words to store the current row of i plus one word for temporary storage. This observation gives rise to the following algorithm that uses a ring structure dd instead of a simple 1-d array d to calculate the values in place.

```
    Algorithm C
    Global inputs: indel, sub
    Inputs: A,B
    Output: z
 1. m := length(A)
 2. n := length(B)
 3. if m > n then begin
 4.    m n := n m
 5.    A B := B A
 6. end
 7. dd[0] := 0
 8. for j := 1 to n do dd[j] := dd[j − 1]+ indel(B[j])
 9. n2 := n + 2
10. r := n
11. for i := 1 to m do begin
12.     r := mod(r + 1, n2)
13.     dd[r] := dd[mod(r + 1, n2)]+ indel(A[i])
14.     for j := 1 to n do begin
15.         r := mod(r + 1, n2)
16.         m1 := dd[r]+ sub(A[i],B[j])
17.         m2 := dd[mod(r + 1, n2)]+ indel(A[i])
18.         m3 := dd[mod(r − 1, n2)]+ indel(B[j])
19.         dd[r] := min(m1,m2,m3)
20.     end
21. end
22. z := dd[mod(r, n2)]
```

The proof of correctness for Algorithm C is again an exercise in index mapping. It is similar to that for Algorithm B and is omitted.

7 Space and Time Analysis of Algorithm C

Apart from storage for the two input strings, the only data structure that depends on the input strings is the ring dd which has exactly n + 2 words. Lines 3 to 6 make sure B is the shorter input string. Thus, the space complexity is O(min(m, n)).

By inspection the time complexity is O(mn). Experiments demonstrate that it takes about the same number of CPU seconds as that of Algorithm A.

Algorithm C takes slightly more CPU time than Algorithm B, but uses less space.

As a special case, this ring idea can also be applied to Hirschberg's Algorithm B [11] for calculating the length of maximal common subsequences. The improved version will reduce the actual local storage requirements from Hirschberg's 2(n + 1) to n + 2 while keeping the actual CPU time about the same.

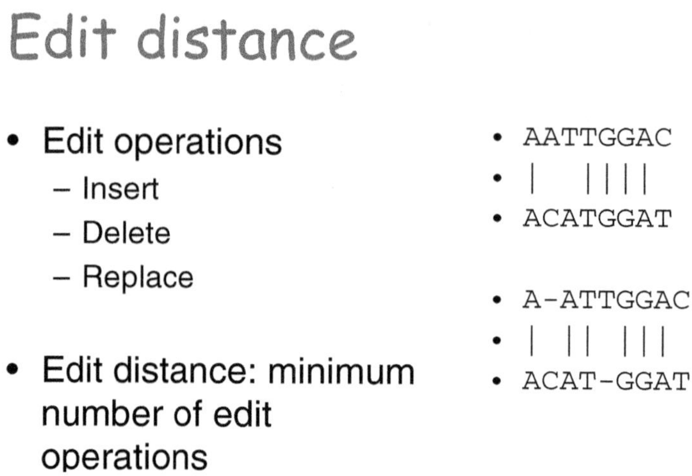

Fig. 2. Two examples of alignment

8 Applications

National Institute of Health (NIH), US Department of Health and Human Services, provides a service called GenBank. It contains all publicly available DNA sequences. As of April 2004, there are over 38,989,342,565 bases in 32,549,400 sequence records. Pekso [12] mentioned that "Genome sequence now accumulate so quickly that, in less than a week, a single laboratory can produce more bits of data than Shakespeare managed in a lifetime, although the latter make better reading."

The following is a sample record. It is the Saccharomyces cerevisiae TCP1-beta gene. There are 1510 a's, 1074 c's, 835 g's, and 1609 t's.

```
   1 gatcctccat atacaacggt atctccacct caggtttaga tctcaacaac gaaccattg
  61 ccgacatgag acagttaggt atcgtcgaga gttacaagct aaaacgagca tagtcagct
 121 ctgcatctga agccgctgaa gttctactaa gggtggataa catcatccgt caagaccaa
 181 gaaccgccaa tagacaacat atgtaacata tttaggatat acctcgaaaa aataaaccg
 241 ccacactgtc attattataa ttagaaacag aacgcaaaaa ttatccacta ataattcaa
       .
       .
       .
4921 ttttcagtgt tagattgctc taattctttg agctgttctc tcagctcctc atatttttc
4981 tgccatgact cagattctaa ttttaagcta ttcaatttct ctttgatc
```

Use edit distance to calculate dissimilarity between AATTGGAC and ACATGGAT. The first alignment requires change the second A from the first string to a C, a T to an A, and finally a C to T. A total of 3 edit operations are needed. The second alignment requires deleting the C from the second string, deleting the second T from the first string and finally change the last C to T. The total costs also is 3 edit operations.

Another application is the parity string problem. Some examples of odd parity strings are 01, 010, 1011. The total number of 1's in the string is odd. Some examples of even parity strings are 000, 1111, 010100. The total number of 1's in the string is even.

The pattern language is simply the set of bit strings. There are exactly two classes: even parity and odd parity. The only edit operation we need is the substitution of a bit by another bit, where a bit can be a 1, a 0, or the empty string.

Step 1: For patterns $p_1, p_2 \in P$,

let $\Delta_1(p_1, p_2) =$ the minimum number of substitutions needed to transform p_1 into p_2.

E.g., 1000 can be transformed into 0001 by deleting

the 1 at the begining and inserting an 1 at the end, so that $\Delta_1(1000, 0001) = 2$.

Step 2: The average intra-group distance for $\breve{Q}_1 = \{\breve{q}_1, \breve{q}_2, ..., \breve{q}_{n_1}\}$ is

$$\rho(\Delta_w) = \frac{2}{n_1(n_1 - 1)} \sum_{i=2}^{n_1} \sum_{j=1}^{i-1} \Delta_w(\breve{q}_i, \breve{q}_j).$$

Step 3: The average inter-group distance between groups $\breve{Q}_1$ and $\breve{Q}_2$ is

$$\upsilon(\Delta_w) = \frac{1}{n_1 n_2} \sum_{i=1}^{n_1} \sum_{j=1}^{n_2} \Delta_w(\breve{q}_i, \breve{r}_j), \text{ where } \breve{Q}_2 = \{\breve{r}_1, \breve{r}_2, ..., \breve{r}_{n_2}\}.$$

Step 4: Stability quotient is $Z(\Delta_w) = \dfrac{\rho(\Delta_w)}{\upsilon(\Delta_w)}$.

Stability optimization minimizes $Z(\Delta_w)$

subject to the constraint that $\sum_{i=1}^{m} w_i = 1$.

Figure 3 shows the first training set consists of two examples: 01 and 10101. The second training set consists of 11 and 00. The distance between the two set is 1/3. The cost for inserting (or deleting because of symmetry) a 0 is 0, for a 1 is 1/3, for a string 00 is 0, for 01 is 1/3, for 10 is 1/3, and for 11 is 0. So we can see that inserting or deleting 0s is free while inserting or deleting a string that contains a 1 costs 1/3 unless there are two 1s. In order to distinguish between even and odd parity strings, it makes sense that if there is an even number of 1s, the cost is free. This scheme will give the optimal objective function value of 0 as shown in Figure 3.

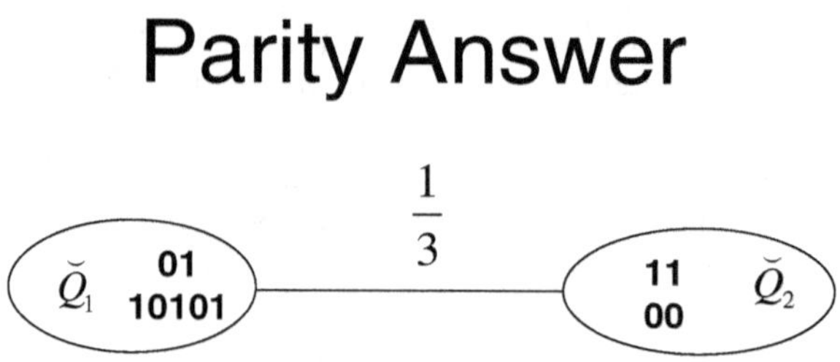

$$S_1 = \begin{Bmatrix} 0 \leftrightarrow \theta \\ 1 \leftrightarrow \theta \\ 00 \leftrightarrow \theta \\ 01 \leftrightarrow \theta \\ 10 \leftrightarrow \theta \\ 11 \leftrightarrow \theta \end{Bmatrix},$$

$$w^* = (0, \frac{1}{3}, 0, \frac{1}{3}, \frac{1}{3}, 0),$$

$$Z^* = 0$$

Fig. 3. Parity problem: optimal weights

9 Conclusion

Two useful algorithms are presented to reduce the quadratic space complexity of Wagner and Fischer's edit distance algorithm to linear. One idea was that instead of calculated the entries of a matrix in a row-after-row order, we did it in a diagonal-after-diagonal order. Another idea to further reduce the space requirement was to treat a diagonal not as a linear array but as a ring structure. These ideas were implemented and tested in the APL language and the resulting programs were applied to the recognition of chromosomes.

References

1. Fu, K. S.: Syntactic Pattern Recognition and Applications. Prentice-Hall, Englewood Cliffs, NJ (1982)
2. Sankoff, D., Kruskal, J. B.: Time Warps, String Edits, and Macromolecules: The Theory and Practice of Sequence Comparison. Addison-Wesley, Reading, Massachusetts (1983)
3. Chan, T. Y. T.: Unifying Metric Approach to The Triple Parity. Artificial Intelligence, 141 (2002) 123-135
4. Chan, T. Y. T.: Inductive Pattern Llearning. IEEE Transactions on Systems, Man, and Cybernetics—Part A: Systems and Humans, 29 (1999) 667-674
5. Myers, E. W.: An O(ND) Difference Algorithm and Its Variations. Algorithmica, 1 (1986) 251-266

6. Myers, E. W., Miller, W.: Optimal Alignments in Linear Space. Computation Application Bioscience, 4 (1988) 11-17
7. Wong, C. K., Chandra, A. K.: Bounds for The String Editing Pproblem. Journal of the ACM, 23 (1976) 13-16
8. Wagner, R. A., Fischer, M. J.: The String to String Correction Problem. Journal of the ACM, 21 (1974) 168-173
9. Chan, T.Y.T.: Unsupervised Classification of Noisy Chromosomes. Bioinformatics, 17 (2001) 438–444
10. Chan, T. Y. T.: Running Parallel Algorithms with APL on A Sequential Machine. APL Quote Quad, 29 (1999) 25-26
11. Hirschberg, D.S.: A Llinear Space Algorithm for Computing Maximal Common Subsequences. Communications of the ACM, 18 (1975) 341-343
12. Pekso, G. A.: Biology's Structurally Sound Foundations. Nature, 401 (1999) 115-116

Prediction of Protein Complexes Based on Protein Interaction Data and Functional Annotation Data Using Kernel Methods

Shi-Hua Zhang[1], Xue-Mei Ning[1], Hong-Wei Liu[2], and Xiang-Sun Zhang[1]

[1] Institute of Applied Mathematics, Academy of Mathematics and Systems Science
Chinese Academy of Sciences, Beijing 100080, China
{zsh, nxm}@amss.ac.cn, zxs@amt.ac.cn

[2] School of Economics, Renmin University of China, Beijing 100872, China
ryuhowell@163.com

Abstract. Prediction of protein complexes is a crucial problem in computational biology. The increasing amount of available genomic data can enhance the identification of protein complexes. Here we describe an approach for predicting protein complexes based on integration of protein-protein interaction (PPI) data and protein functional annotation data. The basic idea is that proteins in protein complexes often interact with each other and protein complexes exhibit high functional consistency/even multiple functional consistency. We create a protein-protein relationship network (PPRN) via a kernel-based integration of these two genomic data. Then we apply the MCODE algorithm on PPRN to detect network clusters as numerically determined protein complexes. We present the results of the approach to yeast *Sacchromyces cerevisiae*. Comparison with well-known experimentally derived complexes and results of other methods verifies the effectiveness of our approach.

1 Introduction

Cellular organization and function are carried out through gene/protein interactions. With ever-increasing different types of genomic data such as DNA sequences, gene expression measurement, protein-protein interaction, and protein phylogenetic profiles, reconstruction of biological machinery from these genomic data is a crucial problem. Protein complex is a group of proteins that often interact with each other, forming a special biological chemical machinery. However, despite recent advances in detection technologies of protein interactions, only a very few of many possible protein complexes has been experimentally determined [1]. Then prediction of protein complexes is a key problem in computational biology. One of such work has been done within the PPI networks [2,3,4]. Proteins in a complex often interact with each other, so protein complexes generally correspond to dense subgraphs in the PPI networks. Recently, three approaches to network clustering including the MCODE (Molecular Complex Detection) algorithm [2], restricted neighborhood search clustering (RNSC) [3], and local clique merging algorithm (LCMA) [4] have been applied to predict protein complexes.

D.-S. Huang, K. Li, and G.W. Irwin (Eds.): ICIC 2006, LNBI 4115, pp. 514–524, 2006.

The MCODE algorithm utilizes connectivity values in PPI networks to identify complexes, a shortcoming of which is that its resulted clusters may be too sparse. While RNSC partitions the PPI networks using a cost function. It is a random algorithm and relatively fewer complexes can be predicted by this algorithm. LCMA algorithm which is based on local clique merging has been shown more efficient than MCODE algorithm preliminarily. It should be noted that proteins in known complexes often correspond to consistent functional annotation [2,3], so the relatively aboundant functional annotation information can be employed to identify protein complexes. In the study of ref.[3], functional homogeneity has been used as a necessary condition for prediction of protein complexes.

Kernel representation of heterogeneous genomic information has already been proven to be very useful tool in computational biology [5,6]. Each type dataset can be represented by means of a kernel function, a real-valued function $K(x, y)$, which defines similarities between pairs of objects (genes, proteins and so on) x and y. Evaluating the kernel on all pairs of data objects yields a symmetric, positive semi-definite matrix $\mathbf{K}$ known as the kernel matrix. The distinguished characteristic is that all types of data are represented in the unified framework even though they might be different in nature. Various kernels have been developed for various genomic data integration [5,6]. For example, the linear kernel and Gaussian kernel are natural choice for datasets which are represented by vectors, while diffusion kernel [7] has proven to be very effective for describing network data. In this study, a simple kernel representation which captures the functional consistency or even multiple functional consistency of protein complexes properly is defined naturally for the protein functional annotation data.

Here we propose an integrated approach that attempts to identify protein complexes using protein interaction data and functional annotation information. We create an integrated protein-protein relationship network (PPRN) by using the kernel methods to integrate these two genomic data. Then the network clustering method called MCODE algorithm is applied to the created PPRN network to detect numerically derived complexes. The MCODE algorithm is developed for detecting complexes in protein interaction networks and it can detect overlapping modules. So they are also suitable for finding complexes in our networks. This approach was applied to yeast *Sacchromyces cerevisiae.* The computed protein complexes show good consistency with well-known yeast protein complexes. Comparison with other methods such as the MCODE algorithm applied on PPI network directly shows the effectiveness of our approach.

2 Systems and Methods

2.1 Materials

We use yeast-related genomic data to predict protein complexes as it is currently the organism with the most comprehensive experimental datasets available publicly.

Protein Interaction Data. A physical network of 4713 yeast *Sacchromyces cerevisiae* proteins containing 14848 protein interactions is used in our work.

The protein-protein interactions were downloaded from the DIP database as of July 2004 and predominantly included data from large-scale experiments [8,9,10,11].

Functional Annotation Data. To employ the functional consistency of protein complexes, we utilize the functional annotation of *Sacchromyces cerevisiae* genes in MIPS Functional Catalog (FunCat) [12] database. FunCat is an annotation scheme for the functional description of proteins from various biology and consists of 28 main functional categories (or branches). The main branches exhibit a hierarchical, tree like structure with up to six levels of increasing specificity and 1307 functional categories are included in total. Here we utilize the functional annotation at the second levels of 68 categories (to the 4713 proteins), so that each protein corresponds to a vector of dimension 68 in which 1 or 0 represents a protein belonging to or not belonging to a category.

Gold Standard Complex Data. To evaluate the effectiveness of our approach for predicting protein complexes, we compare the predicted complexes of the yeast data with known protein complexes in MIPS yeast complex database [13]. In order to removing/filtering the experimentally predicted protein complexes from the dataset to a certain extent, we only use manually annotated complexes derived from literature scanning and the known Gavin benchmark data [10] as our gold standard dataset. Finally, a set M of 439 yeast complexes is used as known complexes set. Its biggest protein complex contains 88 proteins and the average size of it is 9.11.

2.2 Methods

The outline of our method is shown in Figure 1. Two genomic datasets are represented by two kernel matrices respectively. Then a protein-protein relationship network (PPRN) is produced by integration of these two kernels. A powerful tool of detecting network modules is applied to PPRN network. The resulting modules are our numerically detected protein complexes which constitute the predicted complex set P. Validation of these complexes and comparison with related methods verifies our idea that functional annotation information is helpful for the detection of protein complexes.

Kernel Representation and Data Integration. In order to represent each type of genomic information uniformly, kernel representation is an efficient method [5,6]. PPI network can be represented using the diffusion kernel [7]. Let A denote the adjacency matrix of the PPI network and D denote the diagonal matrix of degrees of nodes. So the Laplacian matrix of this network is $L = D - A$. Then the diffusion kernel is defined as

$$K = expm(-\beta L), \tag{1}$$

where $expm$ is a matrix exponential operation and β is a parameter to control

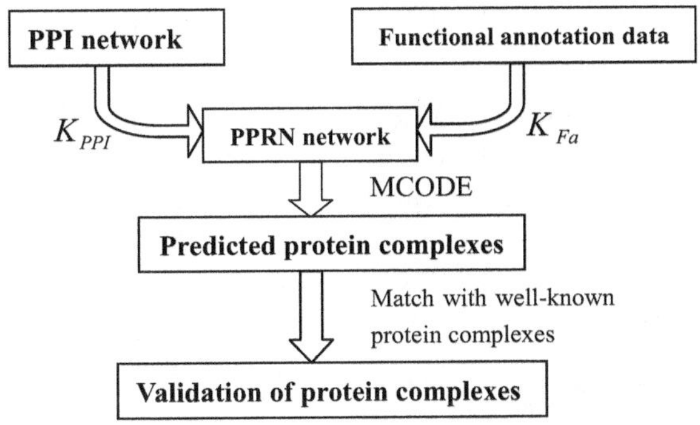

Fig. 1. The schematic diagram of our method for detection of protein complexes

the degree of diffusion. Then the diffusion kernel is normalized so that its all diagonal elements are one:

$$K_{PPI} = \frac{K_{ij}}{\sqrt{K_{ii}K_{jj}}}. \tag{2}$$

Functional annotation data is represented by means of liner kernel:

$$K_{fa}(i,j) = x_i \cdot x_j, \quad K_{Fa} = \frac{K_{fa}}{max(K_{fa})}, \tag{3}$$

where $\cdot$ means the inner product and $max()$ means the maximal value of the matrix.

These two kernels measure the similarity of proteins with respect to every genomic data. A new kernel defined as the sum of the two kernels:

$$K_{Int} = \frac{K_{PPI} + K_{Fa}}{2}, \tag{4}$$

is a simple approach of data integration. Although more complex kernel operation can be employed to create new integrating method, this simple kernel has been used comprehensively [5].

Protein-Protein Relationship Network (PPRN): Kernels describe some implicit similarity of proteins, so any protein kernel matrix K can denote a weighted/unweighted network $G(V, E, W)/G(V, E)$ of protein-protein relationship. The nodes set V consists of all proteins, the matrix W is the value of corresponding kernel matrix which denotes the weights of the edges, and the edge set of such network is defined as:

$$E = \{(i,j)|K_{ij} \geq c\}, \tag{5}$$

where c is a parameter to control the density of the network. We denote the network of kernel K_{Int} as our protein-protein relationship network (PPRN). We believe that a group of proteins which have large enough kernel mutually likely corresponds to a protein complex.

MCODE Algorithm: Bader and Hogue [2] have developed a novel graph theoretic clustering algorithm, i.e., so-called MCODE algorithm (http://cbio.mskcc.org/ bader/software/mcode/index.html), which utilizes connectivity values in PPI networks to detect protein complexes. This algorithm is based on vertex weighting according to its local neighborhood density and then outward traversal from a dense seed protein with a high weighting value to recursively include neighboring vertices whose weight satisfies some given threshold. Here we also apply it on our PPI network and PPRN networks to evaluate our idea that functional annotation can improve the ability of prediction of complexes.

3 Experiments and Results

3.1 Validation of Protein Complexes

We assess the precision of results of applying MCODE algorithm on our PPRN networks by using evaluation metric used in [2,4]. They used the overlap score:

$$OS(p, m) = \frac{k^2}{n_1 \times n_2} \tag{6}$$

to determine matching between a predicted complex $p \in P$ and a known complex $m \in M$, where k is the size of overlap of p and m and n_1,n_2 are the sizes of p and m respectively. Given a predicted complex p and a known complex m, they are considered to be matching if $OS(p, m) \geq 0.2$, where 0.2 is an experientially determined threshold used in [2] firstly and also was used in [4]. And then we refer the notation in [4] to define the set of true positives (TP) as $TP = \{p|\exists m, OS(p, m) \geq 0.2, p \in P, m \in M\}$, and the set of false positives (FP) as $FP = P - TP$. Naturally, the set of false negatives (FN) is defined as $FN = \{m|\forall p, OS(p, m) < 0.2, p \in P, m \in M\}$, and the matched gold-standard complex set MS can be defined as $MS = M - FN$, which contains known complexes matched by predicted complexes. Then the recall (sensitivity) and precision (specificity) are defined as $|TP|/(|TP| + |FN|)$ and $|TP|/(|TP| + |FP|)$ respectively. The so-called F-measure which is defined as

$$F = \frac{2 \times Precision \times Recall}{Precision + Recall} \tag{7}$$

adopted in [4] is used to evaluate the performance of our approach. Just as the authors have pointed that F-measure of every method only can be taken as comparative measures rather than their real values for the incompleteness of known complexes set.

In order to further test our approach, we consider another index which measures the coverage of predicted protein complexes:

$$Cov(p, m) = \frac{k}{n_2}, \quad C_q = \{m|\exists p, Cov(p, m) \geq q, p \in P, m \in M\}, \tag{8}$$

where q is a real number between 0 and 1, the set C_q contains the known complexes whose members appear in a predicted complex above the degree q.

3.2 Experimental Results

Since the noise and incompleteness of known protein interaction data, our approach aims to detect more complexes through integrating functional annotation data to current protein interaction data with high recall and precision. The effectiveness of kernel methods employs the functional consistency of proteins and implicit relationship of interacting proteins. The functional annotation information can complement the absence of existing interactions and correct some false interactions. So integration of the two genomic data can enhance the robustness of network clustering method against only the high noise protein interaction data. Figure 2 shows an example of MIPS complex of size 18 and two matching complexes both of size 13 in PPI and PPRN network respectively by means of MCODE algorithm. Table 1 shows the functional annotation of proteins in Figure 2 (only the functional annotation that is labeled by at least three proteins has been shown). The predicted complex in PPRN network (with $c = 0.24$, see below) is contained within the known complex while for the predicted complex in PPI network only ten proteins are included in it. We can see that all the proteins in the given MIPS complex (the first 18 proteins in table 1) show high multiple functional consistency, while three proteins (the last three proteins labeled in black font) that are included in the predicted complex in PPI network but not included in the given MIPS complex do not show such multiple consistent functional annotation information. This shows our idea that known functional annotation information/functional annotation consistency is helpful for detecting complexes.

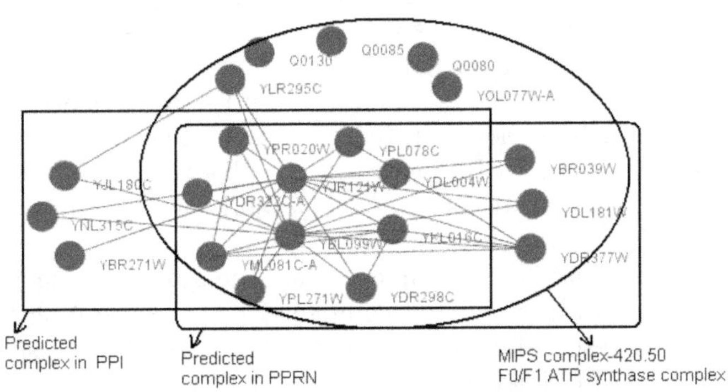

Fig. 2. An example: MIPS complex (MIPS-420.50)-F0/F1 ATP synthase complex and the matching complex predicted in PPI network and PPRN network respectively

In all the study, the diffusion kernel of protein interaction network is computed with $\beta = 3$. And then we choose $c = 0.25$ and 0.24 experientially for producing two PPRN networks with 14423 and 16413 edges respectively. We

Table 1. Functional annotation of proteins in Figure 2

Q0080	02.11	02.13	02.45.15			20.01.01.01	20.01.15	20.03.22	20.09.04	34.01.01.03
Q0085	02.11	02.13	02.45.15	14.10	16.07	20.01.01.01	20.01.15	20.03.22	20.09.04	34.01.01.03
Q0130	02.11	02.13	02.45.15	14.10	16.07	20.01.01.01	20.01.15	20.03.22	20.09.04	34.01.01.03
YOL077W-A	02.11	02.13	02.45.15			20.01.01.01	20.01.15		20.09.04	
YLR295C	02.11	02.13	02.45.15			20.01.01.01	20.01.15	20.03.22	20.09.04	34.01.01.03
YBL099W	02.11	02.13	02.45.15			20.01.01.01	20.01.15	20.03.22	20.09.04	34.01.01.03
YBR039W	02.11	02.13	02.45.15			20.01.01.01	20.01.15	20.03.22	20.09.04	34.01.01.03
YDL004W	02.11	02.13	02.45.15			20.01.01.01	20.01.15	20.03.22	20.09.04	34.01.01.03
YDL181W	02.11		02.45.15				20.01.15			
YDR298C	02.11	02.13	02.45.15		16.07	20.01.01.01	20.01.15	20.03.22	20.09.04	34.01.01.03
YDR322C-A	02.11	02.13	02.45.15	14.10	16.07		20.01.15	20.03	20.09.04	
YDR377W	02.11	02.13	02.45.15	14.10		20.01.01.01	20.01.15			
YJR121W	02.11	02.13	02.45.15			20.01.01.01	20.01.15	20.03.22	20.09.04	34.01.01.03
YKL016C	02.11	02.13	02.45.15	14.10	16.07	20.01.01.01	20.01.15	20.03.22	20.09.04	34.01.01.03
YML081C-A	02.11	02.13	02.45.15			20.01.01.01	20.01.15			
YPL078C	02.11	02.13	02.45.15		16.07	20.01.01.01	20.01.15	20.03.22	20.09.04	34.01.01.03
YPL271W	02.11	02.13	02.45.15			20.01.01.01	20.01.15	20.03.22	20.09.04	34.01.01.03
YPR020W	02.11	02.13	02.45.15		16.07		20.01.15			
YJL180C				14.10						
YNL315C				14.10						
YBR271W										

apply the MCODE algorithm on the two networks to predict protein complexes. For comparison, we also apply it on the original protein interaction network to show the effectiveness of integration of two genomic information using kernel methods. The MCODE algorithm needs two important parameters w and f to control the number and size of resulting clusters. With respect to different networks, the optimal results will be produced by different parameter pairs [2]. We choose some parameter pairs to optimize the biological relevance.

Some of the predicted complexes are of size 3, while complex with size 3 is less statistical significant, since it is easy to produce in a random graph. So we discuss two cases: one is that the predicted complex set includes complexes size of 3 and the other is not. Table 2 and Figure 3 show that the optimal results with respect to the largest F-measures with three groups w in three networks. The results show the integration of functional annotation information with PPI data can enhance the prediction results largely. More complexes are detected using the same network clustering method, and the F-measures of the two PPRN networks are also clearly higher than that of only PPI network used. For example, the F-measures of two PPRN networks are able to achieve 15.85%/22.62% higher than that of PPI networks and 32.66%/39.74% higher in two cases with $w = 0.1$.

We test coverage of predicted complexes, i.e., the degree to which entire complexes appear in the same predicted complexes [16]. Figure 4 shows the large improvement of our results for varying values of q in two cases respectively. For example, in our two PPRN networks, there are 139/140 gold-standard complexes (with $w = 0.1$ and $f = 0$) for which 50% or more of their members appeared in the same predicted complex, compared only 70 in the predicted results of PPI network.

King *et al.*[3] applied RNSC algorithm to predict complexes from protein interaction networks. But they only predicted 45 complexes which match 30 MIPS complexes. A new recent system LCMA algorithm based on local clique merging has been reported to be more efficient than MCODE algorithm. But we do not

Table 2. List of various protein-protein networks and its related results, where P_3 and P_4 represent predicted complexes set with minimum size 3 and 4 of their predicted complexes respectively

Protein-Protein network	para(c, $\langle w, f \rangle$)	$\|P_3\|$	$\|TP\|$	$\|MS\|$	para(c, $\langle w, f \rangle$)	$\|P_4\|$	$\|TP\|$	$\|MS\|$
PPI (MCODE)	(no, $\langle 0.00, 0.00 \rangle$)	157	104	74	(no, $\langle 0.00, 0.20 \rangle$)	147	82	60
PPI (MCODE)	(no, $\langle 0.05, 0.00 \rangle$)	237	143	94	(no, $\langle 0.05, 0.10 \rangle$)	235	121	74
PPI (MCODE)	(no, $\langle 0.10, 0.00 \rangle$)	305	169	105	(no, $\langle 0.10, 0.25 \rangle$)	268	130	92
PPRN (Int+MCODE)	(0.25, $\langle 0.00, 0.00 \rangle$)	371	216	122	(0.25, $\langle 0.00, 0.10 \rangle$)	357	192	112
PPRN (Int+MCODE)	(0.25, $\langle 0.05, 0.00 \rangle$)	510	260	139	(0.25, $\langle 0.05, 0.00 \rangle$)	360	204	95
PPRN (Int+MCODE)	(0.25, $\langle 0.10, 0.00 \rangle$)	591	284	142	(0.25, $\langle 0.10, 0.00 \rangle$)	405	225	97
PPRN (Int+MCODE)	(0.24, $\langle 0.00, 0.30 \rangle$)	377	239	119	(0.24, $\langle 0.00, 0.30 \rangle$)	330	216	104
PPRN (Int+MCODE)	(0.24, $\langle 0.05, 0.00 \rangle$)	526	286	141	(0.24, $\langle 0.05, 0.00 \rangle$)	368	223	93
PPRN (Int+MCODE)	(0.24, $\langle 0.10, 0.00 \rangle$)	633	318	150	(0.24, $\langle 0.10, 0.00 \rangle$)	435	248	105

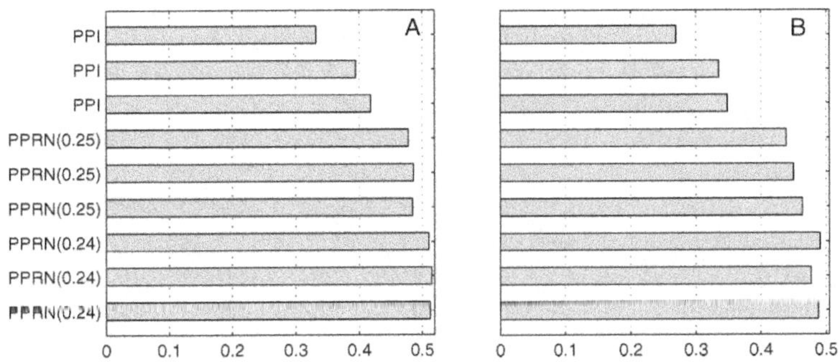

Fig. 3. Comparison of F-measures of applying the MCODE algorithm on various networks with various selective parameters

do direct comparison for the lack of their system, and we emphasize that our contribution is that we complement the incomplete PPI data with functional annotation information by means of kernel methods which well exploit the functional homogeneity of protein complexes or even multiple functional consistency of proteins. Our approach also predicted complexes that do not match current protein complexes set just like other methods have done. Since the known complex set is largely incomplete, these new unmatched complexes could be real complexes likely. So the actual precision of our approach would be higher than current results.

Recent studies have well shown that biological networks (eg. metabolic network, physical interaction networks) show the characteristic of scale-free networks just like many natural networks [14]. Here, we examined the scale-free characteristic of protein-protein relationship networks and size distribution of predicted complexes based on the PPI network and the two PPRN networks. On the top of Figure 5, plots A,B,C show that the probability $P(k)$ of a node with degree in these three networks follows power law: $P(k) \propto k^{-\gamma}$, and at the bottom of Figure 5, plots B,C show that the size distribution of clusters (modules) of two PPRN networks also follow power law clearly, while that of PPI networks has a high slope ($\gamma = 3.56$)(see in plot A).

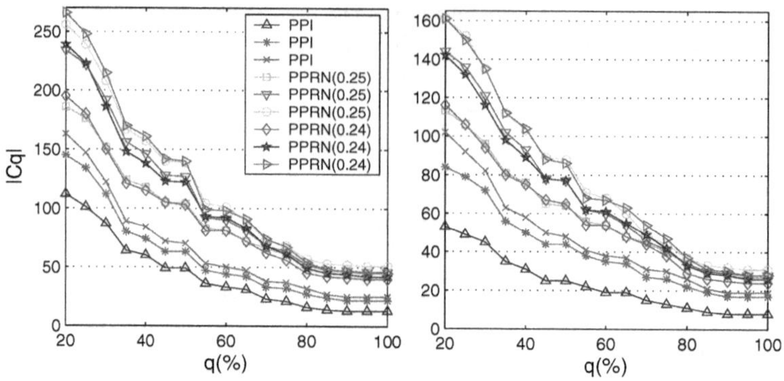

Fig. 4. Complex coverage: $|C_q|$ represents the number of complexes whose member proteins appear in the same predicted complex for various q values. The left figure plots the results with all the predicted complexes including complexes of size 3, while the right removing complexes of size 3.

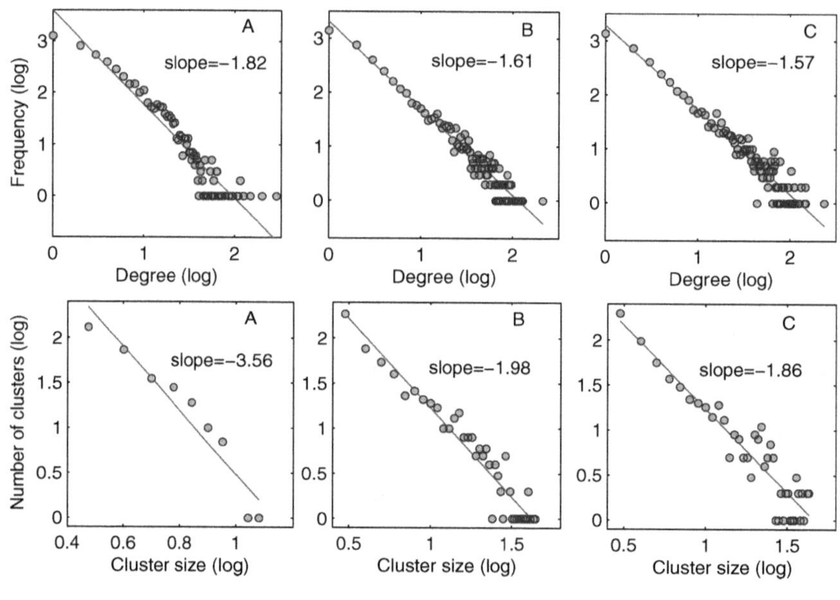

Fig. 5. On the top of the figure, plots show the degree distribution of PPI network and two PPRN networks, and at the bottom of the figure, plots show size distribution of these networks by MCODE algorithm with parameter $w = 0.1$ and $f = 0$

4 Conclusion and Discussion

In this paper, we develop a method of predicting protein complexes based on integration of two important genomic data (physical interaction data and protein functional annotation data) by means of kernel methods. Group of genes/proteins which may correspond to functional modules have been detected comprehensively in physical interaction data [15]. However, it is often hard to conclude that these clusters/modules must have such properties. One reason is that these data is very noisy and incomplete. Prediction of protein complexes has been done based on protein interaction data such as MCODE algorithm [2], RNSC algorithm [3] and recent LCMA algorithm [4]. Detection of molecular pathways/functional modules also have been done based on integration of physical interaction data and another important genomic data—gene expression data [16]. Here, we introduce the functional annotation data to improve the limitation of the physical interaction data and this approach well employ the functional consistency of protein complexes. Kernel representation has been proven to be very useful for various types data, e.g. string, trees, network and so on. Its merit has been comprehensively used in bioinformatics, e.g. inference of biological network [5]. In this study, we well exploit the characteristic of kernel methods and combine these two data. The experimental results with yeast data show the effectiveness of our proposed method. Compared with the results of only using protein interaction data, our predicted complexes match or contain more known experimentally protein complexes. More novel predicted complexes may help biologists to detect new protein complexes experimentally. We can conclude that the combination of these two data sources can produce more better results than only using protein interaction data.

Acknowledgements

This work is partly supported by Important Research Direction Project of CAS "Some Important Problem in Bioinformatics", National Natural Science Foundation of China under Grant No.10471141.

References

1. Sear, R.P.: Specific Protein-Protein Binding in Many-componet Mixtures of Proteinsn. Phys. Biol., 1(2004), 53-60
2. Bader, G.D., Hogue, C.W.: An Automated Method for Finding Molecular Complexes in Large Protein Interaction Networks. BMC Bioinformatics, 4(2003), 2
3. King, A.D., Pržulj, N., Jurisica, I.: Protein Complex Prediction via Cost-based Clustering. Bioinformatics, 20(2004), 3013-3020
4. Li, X.L., Tan, S.H., Foo, C.S., Ng, S.K.: Interaction Graph Mining for Protein Complexes Using Local Clique Merging. Genome Informatics, 16(2005), 260-269
5. Yamanishi. Y, Vert, J.P., Kanehisa, M.: Protein Network Inference from Multiple Genomic Data: a Supervised Approach. Bioinformatics, 20(2004), i363-i370
6. Lanckriet, G.R., De Bie T.D, Cristianini, N., Jordan, M.I., Noble, W.S.: A Statistical Framework for Genomic Data Fusion. Bioinformatics, 20(2004), 2626-2635
7. Kondor, R.I., Lafferty, J.: Diffusion Kernels on Graphs and Other Discrete Input. In Proceedings of the 19th International Conference on Machine Learning, Morgan Kaufmann, University of South Wales, Sydney, Australia, (2002), 315-322

8. Ito, T., Tashiro, K., Muta, S., Ozawa, R., Chiba, T., Nishizawa, M., Yamamoto, K., Kuhara, S. Sakaki, Y.: Toward a Protein-Protein Interaction Map of the Budding Yeast: a Comprehensive System to Examine Two-hybrid Interactions in All Possible Combinations between the Yeast Proteins. Proc. Natl Acad. Sci., USA, 97(2000), 1143-1147
9. Uetz, P., Giot, L., Cagney, G., Mansfield, T.A., Judson, R.S., Knight, J.R., Lockshon, D., Narayan, V., Srinivasan, M., Pochart, P., et al.: A Comprehensive Analysis of Protein-Protein Interactions in Saccharomyces Cerevisiae. Nature, 403(2000), 623-627
10. Gavin, A.C., Bosche, M., Krause, R., Grandi, P., Marzioch, M., Bauer, A., Schultz, J., Rick, J.M., Michon, A.M., Cruciat, C.M., et al.: Functional Organization of the Yeast Proteome by Systematic Analysis of Protein Complexes. Nature, 415(2002), 141-147
11. Ho, Y., Gruhler, A., Heilbut, A., Bader, G.D., Moore, L., Adams, S.L., Millar, A., Taylor, P., Bennett, K., Boutilier, K., et al.: Systematic Identification of Protein Complexes in Saccharomyces Cerevisiae by Mass Spectrometry. Nature, 415(2002), 180-183
12. Ruepp, A., Zollner, A., Maier, D., Albermann, K., Hani, J., Mokrejs, M., Tetko, I., Guldener, U., Mannhaupt, G., Munsterkotter, M. et al.: The FunCat, a Functional Annotation Scheme for Systematic Classification of Proteins from Whole Genomes. Nucleic Acids Res., 32(2004), 5539-5545
13. Mewes, H.W., Frishman, D., Guldener, U., Mannhaupt, G., Mayer, K., Mokrejs, M., Morgenstern, B., Munsterkotter, M., Rudd, S., Weil, B.: MIPS: a Database for Genomes and Protein Sequences. Nucleic Acids Res. 30(2002), 31-34
14. Barabási, A.-L., Oltvai, Z.N.: Network Biology: Understanding the Cell's Functional Organization. Nature Rev. Genet., 5(2004), 101-114
15. Spirin, V., Mirny, L.A.: Protein Complexes and Functional Modules in Molecular Networks. Proc. Natl Acad. Sci., USA, 100(2003), 12123-12126
16. Segal, E., Wang, H., Koller, D.: Discovering Molecular Pathways from Protein Interaction and Gene Expression Data. Bioinformatics, 19(2003), i264-i272

Prediction of Transmembrane Proteins from Their Primary Sequence by Support Vector Machine Approach

C.Z. Cai[1,2], Q.F. Yuan[1,2], H.G. Xiao[1,2], X.H. Liu[1], L.Y. Han[2], and Y.Z. Chen[2]

[1] Department of Applied Physics, Chongqing University, Chongqing 400044, China
caiczh@gmail.com
[2] Department of Pharmacy, National University of Singapore, Singapore 117543

Abstract. Prediction of transmembrane (TM) proteins from their sequence facilitates functional study of genomes and the search of potential membrane-associated therapeutic targets. Computational methods for predicting TM sequences have been developed. These methods achieve high prediction accuracy for many TM proteins but some of these methods are less effective for specific class of TM proteins. Moreover, their performance has been tested by using a relatively small set of TM and non-membrane (NM) proteins. Thus it is useful to evaluate TM protein prediction methods by using a more diverse set of proteins and by testing their performance on specific classes of TM proteins. This work extensively evaluated the capability of support vector machine (SVM) classification systems for the prediction of TM proteins and those of several TM classes. These SVM systems were trained and tested by using 14962 TM and 12168 NM proteins from Pfam protein families. An independent set of 3389 TM and 6063 NM proteins from curated Pfam families were used to further evaluate the performance of these SVM systems. 90.1% and 86.7% of TM and NM proteins were correctly predicted respectively, which are comparable to those from other studies. The prediction accuracies for proteins of specific TM classes are 95.6%, 90.0%, 92.7% and 73.9% for G-protein coupled receptors, envelope proteins, outer membrane proteins, and transporters/channels respectively; and 98.1%, 99.5%, 86.4%, and 98.6% for non-G-protein coupled receptors, non-envelope proteins, non-outer membrane proteins, and non-transporters/non-channels respectively. Tested by using a significantly larger number and more diverse range of proteins than in previous studies, SVM systems appear to be capable of prediction of TM proteins and proteins of specific TM classes at accuracies comparable to those from previous studies. Our SVM systems – SVMProt, can be accessed at http://jing.cz3.nus.edu.sg/cgi-bin/svmprot.cgi.

1 Introduction

Transmembrane (TM) proteins play important roles for signaling, transport, recognition and interaction with extracellular molecules [1,2,3,4]. Many TM proteins, such as G-protein coupled receptors and channels, have been explored as

D.-S. Huang, K. Li, and G.W. Irwin (Eds.): ICIC 2006, LNBI 4115, pp. 525–533, 2006.

therapeutic targets [5,6,7]. Membrane-bound transporters are responsible for absorption and excretion of drugs as well as cellular molecules [8,9]. Thus prediction of TM proteins is important for facilitating functional study of genomes, understanding molecular mechanism of diseases, and for searching new therapeutic targets.

Although TM proteins can be determined by experimental methods such as antibody-binding analysis and C-terminal fusions with indicator proteins [10,11], the number of experimentally determined TM proteins is significantly smaller than the estimated TM proteins in genomes [12,13,14]. Thus computational methods have been developed for facilitating the prediction of TM sequences [13,14,15,16,17,18]. These methods are capable of achieving high prediction accuracy for TM proteins and they can satisfactorily distinguish between TM and globular proteins and between TM and signal peptides. A study of 14 TM protein prediction methods using 270 helical TM chains, 1,418 signal peptides and 616 globular proteins showed that ~95% TM helices are correctly predicted as TM proteins and ~92% of globular proteins are correctly predicted as non-membrane (NM) proteins by the best methods [13]. A more recent study showed that ~95% of the 125 TM proteins and ~99% of the 526 soluble proteins can be correctly predicted by using a modified algorithm [18].

These methods have been developed and tested by using a few hundred to several hundred TM sequences and a slightly higher number of NM proteins. Our search of Swissprot database http://www.expasy.ch/sprot (Swissprot release 44.1, [19]) showed that there are 18358 TM protein sequences and over 134,000 NM proteins. Thus these methods may preferably need to be more adequately tested and trained by using a more diverse set of proteins. Previous studies also revealed that some TM prediction methods tend to predict proteins with more than 5 TM helices at a lower accuracy [13], which affects their prediction capability for such therapeutically and biologically relevant TM proteins as G-protein coupled receptors and certain types of channels and transporters [20]. Some methods have been found to be less capable of distinguishing between signal peptides and membrane helices [13,14]. Therefore, it is useful to evaluate the performance of TM protein prediction methods on specific therapeutically and biologically important classes of TM proteins.

The performance of a statistical learning method, support vector machine (SVM), for the prediction of TM proteins was evaluated in this work by using a diverse set of TM proteins and NM proteins. It was also tested on specific classes of TM proteins. SVM is a relatively new and promising algorithm for binary classification by means of supervised learning and it appears to be more superior than other statistical learning methods [21]. SVM has been applied to the prediction of TM proteins [15,17] and a specific TM class of G-protein coupled receptors [22] as well as other proteins [23,24,25,26,27,28,29,30,31]. These SVM TM protein prediction systems were not trained and tested by using a sufficiently diverse set of TM and NM proteins. Therefore, in this work, a large number of TM and NM proteins were used to train and test a SVM system. SVM systems were also trained and tested for the prediction of therapeutically and biologically important individual classes of TM proteins including G-protein coupled receptors, envelope proteins, outer membrane proteins, and transporters/channels.

Many of the proteins in these four classes either contain more than 5 TM helices or have non-helix TM segments. For instance, G-protein coupled receptors contain 7 TM helices as well as intracellular and extracellular domains. Envelope proteins are located in viral lipoprotein membranes which form the outermost layer of the virion in certain viruses. Outer membrane proteins are located in the outer membrane of organelles like mitochondria, chloroplasts and some eubacteria which are surrounded by a double membrane. Almost all TM transport processes are mediated by integral TM proteins, sometimes functioning in conjunction with extracytoplasmic receptors or receptor domains as well as with cytoplasmic energy-coupling and regulatory proteins or protein domains. Thus the four classes of TM proteins have their own characteristics and they are useful for testing the performance of SVM classification.

2 Methods

2.1 Selection of Transmembrane Proteins and Non-membrane Proteins

All TM proteins used in this study were from a comprehensive search of Swissprot database (Swissprot release 44.1, TrEMBL release 27.1, [19]). A total of 18,358 TM protein sequences were obtained, which include 8,457 G-protein coupled receptors, 450 envelope proteins, 1,492 outer membrane proteins, and 980 transporters and channels. All distinct members in each group were used to construct positive samples for training, testing and independent evaluation of SVM classification system. Multiple entries for a distinct protein were evenly distributed to the training, testing, and independent evaluation set.

The negative samples, i.e. NM proteins, for training and testing our SVM classification systems were selected from seed proteins of the more than 3886 curated protein families in the Pfam database [32] that have no TM protein as a family member. Each negative set contains at least one randomly selected seed protein from each of the Pfam families. For each sub-group of non-G-protein coupled receptor, non-envelope protein, non-out membrane protein, or non-transporter/non-channel, distinct members in the other four sub-groups were added to the negative samples of each of the training, testing and independent evaluation set. For instance, distinct members of envelope proteins, out membrane proteins, transporters and channels are added into the negative samples of the G-protein coupled receptors. It is expected that the number of negative samples in each of these sub-groups may be higher than that in the group of negative samples for all TM proteins.

Training sets of both positive and negative samples were further screened so that only essential proteins that optimally represent each group are retained. The SVM training system for each group was optimized and tested by using separate testing sets of both positive and negative samples composed of all the remaining distinct proteins of a group and those outside the group respectively. The performance of

SVM classification was further evaluated by using independent sets of both positive and negative samples composed of all the remaining proteins of a group and those outside the group respectively. No duplicate protein entry was used in the training, testing and independent evaluation set for each group. For those with sufficient number of distinct members, multiple entries were assigned into each set. For those with less than three distinct members, the proteins were assigned in the order of priority of training, testing and independent evaluation set.

The number of positive and negative samples for each of the training, testing and independent evaluation set for each group of TM proteins is given in Table 1. The training set is composed of 2,105 TM and 2,563 NM proteins, 927 G-protein coupled receptors and 1,320 non-G-protein coupled receptors, 177 envelope proteins and 1,999 non-envelope proteins, 602 outer membrane proteins and 1,539 non-outer membrane proteins, and 485 transporters/channels and 3,511 non-transporters/non-channels. The testing set is comprised of 12,857 TM and 9,605 NM proteins, 4,998 G-protein coupled receptors and 13,216 non-G-protein coupled receptors, 123 envelope proteins and 7,932 non-envelope proteins, 547 outer membrane proteins and 8,385 non-outer membrane proteins, and 335 transporters/channels and 5,632 non-transporters/non-channels. The independent evaluation set is made of 3,389 TM and 6,063 NM proteins, 2,532 G-protein coupled receptors and 7,244 non-G-protein coupled receptors, 150 envelope proteins and 4,952 non-envelop proteins, 343 outer membrane proteins and 4,948 non-outer membrane proteins, and 160 transporters/channels and 3,963 non-transporters/non-channels.

2.2 Feature Vector Construction

Construction of the feature vector for a protein was based on the formula for the prediction of protein-protein interaction [33] and protein function prediction [23,24,25,26]. Details of the formula can be found in the respective publications and references therein. Each feature vector was constructed from encoded representations of tabulated residue properties including amino acids composition, hydrophobicity, normalized van der Waals volume, polarity, polarizability, charge, surface tension, secondary structure and solvent accessibility.

There is some level of overlap in the descriptors for hydrophobicity, polarity, and surface tension. Thus the dimensionality of the feature vectors may be reduced by principle component analysis (PCA). Our own study suggests that the use of PCA reduced feature vectors only moderately improves the accuracy. It is thus unclear to which extent this overlap affects the accuracy of SVM classification. It is noted that reasonably accurate results have been obtained using these overlapping descriptors in various protein classification studies [23,24,25,26,33,34,35,36].

2.3 Support Vector Machine

SVM is based on the structural risk minimization (SRM) principle from statistical learning theory [21]. SVM constructs a hyperplane that separates two different classes of feature vectors. A feature vector xi represents the structural and physico-chemical properties of a protein. There are a number of hyperplanes for an identical group of training data. The classification objective of SVM is to separate the training data with maximum margin while maintaining reasonable computing efficiency. SVM maps feature vectors into a high dimensional feature space using a kernel function $K(\mathbf{x}_i, \mathbf{x}_j)$ followed by the construction of OSH in the feature space [36]. Gaussian kernel function: $K(\mathbf{x}_i, \mathbf{x}_j) = e^{\frac{-\|\mathbf{x}_i - \mathbf{x}_j\|^2}{2\sigma^2}}$ was used in this work because it consistently gives better results than other kernel functions [35]. Linear support vector machine is applied to this feature space and then the decision function is given by:

$$f(\mathbf{x}) = sign[\sum_{i=1}^{l} \alpha_i^0 y_i K(\mathbf{x}, \mathbf{x}_i) + b], \tag{1}$$

where the coefficients α_i^0 and b are determined by maximizing the following Langrangian expression:

$$\sum_{i=1}^{l} \alpha_i - \frac{1}{2} \sum_{i=1}^{l} \sum_{j=1}^{l} \alpha_i \alpha_j y_i y_j K(\mathbf{x}_i, \mathbf{x}_j), \tag{2}$$

under conditions

$$\alpha_i \geq 0 \quad \text{and} \quad \sum_{i=1}^{l} \alpha_i y_i = 0. \tag{3}$$

Positive or negative value from Eq.(1) indicates that the vector $\mathbf{x}$ belongs to the positive or negative class respectively. To further reduce the complexity of parameter selection, hard margin SVM with threshold was used in our own SVM program SVM* [36].

As in the case of all discriminative methods [37], the performance of SVM classification can be measured by the quantity of true positives TP, true negatives TN, false positives FP, false negatives FN, sensitivity $SE = TP/(TP + FN)$, specificity $SP = TN/(TN + FP)$, the overall accuracy (Q) and Matthews Correlation Coefficient (MCC) [25] are given below:

$$Q = (TP + TN)/(TP + FN + TN + FP), \tag{4}$$

$$MCC = \frac{TP \bullet TN - FN \bullet FP}{\sqrt{(TP + FN)(TP + FP)(TN + FN)(TN + FP)}}. \tag{5}$$

3 Results and Discussion

The number of training and testing proteins and prediction results of specific class of TM proteins and the corresponding NM proteins are given in Table 1. In this Table, TP stands for true positive (correctly predicted TM protein of a specific TM class), FN stands for false negative (protein from a specific class of TM proteins incorrectly predicted as a non-class-member), TN stands for true negative (correctly predicted non-class-member), and FP stands for false positive (non-class-member incorrectly predicted as a member of a specific class of TM proteins). The predicted accuracies for TM proteins, G-protein coupled receptors, envelope proteins, outer membrane proteins, and transporters/channel are 90.1%, 95.6%, 90.0%, 92.7% and 73.9% respectively. The predicted accuracies for NM proteins, non-G-protein coupled receptors, non-envelope proteins, non-outer membrane proteins, and non-transporters/non-channels are 86.7%, 98.1%, 99.5%, 86.4%, and 98.6% respectively.

Table 1. Prediction accuracies and the number of positive and negative samples in the training, test, and independent evaluation set of transmembrane proteins (Tr), G-protein coupled receptors (Gp), Envelope proteins (En), Outer Membrane proteins (OM), and Transporters and Channels (TC). Predicted results are given in TP, FN, TN, FP, accuracy for positive samples SE, accuracy for negative samples SP, overall accuracy Q and Matthews correlation coefficient MCC. The number of positive or negative samples in the training set is P or N respectively. The number of positive or negative samples in the test and independent evaluation sets is $TP+FN$ or $TN+FP$ respectively. PF represents Protein Family.

	Training Set		Test Set				Independent Set							
PF	P	N	TP	FN	TN	FP	TP	FN	TN	FP	SE (%)	SP (%)	Q (%)	MCC
Tr	2105	2563	11135	1722	8237	1368	3054	335	5254	809	90.1	86.7	87.9	0.749
Gp	927	1320	4993	5	13212	4	2421	111	7104	140	95.6	98.1	97.4	0.933
En	177	1999	112	11	7904	28	135	15	4927	25	90.0	99.5	99.2	0.867
OM	602	1539	547	0	8384	1	318	25	4276	672	92.7	86.4	86.8	0.499
TC	485	3511	331	4	5628	4	127	33	3909	54	73.9	98.6	97.8	0.735

A direct comparison with results from previous protein studies is inappropriate because of the differences in the specific aspects of proteins classified, dataset, descriptors and classification methods. Nonetheless, a tentative comparison may provide some crude estimate regarding the level of accuracy of our method with respect to those achieved by other studies. With the exception of and transporters/channels, the accuracies for various TM classes are comparable to those of ~95% obtained from previous studies [13,18]. The prediction accuracy for transporters and channels is substantially lower primarily because the collected proteins in this class are not sufficiently diverse to adequately train the corresponding SVM classification system. There are 250 identified families of transporters and 115 families of channels, some of which contain substantial number of distinct proteins [20]. Thus the collected 980 transporters and channels are not enough to fully represent all of the identified families.

The prediction accuracy for the NM proteins and those of negative samples of individual TM classes is comparable to the level of 92%~99% obtained from previous studies. Unlike that of the positive samples, the prediction accuracy for the negative samples of transporters and channels is comparable to those of other classes and those from other studies. This is because the corresponding SVM system was trained by using a diverse set of negative samples that include all representative NM proteins and proteins from other TM classes.

4 Conclusion

SVM appears to be capable of prediction of MP proteins and proteins in specific TM classes from a large number and diverse range of proteins at accuracies comparable to those from other studies. The prediction accuracy of SVM may be further enhanced with the improvement of SVM algorithms particularly the use of multi-class prediction models, more adequate training for distantly related proteins, and the use of the expanded knowledge about specific classes of TM proteins such as transporters and channels. To assist their evaluation and exploration, our SVM classification systems – SVMProt, can be accessed at http://jing.cz3.nus.edu.sg/cgi-bin/svmprot.cgi.

References

1. Stack, J.H., Horazdovsky, B.,Emr, S.D.: Receptor-mediated Protein Sorting to the Vacuole in Yeast: Roles for a Protein Kinase, a Lipid Kinase and GTP-binding Proteins. Annu. Rev. Cell Dev. Biol. **11** (1995) 1–33
2. Le Borgne, R., Hoflack, B.: Protein Transport from the Secretory to the Endocytic Pathway in Mammalian Cells. Biochim. Biophys. Acta **1404** (1998) 195–209
3. Chen, X., Schnell, D.J.: Protein Import into Chloroplasts. Trends Cell Biol. **9** (1999) 222–227
4. Thanassi, D.G., Hutltgren, S.J.: Multiple Pathways Allow Protein Secretion Across the Bacterial Outer Membrane. Curr. Opin. Cell Biol. **12** (2000) 420–430
5. Heusser, C. , Jardieu, P.: Therapeutic Potential of Anti-IgE Antibodies. Curr. Opin. Immunol. **9** (1997) 805–813
6. Saragovi, H.U. , Gehring, K.: Development of Pharmacological Agents for Targeting Neurotrophins and their Receptors. Trends Pharmacol. Sci. **21** (2000) 93–98
7. Sedlacek, H.H.: Kinase Inhibitors in Cancer Therapy: A Look Ahead. Drugs **59** (2000) 435–476
8. Zhang, L., Brett, C.M., Giacommi, K.M.: Role of Organic Cation Transporters in Drug Absorption and Elimination. Annu. Rev. Pharmacol. Toxicol. **38** (1998) 431–460
9. Tamai, I. , Tsuji, A.: Transporter-mediated Permeation of Drugs Across the Blood-brain Barrier. J. Pharmaceut. Sci. **89** (2000) 1371–1388
10. McGovern, K., Ehrmann, M., Beckwith, J.: Decoding Signals for Membrane Proteins using Alkaline Phosphatase Fusions. EMBO J. **10** (1991) 2773–2782
11. Amstutz, P., Forrer, P., Zahnd, C., Pluckthun, A.: In Vitro Display Technologies: Novel Developments and Applications. Curr. Opin. Biotechnol. **12** (2001) 400–405
12. Wallin, E. , von Heijne, G.: Genome-wide Analysis of Integral Membrane Proteins from Eubacterial, Archaean, and Eukaryotic Organisms. Protein Sci. **7** (1998) 1029–1038

13. Chen, C.P., Kernytsky, A., Rost, B.: Transmembrane Helix Predictions Revisited. Protein Sci. **11** (2002) 2774–2791
14. Krogh, A., Larsson, B., von Heijne, G., Sonnhammer, E.L.: Predicting Transmembrane Protein Topology with A Hidden Markov Model: Application to complete genomes. J. Mol. Biol. **305** (2001) 567–580
15. Cai, Y.D., Zhou, G.P., Chou, K.C.: Support Vector Machine for Predicting Membrane Protein Types by using Functional Domain Composition. Biophys. J. **84** (2003) 3257–3263
16. Gromiha, M.M., Ahmad, S., Suwa, M.: Neural Network-based Prediction of Transmembrane -strand Segments in Outer Membrane Proteins. J. Comput. Chem. **25** (2004) 762–767
17. Yuan, Z., Mattick, J.S., Teasdale, R.D.: SVMtm: Support Vector Machines to Predict Transmembrane Segments. J. Comput. Chem. **25** (2004) 632–636
18. Cserzo, M., Eisenhaber, F., Eisenhaber, B., Simon, I.: On Filtering False Positive Transmembrane Protein Predictions. Protein Eng. **15** (2002) 745–752
19. Bairoch, A., Apweiler, R.: The SWISS-PROT Protein Sequence Database And Its Supplement Tremble In 2000. Nucleic Acids Res. **28** (2000) 45–48
20. Saier, M.H.: A functional-phylogenetic Classification System for Transmembrane Solute Transporters. Microbiol. Mol. Biol. Rev. **64** (2000) 354–411
21. Vapnik, V.: The Nature of Statistical Learning Theory. Springer: New York (1999)
22. Karchin, R., Karplus, K., Haussler, D.: Classifying G-protein Coupled Receptors with Support Vector Machines. Bioinformatics **18** (2002) 147–159
23. Cai, C.Z., Han, L.Y., Ji, Z.L., Chen, X., Chen, Y.Z.: SVM-Prot: Web-Based Support Vector Machine Software for Functional Classification of a Protein from Its Primary Sequence. Nucleic Acids Res. **31** (2003) 3692–3697
24. Cai, C.Z., Han, L.Y., Chen, Y.Z.: Enzyme Family Classification by Support Vector Machines. Proteins **55** (2004) 66–76
25. Cai, C.Z., Wang, W.L., Sun, L.Z., Chen, Y.Z: Protein Function Classification via Support Vector Machine Approach. Math. Biosci. **185** (2003) 111–122
26. Cai, C.Z., Han, L.Y., Chen, X. et al.: Prediction of Functional Class of the SARS Coronavirus Proteins by a Statistical Learning Method. J. Proteome Res. **4** (2005) 1855–1862
27. Han, L.Y., Cai, C.Z., Lo, S.L., et al.: Prediction of RNA-binding Proteins from Primary Sequence by a Support Vector Machine Approach. RNA **10** (2004) 355–368
28. Han, L.Y., Cai, C.Z., Ji, Z.L., Chen, Y.Z.: Prediction of Functional Class of Novel Viral Proteins by a Statistical Learning Method Irrespective of Sequence Similarity. Virology **331** (2005) 136–143
29. Han, L.Y., Cai, C.Z., Ji, Z.L., et al.: Predicting Functional Family of Novel Enzymes Irrespective of Sequence Similarity: a Statistical Learning Approach. Nucleic Acids Res. **32** (2004) 6437–6444
30. Cui, J., Han, L.Y., Cai, C.Z., et al.: Prediction of Functional Class of Novel Bacterial Proteins without the Use of Sequence Similarity by a Statistical Learning Method. J. Mol. Microbiol. Biotechnol. **9** (2005) 86–100
31. Lin, H.H., Han, L.Y., Cai, C.Z., Ji, Z.L., Chen, Y.Z.: Prediction of Transporter Family from Protein Sequence by Support Vector Machine Approach. Proteins **62** (2006) 218–231
32. Bateman, A., Birney, E., Cerruti, L. et al.: The Pfam Protein Families Database. Nucleic Acids Res. **30** (2002) 276–280
33. Bock, J.R. and Gough, D.A.: Predicting Protein-protein Interactions from Primary Structure. Bioinformatics **17** (2001) 455–460

34. Lo, S.L., Cai, C.Z., Chen, Y.Z., Chung, M.C.M.: Effect of Training Datasets on Support Vector Machine Prediction of Protein-protein Interactions. Proteomics **5** (2005) 876–884
35. Cai, Y.D., Liu, X.J., Xu, X.B., Chou, K.C.: Support Vector Machines for Predicting HIV Protease Cleavage Sites in Protein. J. Comput. Chem. **23** (2002) 267–274
36. Cai, C.Z., Wang, W.L., Chen, Y.Z.: Support Vector Machine Classification of Physical and Biological Datasets. Inter. J. Mod. Phys. C **14** (2003) 575–585
37. Baldi, P., Brunak, S., Chauvin, Y., Andersen, C.A.F., Nielsen, H.: Assessing the Accuracy of Prediction Algorithms for Classification: An Overview. Bioinformatics **16** (2000) 412–424

Protein Subcellular Location Prediction Based on Pseudo Amino Acid Composition and Immune Genetic Algorithm

Tongliang Zhang, Yongsheng Ding, and Shihuang Shao

College of Information Sciences and Technology,
Donghua University, Shanghai 201620, China
ysding@dhu.edu.cn

Abstract. Protein subcellular location prediction with computational method is still a hot spot in bioinformatics. In this paper, we present a new method to predict protein subcellular location, which based on pseudo amino acid composition and immune genetic algorithm. Hydrophobic patterns of amino acid couples and approximate entropy are introduced to construct pseudo amino acid composition. Immune Genetic algorithm (IGA) is applied to find the fittest weight factors for pseudo amino acid composition, which are crucial in this method. As such, high success rates are obtained by both self-consistency test and jackknife test. More than 80% predictive accuracy is achieved in independent dataset test. The result demonstrates that this new method is practical. And, the method illuminates that the hydrophobic patterns of protein sequence influence its subcellular location.

1 Introduction

In eukaryotic cell, newly proteins play their biological roles only if they are targeted to the correct subcellular compartments. Firstly, Nakia and Kanehisa [1] put forward computational method to predict the protein subcellular location. Recently, several approaches have been introduced on this topic, such as neural networks [2], markov chain models [3], support vector machine [4-7].

Chou and Elrod proposed covariant discriminant algorithm to predict the subcellular locations of protein from amino acid composition. Shortly after, Chou presented the concept of pseudo amino acid composition. Several powerful prediction algorithms are developed based on the concept. The most important difference among these methods is how to construct pseudo amino acid composition. Pan et al [8] used digital signal processing; Cai and Chou [9] employed domain composition; Xiao et al [10] applied complexity measure factor. Wang et al. [11] studied the problem of membrane protein type prediction by using amino acid order effect.

Here, we introduce a new method which based on pseudo-amino acid composition and immune genetic algorithm. The pseudo-amino acid composition is constructed by amino acid composition, hydrophobic patterns and approximate entropy (ApEn) [14] of protein sequence. According to the theory of Lim [12], amino acid residue hydrophobic patterns incline to occur in second structure of a protein sequence. The amino acid

D.-S. Huang, K. Li, and G.W. Irwin (Eds.): ICIC 2006, LNBI 4115, pp. 534 – 542, 2006.

residue hydrophobic value represents the major driving force behind protein folding [13]. Proteins take different functions in different subcellular. So amino acid residue hydrophobic pattern can reveal amino acid sequence. ApEn is used to evaluate the complexity of sequence.

The weight factors of pseudo amino acid composition are crucial to predict subcellular locations. In the reported literature, the method of decision weight factors is test. However, it is difficult to find all appropriate weight factors through just several tests. Here, we apply the immune genetic algorithm (IGA) to find the fittest weight factors. IGA has some new functions than simple genetic algorithm (SGA), for instance, antigen recognition, memory, and adjustment. It promotes the viability of some individuals in population by vaccination. Compared with SGA, IGA not only provides better solution, but also enhances the algorithm convergent speed.

2 Methods

2.1 Datasets

Two datasets constructed by Chou [16] are adopted to test our approach. The training dataset consists of 2191 proteins and independent dataset consists of 2494 proteins. Training dataset includes 145 chloroplast, 571 cytoplasm, 34 cytoskeleton, 49 endoplasmic reticulum, 224 extracellular, 25 golgi apparatus, 37 lysosome, 84 mitochondria, 272 nucleus, 27 peroxisome, 699 plasma membrane, and 24 vacuole. While, Indepen-dent dataset includes 112 chloroplast, 761 cytoplasm, 19 cytoskeleton, 106 endoplasmic reticulum, 95 extracellular, 4 golgi apparatus, 31 lysosome, 163 mitochondria, 418 nucleus, 23 peroxisome, and762 plasma membrane.

2.2 Hydrophobic Value and Hydrophobic Patterns

There are 20 kinds of amino acid in organism which represented by characters as A, C, D, E, F, G, H, I, K, L, M, N, P, Q, R, S, T, V, W, and Y. The characters in protein sequence cannot be used to calculate the characteristic of protein sequence in mathematical computation. The protein sequence has to be described in quantitative way and each element has its corresponding numerical value. The hydrophobic value of amino acid can reflect the amino acid tends to occur in the surface of protein or in the core of protein. We select the hydrophobic value of 20 amino acids to present digitized protein sequence.

Hydrophobic patterns *(i, i+2; i,i+3; i,i+2,i+4; i,i+5; i,i+3,i+4; i,i+1,i+4)*, which based on the theory of Lim[12], are classified into three types to construct pseudo amino acid composition. The hydrophobic patterns *(i ,i+2)* and *(i,i+2,i+4)* occur more frequently in beta-sheets while the patterns*(i,i+3)*, *(i,i+3,i+4)* and *(i,i+1,i+4)* tend to occur more often in alpha-helices. The concept *(i,i+5)* is an extension of the concept of the “helical wheel” or amphipathic alpha-helix[17], where the helix has a hydrophilic side with amino acid side chains extending into the solvent and a hydrophobic side where side chains extend into the non-polar core of the protein. The occurrence frequencies of three types in protein represent the folding pattern.

Table 1. The hydrophobic value of 20 amino acids

Amino acid	Hydrophobic value	Amino acid	Hydrophobic value
A	0.62	M	0.64
C	0.29	S	-0.18
D	-1.05	N	-0.85
E	-0.87	T	-0.05
F	1.19	P	0.12
G	0.48	V	1.08
H	-0.4	Q	-0.78
I	1.38	W	0.81
K	-1.35	R	-1.37
L	1.06	Y	0.26

The selection of hydrophobic amino acids is based on the theory of Rose [18]. The amino acids that have a mean fractional area loss (on transfer from the standard state to the folded protein) of greater than 80% are included. These amino acids tend to be buried in the hydrophobic core of a protein. Seven amino acids meet this criterion: Val (V), Leu (L), Ile (I), Met (M), Cys (C), Phe (F), and Trp (W). The method of calculation occurrence frequencies of three types in protein is described as below:

Step 1: A protein sequence R is expressed by

$$R = R_1 R_2 R_3 \ldots R_N \tag{1}$$

where, R_i is the i-th residue in protein sequence.
Step 2: If the conditions $R_i \in \Psi$ and $R_{(i+2)} \in \Psi$ are satisfied simultaneously, $C_{(i,i+2)} = C_{(i,i+2)} + 1$, where $\Psi = \{V, L, I, M, C, F, W\}$. $C_{(i,i+2,i+4)}$, $C_{(i,i+3)}$, $C_{(i,i+3,i+4)}$, $C_{(i,i+1,i+4)}$, and $C_{(i,i+5)}$ are computed according to the rule mentioned above.

Step 3:

$$F(\alpha) = \sum_{i=1}^{i=N-2} C_{(i,i+2)} + \sum_{i=1}^{i=N-4} C_{(i,i+2,i+4)} \tag{2}$$

$$F(\beta) = \sum_{i=1}^{i=N-2} C_{(i,i+3)} + \sum_{i=1}^{i=N-4} C_{(i,i+3,i+4)} + \sum_{i=1}^{i=N-4} C_{(i,i+1,i+4)} \tag{3}$$

$$F(\gamma) = \sum_{i=1}^{i=N-2} C_{(i,i+5)} \tag{4}$$

where, $F(\alpha)$, $F(\beta)$, $F(\gamma)$ denote the occurrence frequencies of alpha-helices, beta-sheets, and amphipathic alpha-helix, respectively.

2.3 Approximate Entropy (ApEn)

Approximate entropy is a non-negative number that denotes the complexity of time series [14]. It has been successfully applied to the field of EEG and cognition. Pincus

[15] introduced ApEn, a set of measures of system complexity closely related to entropy, which is easily applied to clinical cardiovascular and other time series. The algorithm which is described in his paper is low efficiency and slow speed. Hong et al. [19] improve the algorithm and the speed is almost enhanced by five times. We apply the new algorithm in the study. Suppose a digitized protein sequence X is expressed by $X = u_1 u_2 u_3 .. u_N$.

The approximate entropy can be calculated through the following steps:

Step 1: m-D (dimensional) vector *X(i)* is composed by sequence *u(i)* according to its order. Assume that m=2.

$$X(i) = [u(i), u(i+1), ..., u(i+m-1)] \; i = 1 \sim N-m+1 \tag{5}$$

Step 2: Assume a threshold *r*, if both the distance $dij = u(i) - u(j)$ and the distance $d_{(i+1)(j+1)} = u(i+1) - u(j+1)$ less than *r* then $C_i^m = C_i^m + 1$. Assume r=0.25SD(u) (SD: standard difference) in calculate procedure.

$$C_i^m(r) = \sum_{j=1}^{N-m+1} d_{ij} \cap d_{(i+1)(j+1)} \tag{6}$$

$$C_i^{m+1}(r) = \sum_{j=1}^{N-m} d_{ij} \cap d_{(i+1)(j+1)} \cap d_{(i+2)(j+2)} \tag{7}$$

Step 3: Make the average value of the logarithm of C_i^m and C_i^{m+1}, the results are signed as $\Phi^m(r)$ and $\Phi^{m+1}(r)$

$$\Phi^m(r) = \frac{1}{N-m+1} \sum_{i=1}^{N-m+1} \ln C_i^m(r) \tag{8}$$

$$\Phi^{m+1}(r) = \frac{1}{N-m} \sum_{i=1}^{N-m} \ln C_i^{m+1}(r) \tag{9}$$

Step 4: Approximate entropy is the difference between $\Phi^m(r)$ and $\Phi^{m+1}(r)$

$$\mathrm{ApEn(m,r)} = \Phi^{\mathrm{m}}(\mathrm{r}) - \Phi^{\mathrm{m+1}}(\mathrm{r}) \tag{10}$$

2.4 Immune-Genetic Algorithm

According to the description by Chou [20], a protein can be expressed by a 24-D vector that is composed by 20 amino acid frequency and 4 characteristics of digital amino acid sequence: 3 hydrophobic patterns and ApEn value.

$$X = [x_1, x_2, ..., x_{20}, x_{21}, x_{22}, x_{23}, x_{24}]^T \tag{11}$$

where

$$x_k = \begin{cases} \dfrac{f_k}{\sum_{i=1}^{20} f_i + \sum_{i=1}^{4} w_j p_j} & (1 \le k \le 20) \\ \dfrac{w_k p_k}{\sum_{i=1}^{20} f_i + \sum_{i=1}^{4} w_j p_j} & (21 \le k \le 24) \end{cases} \tag{12}$$

In Eq. (12), f_k is the normalized occurrence frequency of the 20 amino acid in the protein, p_i is the additional characteristic parameter, and w_i is the weight factor for the parameter p_i. In this study, the weight factors are calculated by IGA. The IGA is described as follows:

Step 1: Initialization

Generate N individuals and set search spaces for each parameter, then assign uniform random numbers within search space to each parameter of all individuals. Each individual is an antibody. The fitness function is the antigen.

Step 2: Evaluation

Calculate fitness F for each individual. The results are sorted. If the stop criterion is satisfied, then output the result.

Step 3: Stopping criterion

The implementation will be stopped if the conditions are satisfied:

The value for ΔF does not change for several generations.

$$\Delta F(i) = F(i) - F(i-1) \tag{13}$$

where, $F(i)$ is the fitness value for the i-th generation, $F(i-1)$ is the fitness value for the $(i-1)$-th generation.

2) The number of generation reaches to upper limit G_{max}

Step 4: Update memory set

At the first time, the memory set is empty. m individuals with the highest fitness are saved into the memory set. The memory set is composed of m individuals. In the current population, the d individuals with highest fitness are selected to replace the d individuals in memory set that are the lowest.

Step5: Crossover and mutation

The methods of crossover and mutation are the same as traditional genetic algorithm. After the new N individuals are generated, let the m individuals in the memory set replace the m new individuals. The new population is generated. Return to Step 2.

3 Results and Discussion

Evaluating an algorithm, the success prediction rate and the reliability of prediction should be considered together. Three indexes are applied to evaluate the prediction accuracy: Sensitivity (S_n), personality (S_p), correlation coefficient (CC).

$$S_n = \frac{t_p}{t_p + f_n} \tag{14}$$

$$S_p = \frac{t_p}{t_p + f_p} \tag{15}$$

$$CC = \frac{S_n + S_p}{2} \tag{16}$$

where, t_p is the protein number of right prediction in a subcellular group, f_n is the protein number of wrong prediction in a subcellular group, f_p is the number of the proteins in other groups to be predicted in this group. Sensitivity (S_n) represents the accuracy, and personality (S_p) represents the reliability in procedure of prediction.

Three methods are applied to examine the prediction quality: the self-consistent test, the jackknife test, and independent data set test. In the self-consistent test, the prediction rates for 12 subcellular locations are obtained. S_n, S_p and CC are calculated respectively.

Table 2. Accuracy rate by the self-consistency test for training data set

Subcellular location	Sn	Sp	CC
Chloroplast	0.8690	0.6774	0.7732
Cytoplasmic	0.8266	0.8444	0.8355
Cytoskeleton	1.0000	0.7391	0.8696
Endoplasmic reticulum	0.9388	0.6479	0.7933
Extracellular	0.7009	0.8263	0.7636
Golgi apparatus	1.0000	1.0000	1.0000
Lysosomal	1.0000	0.8810	0.9405
Mitochondrial	0.7976	0.6381	0.7179
Nuclear	0.7794	0.8760	0.8277
Peroxisomal	1.0000	1.0000	1.0000
Plasma membrane	0.9270	0.9614	0.9442
Vacuolar	1.0000	1.0000	1.0000
Total accuracy	1875/2191=85.58%		

IGA is used to calculate the weight factors in Eq. 12. The fitness is a key in IGA. A prediction algorithm cannot be deemed as a good one if the self-consistency is poor. So we select overall accuracy rate of self-consistency as fitness F. Generate 20 individuals and assign uniform random numbers in search space [0, 1.0] for 4 weight factors. The crossover rate is 0.90 and the mutation rate is 0.2. When the number of generations reaches the upper limit 20, the biggest fitness F and the weight factors are obtained: $F=0.8594$

$$\begin{cases} w_1 = 0.0050 \\ w_2 = 0.0933 \\ w_3 = 0.0612 \\ w_4 = 0.0555 \end{cases}$$

The sensitivity (S_n), personality (S_p), and correlation coefficient (CC) of 12 subcellular locations are listed in Table 2. In training data set, 1875 proteins are correctly predicted and 316 proteins are incorrectly done. From the Table 2, we can see that the accuracy (S_n) and the reliability (S_p) are satisfied. It can be testified by the correlation coefficient (CC).

The Jackknife test and independent data set test are the methods often used for cross-validation in statistical prediction. The jackknife is deemed as the most effective and objective method [21, 22]. Chou [23] discussed the mathematical principle of jackknife test. Accordingly, the accuracy rate of jackknife test should be calculated to measure the new predictor algorithm. The independent data set test can be used to examine the practical application of new approach. An independent data set is applied to test the new algorithm. The overall accuracy rate of jackknife on training data set and independent data set test are filled in Table 3, which are 79% and 80.27%, respectively. The results of three other test methods on same data sets are also listed in Table 3. From the table 3, we can see that the overall success rates obtained by current approach are higher than those of Cedano [24] and Pan [8], while almost the same as that of Xiao [10]. However, the accuracy rate of independent data set test is higher than that of Xiao. The accuracy rate of independent data set test obtained by the new algorithm is highest among the algorithm already known.

Table 3. Overall success rates by different test methods and different algorithms

Algorithm	Test method		
	Self-consistency	Jackknife	Independent set
ProtLock	1006/2191=45.9%	971/2191=44.3%	1018/2491=40.8%
Digital signal	1785/2191=81.5%	1483/2191=67.7%	1842/2494=73.9%
Complexity	1884/2191=86.0%	1612/2191=73.6%	1990/2494=79.9%
This work	1875/2191=85.58%	1573/2191=71.80%	2002/2494=80.27%

We test the accuracy rates when the number of subcellular location is reduced from 12 (S^{12}) to seven (S^7) and five (S^5). The results are filled in Table 5. When the small subsets are reduced, the overall accuracy rate decreases in the self-consistency test and independent data set test. The accuracy rates of small subsets are almost 100% in self-consistency and independent data set test, such as cytoskeleton, endoplasmic reticulum, Golgi apparatus, lysosome, peroxisome, and vacuole. But in the jackknife test, the accuracy rates increase with the decreasing of subsets with little proteins number. We can see that the accuracy rate changes from 71.79% (S^{12}) to 76.37% (S^7) and 80.9% (S^5). This because that the protein number in training data set is little. If the training data set is enlarged, the difference between the self-consistency test and jackknife test will be erased.

The aim of this study is to find a new method to predict protein subcellular location based on the concept of pseudo amino acid composition. Hydrophobic of amino acids determine the folding pattern of protein. Therefore, the occurrence frequencies of hydrophobic patterns in protein sequence can reflect the function of protein in some respects. The result obtained in this study proves that the new approach is practical.

Table 4. Overall accuracy rates by different data sets and different test methods

Measures	S12	S7	S5
Self-consistency	1875/2191=85.58%	1731/2044=84.69%	1657/1911=86.71%
Jackknife	1573/2191=71.79%	1561/2044=76.37%	1546/1911=80.9%
Independent set	2002/2494=80.27%	1924/2417=79.60%	1766/2216=79.69%

4 Conclusions

Different compartments of a cell have different physic-chemical environments. So the physic-chemical environments of the compartments of a cell will determine the subcellular location of proteins with special surface physic-chemical characteristic. The amino acid composition might carry a 'signal' in subcellular location. The additional components of pseudo amino acid composition are defined from view of surface physic-chemical characteristic of protein folding. The introduction of frequencies of alpha-helices, beta-sheets, and amphipathic alpha-helix in protein sequence as the pseudo amino acid composition can effectively reflect the sequence feature of a protein. Approximate entropy reflects the complexity of the whole protein sequence. We obtain a higher accuracy rate in predicting the subcellular location by self-consistency test, jackknife test, and independent data set test, especially the result obtained by independent data set test. As such, our new approach is quite promising and practical.

Acknowledgments

This work was supported in part by the National Nature Science Foundation of China (No. 60474037), Program for New Century Excellent Talents in University (NCET-04-4150), and Specialized Research Fund for the Doctoral Program of Higher Education from Educational Committee of China (No. 20030255009). Thanks to Drs. Lei Gao and Wanggen Li for fruitful discussions and valuable comments on the manuscript.

References

1. Nakai, K., Kanehisa, M.: A Knowledge Base for Predicting Protein Localization Sites in Eukaryotic Cells. Genomics, **14** (1992) 897–911
2. Reinhardt, A., Hubbard, T.: Using Neural Networks for Prediction of the Subcellular Location of Proteins. Nucleic Acids Res., **26** (1998) 2230–2236
3. Yuan Z.: Prediction of Protein Subcellular Locations using Markov Chain Models. FEBS Letter, **451** (1999) 23–26
4. Park, K.J., Kanehisa, M.: Prediction of Protein Subcellular Locations Support Vector Machines using Compositions Amino Acids and Amino Acid Pairs. Bioinformatics, **19** (2003) 1656–1663
5. Chou, K.C., Cai, Y.D.: Using Functional Domain Composition and Support Vector Machines for Prediction of Protein Subcellular Location. Journal of biological chemistry, **277** (2002) 45765–45769

6. Hua, S., Sun, Z.: Support Vector Machine Approach for Protein Subcellular Localization Prediction. Bioinformatics, **17** (2001) 721–728
7. Cai, Y.D., Liu, X.J., Xu, X.B., Chou, K.C.: Support Vector Machines for Prediction of Protein Subcellular Llocation by Incorporating Quasi-sequence-order Effect. J. Cell Biochemistry, **84** (2002) 343–348
8. Pan, Y.X., Zhang, Z.Z., Guo, Z.M., Feng, G.Y., Huang, Z.D., He, L.: Application of Pseudo Amino Acid Composition for Predicting Protein Subcellular Location: Stochastic Signal Processing Approach. Journal of Protein Chemistry, **22** (2003) 395-402
9. Cai, Y.D., Chou, K.C.: Nearest Neighbor Algorithm for Predicting Protein Subcellular Location by Combining Function Domain Composition and Pseudo Amino Acid Composition. Biochem. Biophys. Res. Comm., **305** (2003) 407-411
10. Xiao, X., Shao, S.H., Ding, Y.S., Huang, Z.D., Huang Y.S., Chou K.C.: Using Complexity Measure Factor to Predict Protein Subcellular Location. Amino Acid, **28** (2005) 57-61
11. Wang, M., Yang, J., Xu, Z.J., Chou, K.C.: Weight-support Vector Machines for Prediction Membrane Protein Type Based on Pseudo-amino Acid Composition. Protein Engineering Design and Selection, **17** (2004) 509-516
12. Lim, V.I.: Algorithms for the Prediction of A-helical and 0-structural Regions in Globular Proteins. J. Mol. Biol., **88** (1974) 873-894
13. Dill K.A.: Dominant Forces in Protein Folding. Biochemistry, **29** (1990) 7133-7155
14. Sadovsky M.G.: The Method to Compare Nucleotide Sequence based on Minimum Entropy Principle. Bull Math Biol., **65** (2003) 309-322
15. Pincus, S.M.: Approximate Entropy as a Measure of System Complexity. PNAS, **88** (1991) 2297–2301
16. Chou, K.C., Elrod, D.W.: Protein Subcellular Location Prediction. Protein Eng., **12** (1999) 183-190
17. Schiffer, M., Edmundson, A.: Use of Helical Wheels to Represent the Structures of Proteins and to Identify Segments with Helical Potential. Biophys. J., **7** (1967) 121-133
18. Rose, G.D., Geselowitz, A.R., Lesser, G.J., Lee, R.H., Zehfus M.H.: Hydrophobic of Amino Acid Residue in Globular Proteins. Science, **229** (1985) 834-838
19. Hong, B., Tang, Q.Y., Yang, F.S.: Apen and Cross-ApEn: Property, Fast Algorithm and Preliminary Application to the Study of EEG and Cognition, Signal Process, **15** (1999) 100-108
20. Chou, K.C.: Prediction of Protein Cellular Attributes using Pseudo-amino-acid Composition. Protein: struct. Funct. Genet., **43** (2001) 246-255
21. Zhou, G.P.: An Intriguing Controversy over Protein Structural Class Prediction. J. Protein Chem., **17** (1998) 729-738
22. Zhou, G.P., Assa-Munt N.: Some Insight into Protein Structural Class Prediction. Protein: Structure, Function, and Genetics, **50** (2001) 44-48
23. Chou, K.C., Zhang, C.T.: Review: Prediction of Protein Structural Classes. Crit. Rev. Biochem. Mol Biol., **30** (1995) 275-349
24. Cedano, J., Aloy, P., P'erez-Pons, J.A., Querol, E.: Relation between Amino Acid Composition and Cellular Location of Protein. J. Mol. Biol., **266** (1997) 594-600

SPDBS: An SBML-Based Biochemical Pathway Database System*

Tae-Sung Jung[1], Kyoung-Ran Kim[1], Seung-Hyun Jung[1], Tae-Kyung Kim[1], Myung-Sang Ahn[2], Jong-Hak Lee[3], and Wan-Sup Cho[2]

[1] Dept. of Information Industrial Engineering, Chungbuk National University, 361763 Cheongju, Chungbuk, Korea
{mispro, k2ran, sane7142, tkkim}@chungbuk.ac.kr

[2] Dept. of Management Information Systems, Chungbuk National University, 361763 Cheongju, Chungbuk, Korea
{epita55, wscho}@chungbuk.ac.kr

[3] Division of Computer & Information Communications Engineering, Catholic University of Daegu, 712702 Gyeongsan, Gyeongbuk, Korea
jhlee11@cu.ac.kr

Abstract. Biochemical pathways such as metabolic, regulatory, or signal transduction pathways can be viewed as a complicated network of interactions among molecular species in the cell. As the amount of pathway information for various organisms is increasing very rapidly, performing various analyses on the full network of pathways for even multiple organisms can be possible and therefore developing an integrated database for storing and analyzing pathway information is becoming a critical issue. However, analyzing these networks is not easy because of the nature of the existing pathway databases, which are often heterogeneous, incomplete, and/or inconsistent. To solve this problem, SBML(Systems Biology Markup Language) – a computer-readable format for representing various biochemical pathways – has been adopted by the most of the SW packages in systems biology. We developed an SBML-based Biochemical Pathway Database System (SPDBS) that supports (1) efficient integration of the heterogeneous and distributed pathway databases, (2) prediction of the metabolic pathways for a given (non-annotated) genome sequence, (3) dynamic visualization/ simulation of the pathways, (4) starting from the detailed pathway graph, build networks at different levels of representation, (5) imports/exports of SBML documents for the simulation and/or exchange of the biochemical pathways in various applications. To evaluate the system, we applied the system to the construction of pathways from its genome sequences. For the E. coli genome sequence, SPDBS estimates the same metabolic pathways as the original well-known E. coli pathway. We are applying our system to the pathway prediction of S. chungbukensis DJ77 and Vibrio vulnificus CMCP6.

* This work was supported by the Regional Research Centers Program of the Ministry of Education & Human Resources Development and the Program for the Training of Graduate Students in Regional Innovation which was conducted by the Ministry of Commerce, Industry and Energy in Korea.

D.-S. Huang, K. Li, and G.W. Irwin (Eds.): ICIC 2006, LNBI 4115, pp. 543 – 550, 2006.

1 Introduction

There are three kinds of *Biochemical pathways,* metabolic, regulatory, or signal transduction pathways. The pathways can be viewed as interconnected processes including an intricate network of interactions between molecular compounds in the cell[1]. *Metabolic pathways* are responsible for carrying out the chemical reactions that provide basic biological functions (DNA, RNA, protein synthesis and degradation, energy metabolism, fatty acid synthesis, and many others). *Regulatory pathways* are responsible for converting genetic information into proteins (gene products). *Signal transaction pathways* are concerned with coordinating metabolic processes with transcription and protein synthesis. Each of these pathways has been kept in a separate (independent) database with distinct attributes even though they are related each other.

Pathway databases contain data of biochemical pathways which consist of two kinds of information: *biochemical components* (e.g., substrates, enzymes, products) and *their interactions* [1]. Most existing pathway databases focus on specific types of pathways rather than an integrated one: e.g., *Transpath*[2] for protein-DNA interactions, *KEGG*[5] for metabolic pathways, *EcoCyc* and *MetaCyc*[6] for *E. coli* and other organisms' metabolic pathways, *BIND*[2] for signal transduction pathways, *PathFinder*[3], *WIT*[14], *PathMAPA*[15], *BioJAKE*[17], and *MPW*[18] for metabolic pathways. Pathway databases raise many important and challenging computational and bioinformatics issues, such as querying, navigation, and visualization of the pathways; seamless integration/analysis of the heterogeneous pathway data distributed in diverse sources.

Systems Biology Markup Language (SBML) is a computer-readable format for representing models of various biochemical pathways[4,21]. SBML has been evolving since mid-2000 through the efforts of an international group of software developers and users. Various knowledge on the metabolic, regulatory, and signal transduction pathways can be represented in the unified SBML data model[4,21]. Recently, KEGG and EcoCyc export SBML files for the metabolic pathways, and more than 90 software products support SBML as a standard data format.

We propose an SBML-based integrated pathway database system, called *SPDBS(SBML-based Biochemical Pathway Database System,* can access the system from http://database.cbnu.ac.kr/SPDBS/CBPathway.jnlp*),* for the integration and management of heterogeneous biochemical pathways. We adopt an object database for implementing SBML data model. Object model has been regarded as a strong tool for integrating metabolic, regulatory, and signal transduction pathways, removing the (artificial) barrier between 3 kinds of biochemical pathways, and allowing analysis of the whole pathways[1]. In SPDBS, the result of a query can be visualized dynamically and the users can edit or simulate the result pathways by using various simulation tools. For gene list or (non-annotated) sequence data, SPDBS provides dynamic pathway reconstruction or estimation by using an orthologous database[12]. Orthologous database provides useful information for estimating the function of unknown genes. To remove ambiguity in the data integration or query, gene ontology(GO)[23] has been utilized. GO guarantees consistent terminologies in the description of biological entities.

To evaluate SPDBS, we applied the system to the reconstruction of various pathways from their genome sequences or particular genes. In Section 4, we describe pathway reconstruction procedure for *E. coli* genome sequence. The result shows that SPDBS estimates the same metabolic pathways as the original one. And we reconstructed TCA cycle of *S. chungbukensis* DJ77 from non-annotated sequences using SPDBS. We verify the reconstructed pathway biologically with biologists.

The paper is organized as follows. In Section 2, we present related work. In Section 3, we describe the system architecture of the SPDBS. In Section 4, we present key functions of SPDBS and evaluation results for E. coli genome sequence. In Section 5, we conclude our paper.

2 Related Work

In this section, we discuss SBML documents, conventional pathway database systems, and orthologous databases as related issues.

The *SBML(Systems Biology Markup Language)* has been developed by the systems biology community[4]. It is a free and open language designed as computer readable format for representing models of biochemical reaction networks. It uses XML as its description language and UML for modeling the components. Recently, the utility *KEGG2SBML* has been released for converting metabolic pathways in KEGG (text file with pathway image) to the SBML documents[4]. SBML will be popular in the description of the pathways and protein interaction databases because there are over 90 software packages supporting SBML including KEGG and EcoCyc[4].

A wide variety of *pathway database systems* exist for the storage and retrieval of pathway information. The most widely used system is KEGG [5]. While KEGG uses static approaches for the pathway data visualization, its value lies primarily in the breadth of its content coverage. The benefits of the first-generation systems such as WIT [13], MPW [17] and EcoCyc [6], which perform limited dynamic visualization, comes mainly from their content. General purpose systems for pathway rendering include BioJake [15], PathDB [9], PathFinder [3], Pathways Database System [7], PaVESy [10], and VitaPad [11]. However, none of these systems import/export SBML documents into/from the database. Furthermore, they do not provide useful tools such as integration of the various pathways into a database, query/navigation/ modification of the pathway database, estimation/ reconstruction of the pathways from gene-list or (non-annotated) genome sequence, and visualization of the pathways.

The genes that have evolved directly from an ancestral gene [12], [13] are called *orthologous genes.* They are most likely to share the function and similar sequence [18]. Therefore, orthologous genes are useful to predict the functions of unknown genes [19]. In SPDBS, orthologous genes are used in the estimation of the pathways from genome sequence in which unknown genes are included. For this purpose, we have developed an orthologous database for 88 species [12].

3 SPDBS: An SBML-Based Biochemical Pathway Database System

In this section, we present the system architecture of the SBML-based biochemical pathway database system(SPDBS). SPDBS has 4 layers as shown in Fig. 1; a database construction system, a local pathway database, an analysis system, and a user interface. We will introduce each layer one by one.

(a) *Database Construction System (DCS)*: DCS constructs a unified integrated database from external pathway databases. It has two kinds of tools – *SBML Converter* and *Mediators* - for the integration of heterogeneous pathway databases. The *SBML Converter* transforms SBML documents from external sources into the local object database. The *Mediator* receives various kinds of biochemical pathway data and converts them into the local object database. We have developed several mediators for various external data sources.

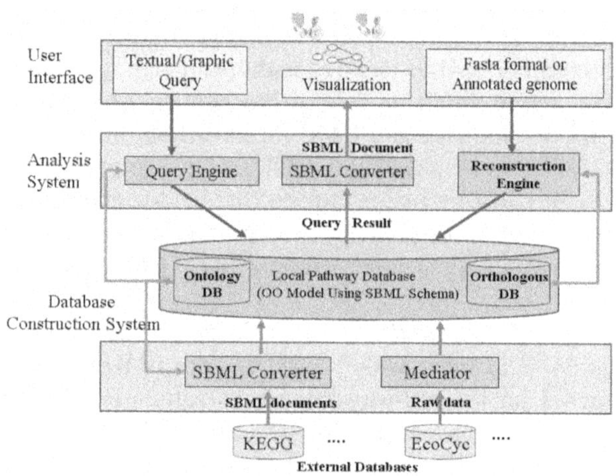

Fig. 1. System Architecture

(b) *Local Pathway Database*: It consists of an *ontology database*[1], an *orthologous database* [12], and *a biochemical pathway database*. The *ontology database* is used for removing terminology ambiguity in the database construction and queries. The *orthologous database* is used for estimating the function of the unknown genes in the genome sequence. The *biochemical pathway database* stores and manages pathway information in an object data model. Although various graph-based data models (such as compound graphs, reaction graphs, bipartite graph, hyper-graphs etc.) have been proposed for representing pathway information, object data model can be seen as a generalization of the graph-based data models [1], and thus it can be regarded as a unified data model for

[1] Gene ontology solves terminology ambiguity problems in the database integration and query. For example, *'EC number 4.1.2.13'* is represented as *'FbaA'* in KEGG. But, it is represented as *'FBABSYN-MONOMER'* in EcoCyc. This leads to ambiguity.

representing various kinds of pathway information. For the pathway information, we developed an object database according to the schema in the SBML[4], [21].

(c) *Pathway Analysis System*: It provides services for gene function prediction, dynamic pathway prediction/reconstruction, and the SBML interface. Pathway reconstruction can be done either from gene list or sequence data; In the case of non-annotated sequence data, the reconstruction engine uses orthologous database for estimating the function of the unknown genes. Result pathways can be represented in an SBML format. For simulation, biochemists can modify the initial quantities and the kinetic laws of the result pathways.

(d) *Graphic User Interface (GUI):* GUI shown in Fig. 2 provides four panels with menu and toolbar: *the left panel* (left top) for the tree-like view of the pathways, *an input query panel* (left bottom) and *an edit panel* (show when it need) for querying and editing the property of an entity. There are a *visualization panel* (right top) and a *result panel* (right bottom) at the right side. Fig. 2 shows a snapshot of the user interface for the 'Pentose and glucuronate interconversions' pathway of Vibrio cholerae.

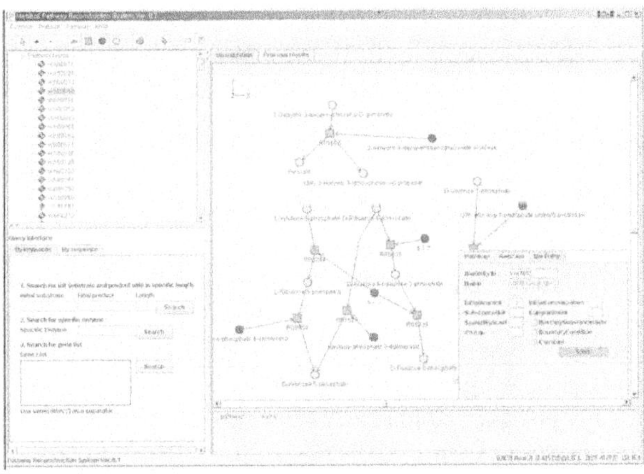

Fig. 2. GUI: Visualization of the 'Pentose and glucuronate interconversions' pathway of Vibrio cholerae

4 Useful Functions and Evaluation of SPDBS

In this section, various functions of SPDBS are presented in detail. We then describe a real application of SPDBS for the reconstruction of the *E. coli* pathway.

4.1 Query and Retrieval Service

SPDBS provides various querying, analysis services and functions. First, given specific substrate and product, SPDBS provides corresponding pathway information to the users. For example, let assume that the user want to find the pathways which consists of ' D-Ribulose ' substrate and ' D-Xylulose 5-phosphate' product. To do this, the initial substrate and the final product with the length (limitation of reaction step) are provided to the system, then the result of the query is displayed on the result panel.

Second, user can find the pathways for given enzyme. If a user input the enzyme 'D-ribulokinase', then SPDBS provides corresponding pathways. User can get visualized pathway by selecting the pathway from the result panel.

Third, SPDBS constructs pathways for a given gene-list. Since multiple pathways can be generated for a single gene-list, SPDBS provides multiple visualization panels. Comparison of the result pathways can be done in the multiple visualization panels.

Fourth, SPDBS estimates pathways for genome sequence in which unknown genes are included (i.e., non-annotated genome sequence). For the genome sequence, the system searches the top-hit scored gene by using the BLAST. For the selected genes, orthologous genes are used for the prediction of their pathways. In the next section, we will describe this service in detail.

Fifth, A metabolic pathway for an increasing number of organisms reveals that even small networks can contain thousands of reactions and chemical compounds. The intimate connectivity between components complicates their decomposition into biologically meaningful sub-pathway. So, SPDBS provides a graph abstraction function. Users can make a group some nodes to an abstract node.

Sixth, SPDBS can export the result pathways into the SBML documents. From the SBML documents, the user can modify the initial quantities and kinetic laws for the pathway, and the result can be exported into the SBML documents. For the result SBML documents, simulation can be done with the help of public simulation software such as Gepasi[22] or COPASI[24].

4.2 Pathway Reconstruction System

For a non-annotated genome sequence, SPDBS estimates the functions of the unknown genes by using orthologous database [12] and dynamically generates metabolic pathways in a graphical form (dynamic visualization). Note that most of the existing systems including KEGG and Boehringer Mannheim support static visualization of metabolic pathways [3], [5]. This not only limits the usefulness of the systems, but also increases maintenance cost.

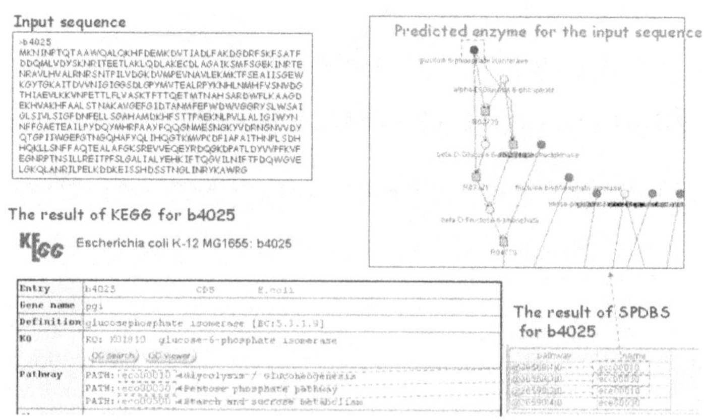

Fig. 3. Evaluation of SPDBS through comparison the results with KEGG

We evaluate the SPDBS through comparison with KEGG. We input the sequence of a gene 'b4025' in *E.coli* K12 and searched for a pathway which modified by the gene 'b4025'. Fig. 3 shows the result for 'b4025' sequence. We can know the results are same.

Also, we reconstructed a metabolic pathway of *S. chungbukensis* DJ77 using SPDBS. Fig. 4 shows the result of reconstructed TCA cycle of *S. chungbukensis* DJ77. We input the non-annotated sequences (about for 10 genes) to SPDBS. And then SPDBS predict the gene for each sequence. Finally, we reconstruct the metabolic pathway through the predicted genes. Now, we verify the reconstructed pathway biologically with biologist.

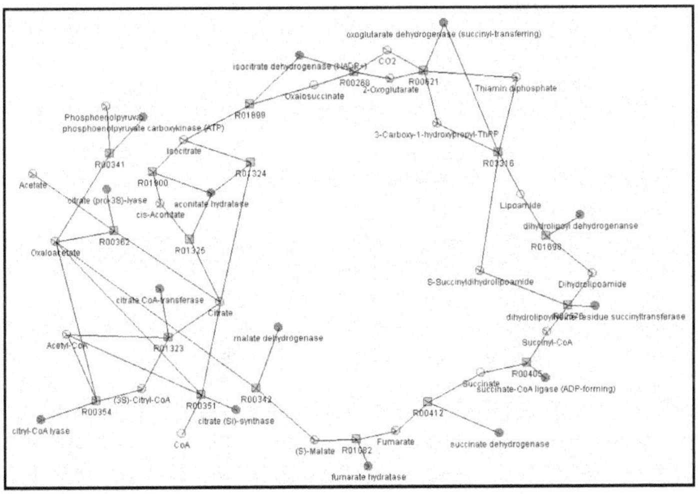

Fig. 4. The reconstructed Citrate Cycle (TCA Cycle) of S. chungbukensis DJ77 using SPDBS system

5 Conclusions and Future Research

The Systems Biology Markup Language (SBML) is a computer-readable format for representing models of biochemical reaction networks. SBML is applicable to metabolic networks, cell-signaling pathways, regulatory networks, and many others such as simulation and visualization of the pathways. SPDBS is a biochemical pathway data-base system for integration, dynamic retrieval, reconstruction/prediction, visualiza-tion, and simulation of the pathways. The system provides SBML import/export for the exchange of the biochemical pathways. To solve the terminology ambiguity, the system uses gene ontology during data integration and query processing. For a genome sequence (gene-list or Fasta format file), the system provides pathway reconstruction service. We are applying our system to the pathway prediction of *Vibrio vulnificus* CMCP6 and *Sphingomonas chungbukensis DJ77*[9] genome sequences whose annota-tions are not completed yet. For the incomplete pathways, we will devise a data mining technique to recommend candidate sub-pathways. We also accommodate other kinds of biochemical pathways data in the SPDBS, and provide integrated pathway analysis services to the users.

References

1. Deville Y., et al.: An Overview of Data Models for the Analysis of Biochemical Pathways. Briefings in Bioinformatics, 4(3). (2003) 246–259
2. Frank Schacherer, et al.: The Transpath Signal Transduction Database: A Knowledge Base on Signal Transduction Networks. Bioinformatics, 17(11) (2001)1053–1057
3. Goesmann, A., et al.: PathFinder: Reconstruction and Dynamic Visualization of Metabolic Pathways, Bioinformatics, 18. (2002) 124–129
4. Hucka, M., et al.: The Systems Biology Markup Language (SBML): A Medium for Representation and Exchange of Biochemical Network Models. Bioinformatics, 19(4). (2003) 524–531
5. Kanehisa, M. and Goto, S.: KEGG: Kyoto Encyclopedia of Genes and Genomes. Nucleic Acids Res., 28. (2002) 27–30
6. Karp, P. D., et al.: The MetaCyc database. Nucleic Acids Res., 30. (2000) 59–61
7. Krishnamurthy, L., et al.: Pathways Database System: An Integrated System for Biological Pathways. Bioinformatics, 19. (2003) 930–937
8. Kuffner, R. M., et al.: PathDB. The Molecular Biology Database. 2004
9. Kwon, H. R., et al: Gene List of Sphingomonas Chungbukensis DJ77. In Proc. International Conference on Genome Informatics, Japan. 2003
10. Ludemann, A., et al.: PaVESy: Pathway Visualization and Editing System. Bioinformatics, 20. (2004) 2841–2844
11. Matthew, H., et al.: VitaPad: Visualization Tools for the Analysis of Pathway Data. Bioinformatics, 21. (2004) 1596–1602
12. Oh, J. S. et al.: Othologous Group Clustering System based on the Grid Computing. In Proc. of the International Joint Conference on InCoB, AASBi, and KSBI (BioInfo 2005). (2005) 72–77
13. Overbeek, R., et al.: WIT: Integrated System for High Throughput Genome Sequence Analysis and Metabolic Reconstruction. Nucleic Acids Res., 28. (2000) 123–125
14. Pan, D., et al.: PathMAPA: A Tool for Displaying Gene Expression and Performing Statistical Tests on Metabolic Pathways at Multiple Levels for Arabidopsis. BMC Bioinformatics, 4, 56. (2003)
15. Paul, Shannon, et al.: Cytoscape: A Software Environment for Integrated Models of Biomolecular Interaction Networks. Genome Research, 13. (2003) 2498–2504
16. Salamonsen, W., et al: BioJAKE: A Tool for the Creation, Visualization and Manipulation of Metabolic Pathways. In Proc. Pac. Symp. Biocomput. (1999) 392–400
17. Selkov, E., et al.: MPW: The Metabolic Pathways Database. Nucleic Acids Res.(1998) 43–45
18. Tatusov, R. L., et al.: A genomic Perspective on Protein Families. Science 278. (1997) 631–637
19. Trost, E., et al.: Java Editor for Biological Pathways. Bioinformatics, 19. (2003) 786–787
20. Reshetnikoy, V., et al.: Vector PathBlazer: A New Pathway Analysis And Visualization Tool. In Proc. ISMB. (2003)
21. http://sbml.org/index.psp
22. http://www.gepasi.org
23. http://www.geneontology.org
24. COPASI, http://www.copasi.org

Supervised Inference of Gene Regulatory Networks by Linear Programming

Yong Wang[1,2], Trupti Joshi[3], Dong Xu[3], and Xiang-Sun Zhang[2], and Luonan Chen[1,4]

[1] Osaka Sangyo University, Nakagaito 3-1-1, Daito, Osaka 574-8530, Japan
ywang@ctex.org, chen@elec.osaka-sandai.ac.jp
[2] Academy of Mathematics and Systems Science, CAS, Beijing 100080, China
zxs@amt.ac.cn
[3] Computer Science Department and Christopher S. Bond Life Sciences Center, University of Missouri, Columbia, MO 65211, USA
{joshitr, xudong}@missouri.edu
[4] Institute of systems biology, Shanghai University, 200444, China

Abstract. The development of algorithms for reverse-engineering gene regulatory networks is boosted by microarray technologies, which enable the simultaneous measurement of all RNA transcripts in a cell. Meanwhile the curated repository of regulatory associations between transcription factors (TF) and target genes is available based on bibliographic references. In this paper we propose a novel method to combine time-course microarray dataset and documented or potential known transcription regulators for inferring gene regulatory networks. The gene network reconstruction algorithm is based on linear programming and performed in the supervised learning framework. We have tested the new method using both simulated data and experimental data. The result demonstrates the effectiveness of our method which significantly alleviates the problem of data scarcity and remarkably improves the reliability.

Keywords: Systems biology, gene regulatory network, linear programming.

1 Introduction

Microarray technologies have produced tremendous amounts of gene expression data [1]. For example the Stanford Microarray Database (SMD) has deposited data for 60,222 experiments, from 302 labs and 36 organisms, as of March, 2006. It is necessary and important to understand gene expression and regulation through mining these data. A straightforward way on microarray data analysis is the reconstruction of gene regulatory network, which aims to find the underlying network of gene-gene interactions from the measured dataset of gene expression [2]. A wide variety of approaches have been proposed to infer gene regulatory networks from time-course data [3, 4, 5] or perturbation experiments [6], such as discrete models of Boolean networks and Bayesian networks, and

D.-S. Huang, K. Li, and G.W. Irwin (Eds.): ICIC 2006, LNBI 4115, pp. 551–561, 2006.

continuous models of neural networks, difference equations [1] and differential equations [7, 8].

The common problem for all these models is scarcity of data [9]. Since a typical gene expression dataset consists of relatively few time points (often less than 20) with respect to a large number of genes (generally in thousands). In other words, the number of genes far exceeds the number of time points for which data are available, making the problem of determining gene regulatory network structure a difficult and ill-posed one.

In this paper we propose a novel method to combine computational analysis of microarray dataset and biological experiment results together for inferring gene regulatory network with the consideration of sparsity of connections. We develop a supervised gene network reconstruction algorithm by linear programming based on the differential equation model [9]. The original idea of supervised learning is to incorporate the known documented interactions between genes into the parameter estimation by keeping sparsity, because many regulatory associations have been identified and recorded in literatures or databases, which are valuable information for the inference of gene networks. For example, YEASTRACT (Yeast Search for Transcriptional Regulators And Consensus Tracking, `http://www.yeastract.com/`) is a curated repository of more than 12,500 regulatory associations between transcription factors (TF) and target genes in *Saccharomyces cerevisiae*, based on more than 900 bibliographic references. All the information in YEASTRACT will be updated regularly to match the recent literature on yeast regulatory networks. In this paper, the supervised information is adopted in the inference procedure by exploiting the general solution form of arbitrary Jacobian matrix for gene regulatory network. The proposed method theoretically ensures the derivation of feasible network structure with respect to the used dataset, thereby not only significantly alleviating the problem of data scarcity but also remarkably improving the reliability. Specifically, it can be expected that the information of documented regulatory interactions considered will lead to biologically plausible results.

2 Methods

In general, one can represent a gene regulatory network model as a directed graph. In this paper the graph is mathematically expressed by a set of linear differential equations. The regulatory influence between genes are formulated as a matrix and obtained as the general solution of differential equations [9]. The supervised learning information is added in the procedure of seeking general solution from particular solution by solving a linear programming problem. Fig. 1 illustrates the schematic of the proposed method. The experiment data obtained by microarray technology is analyzed and normalized, then the time course data of gene expression are collected as a matrix. The dynamic properties of the gene regulatory network are described by ordinary differential equation model. To infer the relationships between genes, previously known regulatory interactions in the network are picked as supervised information. Then novel influences (or

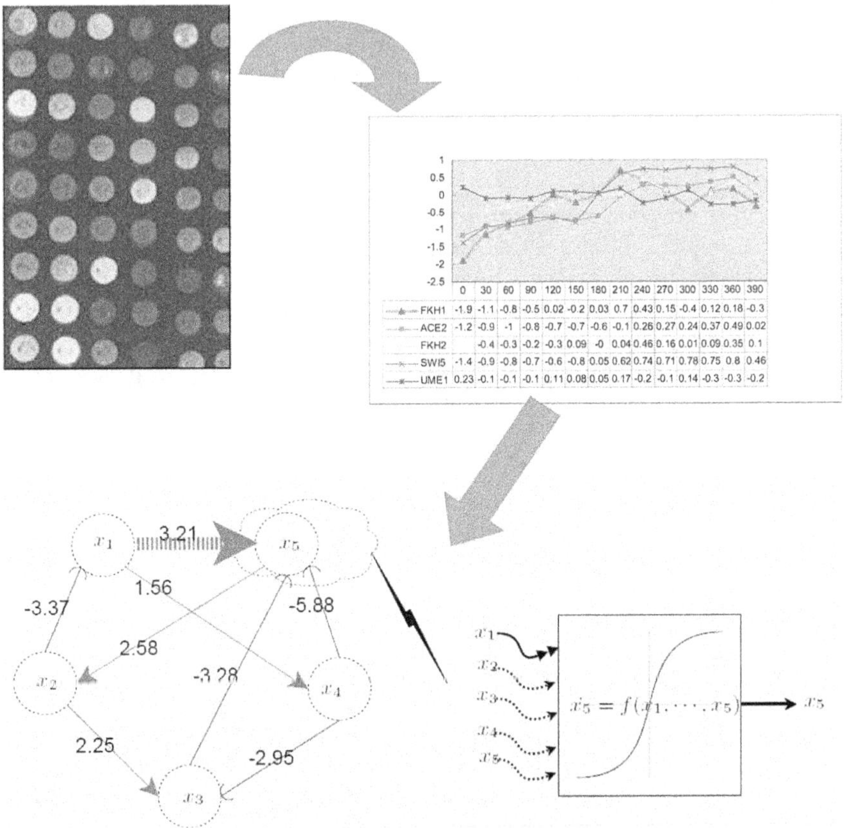

Fig. 1. Graph depiction of the strategy to construct the network by supervised learning. The experimenter measures the expression (concentration) of many or all RNA transcripts in the cells by microarray. Then time course data are collected as a matrix by normalization. The ordinary differential equation is used to inference the relationship between genes. In the network previously known regulatory influences are marked in bold and special line, novel influences (or false positives) are marked in common line. Arrows and arcs denote activation and repression, respectively.

false positives) are obtained by linear programming inference algorithm as a result. Noticed that the network structure inferred with or without supervision are all solutions of the ordinary differential equation. They differ only in that the supervised one learns the known correct information in the inference procedure and the results tend to be more consistent with the known literatures.

2.1 Gene Regulatory Network

The model is based on relating the changes in gene transcript concentration to each other and to the external perturbation [10]. The external perturbation means an experimental treatment that can alter the transcription rate of the genes in the cell. An example of perturbation is the treatment with a chemical compound, or a genetic perturbation involving over-expression or down-regulation of particular genes. The ordinary differential equation is used to represent the rate of synthesis of a transcript as a function of the concentrations of every other transcript in a cell and the external perturbation:

$$\dot{x}(t) = Jx(t) + pc(t), \quad t = t_1, ..., t_m \tag{1}$$

Where $x(t)$ is expression level (mRNA concentrations) of gene at time point t, $J = (J_{ij})_{n\times n} = \partial f(x)/\partial x$ is an $n \times n$ Jacobian matrix or connectivity matrix. Here p represents the effort of perturbation on x and $c(t)$ represents the external perturbation at time t. By introducing p as a variable in inference, the approach can be a powerful methodology for the drug discovery process. Since it would be able to identify the compound mode of action via a time course gene expression profile and the feasibility is proved by reconstruct the SOS system in the bacteria *Escherichia coli* [6,10].

Writing the equation (1) in a compact form for all time points using matrix notation as

$$\dot{X} = JX + PC \tag{2}$$

where $X = (x(t_1), ..., x(t_m))$ and $\dot{X} = (\dot{x}(t_1), ..., \dot{x}(t_m))$ are all $n \times m$ matrices with the first derivative of mRNA concentration $\dot{x}_i(t_j) = [x_i(t_{j+1}) - x_i(t_j)]/[t_{j+1} - t_j]$ for $i = 1, ..., n; j = 1, ..., m$. Suppose that there are s times external perturbation, then $C = (c(t_1), ..., c(t_m))$ is a $s \times m$ matrix representing the s perturbations. The unknowns to calculate are connectivity matrix J and P. J is an $n \times n$ connectivity matrix, composed of elements J_{ij}, which represents the influence of gene j on gene i with a positive, zero or negative sign indicating activation, no interaction and repression respectively. P is an $n \times s$ matrix representing the effect of s perturbations on the n gene system. An non-zero element P_{ij} of P implies that the ith gene is a direct target of the jth perturbation. The equation (2) can be reformed as:

$$\dot{X} = [J \ P] \begin{bmatrix} X \\ C \end{bmatrix} \tag{3}$$

By adopting SVD to $\begin{bmatrix} X \\ C \end{bmatrix}$, i.e.,

$$\begin{bmatrix} X \\ C \end{bmatrix}^T_{m\times(n+s)} = U_{m\times(n+s)} E_{(n+s)\times(n+s)} V^T_{(n+s)\times(n+s)} \tag{4}$$

where U is a unitary $m \times (n+s)$ matrix of left eigenvectors, $E = diag(\ e_1, \cdots, e_{(n+s)})$ is a diagonal $(n+s) \times (n+s)$ matrix containing the $(n+s)$ eigenvalues and $(V)^T$ is the transpose of a unitary $(n+s) \times (n+s)$ matrix of right eigenvectors. Then we can have a particular solution with the smallest L_2 norm for the Jacobian matrix $\hat{J} = (\hat{J}_{ij})_{n\times n}$ and $\hat{P} = (\hat{P}_{ij})_{n\times s}$ as

$$[\hat{J} \ \hat{P}] = (\dot{X})U(E)^{-1}V^T \tag{5}$$

where $E^{-1} = diag(1/e_i)$ and $1/e_i$ is set to be zero if $e_i = 0$. Thus, the general solution of the Jacobian matrix $J = (J_{ij})_{n\times n}$ and $P = (P_{ij})_{n\times s}$ are

$$[J\ P] = [\hat{J}\ \hat{P}] + YV^T \tag{6}$$

$Y = (y_{ij})$ is an $n \times (n+s)$ matrix. Solutions of (6) represent all of the possible networks that are consistent with the perturbation microarray dataset, depending on arbitrary Y. Then given the gene of interest, the supervised information can be incorporated during the process of getting the sparse structure of J by similarly solving linear programming as described in 2.2.

2.2 Supervised Learning by Linear Programming

With the general solution expression of (6), the next step is to pick a biologically meaningful solution by determining variable Y. In [11], the objective is to make the zero elements of J as much as possible. Though the inferred network is sparse in some sense, it is still a heuristic idea and cannot be biologically ensured. In this paper we add the supervised information during the inference process of the network structure. This strategy will drive the network toward the direction in a more biological meaning way.

Suppose that the known information of gene regulatory network is expressed by K, which is an $n \times n$ sparse matrix. If the element K_{ij} is nonzero, it means that gene j has regulatory influence (the activation or depression depends on the sign of K_{ij}) on gene i and this interaction is assumed to be revealed by biological experiments. We will discuss how to incorporate these valuable information in the inference of whole network by linear programming formulation.

The $\hat{J}$ is the particular solution of the microarray dataset by SVD. The information of K is incorporated in the general solution $J = (J_{ij})_{n\times n}$ by a proper Y such that

$$[J\ P] = [\hat{J}\ \hat{P}] + YV^T, \tag{7}$$

and at the same time the following conditions are satisfied

$$(J_{ij})_{n\times n} = (K_{ij})_{n\times n}, \quad K_{ij} \neq 0 \tag{8}$$

Considering the L_1 problem:

$$\min_Y |[(K - \hat{J})\ \ \hat{P}] + Y^k V^T| \tag{9}$$

where $\hat{J}$, K, $\hat{P}$ and V^T are given. Without loss of generality, the problem is expressed as the standard form

$$\min_X |AX - B| \tag{10}$$

where the coefficient A is an $n \times (n+s) \times nl$ matrix, and l is the number of zero elements in E^{-1} in (4). X and B are $l \times n$ and $(n+s) \times n$ matrices, respectively. In fact we Do not need to solve such a large LP problem. Observing that A has a special structure as

$$A = \begin{bmatrix} A_1 & & & \\ & A_2 & & \\ & & \cdots & \\ & & & A_n \end{bmatrix}$$

The size of $A_i, i = 1, 2, \cdots, n$ is $(n + s) \times l$. With the correspondingly decomposing $X = [X_1, X_2, \cdots, X_n]$ and $B = [B_1, B_2, \cdots, B_n]$, the solution of the raw problems (10) can be obtained by solving the following n small problems:

$$\min_{X_i} |A_i X_i - B_i| \quad i = 1, 2, \ldots, n \tag{11}$$

where the dimensions of A_i, X_i and B_i are $n \times l$, $l \times 1$ and $n \times 1$, respectively. Clearly the decomposition scheme reduces the storage space from about $O(n^4)$ to $O(n^2)$ and requires $O(n^4)$ computations. Furthermore, it provides the formulation for parallel computation so as to rapidly find regulatory relationship of the interested gene in a high priority. i. e. the regulatory influence relationship of a designated gene can be obtained independently by solving a small scale L1 problem. Without loss of generality, only gene i is to be considered in the following inference algorithm.

Noticing that $B_i = -\hat{J}_i + K_i(B_i, \hat{J}_i$ and K_i are the ith row of the matrix B, $\hat{J}$ and K, respectively), let $I = \{j | K_{ij} \neq 0\}$ and $|I| = q$, by introducing $2(n-q)$ slack variables u_j, v_j, $j \in \{1, 2, \cdots, n\} \setminus I$, the L1 problem

$$\min_{x_{ij}} \sum_{j=1}^{n} |\sum_{k=1}^{l} a_{ik} x_{ik} - \hat{J}_{ij} + K_{ij}| \tag{12}$$

is equivalent to the linear programming model as follows:

$$\begin{aligned} \min_{x_{ij}, u_j, v_j} \quad & \textstyle\sum_{j \in I} (u_j + v_j) && \\ s.t. \quad & \textstyle\sum_{k=1}^{l} a_{ik} x_{ik} - \hat{J}_{ij} + K_{ij} = u_j - v_j, && j \in \{1, 2, \cdots, n\} \setminus I \\ & \textstyle\sum_{k=1}^{l} a_{ik} x_{ik} = \hat{J}_{ij}, && j \in I \\ & u_j, v_j \geq 0, && j \in \{1, 2, \cdots, n\} \setminus I \\ & x_{ij} \in R, && j \in \{1, 2, \cdots, n\} \end{aligned} \tag{13}$$

The variables need to be solved are $x_{i1}, x_{i2}, \cdots, x_{il}, u_1, v_1, \cdots, u_{n-q}, v_{n-q}$ and the total number is $2n - 2q + l$. There are n equality constraints and $2(n - q)$ inequality constraints. The LP problem for such a scale can be dealt directly by simplex method.

In the above linear programming formulation, the supervised information is incorporated through the first $n-q$ equality constraints. If a little supervised information is added (q is small), the LP will generally give the unique solution which is a particular solution and satisfies $(J_{ij})_{n \times n} = (K_{ij})_{n \times n}, \quad K_{ij} \neq 0$. But when much supervised information is applied to the gene regulatory system (q is large), the LP will give an approximate solution due to redundant constraints. In this case the system is over-determined and not all $(J_{ij})_{n \times n} = (K_{ij})_{n \times n}, \quad K_{ij} \neq 0$ are satisfied. The answer of how much supervised information should be incorporated in the inference depends on the inherent degree of freedom depicted by

l. For example in a large-scale gene network, the time course data is relatively scarce ($l \approx n$) and much supervised information can be added. As reported in [11] for a large system, the smallest number of time points needed is $O(\log n)$ to reconstruct the $n \times n$ connectivity matrix for an n-gene network. Hence, the proper incorporation of known information of interactions in the inference algorithm will alleviate the requirement and dependence on high quality microarray data greatly.

Next we want to further address how to set values for matrix K from know information. In our model the knowledge of gene regulatory network is expressed and incorporated by matrix K. If the element K_{ij} is nonzero, it means that gene j has regulatory influence on gene i. The activation or depression role differs on the sign of K_{ij}. It is better to provide the quantitative strength of the know regulatory interactions, but many constraints are basically all qualitative instead of quantitative in the databases or literatures though it is true that they are readily available . You may know gene i activates gene j, but the quantitative relationship as described by the K_{ij} is not known. The feasible way is to assign a constant to K_{ij} and use its sign to indicate activation or depression relationship. Since in some meaning, the major function of element k_{ij} is to introduce constraints in the linear programming and this function can be preformed by keeping it a constant. In the long run, a systematic search/protocol or a soft (boundary) constraint can be developed. In this paper, The $k_{ij} = 1$ or $k_{ij} = -1$ are simply set to know activation or regression regulatory interaction in our experiments using biological microarray data.

3 Results

In this section, we first report on simulated numerical test that we have designed to benchmark our method by using supervised inference strategy. Then we will describe the gene regulatory network inference using yeast microarray gene expression data. As analyzed in Methods section, when no supervised information is applied, our method is similar to the method of [11], which can recover the network connectivity from gene expression measurements in the presence of noise by singular value decomposition (SVD) and regression. With supervised information, we can further infer the network structure in a more accurate and robust manner. In this section, only preliminary results are given for incorporating supervised information by single microarray dataset, the experiments of gene network inference by perturbation dataset and multiple datasets can be conducted similarly.

3.1 Simulated Data

The first example is a small simulated network with five genes governed by

$$\begin{aligned}
\dot{x}_1(t) &= -x_1(t) - 0.2x_2(t) + 0.5x_4(t) + \xi_1(t),\\
\dot{x}_2(t) &= -0.8x_1(t) - 1.5x_2(t) + x_3(t) - 0.5x_5(t) + \xi_2(t),\\
\dot{x}_3(t) &= 0.6x_1(t) + 0.2x_2(t) - x_3(t) - 0.3x_4(t) + \xi_3(t),\\
\dot{x}_4(t) &= 0.9x_2(t) - x_4(t) - 1.5x_5(t) + \xi_4(t),\\
\dot{x}_5(t) &= -0.2x_1(t) + 0.7x_4(t) - 1.5x_5(t) + \xi_5(t),
\end{aligned}$$

where x_i reflects the expression level of the gene-i and $\xi_i(t)$ represents noise for $i = 1, 2, 3, 4, 5$.

To test our method, we randomly choose the initial condition of the system and take several points of x as a measured time-course dataset. In our simulated example, we obtained a dataset with 4 time points, and applied our method to reconstruct the connectivity matrix or the Jacobian matrix J. We set all of noises $\xi_i(t), i = 1, 2, 3, 4, 5$ obeying normal distribution in the simulated example with noise level of $N(0, 0.005)$. And the supervised information K is incorporated by fixing $K_{2,5} = -0.5$ and $K_{5,4} = 0.7$, which means the depression influence of gene 5 → gene 2 and the activation influence of gene 4 → gene 5 are measured and known.

The numerical results are depicted in Figure 2, which shows the reconstructed networks without and with supervised information, respectively. Clearly with the supervised information, it infers the network more accurately. Without supervised information (Fig. 2 (b)), the strong self repressive relation of gene x_5, the activation relation between gene x_4 and gene x_5 are neglected. When the supervised information is used, the topology of the network becomes correct and the predicted connectivity matrix, which represents the strengths among gene interactions, is very close to the true one (Fig. 2 (c)). Such results imply that our method is able to infer the solution of the highly under-determined problem in an accurate manner when a sufficient number of known interactions are available even although the microarray dataset has only a few time points.

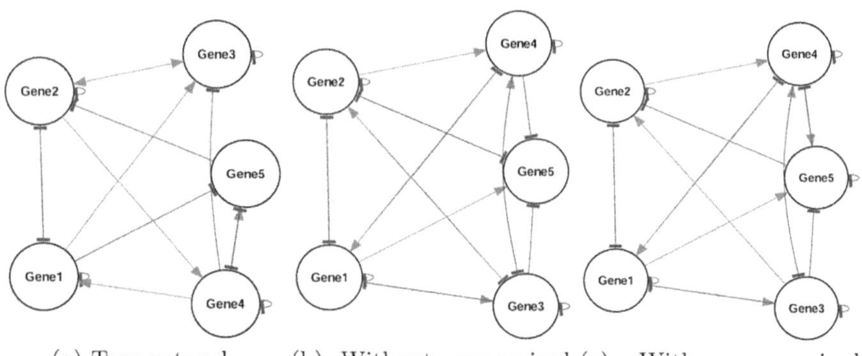

(a) True network (b) Without supervised learning (c) With supervised learning

Fig. 2. Regulatory network reconstruction for the simulated example with 5 genes. Activation is shown in red arrow and repression in blue arc.

3.2 Application to Experimental Data

We applied our method using experimental data. To ensure high quality of the data, we only used microarray experimental data generated from whole genome Affymetrix chips, instead of any oligo or cDNA array data. The experiments are conducted to test our method using a small number of genes of Heat-Shock

Response data for yeast. We created an input dataset for 10 transcription factors related to heat-shock response in yeast *Saccharomyces cerevisiae.* 2 out of the 10 transcription factors (HSF1 and SKN7) are known to be directly involved in heat shock response. HSF1 and SKN7 each are known to regulate 4 other transcription factors among the ten. TYE7 → HSF1 and RPN4 → HSF1 are documented known regulation. This information was obtained from YEASTRACT (`http://www.yeastract.com/index.php`). For the 10 transcription factors, we used microarray dataset at the Stanford Microarray Database (`http://smd.stanford.edu/`) (y14, with 5 time points) for gene expression under heat shock conditions. We applied the proposed method to this dataset. As indicated in 2.2 subsection, the $k_{ij} = 1$ or $k_{ij} = -1$ are simply set to know regulatory interaction in our real data experiments. As shown in Fig. 3, the prediction succeeded in reconstructing 2 edges of the network which are identical with the documented known regulation (see subfigure (b) in Fig. 3). But these regulation information can not be obtained without the supervised learning procedure (see the subfigure (a) in Fig. 3).

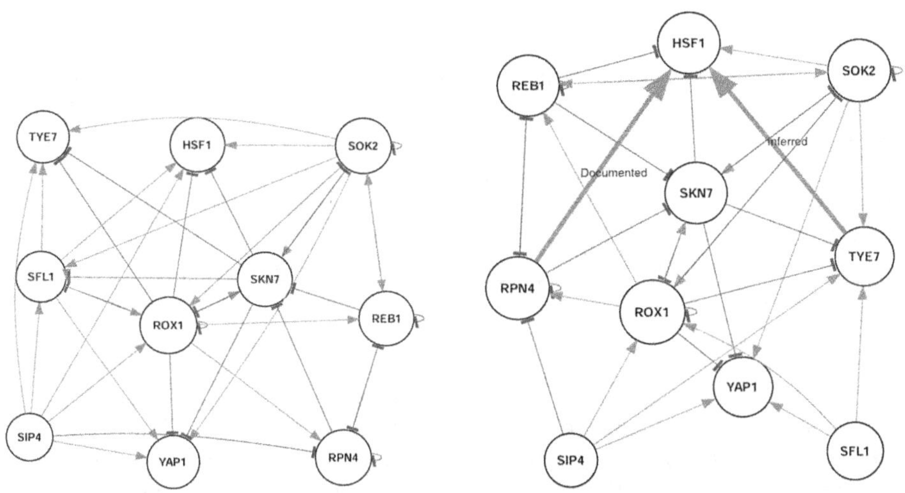

(a) Without supervised learning (b) With supervised learning

Fig. 3. Regulatory network reconstruction for a set of 10 transcription factors for heat shock response microarray data y14. Activation is shown in red arrow and repression in blue arc. The confirmed edges are shown in bold arrows with labels.

4 Discussion and Conclusion

Microarray gene expression data become increasingly common data source that can provide insights into biological processes at a system-wide level. In contrast to the conventional methods suffering the dimensionality problem, the main contribution of this paper is development of a methodology to reconstruct gene regulatory network by using existing interaction information of genes from database.

In other words, we provide a general framework to handle the microarray data by fully considering the accumulative biological experiment results, which not only makes the inferred results more accurate, but also alleviates the problem of dimensionality or data scarcity greatly. We have tested our approach to both simulated small-size problems and experimental biological data in this paper, which verified the efficiency and effectiveness of our algorithm.

In this paper, the supervised learning strategy is incorporated in the network inference. These approaches are called "supervised" because the model is learned from a training data consisting of a set of system responses. In our method, the training data are collected from the known transcription relationship from YEASTRACT, which allows the identification of documented or potential transcription regulators of a given gene and documented or potential regulatory for each transcription factor. Compared with the supervised learning strategy in [6], our method has advantages in simplifying the design of experiments and requiring less expensive computation. In our method, only time-course data of gene expression is utilized. While in [6] the perturbations around steady state are used. The reverse engineering algorithm implemented by linear programming in our method simplifies the complexity of network inference greatly. The strategy in [6] to fix the number of interaction of every gene and to enumerate all the cases by greedy algorithm is expensive and time consuming compared with the simplex algorithm for linear programming.

To examine causal relation among genes, a major source of errors comes from the noises of the gene expression data intrinsic to microarray technologies. To reduce the defect of unreliable data, a feasible method is to combine multiple microarray gene expression datasets together to get the more consistent network. Our supervised strategy in inferring gene regulatory network is ready to generalized in multiple datasets case. The more simulations and experiments are in progress.

The tendency is clearly that more and more data of gene interaction will be experimentally measured, reported and deposited in the literatures or databases. Meanwhile the quality of these data and microarray data keeps improving. It is reasonable to expect that our inference method will continue to prove valuable in analyzing and predicting the behavior or gene regulatory networks.

References

1. van Someren, E. P., Wessels, L. F. A., Reinders, M. J. T., Backer, E.: Robust Genetic Network Modeling by Adding Noisy Data. In: Proceedings of 2001 IEEE-EURASIP Workshop on Nonlinear Signal and Image Processing, Japan, (2001) 234-246
2. Hartemink, A. J.: Reverse Engineering Gene Regulatory Networks. Nature Biotechnology, 23 (5) (2005) 554-555
3. Holter, N. S., Maritan, A., Fedoroff, M. C. N. V., Banavar, J. R.: Dynamic Modeling of Gene Expression Data. Proc. Natl. Acad. Sci. USA, 98 (2001) 1693-1698
4. Tegner, J., Yeung, M. K. S., Collins, J.: Reverse Engineering Gene Networks: Integrating Genetic Perturbations with Dynamical Modeling. Proc. Natl. Acad. Sci. USA, 100 (2003) 5944-5949

5. Dewey, T. G., Galas, D. J.: Dynamic Models of Gene Expression and Classification. Functional & Integrative Genomics, 1 (4) (2001) 269-278
6. Gardner, T. S., Di Bernardo, D., Lorenz, D., Collins, J. J.: Inferring Genetic Networks and Identifying Compound Mode of Action Via Expression Profiling. Science, 301 (5629) (2003) 102-105
7. Chen, L., Aihara, K.: Stability and Bifurcation Analysis of Differential-difference-algebraic Equations. IEEE Trans. on Circuits and Systems - I, 48 (3) (2001) 308–326
8. Chen, L., Aihara, K.: Stability of Genetic Regulatory Networks with Time Delay. IEEE Trans. on Circuits and Systems-I,49 (5) (2002) 602-608
9. Wang, Y., Joshi, T., Zhang, X. S., Xu, D., Chen, L.: Inferring Gene Regulatory Networks from Multiple Microarray Datasets (2006) (In submission)
10. Bansal, M., Gatta, G. D., di Bernardo, D.: Inference of Gene Regulatory Networks and Compound Mode of Action from Time Course Gene Expression Profiles. Bioinformatics, 10 (1093) (2006) 1-8
11. Yeung, M. K. S., Tegner, J., Collins, J.: Reverse Engineering Gene Networks Using Singular Value Decomposition and Robust Regression. Proc. Natl. Acad. Sci., USA, 99 (2002) 6163-6168

Double Optimization for Design of Protein Energy Function

Seung-Yeon Kim[1] and Julian Lee[2]

[1] Computer Aided Molecular Design Research Center, Soongsil University, 156-743 Seoul, Korea
sykim@ssu.ac.kr
[2] Department of Bioinformatics and Life Science, Soongsil University, 156-743 Seoul, Korea
jul@ssu.ac.kr
http://bioinfo.ssu.ac.kr/~jul/jul_eng.htm

Abstract. We propose an automated protocol for designing the energy landscape suitable for the description of a given set of protein sequences with known structures, by optimizing the parameters of the energy function. The parameters are optimized so that not only the global minimum-energy conformation becomes native-like, but also the conformations distinct from the native structure have higher energies than those close to the native one, for each protein sequence in the set. In order to achieve this goal, one has to sample protein conformations that are local minima of the energy function for given parameters. Then the parameters are optimized using linear approximation, and then local minimum conformations are searched with the new energy parameters. We develop an algorithm that repeats this process of parameter optimization based on linear approximation, and conformational optimization for the current parameters, ultimately leading to the optimization of the energy parameters. We test the feasibility of this algorithm by optimizing a coarse grained energy function, called the UNRES energy function, for a set of ten proteins.

1 Introduction

According to the thermodynamic hypothesis [1], proteins adopt native structures that minimize their free energies. Therefore, obtaining energy function that accurately describes proteins would lead not only to the prediction of three-dimensional structures, but also to the understanding of folding mechanism [2,3].

Energy functions are generally parameterized from quantum mechanical calculations and experimental data on model systems. However, such calculations and data do not determine the parameters with perfect accuracy. The residual errors in energy functions may have significant effects on simulations of macromolecules such as proteins where the total energy is the sum of a large number of interaction terms. Moreover, these terms are known to cancel each other to

D.-S. Huang, K. Li, and G.W. Irwin (Eds.): ICIC 2006, LNBI 4115, pp. 562–570, 2006.

a high degree, making their systematic errors even more significant. Thus it is crucial to refine the parameters of a energy function before it can be successfully used to study protein folding.

In this work, we develop an automated protocol for the parameter optimization, where the parameters are modified so as to make conformations with larger values of root-mean-square deviation (RMSD) have higher values of energy relative to those with smaller values of RMSD. This goal is achieved by repeating two distinct optimization procedures, sampling local minimum-energy conformations for a given parameter set, and parameter optimization using linear approximation. The parameter optimization based on linear approximation is performed by supernodal Cholesky factorization method [4], a Linear Programming algorithm, and the local minimum-energy conformations with low energies for a given parameter are sampled using the conformational space annealing method [5,6,7,8].

We show the feasibility of this algorithm by successfully optimizing the UNRES energy function [9,10], for a set of ten proteins. Our work is an improvement over previous works [3,11,12,13,14,15,16], where either primitive methods of parameter optimization were used, or the optimization was performed for the training set of a small number of protein sequences, at most four of them.

2 Methods

2.1 Constrained and Unconstrained Conformational Searches

In order to check the performance of a energy function for a given set of parameters, one has to perform two types of conformational search, the constrained and unconstrained conformational searches. In the constrained search, the backbone angles of the conformations are fixed to the values of the native conformations, and only the side-chain angles are minimized with respect to the energy. We call the resulting conformations the super-native. The other conformations are obtained from unconstrained conformational search. The conformations obtained from the constrained and unconstrained searches are added to the structural database of local minimum-energy conformations for each protein. The search algorithm we use is the conformational space annealing (CSA) method [5,6,7,8]. The CSA method can be considered as a genetic algorithm that enforces a broad sampling in its early stages and gradually allows the conformational search to be focused into narrow conformational space in its later stages. As a consequence, many low-energy local minima including the GMEC of the benchmark protein can be identified for a given parameter set.

2.2 Parameter Refinement Using Linear Programming

The changes of energy gaps are estimated by the linear approximation of the energy function in terms of parameters. Among the conformations with non-zero RMSD values in the structural database, 50 (an arbitrary number) conformations with the smallest RMSD values are selected as the native-like conformations, while the rest are considered as the non-native ones. Since an energy function can be considered to describe the nature correctly if native-like structures have lower energies than non-native ones, the parameters are optimized to minimize the energy gaps $E_{\text{gap}}^{(1)}$ and $E_{\text{gap}}^{(2)}$,

$$
\begin{aligned}
E_{\text{gap}}^{(1)} &= E^{\text{N}} - E^{\text{NN}} \\
E_{\text{gap}}^{(2)} &= E^{\text{SN}} - E^{\text{NN}}
\end{aligned}
\tag{1}
$$

for each protein in the training set, where E^{N} and E^{SN} are the highest energies of the native-like and super-native conformations, respectively, and E^{NN} is the lowest energy of the non-native conformations. The energies are the modified ones that are weighted with the RMSD values of the conformations:

$$E_{\text{modified}} = E + 0.3\ \text{RMSD}. \tag{2}$$

Weighting the energies with the RMSD values makes the large RMSD conformations have high energies compared to ones with small RMSD values. The parameter optimization is carried out by minimizing the energy gaps $E_{\text{gap}}^{(1)}$ and $E_{\text{gap}}^{(2)}$ of each protein in turn, while imposing the constraints that all the other energy gaps, including those from the other proteins, do not increase.

In this work, we adjust the 715 linear parameters of the UNRES energy function [9,10], a coarse-grained protein energy function. The energy of a local minimum-energy conformation can be written as:

$$E = \sum_j p_j e_j(\mathbf{x}_{\text{min}}) \tag{3}$$

where e_i's are the energy components evaluated with the coordinates $\mathbf{x}_{\text{min}}$ of a local minimum-energy conformation. Since the positions of local minima also depend on the parameters, the full parameter dependence of the energy gaps are nonlinear. However, if the parameters are changed by small amounts, the energy with the new parameters can be estimated by the linear approximation:

$$E^{\text{new}} \approx E^{\text{old}} + \sum_i (p_i^{\text{new}} - p_i^{\text{old}}) e_i(\mathbf{x}_{\text{min}}) \tag{4}$$

where the p_i^{old} and p_i^{new} terms represent the parameters before and after the modification, respectively. The parameter dependence on the position of the local minimum can be neglected in the linear approximation, since the derivative in the conformational space vanishes at a local minimum. The additional term 0.3 RMSD of Eq.(2) vanishes in these expressions due to the same reason. The changes of the energy gaps are estimated as:

$$
\begin{aligned}
\Delta E_{\text{gap}}^{(1)} &= E_{\text{gap}}^{(1)}(\{p_j^{\text{new}}\}) - E_{\text{gap}}^{(1)}(\{p_j^{\text{old}}\}) \\
&= (E^{\text{N}}(\{p_j^{\text{new}}\}) - E^{\text{NN}}(\{p_j^{\text{new}}\})) - (E^{\text{N}}(\{p_j^{\text{old}}\}) - E^{\text{NN}}(\{p_j^{\text{old}}\})) \\
&= \sum_j (e_j^{\text{N}} - e_j^{\text{NN}})(p_j^{\text{new}} - p_j^{\text{old}})
\end{aligned}
\tag{5}
$$

$$\begin{aligned}\Delta E_{\text{gap}}^{(2)} &= E_{\text{gap}}^{(2)}(\{p_j^{\text{new}}\}) - E_{\text{gap}}^{(2)}(\{p_j^{\text{old}}\}) \\ &= (E^{\text{SN}}(\{p_j^{\text{new}}\}) - E^{\text{NN}}(\{p_j^{\text{new}}\})) - (E^{\text{SN}}(\{p_j^{\text{old}}\}) - E^{\text{NN}}(\{p_j^{\text{old}}\})) \\ &= \sum_j (e_j^{\text{SN}} - e_j^{\text{NN}})(p_j^{\text{new}} - p_j^{\text{old}}) \end{aligned} \tag{6}$$

The magnitude of the parameter change $\delta p_j \equiv p_j^{\text{new}} - p_j^{\text{old}}$ is bounded by a certain fraction ϵ of p_j^{old}. We use $\epsilon = 0.01$ in this study. First, the vector δp_j is chosen within the bound to decrease the energy gap $\Delta E_{\text{gap}}^{(1)}$ of the selected protein as much as possible while imposing the constraints that any positive values among $E_{\text{gap}}^{(2)}$ and the energy gaps of the other proteins do not increase and negative values do not become positive. Denoting the energy gaps of the k-th protein as $E_{\text{gap}}^{(p=1,2)}(k)$ and assuming the i-th protein is selected for the decrease of the energy gap, this problem can be phrased as follows:

Minimize

$$\Delta E_{\text{gap}}^{(1)}(i) = \textstyle\sum_j (e_j^{\text{N}}(i) - e_j^{\text{NN}}(i))(p_j^{\text{new}} - p_j^{\text{old}})$$

with constraints

$$|\delta p_i| \leq \epsilon$$

$$\Delta E_{\text{gap}}^{(2)}(i) = \sum_j (e_j^{\text{SN}}(i) - e_j^{\text{NN}}(i))(p_j^{\text{new}} - p_j^{\text{old}}) \leq \begin{cases} 0 & \text{if } E_{\text{gap}}^{(2)}(i) > 0 \\ -E_{\text{gap}}^{(2)}(i) & \text{otherwise} \end{cases}$$

$$\Delta E_{\text{gap}}^{(p=1,2)}(k \neq i) = \sum_j (e_j^{\text{(S)N}}(k) - e_j^{\text{NN}}(k))(p_j^{\text{new}} - p_j^{\text{old}}) \leq \begin{cases} 0 & \text{if } E_{\text{gap}}^{(p)}(k) > 0 \\ -E_{\text{gap}}^{(p)}(k) & \text{otherwise} \end{cases}$$

This is a global optimization problem where the linear parameters p_j are the *variables*. The object function to minimize, and the constraints, are all linear in p_j. This type of the optimization problem is called the Linear Programming. It can be solved exactly, and many algorithms have been developed for solving the Linear Programming problem. We use the primal-dual method with supernodal Cholesky factorization [4] in this work, which finds an accurate answer with reasonably computational costs.

After minimizing $\Delta E_{\text{gap}}^{(1)}(i)$, we solve the same form of linear programming where now $\Delta E_{\text{gap}}^{(2)}(i)$ is the objective function and the other energy gaps become constrained. Then we select another protein and repeat this procedure (300 times in this work) of minimizing $\Delta E_{\text{gap}}^{(1)}$ and $\Delta E_{\text{gap}}^{(2)}$ in turn.

The algorithm of parameter optimization based on linear approximation, using repeated Linear Programming, is summarized in algorithm 1.

2.3 Reminimization and New Conformational Search

Since the procedure of the previous section was based on the linear approximation Eqs.(5) and (6), we now have to evaluate the true energy gaps using the newly obtained parameters. The breakdown of the linear approximation may

Algorithm 1. Parameter optimization for conformations in the structural database, using linear programming

N_p = Number of protein sequences in the dataset (constant)
N_{it} = Maximum number of iteration (300 in this work)
$\mathbf{p}(i)$ = The initial parameters at the start of this sub-algorithm
for $i = 0$ to N_{it} **do**
  Calculate energy gaps $E^{(1)}_{\text{gap}}(p)(p = 1, \cdots, N_p)$
  for $p = 1$ to N_p **do**
    minimize $\Delta E^{(1)}_{\text{gap}}(p)$, while constraining the other energy gaps (Linear Programming)
    minimize $\Delta E^{(2)}_{\text{gap}}(p)$, while constraining the other energy gaps (Linear Programming)
  end for
  if Energy gaps are negative for all of the N_p proteins **then**
    End this sub-algorithm, since parameters are optimized for the conformations in the structural database.
  end if
end for

come from two sources. First, the conformations corresponding to the local minima of the energy for the original set of parameters are no longer necessarily so for the new parameter set. For this reason, we reminimize the energy of these conformations with the new parameters. Since super-native conformations are not local minimum-energy conformations, even with the original parameters, the unconstrained reminimization of these conformations with the new parameters may furnish low-lying local minima with small values of RMSD. Second, the local minima obtained from conformational searches with the original parameter set may constitute only a small fraction of low-lying local minima. After the modification of the parameters, some of the local minima which were not considered due to their relatively high energies, can now have low energies for the new parameter set. It is even possible that entirely distinct low-energy local minima appear. Therefore these new minima are taken into account by performing subsequent conformational searches with the newly obtained parameter set.

2.4 Update of the Structural Database and Iterative Refinement of Parameters

The low-lying local energy minima found in the new conformational searches are added into the energy-reminimized conformations to form a structural database of local energy minima. The conformations in the database are used to obtain the energy gaps, which are used for the new round of parameter refinement. As the procedure of [conformational search → parameter refinement → energy reminimization] is repeated, the number of conformations in the structural database increases. This iterative procedure is continued until sufficiently good native-like conformations are found from the unconstrained conformational search. The whole procedure is summarized in algorithm 2.

Algorithm 2. The algorithm for protein energy function parameter optimization

N_p = Number of protein sequences in the dataset (constant)
N_{it} = Maximum number of iteration (constant)
$\mathbf{p}(i)$ = the initial parameters
for $i = 0$ to N_{it} **do**
  for $p = 1$ to N_p **do**
    For the protein sequence p, sample low-lying local minimum-energy conformations with no constraints, and save their coordinates and energy components.
    For the protein sequence p, sample low-lying local minimum-energy native-like conformations (constrained sampling), and save their coordinates and energy components
  end for
  if Low-lying conformations found from unconstrained search are native-like for all of the N_p sequences **then**
    Parameter optimization accomplished. End the algorithm.
  end if
  if $i == 0$ **then**
    structural database $\Leftarrow$ low-lying conformations obtained from unconstrained search + low-lying native-like conformations obtained from constrained search (coordinates and energy components)
  else
    structural database $\Leftarrow$ structural database + low-lying conformations obtained from unconstrained search + low-lying native-like conformations obtained from constrained search (coordinates and energy components)
  end if
  Optimize parameters using Linear Programming, so that the energy gap $E_g = E_{na} - E_{nn}$ between the maximum energy among native-like conformations, E_{na}, and the minimum energy among non-native conformations, E_{nn} (See alg. 1), is minimized for each of the N_p sequences in the training set $\Rightarrow \mathbf{p}(i+1)$ (new parameters)
  Reminimize structural database with respect to the energy function with the new parameters
end for

3 Results and Discussions

3.1 Ten Proteins in the Training Set and Two Proteins in the Test Set

We apply our protocol to the optimization of UNRES energy function, for a training set consisting of ten proteins, that belong to the structural class of α proteins. The PDB codes of these proteins are, 1BBA(36), 1BDD(60), 1EDI(56), 1EDK(56), 1HNR(47), 1IDY(54), 1PRB(53), 1PRU(56), 1VII(36), and 1ZDB(38), where the number inside parentheses are the lengths of these proteins. The initial parameter set is the one used in CASP3[17,18]. The optimized parameter set obtained using the training set above, is useful for the protein folding study of an α protein, including the tertiary structure prediction, where the secondary structure content

Table 1. The C_α RMSD of GMEC, for the ten proteins in the training set and the two proteins in the test set. The number in parantheses are the smallest RMSD values found in the fifty low-energy conformations.

	1BBA	1BDD	1EDI	1EDK	1HNR	1L2Y
initial parameters	8.9 (8.1)	9.8 (7.2)	7.8 (5.0)	7.6 (4.9)	9.9 (6.6)	6.5 (4.6)
optimized parameters	9.3 (4.3)	3.9 (2.9)	3.8 (2.4)	3.9 (2.4)	9.4 (5.3)	3.6 (3.1)
	1IDY	1PRB	1PRU	1VII	1ZDB	1F4I
initial parameters	11.2 (6.9)	10.1 (7.5)	11.5 (6.9)	6.3 (4.9)	7.7 (6.7)	6.8 (5.1)
optimized parameters	10.1 (4.9)	6.2 (5.9)	6.7 (5.8)	5.3 (3.5)	7.6 (3.0)	5.4 (4.2)

can be determined relatively easily using experimental methods such as Circular Dichroism (CD) or Nuclear Magnetic Resonances (NMR). It is of course possible to obtain energy function parameters suitable for the general description of proteins regardless of their structural classes, using training set consisting of proteins that belong to diverse structural classes[13,14].

The performance of UNRES energy function with the optimized parameters was tested, by sampling low-energy conformations of two α proteins not included in the training set, 1F4I(40) and 1L2Y(20).

For proteins both in training and test sets, fifty conformations were sampled in each conformational search. The RMSDs of C_α coordinates from those of native structures for the global minimum-energy conformations (GMEC) are shown in Table 1, along with the smallest values of RMSD found among the fifty low-energy conformations, obtained with the initial and optimized parameters. Five iterations of linear optimization were performed in order to obtain the optimized parameter set. The energies are not displayed since their numerical

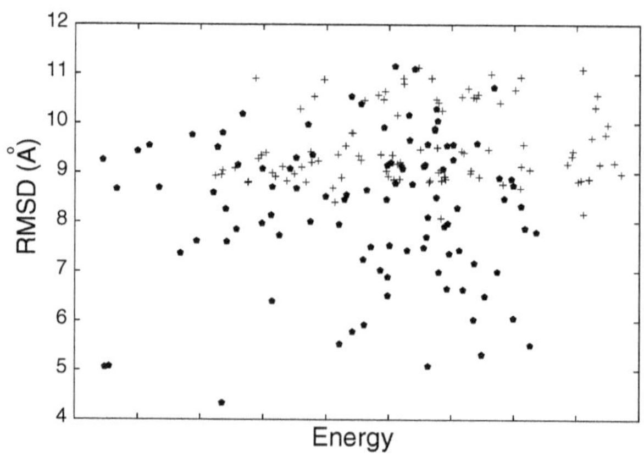

Fig. 1. Plots of the energy and C^α RMSD values of fifty low-energy conformations for the protein 1BBA, obtained using the initial (*plus signs*) and the optimized parameters (*filled circles*). Although the RMSD of the GMEC is smaller for conformations obtained with the initial parameters, much more native-like low-energy conformations are obtained with the optimized parameters.

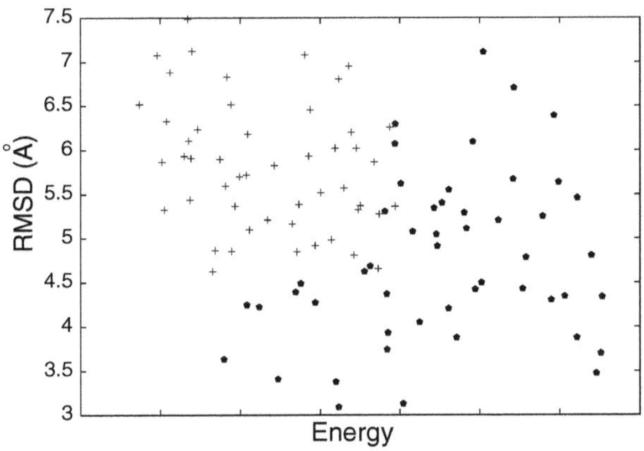

Fig. 2. Plots of the energy and C$^\alpha$ RMSD values of fifty low-energy conformations for the protein 1L2Y, obtained using the initial (*plus signs*) and the optimized parameters (*filled circles*). The GMEC found with the optimized parameters has smaller RMSD value than that found with the initial parameters. Also, much more native-like low-energy conformations are obtained with the optimized parameters.

values have no physical meaning, due to the fact that the overall scale of the linear parameters is not fixed in our protocol. We see from the data that after five iteration, the parameters are indeed optimized for the ten proteins in the training set. The smallest RMSD values found among the fifty low-energy conformations decreased for all of the proteins in the training set, and the RMSD values of the GMECs also decreased for all of them except 1BBA. Since the native structure of a protein is usually predicted by constructing several models, rather than selecting just one GMEC, GMEC not being native-like is not a serious problem, as long as there are sufficient number of native-like conformations in the final set of low-energy conformations obtained by the sampling algorithm. In fact, even for 1BBA where RMSD value of the GMEC increased, those of the second and third lowest energy conformations all decreased, and there are much more native-like low-energy conformations obtained with the optimized parameters, as can be seen in Fig. 1 where the RMSD values of the fifty low-energy conformations are plotted against their energy values.

We see that also for the two proteins in the test set, the lowest RMSD value found from the fifty low-energy conformations, as well as that of GMEC, decrease as the parameters are optimized. Again, the low-energy conformations become more native-like overall, as can be seen from the plot of RMSD against energy values for 1L2Y(Fig. 2). The result suggests that the optimized parameters are transferable to α proteins not included in the training set, and can be used for the study of protein folding of such proteins.

Acknowledgement

This work was supported by the Soongsil University Research Fund.

References

1. Anfinsen, C. B.: Principles that govern the folding of protein chains. Science vol. 181 (1973) 223–230
2. Kim, S.-Y., Lee, J., Lee, J.: Folding of small proteins using a single continuous potential. J. Chem. Phys. vol.120 (2004) 8271–8276
3. Lee, J., Kim, S.-Y., Lee, J.: Optimization of potential-energy parameters for folding of several proteins. J. Kor. Phys. Soc. vol. 44 (2004) 594–598
4. Mészáros, C. A.: Fast Cholesky Factorization for Interior Point Methods of Linear Programming. Computers & Mathematics with Applications vol. 31 (1995) 49–54
5. Lee, J., Scheraga, H. A., Rackovsky. S.: New optimization method for conformational energy calculations on polypeptides: Conformational space annealing. J. Comput. Chem. vol. 18 (1997) 1222–1232
6. Lee, J., Lee, I.H., Lee, J.: Unbiased Global Optimization of Lennard-Jones Clusters for $N >= 201$ Using the Conformational Space Annealing Method. Phys. Rev. Lett. vol. 91 (2003) 080201-1–4
7. Kim S.-Y., Lee S.J., Lee J.: Conformational space annealing and an off-lattice frustrated model protein Conformational space annealing and an off-lattice frustrated model protein. J. Chem. Phys. vol. 119 (2003) 10274–10279
8. Kim S.-Y., Lee S.B., Lee J.: Structure optimization by conformational space annealing in an off-lattice protein model. Phys. Rev. E vol. 72 (2003) 011916-1–6
9. Liwo, A., Oldziej, S., Pincus, M. R., Wawak, R. J., Rackovsky, S., Scheraga, H. A.: A united-residue force field for off-lattice protein-structure simulations. I: Functional forms and parameters of long-range side-chain interaction potentials from protein crystal data. J. Comput. Chem. vol. 18 (1997) 849–873
10. Liwo, A., Pincus, M. R., Wawak, R. J., Rackovsky, S., Oldziej, S., Scheraga, H. A.: A united-residue force field for off-lattice protein-structure simulations. II: Parameterization of local interactions and determination of the weights of energy terms by Z-score optimization. J. Comput. Chem. vol. 18 (1997) 874 – 887
11. Lee, J., Ripoll, D. R., Czaplewski, C., Pillardy, J., Wedemeyer, W. J., Scheraga, H. A.: Optimization in Macromolecular Potential Enrgy Functions by Conformational space Annealing. J. Phys. Chem. B vol. 105 (2001) 7291–7298
12. Liwo, A., Arlukowicz, P., Czaplewski, C., Oldziej, S., Pillardy, J, Scheraga, H. A. : A method for optimizing potential-energy functions by a hierarchical design of the potential-energy landscape: Applicaiton to the UNRES for field. Proc. Natl. Acad. Sci. U.S.A. vol. 99 (2002) 1937–1942
13. Lee, J.; Park, K.; Lee, J.: Full Optimization of Linear Parameters of a United Residue Protein Potential. J. Phys. Chem. B vol. 106 (2002) 11647–11657
14. Lee, J., Kim, S-.Y., Lee, J.: Design of a protein potential energy landscape by parameter optimization. J. Phys. Chem. B vol. 108 (2004) 4525–4534
15. Lee, J., Kim, S.-Y., Lee, J. Protein structure prediction based on fragment assembly and parameter optimization. Biophys. Chem. vol. 115 (2005) 209–214
16. Lee, J., Kim, S.-Y., Lee, J. Protein structure prediction based on fragment assembly and beta-strand pairing energy function. J. Korean Phys. Soc. vol. 46 (2005) 707–712
17. Lee, J., Liwo, A., Ripoll, D. R., Pillardy, J., Scheraga, H. A.: Calculation of Protein Conformation by Global Optimization of a potential energy function. Proteins. Suppl. **3** (1999) 204–208
18. Third Community Wide Experiment on the Critical Assessment of Techniques for Protein Structure Prediction; Asilomar Conference Center, December 13-17, 1998; http://predictioncenter.org/casp3/Casp3.html.

Efficient Solution of Bidomain Equations in Simulation of Cardiac Excitation Anisotropic Propagation

Yu Zhang[1], Ling Xia[1], and Guanghuan Hou[2]

[1] Department of Biomedical Engineering, Zhejiang University, Hangzhou 310027, China
xialing@zju.edu.cn
[2] Department of Mathematics, Zhejiang University, Hangzhou 310027, China

Abstract. Bidomain equations are used to characterize myocardial physiological activation and propagation. Their numerical solution is costly computation because of the higher temporal and spatial discretization requirements, especially in three dimensions. In most previous studies, the heart was supposed to be homogeneous isotropic medium, and thus can use the mondomain equation in stead of the bidomain equations to simulate the heart excitation propagation. Simulation of heart excitation anisotropic propagation in three-dimensional large tissue size by solving bidomain equations has not been implemented yet. In this paper, we present an efficient solution of bidomain equations in simulation of heart excitation anisotropic propagation by combining some numerical techniques such as non-standard finite difference (NSFD), domain decomposition and multigrid methods. The results show that the proposed method can successfully be used to simulate heart excitation anisotropic propagation in three-dimensional large tissue size, and it suggests that such method may provide a good basis for heart simulation research in a more physiologically way.

1 Introduction

Computational models of activation and propagation in virtual cardiac tissues are now becoming an established research tool for investigating mechanisms of normal and abnormal heart activation [1]. More recently, on the basis of anisotropic bidomain theory [2], reaction-diffusion equations have been used to characterize myocardial physiological propagation and electrical behavior generated by ionic movements in myocardial cell by coupling with myocardial cell model which is composed of a large number of ordinary differential equations (ODEs).

However, the bidomain equations are computationally very expensive. Most researchers simplify heart as homogeneous isotropic medium, thus the bidomain equations can be replaced by the mondomain equation in which an operator splitting approach and adaptive time step can be used to solve parabolic equation and avoid the solution of a large non-linear PDE system (PDEs) at every time step. Huiskamp [3] created a model of the ventricles that was composed of 810546 elements and used a biophysically based membrane model in a monodomain framework to simulate the spread of excitation. Gulrajani et al [4] created a cardiac model with 12 million cell

D.-S. Huang, K. Li, and G.W. Irwin (Eds.): ICIC 2006, LNBI 4115, pp. 571 – 581, 2006.

units by using an equal anisotropy assumption to simplify the system to a monodomain model and performed the computations on a large parallel shared memory computer. All these models of cardiac propagation are far less expensive computationally than the bidomain frameworks, but they fail to provide an avenue for introducing extracellular stimulation nor do they produce extracellular potentials that can be compared directly with experimental recordings. Recently, some researchers have adopted bidomain framework to model different aspects of cardiac activation, but only implement in simplified and segmental 3D anatomical heart geometry for limitation of computational capacity. Most bidomain implementations are based around a finite difference solution technique but the finite element method has also been used [5], as well as the finite volume technique [6].

Since the heart is a heterogeneous 3-D object with anisotropic properties, we have to solve another elliptic equation, which is actually to solve an algebra equation Ax=b. Since the number of the grid units in heart propagation model maybe millions, inverting the matrix A is infeasible for the common iterate methods, such as Gauss-Sediel iteration or C-G iteration, and even for the multigrid method [7], and that elliptic equations are known to yield matrices that become progressively more ill-conditioned as the spatial resolution is increased. In this paper, we present an efficient solution of bidomain equations in simulation of heart excitation anisotropic propagation by combining some numerical techniques such as non-standard finite difference (NSFD) [8], domain decomposition [9] and multigrid methods, which can drastically reduce the overall time of simulation.

2 Methods

2.1 The Bidomain Model and Cardiac Cell Model

The bidomain model applied in heart is:

$$\frac{\partial V_m}{\partial t}(\underline{x},t)=\frac{1}{C_m}\left[-\sum I_{ion}(\underline{x},t)-I_{app}(\underline{x},t)-\frac{1}{\beta}\nabla\cdot(D_e(\underline{x})\nabla\phi_e(\underline{x},t))\right] \tag{1}$$

$$\nabla\cdot\big((D_i(\underline{x})+D_e(\underline{x}))\nabla\phi_e(\underline{x},t)\big)=-\nabla\cdot\big(D_i(\underline{x},t)\nabla V_m(\underline{x},t)\big) \tag{2}$$

The parabolic (1) and elliptic (2) equations are the governing equations of the bidomain model. The main variables V_m and Φ_e are the transmembrane potential and the extracellular potential, I_{ion} is the current density flowing through the ionic channels, $I_{app}(\underline{x},t)$ is an applied stimulus current per unit area, β is a geometrical constant, C_m is the capacitance of the cell membrane, D_i and D_e are conductivity tensors which is used to describe the anisotropic nature of the cardiac microstructure, $\underline{x}$ is the spatial position. They can be simplified by the assumption of equal anisotropy, treating heart as homogeneous isotropic medium.

$$D_i(\underline{x}) = \frac{1}{k} D_e(\underline{x}) \tag{3}$$

$$\frac{\partial V_m}{\partial t}(\underline{x},t) = \frac{1}{C_m}\left[-\sum I_{ion}(\underline{x},t) - I_{app}(\underline{x},t) + \frac{1}{\beta}\left(\frac{k}{k+1}\right)\nabla\cdot\left(D_i(\underline{x})\nabla V_m(\underline{x},t)\right)\right] \tag{4}$$

where k is the anisotropy ratio. The resulting parabolic equation is known as the monodomain equation (4), details about bidomain theory can see Ref. [1].

The electrophysiological behavior of single cell can be described by the following differential equations which constitute the non-linear parts of the PDEs:

$$\begin{aligned} dV_m / dt = &-(1/C_m)(I_{Na} + I_{NaL} + I_{CaL} + I_{to} + I_{Kr} + I_{Ks} + I_{K1} \\ &+ I_{NaK} + I_{NaCa} + I_{pCa} + I_{pK} + I_{bCa} + I_{bNa} + I_{stim}) \end{aligned} \tag{5}$$

$$I_x = G_x \cdot (V_m - E_x) \tag{6}$$

where I_x is the current flow through the ion channel x, G_x is the conductance of the channel and E_x is the reversal potential for the channel which is the Nernst potential for the ion or ions. The conductance is the maximal conductance g_x multiplied by the state of the channel which is modeled with a set of gating variables. For a channel with m different gates, each with n_i sub-units per gate, the model of the channel conductance is given by ODEs of the form:

$$G_x = g_x \prod_{i=1}^{m} x_i^{ni} \tag{7}$$

$$\frac{dx_i}{dt} = \alpha_{x,i}(V_m)\cdot(1-x_i) - \beta_{x,i}(V_m)\cdot x_i, \quad or \quad \frac{dx_i}{dt} = \frac{x_{i,\infty}(V_m) - x_i}{\tau_{x,i}(V_m)} \tag{8}$$

$$\tau_{x,i}(V_m) = \frac{1}{\alpha_{x,i}(V_m) + \beta_{x,i}(V_m)}, \; x_{i,\infty}(V_m) = \frac{\alpha_{x,i}(V_m)}{a_{x,i}(V_m) + \beta_{x,i}(V_m)} \tag{9}$$

where $x_{i,\infty}$ represents the steady state behavior of the gate. The $\alpha_{x,i}(V_m)$ and $\beta_{x,i}(V_m)$ coefficients are known as rate constants and are voltage dependent.

In this study, the Tusscher et al.'s ionic model of human ventricular tissue [10] is used since it includes a high level of electrophysiological detail while maintaining a low computational cost. There are thirteen ionic currents in the ten Tusscher et al.'s ionic model [10], fourteen gating parameters in total. These ODEs for gating parameters are coupled with an ODE for the calcium concentration in the cell and an ODE for the membrane voltage, making sixteen ODEs in total.

2.2 Boundary Conditions

In order to specify the boundary conditions, let the closed surface S_H be a boundary separating bidomain region H and surrounding volume conductor B, and let the closed surface S_B bound region B. $\boldsymbol{n}$ denotes the unit outward normal to S_H and S_B. Φ_o is the extracardiac potential and D_i is the isotropic scalar conductivity outside of H.

$$\phi_o = \phi_e,\ n \cdot D_o(\underline{x})\nabla\phi_o = n\big(D(\underline{x})\nabla\phi_e(\underline{x},t) + D_i(\underline{x})\nabla V_m(\underline{x},t)\big) \quad on \quad S_H \tag{10}$$

$$n \cdot \sigma_o \nabla\phi_o = 0 \quad on \quad S_B \tag{11}$$

Since the sources in H are related to the presence of intracellular medium, which is absent in B, we may assume that the vector $D_i \nabla V_m$ in Eq.(10) is tangent to the surface S_H; this becomes an boundary condition:

$$n \cdot D_i(\underline{x})\nabla V_m(\underline{x},t) = 0 \quad on \quad S_H \tag{12}$$

$$n \cdot D_i(\underline{x})\nabla\phi_e(\underline{x},t) = 0 \quad on \quad S_H \tag{13}$$

2.3 Numerical Considerations

The equations of bidomain model can be solved with different time steps and at different spatial resolutions as a decoupled system, a system of a parabolic PDE, an elliptic PDE and a nonlinear system of ODEs. First, we introduce a temporal discretization of the time intervals $t \in (0, t_N]$. Let Δt be the global time step or the time interval between each integration process $\Delta t = t_N/N$ and M be an even number of fractional time steps. Define $\Delta t' = \Delta t/M$ as the fractional diffusion time step. In the spatial direction, we use rectangular blocks the domain (H), which can be partitioned into a set of smaller, simpler elements to generate computational matrices. Both the parabolic PDE and the nonlinear ODEs system are solved via the explicit forward-Euler scheme with an operator splitting technique [1]. The equations can be solved in the following steps:

1. Assume we know the value of x^n, y^n, $V_m{}^n$ and $\Phi_e{}^n$ at t^n.
2. Compute $V_m{}^{n+1/2}$ in H by solving the parabolic equation (Eq.(1), or (4)) using a step size $\Delta t'$ repeated $M/2$ times. (Eq.(14) for bidomain, Eq.(15) for monodomain)

$$V_m^{\ n+1/2} = V_m^{\ n} + \frac{\Delta t'}{C_m}\left[-\sum I_{ion}^{\ n} - I_{app}^{\ n} - \frac{1}{\beta}\nabla\cdot\left(D_e\nabla\phi_e^{\ n}\right)\right] \tag{14}$$

$$V_m^{\ n+1/2} = V_m^{\ n} + \frac{\Delta t'}{C_m}\left[-\sum I_{ion}^{\ n} - I_{app}^{\ n} + \frac{1}{\beta}\left(\frac{k}{k+1}\right)\nabla\cdot\left(D_i\nabla V_m^{\ n}\right)\right] \tag{15}$$

3. Compute x^{n+1}, y^{n+1} by solving the ODEs of the cell model using nonstandard finite difference scheme, in which x^{n+1} represents gating variables, y^{n+1} represents I_{ion}.

$$x^{n+1} = x_\infty\left(V_m^{\ n+1/2}\right) + \left(x^n - x_\infty\left(V_m^{\ n+1/2}\right)\right)e^{\left(-\frac{\Delta t}{\tau_x\left(V_m^{\ n+1/2}\right)}\right)} \tag{16}$$

$$y^{n+1} = y^n + \Delta t \cdot f\left(V_m^{\ n+1/2}, x^n\right) \tag{17}$$

4. Compute V_m^{n+1} in H by solving the parabolic equation (Eq.(1), or (4)) using a step size $\Delta t'$ repeated $M/2$ times again. (same as step 2)
5. Compute Φ_e^{n+1} in H by solving the elliptic equation (Eq.(2)) using domain decomposition and multigrid methods:

$$\nabla\cdot\left((D_i+D_e)\nabla\phi_e^{n+1}\right)=-\nabla\cdot\left(D_i\nabla V_m^{n+1}\right) \qquad \textit{only for bidomain model} \tag{18}$$

2.4 Efficient ODE-Solvers

Generally, the ODEs system is solved using forward Euler method, but it has to take small step-size to maintain the stable computation. Here we take advantage of the linear equations (Eq.(8)) of the ODEs governing the gating variables by making use of discrete derivative (Eq.(19)) from the exact solution as part of a NSFD scheme [8], to update gating variables,

$$x_{n+1}=x_\infty+\left(x_n-x_\infty\right)e^{\left(-\frac{\Delta t}{\tau_x}\right)} \tag{19}$$

where x_∞ *and* τ_x, are defined previously and can be evaluated explicitly using the value of V_m from the previous time step. It leaves the ODEs only for membrane potential (V_m) and calcium concentration ($[Ca^{2+}]_i$) which can solved numerically by forward Euler method. In this way, the system is no longer stiff. A larger step-size can now be used.

2.5 Efficient Parabolic Equation-Solvers

Most studies [1] used the conventional Euler method to integrate the PDEs by using a very small time step to keep numerical stability and accuracy, which makes the computation much slow. Here, we solve the PDEs for cardiac excitation propogation by using the well-known operator splitting and adaptive time step methods, which can speed up the computation and maintain the accuracy as well. The operator-splitting scheme can then be symbolically represented as:

$$y^{n+1}=D_{\Delta t'}^{M}R_{\Delta t}D_{\Delta t'}^{M}y^{n} \tag{20}$$

where $y^{n+1}$and y^n are solutions at t^{n+1} and t^n respectively, $R_{\Delta t}$ represents the stiff integration of the reaction term over a step size Δt (see step.3). $D_{\Delta t'}$ represents the integration of the diffusion term using a step size $\Delta t'$ and the superscript indicates this integration is to be repeated M times (see step.2, step.4). This construct enables the global time step ($\Delta t = M\Delta t'$) to become significantly larger than the diffusion stability limit, resulting in an improved computational performance.

2.6 Efficient Elliptic Equation-Solvers

We decompose H into some subdomains: $H:=\bigcup_{i=1}^{n}H_i$ with non-overlapping decomposition. Then a smaller linear system can be computed in each subdomain separately:

$$\nabla \cdot \left((D_i + D_e) \nabla \phi_{e,i}^{\ n+1,k} \right) = -\nabla \cdot \left(D_i \nabla V_{m,i}^{\ n+1} \right) \quad in \quad H_i \tag{21}$$

$$B_{i,j}\phi_{e,i}^{n+1,k} = B_{i,j}\phi_{e,j}^{n+1,k-1} \quad on \quad \partial H_i \cap \bar{H}_j, \quad B_{i,j} = \frac{\partial}{\partial n_i} + c_1 + c_2 \frac{\partial}{\partial \tau_i} + c_3 \frac{\partial^2}{\partial \tau_j^2} \tag{22}$$

$$\partial_n \phi_{e,i}^{\ n+1,k} = 0 \quad on \quad \partial H_i \cap \partial H \tag{23}$$

where $B_{i,j}$ is artificial boundary condition (ABC) on the boundary of H_i intercepted by H_j.τ_i is the tangential derivative along the interface, n_i is the normal, and the coefficient c_1, c_2, c_3 is used to minimize the convergence rate of the Schwarz algorithm [11, 12].

It is well known that classical relaxation methods such as the Jacobi or Gauss-Seidel methods, applied to the iterative solution of the resulting linear system, can quickly damp the high frequencies of the error. However, they do not allow for an efficient treatment of the low frequency components, so we use the multigrid method in subdomain. The basic idea of the multigrid grid is to transfer the partially solved solution on a coarser grid in order to transform the low frequencies of the fine grid solution in high frequencies, which can be efficiently damped by the standard relaxation methods. In the subdomain, let $\tilde{\phi}_{e,j,h}^{n+1,k}$ (subscript h denotes in fine grid) be the approximate iterative solution by classical relaxation methods:

$$\nabla \cdot \left((D_i + D_e) \nabla v_h \right) = r_h \tag{24}$$

where $r_h = -\nabla \cdot \left(D_i \nabla V_{m,i,h}^{n+1} \right) - \nabla \cdot \left((D_i + D_e) \nabla \tilde{\phi}_{e,i,h}^{n+1,k} \right)$ is the residual; since v_h is smooth, Eq.(24) can be approximated on the coarser grid with mesh spacing $H>h$ by

$$\nabla \cdot \left((D_i + D_e) \nabla v_H \right) = I_h^H r_h \tag{25}$$

The function I_h^H is to interpolate the value in fine grid to coarser grid (I_H^h is the opposite). Then after working out v_H from Eq.(25), we can re-interpolate it to the finer grid, and added to the previous approximate solution $\tilde{\phi}_{e,i,h}^{n+1,k+1} = \tilde{\phi}_{e,i,h}^{n+1,k} + I_H^h v_H$. Actually, the higher level multigrid method may be more feasible, since it brings more accuracy while only increasing a little more computation (see [7]). The procedure is similar to the two level multigrid, just to reuse the two level multigrid when solve the equation in coarser grid and then resort it for the more coarser grid. In addition, there are many cases of the multi-level multigrid method. The most important cases in practice are V-cycle, and W-cycle. For more details, see [7, 13].

3 Results

3.1 Numerical Results for Single Cell Experiments with NSFD Scheme

By using of Matlab7.0's ode15s function with setting both absolute and relative tolerances (*atol* and *rtol*, respectively) to 10^{-10} which is a variable step-size stiff solver for ODEs, which is based on a family of linear multistep methods, we get the reference solution of the membrane voltage V_m^{ref} which is shown in Fig.1.

The relative root-mean square error (E_{RMS}) between the numerical solution V_m and the reference V_m^{ref} can be computed using Eqs. (26) to give a more intuitive sense of the error between solutions (Table 1). As we know, the ODEs governing the gating variables are linear, so we make use of their known exact solution as part of a non-standard finite difference scheme to update gating variables, so that the stiffness of the ODEs system of ionic model is reduced and a larger step-size can be used.

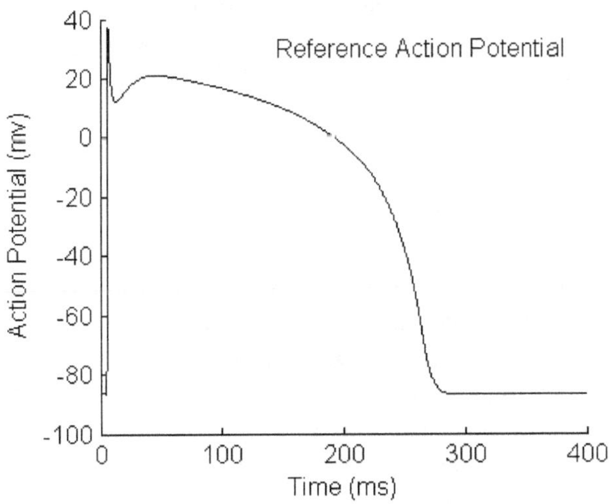

Fig. 1. Reference solution generated for the membrane voltage V_m^{ref}

$$E_{RMS} = \sqrt{\frac{\sum_{i=1}^{N}\left(V_i - V_i^{ref}\right)^2}{\sum_{i=1}^{N} V_i^{ref}}} \tag{26}$$

3.2 Validation of the Model on Simple and Anatomical Geometry

First, we simulate isotropy and anisotropic propagation inside a homogeneous virtual tissue composed of uniform cell units (40×40), in which the fiber direction was kept constant and anisotropic conductivity tensors Dl =0.45cm2/ms (l for longitudinal),

Table 1. R_{RMS} values for the NSFD scheme for some step sizes Δt

Δt(ms)	E_{RMS}
0.005	0.0086
0.01	0.0236
0.05	0.0392
0.1	0.0423
0.2	0.0496
0.3	0.0654

Dt =0.13 cm2/ms (t for transversal). The stimulus current is applied at the center of the region and Δt =10μs, M=20, Dx=0.39mm. Fig.2 shows the equal anisotropy propagated isochrones agree with theoretical predictions very well, i.e., the ratio of the longitudinal and transversal axes of isochrone is proportional to the square-root of the conductivities in the corresponding directions [14].

Second, we test the proposed method on the heart model with real anatomical structure. A dog ventricle data set, composed of the volume image and tensor eigenvector reconstructed from diffusion tensor magnetic resonance imaging developed by Edward [15], is used as the domain for simulating normal activation. We chose ten slices of the cardiac tissue for simulation test. The spatial matrix size is 12×130×130 which corresponds to pixel size of 0.78 × 0.78 × 0.78 mm. Initial activation points (Purkinje–myocardial junction sites) and Fiber angles are shown in Fig. 3 at a representative section perpendicular to the long axis of our heart model. We set the discretization parameters as Δt =10μs, M=20, Dx=0.39mm and the tolerance for the elliptic solve is set to 10-6. All the simulation are performed on Dual-Core Intel® Xeon® processors Dell Precision WorkStation 670 System (2R) with 4Gb physical RAM, cost 17 hrs and 40 mins for 55ms excitation propagation simulation. Fig. 4 shows the activation isochrones, the wave front radiates from the points of stimulation, initially from the two sites in the left ventricle. This initiates a wave of depolarization that spreads concentrically around the left ventricle. After about 50 ms the wave fronts merge in the posterior part of the right ventricle, which is a little different from the real physiological condition. The reason is that, there are more activation points in a real heart, but here it is a segmental 3D model and only activation points lying in the vicinity of the tissue were included. In physiological condition wave fronts originating from other sites lying outside the selected segment could depolarize tissue inside the segment before the wave front comes from our activation points. This possible effect is not included in the present simulation.

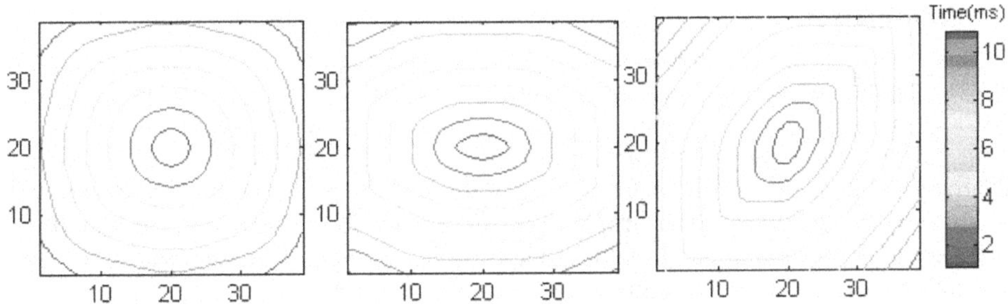

Fig. 2. Activation isochrones from simulations in uniformly isotropy and anisotropic virtual slab tissue. The left panel shows view of isotropy propagation, the middle and right display views of anisotropic propagation simulation with fiber directions of 0°and 60°; the slab are 40 cells in each direction.

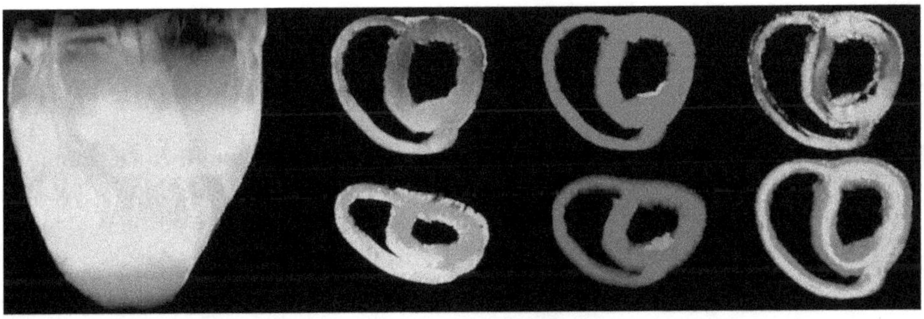

Fig. 3. Anatomical model of the dog ventricle. The left panel shows 3D view of the whole ventricle, and the next displays view of the selected part of the cardiac tissue for simulation. After that displays the initial activation points (*Purkinje–myocardial junction sites*). The right view displays the fiber angles (*above is azimuth angles, bottom is colatitude angles*).

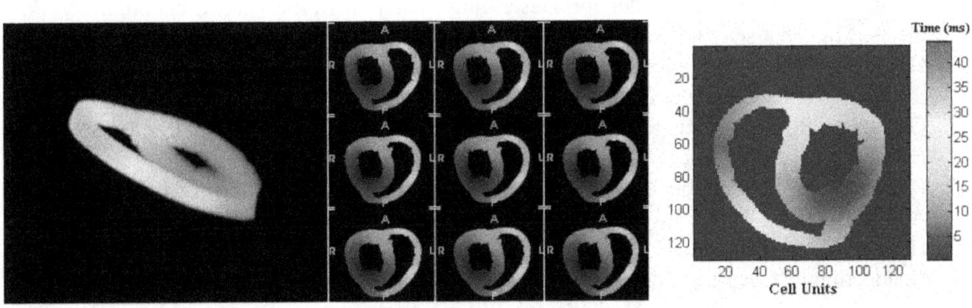

Fig. 4. Isochrones of simulated normal activation of the ventricles. The left panel shows 3D isochrones in selected part of ventricle, and the right view displays the activation isochrones on the first slice of the tissue. The middle panels display views of the activation in other slices.

5 Conclusion

In this paper, we combine the advantage of domain decomposition and multigrid methods to give a better scheme for solving large scale algebra equation that arise from elliptic problems, which can significantly speed up integration while preserving accuracy. The results show that the proposed method can successfully be used to simulate heart excitation anisotropic propagation in three-dimensional large tissue size, and it suggests that such method may provide a good basis for heart simulation research in a more physiologically way.

Since the action potential varies greatly on the wave front of excitation propagation, the adaptive multigrid may supply an inexpensive switching criterion, with which adaptive smoother terminates when the (scaled) residuals have become small, thus effectively decease the iteration steps. Furthermore, the multilevel scheme of elliptic partial differential equations can be combined with adaptive mesh refinement. However, the adaptive choice of the discretization order is difficult, which will be a good topic for future research.

Acknowledgments

This project is supported by the 973 National Key Basic Research & Development Program (2003CB716106), the National Natural Science Foundation of China (30370400, 30570484) and the Program for New Century Excellent Talents in University (NCET-04-0550). The authors are also grateful to Dr. Edward Hsu of Duke University for his providing DTMRI data.

References

1. Lines, G.T., Buist, M.L., et al.: Mathematical Models and Numerical Methods for the Forward Problem in Cardiac Electrophysiology. Comput Visual Sci. 5 (2003) 215–239
2. Tung, L.: A Bidomain Model for Describing Ischemic Myocardial d-c Potentials. PhD thesis, Massachusetts Institute of Technology (1978)
3. Huiskamp, G.J.: Simulation of Depolarization in a Membrane-equations-based Model of the Anisotropic Ventricle. IEEE Trans. Biomed. Eng. 45 (1998) 847–855
4. Gulrajani, R.M., Trudel, M.C., Leon, L.J.: A Membrane-Based Computer Heart Model Employing Parallel Processing. Biomedizinische Technik, 46 (2001) 20–22
5. Rogers, J.M., McCulloch, A.D.: A Collocation-Galerkin Finite Element Model of Cardiac Action Potential Propagation. IEEE Trans. Biomed. Eng. 41 (1994) 743–757
6. Harrild, D.M., Penland, C., Henriquez, C.S.: A Flexible Method for Simulating Cardiac Conduction in Three-Dimensional Complex Geometries. J. Electrocardiol. 33 (2000) 241–251
7. Scott, R. et al.: Multigrid Methods for Elliptic Problems: A Review. Mon. Wea. Rev, (1986) 943-959
8. Mickens, R.E.: Applications of Nonstandard Finite Difference Schemes, World Scientific, Singapore (2000)
9. Lions, P.L.: On the Schwarz alternating method III: A Variant for Nonoverlapping Subdomains, Third International Symposium on Domain Decomposition Methods for Partial Differential Equations, SIAM, Philadelphia, PA, (1990) 202-223

10. ten Tusscher, K.H.W.J., Noble, D., Noble, P.J., Panfilov, A.V.: A Model for Human Ventricular Tissue. Am J Physiol Heart Circ Physiol. 286 (2004) 1573-1589
11. Japhet, C. et al.: The Best Interface Conditions for Domain Decomposition Methods: Absorbing Boundary Conditions, (2003)
12. Japhet, C., Nataf, F., Roux, F.X.: Extension of a Coarse Grid Preconditioner to non-symmetric Problems. Contemporary Mathematics, 218 (1998) 279-286
13. Rude, U.: Fully Adaptive Multigrid Methods, SIAM Journal on Numerical Analysis, 30 (1993) 230-248
14. Gulrajani, R.M.: Models of the Electrical Activity of the Heart and Computer Simulation of the Electrocardiogram. CRC Crit Rev Biomed Eng, 16 (1988) 1–61
15. Edward, W.H.: Three-dimensional Diffusion Tensor Microscopy of Fixed Mouse Hearts, Magnetic Resonance in Medicine, 52 (2004) 453-460

Analysis of Numerical Solutions to Stochastic Age-Dependent Population Equations

Qimin Zhang and Xining Li

School of Mathematics and Computer Science Ningxia University, Yinchuan Ningxia 750021, P.R. China
zhangqimin64@sina.com

Abstract. In this paper, stochastic age-dependent population equations, one of the important classes of hybrid systems, are studied. In general, most of stochastic age-dependent population equations do not have explicit solutions. Thus numerical approximation schemes are invaluable tools for exploring their properties. The main purpose of this paper is to develop a numerical scheme and show the convergence of the numerical approximation solution to the true solution.

1 Introduction

We consider the convergence of stochastic age-dependent population equation of the form

$$\frac{\partial P}{\partial t} + \frac{\partial P}{\partial a} = -\mu(t,a)P + f(t,P) + g(t,P)\frac{dW_t}{dt},$$

or

$$d_tP = [-\frac{\partial P}{\partial a} - \mu(t,a)P + f(t,P)]dt + g(t,P)dW_t, \qquad (1)$$

where $P(t,a)$ denotes the population density of age a at time t, $\beta(t,a)$ denotes the fertility rate of females of age a at time t, $\mu(t,a)$ denotes the mortality rate of age a at time t. $f(t,P)$ denotes effects of external environment for population system. $g(t,P)$ is a diffusion coefficient.

There has been an increasing interest in the study of age-dependent population equations. For example, Cushing [1] investigated hierarchical age-dependent populations with intra-specific competition or predation when $g(t,P) = 0$. Allen and Thrasher [2] considered vaccination strategies in age-dependent populations in the case of $g(t,P) = 0$. Existence, uniqueness and stability of strong solutions equation (1) were discussed in [3]. Pollard [4] studied the effects of adding stochastic terms to discrete-time age-dependent models that employ Leslie matrices. Using Lyapunov functional, numerical analysis for stochastic age-dependent population equation (1) has been studied by Zhang[5].Abia [6] gave the numerical solution for non-linear age-dependent population models basing second-order finite difference schemes; explicit schemes of first-order were considered by Kostova [7] for a model describing interacting population dynamics; explicit two-step

D.-S. Huang, K. Li, and G.W. Irwin (Eds.): ICIC 2006, LNBI 4115, pp. 582–592, 2006.

second-order methods based on the central-difference operator along characteristics were studied by Milner and Rabbiolo [8] for the Gurtin- MacCamy model but assuming that the fertility rate in Eq.(1) is independent of the total size of the population. In this paper, by virtue of a direct approach quite different from those mentioned above, we shall prove that the numerical solutions converge to the exact solutions and provide the order of convergence. We give an example to show the efficiency of our numerical method.

2 Preliminaries and Approximation

Let

$$V = H^1([0,A]) \equiv \{\varphi | \varphi \in L^2([0,A]), \frac{\partial \varphi}{\partial x_i} \in L^2([0,A]),$$
$$\text{where} \frac{\partial \varphi}{\partial x_i} \text{is generalized partial derivatives}\}.$$

V is a Sobolev space. $H = L^2([0,A])$ such that

$$V \hookrightarrow H \equiv H' \hookrightarrow V'.$$

V' is the dual space of V. We denote by $\|\cdot\|$, $|\cdot|$ and $\|\cdot\|_*$ the norms in V, H and V' respectively; by $\langle\cdot,\cdot\rangle$ the duality product between V, V', and by $(\cdot,\cdot)$ the scalar product in H. For an operator $B \in \mathcal{L}(M,H)$ be the space of all bounded linear operators from M into H, we denote by $\|B\|_2$ the Hilbert-Schmidt norm, i.e.

$$\|B\|_2^2 = tr(BWB^T).$$

Let $(\Omega,\mathcal{F},P)$ be a complete probability space with a filtrations $\{\mathcal{F}_t\}_{t\geq 0}$ satisfying the usual conditions (i.e., it is increasing and right continuous while $\mathcal{F}_0$ contains all P-null sets).

Let $C = C([0,T];H)$ be the space of all continuous function from $[0,T]$ into H with sup-norm $\|\psi\|_C = \sup_{0\leq s\leq T}|\psi|(s)$, $L_V^p = L^p([0,T];V)$ and $L_H^p = L^p([0,T];H)$.

Consider stochastic age-dependent population equation of the form

$$\begin{cases} d_tP = [-\frac{\partial P}{\partial a} - \mu(t,a)P + f(t,P)]dt + g(t,P)dW_t, & in \quad Q = (0,T)\times(0,A), \\ P(0,a) = P_0(a), & in \quad [0,A], \\ P(t,0) = \int_0^A \beta(t,a)P(t,a)da, & in \quad [0,T]. \end{cases} \tag{2}$$

where $T>0$, $A>0$, $f(t,\cdot): L_H^2 \to H$ be a family of nonlinear operators, $\mathcal{F}_t$-measurable almost surely in t. $g(t,\cdot): L_H^2 \to \mathcal{L}(M,H)$ is the family of nonlinear operator, $\mathcal{F}_t$-measurable almost surely in t.

The integral version of Eq.(2) is given by the equation

$$P_t = P_0 - \int_0^t \frac{\partial P_s}{\partial a}ds - \int_0^t \mu(s,a)P_sds + \int_0^t f(s,P_s)ds + \int_0^t g(s,P_s)dW_s, \tag{3}$$

here $P_t = P(t,a)$.

For system (2) the discrete approximate solution on $t = 0, \Delta t, 2\Delta t, \cdots, N\Delta t$ is defined by the iterative scheme

$$Q_t^{n+1} = Q_t^n - \frac{\partial Q_t^n}{\partial a}\Delta t - \mu(t,a)Q_t^n\Delta t + f(t, Q_t^n)\Delta t + g(t, Q_t^n)\Delta W_n, \quad (4)$$

with initial value $Q_t^0 = P(0,a)$, $Q^n(t,0) = \int_0^A \beta(t,a)Q_t^n da$, $n \geq 1$. Here, Q_t^n is the approximation to $P(t_n, a)$, for $t_n = n\Delta t$, the time increment is $\Delta t = \frac{T}{N} << 1$, Brownian motion increment is $\Delta W_n = W(t_{n+1}) - W(t_n)$.

For convenience, we shall extend the discrete numerical solution to continuous time. We first define the step function

$$Z_t = Z(t,a) = \sum_{k=0}^{N-1} Q_t^k \mathbf{1}_{[k\Delta t,(k+1)\Delta t)}, \quad (5)$$

where $\mathbf{1}_G$ is the indicator function for the set G. Then we define

$$Q_t = P_0 - \int_0^t \frac{\partial Q_s}{\partial a}ds - \int_0^t \mu(s,a)Z_s ds + \int_0^t f(s, Z_s)ds + \int_0^t g(s, Z_s)dW_s, \quad (6)$$

with $Q_0 = P(0,a)$, $Q(t,0) = \int_0^A \beta(t,a)Q_t da$, $Q_t = Q(t,a)$. It is straightforward to check that $Z(t_k, a) = Q_t^k = Q(t_k, a)$.

The key contribution of this paper is to show that the approximate solution Q_t will converge to the true solution P_t of stochastic age-dependent population equation (2) under the given conditions. Because Q_t interpolates the discrete numerical solution, this will immediately give a convergence for Q_t^k.

As the standing hypotheses we always assume that the following conditions are satisfied:

(i) $f(t,0) = 0$, $g(t,0) = 0$;

(ii)(Lipschitz condition) there exists a positive constant K such that $x, y \in C$

$$|f(t,y) - f(t,x)| \vee \|g(t,y) - g(t,x)\|_2 \leq K\|y - x\|_C, \ a.e.t; \quad (7)$$

(iii) $\mu(t,a)$, $\beta(t,a)$ are continuous in $\bar{Q}$ such that

$$0 \leq \mu_0 \leq \mu(t,a) \leq \bar{\alpha} < \infty, \quad 0 \leq \beta(t,a) \leq \bar{\beta} < \infty. \quad (8)$$

Notes It is known (cf.[3]) that Eq.(2) has a unique strong solution on $t \in [0, T]$ under conditions (i)–(iii).

3 The Main Results

In this section, we will prove the convergence of Q_t. Throughout the following analysis we use $C_1, C_2, \cdots$ to denote generic constants that depend upon K, T, but not upon Δt. The precise values of these constants may be determined via the proofs.

In order to prove the convergence of Q_t, we first give several lemmas. our first lemma shows that the continuous approximation has bounded second moments in a strong sense.

Lemma 1. *Under the assumptions above, then*

$$E \sup_{t\in[0,T]} |Q_t|^2 \le C_1. \tag{9}$$

Proof. From (6), one can obtain

$$dQ_t = -\frac{\partial Q_t}{\partial a}dt - \mu(t,a)Z_t dt + f(t,Z_t)dt + g(t,Z_t)dW_t. \tag{10}$$

Applying Itô formula to $|Q_t|^2$ yields

$$\begin{aligned}
|Q_t|^2 &= |Q_0|^2 + 2\int_0^t \langle -\frac{\partial Q_s}{\partial a} - \mu(s,a)Z_s, Q_s\rangle ds + 2\int_0^t (f(s,Z_s),Q_s)ds \\
&\quad +2\int_0^t (Q_s, g(s,Z_s)dW_s) + \int_0^t \|g(s,Z_s)\|_2^2 ds \\
&\le |Q_0|^2 - 2\int_0^t \langle \frac{\partial Q_s}{\partial a}, Q_s\rangle ds - 2\mu_0\int_0^t (Z_s,Q_s)ds + 2\int_0^t (f(s,Z_s),Q_s)ds \\
&\quad + \int_0^t \|g(s,Z_s)\|_2^2 ds + 2\int_0^t (Q_s, g(s,Z_s)dW_s).
\end{aligned}$$

Since

$$\begin{aligned}
&-\langle \frac{\partial Q_s}{\partial a}, Q_s\rangle \\
&= -\int_0^A Q_s d_a(Q_s) = \frac{1}{2}(\int_0^A \beta(s,a)Q_s da)^2 \\
&\le \frac{1}{2}\int_0^A \beta^2(s,a)da \int_0^A Q_s^2 da \\
&\le \frac{1}{2}A^2\bar{\beta}^2|Q_s|^2.
\end{aligned}$$

Therefore, we get that

$$\begin{aligned}
|Q_t|^2 &\le |Q_0|^2 + A^2\bar{\beta}^2\int_0^t |Q_s|^2 ds + 2\int_0^t |Q_s||f(s,Z_s)|ds \\
&\quad +2\mu_0\int_0^t |Q_s||Z_s|ds + \int_0^t \|g(s,Z_s)\|_2^2 ds + 2\int_0^t (Q_s, g(s,Z_s)dW_s).
\end{aligned}$$

Now, it follows that for any $t \in [0, T]$

$$
\begin{aligned}
&E \sup_{0 \le s \le t} |Q_s|^2 \\
&\le E|Q_0|^2 + (A^2\bar{\beta}^2 + 2 + \mu_0) \int_0^t E \sup_{0 \le s \le t} |Q_s|^2 ds + \int_0^t E|f(s, Z_s)|^2 ds \\
&\quad + \int_0^t E\|g(s, Z_s)\|_2^2 ds + 2E \sup_{0 \le s \le t} \int_0^s (Q_r, g(r, Z_r) dW_r).
\end{aligned}
$$

Using condition (ii) yields

$$
\begin{aligned}
&E \sup_{0 \le s \le t} |Q_s|^2 \\
&\le E|Q_0|^2 + (A^2\bar{\beta}^2 + 2 + \mu_0) \int_0^t E \sup_{0 \le s \le t} |Q_s|^2 ds \\
&\quad + 2K^2 \int_0^t E\|Z_s\|_C^2 ds + 2E \sup_{0 \le s \le t} \int_0^s (Q_r, g(r, Z_r) dW_r).
\end{aligned}
\tag{11}
$$

We can argue in a similar manner as Zhang did in paper[5]. Hence our claim is proved.

Next, we show that the continuous time approximation remains close to the step function Z_t in a strong sense.

Lemma 2. *Assume the preceding hypotheses hold, then*

$$
E \sup_{0 \le t \le T} |Q_t - Z_t|^2 \le C_3 \Delta t. \tag{12}
$$

Proof. For $\forall t \in [0, T]$, there exists an integer k such that $t \in [k\Delta t, (k+1)\Delta t)$. We have

$$
\begin{aligned}
&Q_t - Z_t \\
&= Q_t - Q_t^k \\
&= -\int_{k\Delta t}^t \frac{\partial Q_s}{\partial a} ds - \int_{k\Delta t}^t \mu(s, a) Z_s ds + \int_{k\Delta t}^t f(s, Z_s) ds \\
&\quad + \int_{k\Delta t}^t g(s, Z_s) dW_s.
\end{aligned}
$$

Thus,

$$
\begin{aligned}
&|Q_t - Z_t|^2 \\
&\le 4\left|\int_{k\Delta t}^t \frac{\partial Q_s}{\partial a} ds\right|^2 + 4\left|\int_{k\Delta t}^t \mu(s, a) Z_s ds\right|^2 + 4\left|\int_{k\Delta t}^t f(s, Z_s) ds\right|^2 \\
&\quad + 4\left|\int_{k\Delta t}^t g(s, Z_s) dW_s\right|^2.
\end{aligned}
$$

Now, the Cauchy-Schwarz inequality and the assumptions give

$$\begin{aligned}
&|Q_t - Z_t|^2 \\
&\leq 4\Delta t \int_{k\Delta t}^{t} \|Q_s\|^2 ds + 4\bar{\alpha}^2 \int_{k\Delta t}^{t} |Q_s|^2 ds + 4\Delta t \int_{k\Delta t}^{t} |f(s, Z_s)|^2 ds \\
&\quad +4 \left| \int_{k\Delta t}^{t} g(s, Z_s) dW_s \right|^2 \\
&\leq 4\Delta t \int_{0}^{T} \|Q_s\|^2 ds + 4\bar{\alpha}^2 \int_{k\Delta t}^{t} |Q_s|^2 ds + 4K^2 \Delta t \int_{0}^{T} |Q_s|^2 ds \\
&\quad +4 \left| \int_{k\Delta t}^{t} g(s, Z_s) dW_s \right|^2 .
\end{aligned}$$

Hence

$$\begin{aligned}
&E \sup_{t\in[0,T]} |Q_t - Z_t|^2 \\
&\leq 4\Delta t \int_{0}^{T} E\|Q_s\|^2 ds + 4\bar{\alpha}^2 C_1 \Delta t + 4K^2 \Delta t T C_1 \\
&\quad +4E \sup_{t\in[0,T]} \max_{k=0,1,\cdots,N-1} \left| \int_{k\Delta t}^{t} g(s, Z_s) dW_s \right|^2 .
\end{aligned}$$

Using the Doob inequality and (7), (9) yield

$$\begin{aligned}
&E \sup_{t\in[0,T]} |Q_t - Z_t|^2 \\
&\leq 4\Delta t \int_{0}^{T} E\|Q_s\|^2 ds + 4\bar{\alpha}^2 C_1 \Delta t + 4K^2 \Delta t T C_1 \\
&\quad +4 \max_{k=0,1,\cdots,N-1} \int_{k\Delta t}^{(k+1)\Delta t} E|g(s, Z_s)|^2 ds \\
&\leq 4\Delta t \int_{0}^{T} E\|Q_s\|^2 ds + 4\bar{\alpha}^2 C_1 \Delta t + 4K^2 \Delta t T C_1 + 4K^2 C_1 \Delta t,
\end{aligned}$$

Invoking lemma 3.2 of Zhang[5], we obtain the result (12).

We are now in a position to prove a strong convergence result.

Theorem 1. *Under the assumptions in Lemma 2, then*

$$E \sup_{0\leq t\leq T} |P_t - Q_t|^2 \leq C_4 \Delta t. \tag{13}$$

Proof. Combining (3) with (6) has

$$\begin{aligned}
&P_t - Q_t \\
&= -\int_{0}^{t} \frac{\partial (P_s - Q_s)}{\partial a} ds - \int_{0}^{t} \mu(s, a)(P_s - Z_s) ds \\
&\quad + \int_{0}^{t} (f(s, P_s) - f(s, Z_s)) ds + \int_{0}^{t} (g(s, P_s) - g(s, Z_s)) dW_s .
\end{aligned}$$

Therefore using Itô formula, along with the Cauchy-Schwarz inequality and (7) yields,

$$\begin{aligned}
&d|P_t - Q_t|^2 \\
&= -2\langle P_t - Q_t, \frac{\partial(P_t - Q_t)}{\partial a}\rangle dt - 2(P_t - Q_t, \mu(t,a)(P_t - Z_t))dt \\
&\quad +2(P_t - Q_t, f(t,P_t) - f(t,Z_t))dt + \|g(t,P_t) - g(t,Z_t)\|_2^2 dt \\
&\quad +2(P_t - Q_t, (g(t,P_t) - g(t,Z_t))dW_t) \\
&\le A^2\bar{\beta}^2|P_t - Q_t|^2 dt + 2\bar{\alpha}|P_t - Q_t||P_t - Z_t|dt \\
&\quad +2K|P_t - Q_t|\|P_t - Z_t\|_C dt + K^2\|P_t - Z_t\|_C^2 dt \\
&\quad +2(P_t - Q_t, (g(t,P_t) - g(t,Z_t))dW_t).
\end{aligned}$$

Hence, for any $t \in [0,T]$,

$$\begin{aligned}
&E\sup_{s\in[0,t]}|P_s - Q_s|^2 \\
&\le (A^2\bar{\beta}^2 + \bar{\alpha} + K)\int_0^t E\sup_{r\in[0,s]}|P_r - Q_r|^2 ds \\
&\quad +\bar{\alpha}E\int_0^t |P_s - Z_s|^2 ds + K(K+1)E\int_0^t \|P_s - Z_s\|_C^2 ds \\
&\quad +2E\sup_{s\in[0,t]}\int_0^s (P_s - Q_s, (g(s,P_s) - g(s,Z_s))dW_s) \qquad (14)\\
&\le (A^2\bar{\beta}^2 + 3\bar{\alpha} + 3K + 2K^2)\int_0^t E\sup_{r\in[0,s]}|P_r - Q_r|^2 ds \\
&\quad +2\bar{\alpha}E\int_0^t |Q_s - Z_s|^2 ds + 2K(K+1)E\int_0^t \|Q_s - Z_s\|_C^2 ds \\
&\quad +2E\sup_{s\in[0,t]}\int_0^s (P_s - Q_s, (g(s,P_s) - g(s,Z_s))dW_s).
\end{aligned}$$

By Burkholder-Davis-Gundy's inequality, we have

$$\begin{aligned}
&E\sup_{s\in[0,t]}\int_0^s (P_s - Q_s, (g(s,P_s) - g(s,Z_s))dW_s) \\
&\le k_1 E[\sup_{0\le s\le t}|P_s - Q_s|(\int_0^t \|g(s,P_s) - g(s,Z_s)\|_2^2 ds)^{1/2}] \qquad (15)\\
&\le \frac{1}{4}E[\sup_{0\le s\le t}|P_s - Q_s|^2 + k_2\int_0^t E\|P_s - Z_s\|_C^2 ds,
\end{aligned}$$

where k_1, k_2 are two positive constants. Therefore inserting (15) into (14) has

$$\begin{aligned}
&E \sup_{s\in[0,t]} |P_s - Q_s|^2 \\
&\leq (A^2\bar{\beta}^2 + 3\bar{\alpha} + 3K + 2K^2 + 2k_2) \int_0^t E \sup_{r\in[0,s]} |P_r - Q_r|^2 ds \\
&\quad +2(\bar{\alpha} + K + K^2 + k_2) \int_0^t E \sup_{s\in[0,t]} |Q_s - Z_s|^2 ds \\
&\quad +\frac{1}{2} E \sup_{s\in[0,t]} |P_s - Q_s|^2.
\end{aligned}$$

Applying Lemma 2 we obtain a bound of the form

$$E \sup_{s\in[0,t]} |P_s - Q_s|^2 \leq D_1 \Delta t + D_2 \int_0^t E \sup_{r\in[0,s]} |P_r - Q_r|^2 ds,$$

where

$$D_1 = 4(\bar{\alpha} + K + K^2 + k_2) T C_3,$$

$$D_2 = 2(A^2\bar{\beta}^2 + 3\bar{\alpha} + 3K + 2K^2 + 2k_2).$$

The result (14) then follows from the continuous Gronwall inequality with

$$C_4 = D_1 \exp(D_2 T).$$

It is easy to deduce that the following theorem is satisfied.

Theorem 2. *Assume the preceding hypotheses hold, the approximate solution (6) will converge to the true solution of Eq. (2) in the sense*

$$\lim_{\Delta t\to 0} E\left[\sup_{0\leq t\leq T} |P_t - Q_t|^2\right] = 0. \tag{16}$$

4 One Example

Let us consider a stochastic age-dependent population equation of the form

$$\begin{cases}
d_t P = \left[-\frac{\partial P}{\partial a} - \frac{1}{2(1-a)^2} P - \frac{1}{2} t P\right] dt + P dB_t, & in \;\; Q = (0,T) \times (0,1), \\
P(0,a) = \frac{1}{2} \exp(-\frac{1}{1-a}), & in \;\; [0,1], \\
P(t,0) = \int_0^1 \frac{1}{2(1-a)^2} P(t,a) da, & in \;\; [0,T].
\end{cases} \tag{17}$$

Here B_t is a real standard Brownian motion (so, $M = R$ and $W = 1$). We can set this problem in our formulation by taking $H = L^2([0,1])$,

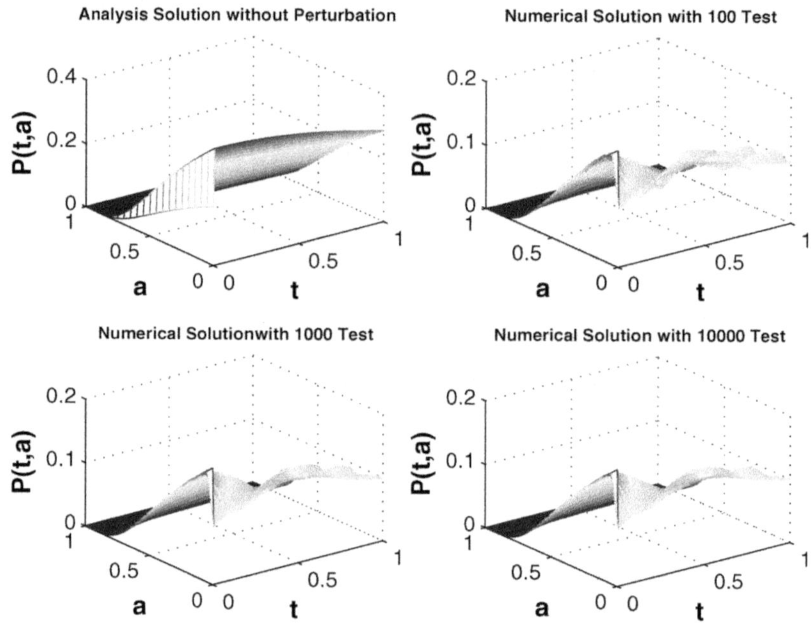

Fig. 1. Numerical Simulations of Stochastic Population Equation

$V = W_0^1([0,1])$ (a Sobolev space with elements satisfying the boundary conditions above),$\mu(t,a) = \beta(t,a) = \frac{1}{2(1-a)^2}$, $f(t,P) = -\frac{1}{2}tP$, and $g(t,P) = P$, $P_0(a) = \frac{1}{2}\exp(-\frac{1}{1-a})$.

Clearly, the system (17) satisfy conditions (i)-(iii). Consequently, the approximate solution will converge to the true solution of (17) for any $(t,a) \in (0,T) \times (0,1)$ in the sense of Theorem 2

Take $T = 1$ in Eq.(17). Fig.1 gives the pictures above with fixed step sizes $\Delta t = 0.005$, $\Delta a = 0.05$. The Upper in the left is the explicit solution to Eq.(17) without perturbation, that is $EP(t,a) = \exp(-\frac{1}{1-a} - \frac{t^2}{4})$. The other three pictures are numerical simulations of the stochastic age-dependent population equations with 100, 1000 and 10000 experiments respectively, where $EQ(t,a) = \frac{1}{n}\sum_{k=1}^{n} Q_k(t,a)$. It clearly reveals the fact that the numerical approximation will tend to the true solution in the mean sense.

It is difficult to obtain the true explicit solution to Eq.(17), so the explicit solution $P(t,a)$ to Eq.(17) can be replaced by $\exp(-\frac{1}{1-a} - \frac{t^2}{4})(1+\Delta B_t)$. The Fig.2 shows the computer simulation for the difference between $\exp(-\frac{1}{1-a} - \frac{t^2}{4})(1 + \Delta B_t)$ and the Euler approximation solution $Q(t,a)$. The maximum value of the error square is not greater than 0.05. Clearly the numerical approximation will tend to the true solution in the mean square sense.

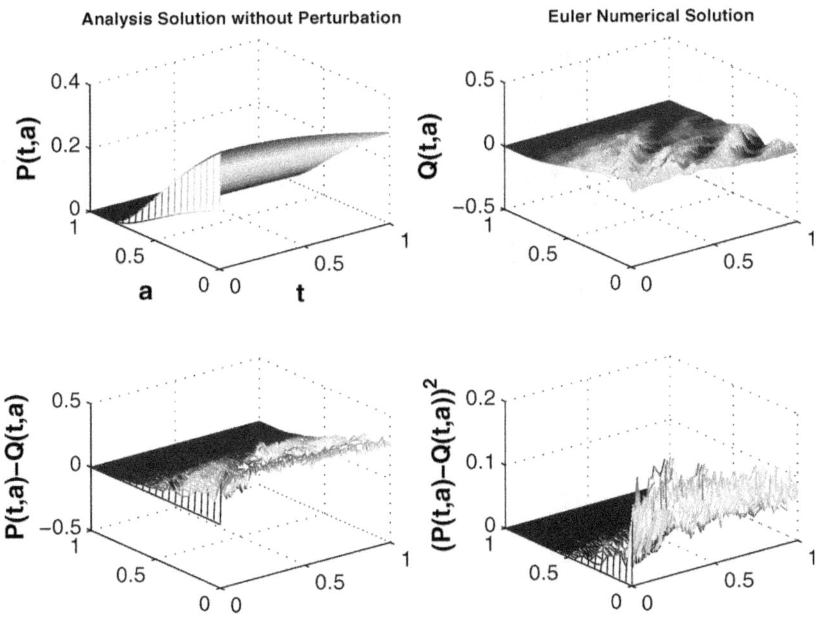

Fig. 2. Error Simulation of Stochastic Population Equation

5 Conclusion

In general, most of stochastic age-dependent population equations do not have explicit solutions, thus numerical approximation schemes are invaluable tools for exploring their properties. Using the Doob inequality and Burkholder-Davis-Gundy's inequality, the main purpose of this paper is to develop a numerical scheme and show the convergence of the numerical approximation solution to the true solution. We give an example to show the efficiency of our numerical method.

Acknowledgements

The research was supported by The National Social Science Foundation (No.05X RK008)(China); also was supported by NingXia Natural Science Foundation(No. G002)(China).

References

1. Cushing, J.: The Dynamics of Hierarchical Age-structured Populations. J. Math. Biol. **32**(1994) 705–729
2. Allen, L.,Thrasher, D.: The Effects of Vaccination in An Age-dependent Model for Varicella and Herpes Zoster. IEEE Tran. Auto. Contr. **43** (1998) 779–789

3. Zhang, Q., Liu, W., Nie Z.: Existence, Uniqueness and Exponential Stability for Stochastic Age-dependent Population. Appl. Math. Compu. **154** (2004) 183–201
4. Pollard, J.: On The Use of The Direct Matrix Product in Analyzing Certain Stochastic Population Model. Biom. **53** (1966) 397–415
5. Zhang, Q., Han, C.: Numerical Analysis for Stochastic Age-dependent Population Equations. Appl. Math. Compu. **169**(2005) 278–294
6. Abia, L., López-Marcos, J.: On The Numerical Integration of Non-local Terms for Age-structured Population Models. Math. Biosc. **157** (1999) 147–167
7. Kostova, T.: Numerical Solutions to Equations Modelling Nonlinearly Interacting Age-dependent Populations. Comput. Math. Applic. **19** (1990) 95–103
8. Milner, F., Rabbiolo, G.: Rapidly Converging Numerical Algorithms for Models of Population Dynamics. J. Math. Biol. **30** (1992) 733-753

Classifying G-Protein Coupled Receptors with Hydropathy Blocks and Support Vector Machines

Xing-Ming Zhao[1], De-Shuang Huang[1], Shiwu Zhang[2], and Yiu-ming Cheung[3]

[1] Institute of Intelligent Machines, CAS
P.O. Box.1130, Hefei, Anhui, 230031, China
{x_mzhao, dshuang}@iim.ac.cn

[2] Department of Precision Machinery and Instrumentation
University of Science and Technology of China
Hefei, Anhui, China
swzhang@ustc.edu.cn

[3] Department of Computer Science, Hong Kong Baptist University,
Hong Kong, China
ymc@comp.hkbu.edu.hk

Abstract. This paper developes a new method for recognizing G-Protein Coupled Receptors (GPCRs) based on features generated from the hydropathy properties of the amino acid sequences. Using the hydropathy characteristics, namely hydropathy blocks, the protein sequences are converted into fixed-dimensional feature vectors. Subsequently, the Support Vector Machine (SVM) classifier is utilized to identify the GPCR proteins belonging to the same families or subfamilies. The experimental results on GPCR datasets show that the proteins belonging to the same family or subfamily can be identified using features generated based on the hydropathy blocks.

1 Introduction

G-protein coupled receptors (GPCRs) is a superfamily of proteins that play key role in the signaling network which regulates the basic cellular processes: neurotransmission, cellular metabolism, secretion, inflammatory and immune response, taste and vision, etc [1]. Because of their crucial role in signal transduction, many researchers are focusing on understanding their structure and function. At present, more than 50% of drugs available on the market act through GPCRs [2]. However, their tertiary structures remain mostly unresolved, because of their difficult crystallization. To date, only the structure of one GPCR (bovine rhodopsin) has been solved [3]. On the other hand, the amino acid sequences of more than 1000 GPCR-related proteins have been determined. Therefore, it is strongly desired to develop an efficient and reliable method to predict the structure and function of GPCRs from sequence data.

One way towards this task is to classify the GPCRs into known families to induce its structural and functional features. In the literature, a variety of

D.-S. Huang, K. Li, and G.W. Irwin (Eds.): ICIC 2006, LNBI 4115, pp. 593–602, 2006.

approaches, e.g. PSI-BLAST [4], Hidden Marked Models (HMM) [5], and so forth, have been developed for classifying protein sequences by searching against protein databases with annotations of structure and function. However, these methods will not work when query proteins lack significant sequence similarity to the database sequences [2]. Recently, a machine learning method, namely Support Vector Machine (SVM) [6] [7], has been successfully applied to classifying GPCRs and shown the superiority to other methods [3] [2].

Most of the methods described above for classifying GPCR proteins depend on alignment or the similarities between protein sequences. Generally, the similarities can be characterized by the physicochemical properties of amino acids in protein sequences, such as polarity, solubility, hydropathy, and so forth. Among these physicochemical properties, hydropathy (the patterns of hydrophilicity and hydrophobicity) is known to be well conserved. In the literature, many methods have been developed for explaining the folding of proteins. Since the hydrophobicity profile of a membrane protein is conserved much better than the amino acid sequence itself during the evolution process, the hydropathy profile has been used to identify the membrane proteins in various classical methods [8] [9] [10]. Furthermore, the hydropathy patterns presenting in the protein sequences are used to develop reduced amino acid alphabets for protein secondary structure prediction [11] [12]. Recently, Panek et al. has developed a new method for identification of protein families based on hydropathy distributions in proteins [13]. In the method developed by Panek et al., the number of segments must be optimally estimated in advance and the length of the protein sequence is not taken into account.

In this paper, we present a new method for classifying the GPCRs based on the hydropathy blocks. Using the hydropathy characteristics within the three hydropathy subsets (strongly hydrophilic, strongly hydrophobic and weakly hydrophilic or weakly hydrophobic), we generate a number of $k - tuples$, called hydropathy blocks here. Instead of dividing the protein sequences into segments and counting the number of hydropathy characters in each segment, the frequencies of the hydropathy blocks occurring in the protein sequence are calculated and used to generate fixed-dimensional vectors. Having obtained the feature vectors, the Support Vector Machine (SVM) classifier is utilized to classify the protein sequences into known families. Experiments on classifying GPCRs show the promising results.

The remainder of this paper is organized as follows. Section 2 describes the methods for feature extraction and presents an overview of Support Vector Machine. The experimental results and discussions are reported in Section 3. Finally, a conclusion is drawn in Section 4.

2 Methods

2.1 Feature Extraction

A protein sequence S, of length n, is defined as a linear succession of n symbols from the 20-letter amino acid alphabet {A, C, D, E, F, G, H, I, K, L, M, N, P, Q,

R, S, T, V, W, Y }. According to the hydropathy scale, the 20 amino acids can be grouped into 3 hydropathy sets: strongly hydrophilic (POL), strongly hydrophobic (HPO) and weakly hydrophilic or weakly hydrophobic (Ambi) (Table 1) [2]. Here, the abbreviations, i.e. POL, HPO and Ambi, for hydropathy characteristics used by Panek et al. are adopted. Proline and glycine are not classified into any hydropathy set because of their unique backbone properties, and Cysteine is excluded from any set because it has polarizable properties [14].

Table 1. Hydropathy Characteristics [13]

Hydropathy characteristics	Abbreviation	Amino acids
Strongly hydrophilic (polar)	POL	r,d,e,n,q,k,h
Strongly hydrophobic	HPO	l,i,v,a,m,f
Weakly hydrophilic or weakly hydrophobic (ambiguous)	Ambi	s,t,y,w

In this paper, the $2-tuples$ [15] generated from the 3 hydropathy sets instead the alphabet of 20 amino acids are used to generate feature vectors. The $2-tuples$ extract various patterns of two consecutive residues with the same hydropathy characteristics in the protein sequence S and counts the number of occurrences of the $2-tuples$ with overlapping. Here, the 2-*tuples* generated from the 3 hydropathy sets are called hydropathy blocks (abbreviated as 2-HBs) because they consist of successive residues with the same hydropathy characteristics just like the BLOCKS presenting in the protein sequences. Since the amino acids are grouped into 3 sets: POL, HPO and Ambi, we need to generate the 2-HBs for each set, respectively. The union of 2-HBs belonging to the POL set is noted as POL2, and the size of POL2 will be $7^2 = 49$. Similarly, the union of the 2-HBs belonging to the HPO set is noted as HPO2 and the union of the 2-HBs belonging to the Ambi set is noted as Ambi2. The size of HPO2 is $6^2 = 36$, and the size of Ambi2 is $4^2 = 16$.

One can count the number of the occurrences of 2−HBs by taking a $2-letter$ sliding window that is run through the sequences, from position 1 to $n-2+1$. Consequently, the number of occurrences of the 2-HBs in protein sequence S can be represented as:

$$\begin{aligned} POL_2 &= \{POL2_1^S, POL2_2^S, \cdots, POL2_i^S, \cdots, POL2_{49}^S\} \\ HPO_2 &= \{HPO2_1^S, HPO2_2^S, \cdots, HPO2_j^S, \cdots, HPO2_{36}^S\} \\ Ambi_2 &= \{Ambi2_1^S, Ambi2_2^S, \cdots, Ambi2_m^S, \cdots, Ambi2_{16}^S\} \end{aligned}$$

where $POL2_i^S$ represents the number of the ith 2-HB of POL2 occurring in sequence S, $HPO2_j^S$ represents the number of the jth 2-HB of HPO2 occurring in sequence S, and $Ambi2_m^S$ represents the number of the mth 2-HB of Ambi2 occurring in sequence S. For example, for protein sequence

$S = \{STWPVERKHLVIVITN\}$, n=16, the counting determined by taking a $2 - letter$ sliding window for $n - 2 + 1 = 15$ times would be:

$$W_2^{POL} = \{ER, RK, KH\}$$
$$W_2^{HPO} = \{LV, VI, IV, VI\}$$
$$W_2^{Ambi} = \{ST, TW\}$$

and the number of 2-HBs occurring in S is:

$$POL_2 = \{1, 1, 1\}$$
$$HPO_2 = \{1, 2, 1\} Ambi_2 = \{1, 1\}$$

In the above countings, those HBs not presented in S are not listed. Hence, the protein sequence can be represented as a linear succession of the 2-HBs, i.e., $V = [POL_2,\ HPO_2,\ Ambi_2]$.

For a protein sequence S, suppose the size of $POL2$ be L_{POL2}, the size of $HPO2$ be L_{HPO2}, the size of $Ambi2$ be L_{Ambi2}, and $X = [X^S_{POL2}, X^S_{HPO2}, X^S_{Ambi2}]$, with

$$X^S_{POL2_i} = \frac{C_i^{POL}}{\sum_{i=1}^{L_{POL2}} C_i^{POL}},\ i = 1, \cdots, L_{POL2} \tag{1}$$

$$X^S_{HPO2_j} = \frac{C_j^{HPO}}{\sum_{j=1}^{L_{HPO2}} C_j^{HPO}},\ j = 1, \cdots, L_{HPO2} \tag{2}$$

$$X^S_{Ambi2_m} = \frac{C_m^{Ambi}}{\sum_{m=1}^{L_{Ambi2}} C_m^{Ambi}},\ m = 1, \cdots, L_{Ambi2} \tag{3}$$

where $X^S_{POL2_i}$ is the frequency of the i-th 2-HB of $POL2$ occurring in S, and C_i^{POL} is the times of the i-th 2-HB of $POL2$ occurring in S, and so forth. Consequently, the protein sequence S can be converted into a 101 dimensional vector:

$$v = \frac{X}{\sqrt{length(S) - 1}} \tag{4}$$

where $length(S)$ is the length of protein sequence S.

2.2 Construction of Support Vector Machine

After converting the protein sequences into feature vectors, the next step is to classify the protein sequences into certain families. Many pattern recognition methods have been developed for protein sequence classification, among which the Support Vector Machine shows superior to other methods because of its good generality and strong statistical theory ground.

Support Vector Machine (SVM) is a class of supervised learning algorithms introduced by Vapnik [6], which is based on the well developed statistical learning theory. An important feature of SVM is that due to Mercer's conditions on the kernels, the corresponding optimization problems are convex and hence have no local minima [16].

Given a set of labelled training vectors $\mathbf{x_i}, i = 1, \ldots, l$, from two classes, labelled by $y_i \in \{-1, +1\}, i = 1, \ldots, l$, SVM maps the input vectors $\mathbf{x} \in R^n$ into a high dimensional feature space H by a mapping function $\Phi(\cdot)$ and finds a hyperplane, which maximizes the margin, i.e., the distance between the hyperplane and the nearest data points of each class in the space H. The decision function implemented by SVM can be written as:

$$f(\mathbf{x}) = \mathrm{sgn}(\sum_{i=1}^{l} y_i \alpha_i \cdot K(\mathbf{x_i}, \mathbf{x}) + b) \tag{5}$$

with

$$K(\mathbf{x_i}, \mathbf{x}) = < \Phi(\mathbf{x_i}), \Phi(\mathbf{x}) > \tag{6}$$

where α_is are the coefficients obtained by solving the following optimization problem:

$$\begin{aligned} &\text{maximize} \quad \sum_{i=1}^{l} \alpha_i - 1/2 \sum_{i=1}^{l} \sum_{j=1}^{l} \alpha_i \alpha_j \cdot y_i y_j \cdot K(\mathbf{x_i}, \mathbf{x_j}) \\ &\text{subject to } 0 \leq \alpha_i \leq C \\ &\qquad \sum_{i=1}^{l} \alpha_i y_i = 0 \quad i = 1, 2, \cdots, l. \end{aligned}$$

where C is a regularization parameter, which controls the trad-off between the margin and the misclassification error.

To construct the SVM classifier, we need to tune two parameters: the capacity, and the choice of kernel. The capacity generally allows us to control how much tolerance for the errors in the classification we allow. Hence, the capacity will affect the generalization ability of the SVM classifier and prevent it from overfitting the training set. On the other hand, the kernel function allows the SVM classifier to create a hyperplane in high dimensional space that effectively separates the training data. In this paper, the Gaussian kernel is employed for all the classifiers and the capacity is set to 1000.

3 Experimental Results and Discussions

We investigated the performance of our proposed method on identifying the families(classes) and subfamilies of the G protein-coupled receptors (GPCRs). Following the same three steps described in [3] and [2], we first discriminate the GPCRs from the non-GPCRs, then discriminate the five major classes of GPCR, and finally identify the subfamilies of ClassA of GPCRs.

Initially, we applied our proposed method (note as SVM-HB hereinafter) to identifying the GPCRs from the non-GPCRs. The dataset used here was obtained from http://www.soe.ucsc.edu/research/compbio/gpcr/. It consisted of 778 GPCRs from the five major GPCR classes and was used as the positive examples. Another dataset, which consisted of 99 decoy negative examples and 2425 negative examples obtained from SCOP version 1.37 PDB90 domain data, was used as the negative examples. Our proposed method was compared with GPCRpred developed by Raghava et al. [2], where SVMs were trained and tested using features generated from the dipeptide compositions. The performance was evaluated by 5-fold cross validations. Furthermore, four parameters - sensitivity, specificity, accuracy and Matthews' correlation coefficient (MCC)- were used to measure the performance of the two methods. The results were summarized in Table 2.

Table 2. The performance of SVM-HB and GPCRpred in differentiating GPCRs from non-GPCRs by 5-fold cross-validation

Method	Sensitivity (%)	Specificity (%)	ACC (%)	MCC
SVM-HB	100.00	100.00	99.96	0.99
GPCRpred[a]	98.60	99.80	99.50	0.99

a. The results were obtained from [2].

It can be seen from Table 2 that our proposed SVM-HB can differentiate GPCRs from non-GPCRs with accuracy of 99.96% and MCC of 0.9992, outperforms GPCRpred. The results suggest that using features generated from the hydropathy blocks, the SVM classifier can get better results in differentiating GPCRs from non-GPCRs than the one using features generated from the dipeptide components.

Subsequently, our proposed method was used to identify the five major classes of the GPCRs. According to the GPCRDB information system, the GPCR superfamily can be divided into five major classes: Class A (receptors related to rhodopsin and andrenergic receptors), Class B (receptors related to calcitonin and parathyroid hormone (PTH) receptors), Class C (receptors related to metabotropic receptors), Class D (receptors related to pheromone receptors) and Class E (receptors related to cAMP receptors). The datasets for the five classes were obtained from the work of Karchin et al. [3], and they were also used by Raghava et al. [2].

To predict the class of GPCRs, a series of binary SVMs were constructed, which were trained and tested using the feature vectors generated from hydropathy blocks. Our proposed SVM-HB was also compared with GPCRpred. Both SVM-HB and GPCRpred were evaluated by 2-fold cross-validation because of the low number of sequences. The performance was measured by the recognition rates (ACC) and MCC. The performance of the SVM-HB and GPCRpred is summarized in Table 3. It can be seen that SVM-HB can discriminate between all the GPCR classes exactly except for Class D. As pointed out in [2], the poor results were obtained due to the lower number of sequences for training. It was

Table 3. The performance of SVM-HB and GPCRpred in discriminating the five major GPCR classes

GPCR classes	Seq	SVM-HB		GPCRpred[b]	
		ACC(%)	MCC	ACC(%)	MCC
Rhodopsin and andrenergic-like receptors (Class A)	692	100.00	1.00	98.10	0.80
Calcitonin and PTH-like receptors (Class B)	56	100.00	1.00	85.70	0.84
Metabotropic-like receptors (Class C)	16	100.00	1.00	81.30	0.81
Pheromone-like receptors (Class D)	12	94.90	0.88	36.40	0.49
cAMP-like receptors (Class E)	3	100.00	1.00	100.00	1.00

b. The results were obtained from [2].

readily found from Table 3 that our proposed SVM-HB performs significantly better than the GPCRpred method.

Generally, the determination of GPCR function is crucial for pharmaceutical research. To determine GPCR function from sequence information, we need to discriminate the GPCR subfamilies instead of superfamilies. Although superfamily discrimination can be achieved by various methods that generalize features shared by a diverse group of examples, it is usually difficult to discriminate the examples belonging to different subfamilies because they may differ only slightly [3]. The problem of recognizing GPCR subfamilies lies in the fact that the subfamily classification in GPCRDB is defined chemically (according to which ligand the receptor binds) rather than by sequence homology [3]. Here, we applied our proposed SVM-HB to classifying the subfamilies of the rhdopsin-like family. The datasets were obtained from http://www.soe.ucsc.edu/research/compbio/gpcr/, and the 14 datasets used in this experiment were the same ones used by Raghava et al.

The performance of our proposed method was evaluated using 2-fold cross-validation and compared with CPGRpred. The performance of SVM-HB and GPCRpred in terms of recognition rates (ACC) and MCC is shown in Table 4. It can be seen from that all subfamilies can be discriminated exactly using SVM-HB except Amine and Rhodopsin, and the performance of SVM-HB is much better than GPCRpred. The overall recognition rate and MCC for the 14 subfamilies of the rhodopsin-like family using SVM-HB are 99.93% and 0.9986, respectively. Poor results were obtained due to the lower number of sequences for training.

Furthermore, the ROC scores were also used to estimate the performance of the GPCRpred and SVM-HB methods in identifying subfamilies of the rhodopsin-like family of GPCR. The experimental results are shown in Fig.1, where a higher curve corresponds to a more accurate classification result. It can

Table 4. The performance of SVM-HB and GPCRpred in identifying subfamilies of the rhdopsin-like family of GPCR

GPCR classes	SVM-HB		GPCRpred[c]	
	ACC(%)	MCC	ACC%)	MCC
Amine	99.10	0.99	99.10	0.99
Peptide	100.00	1.00	99.70	0.95
Hormone proteins	100.00	1.00	100.00	1.00
Rhodopsin	99.96	0.99	98.90	0.99
Olfactory	100.00	1.00	100.00	0.99
Prostanoid	100.00	1.00	100.00	0.99
Nucleotide-like	100.00	1.00	85.40	0.92
Cannabis	100.00	1.00	100.00	1.00
Platelet activating factor	100.00	1.00	100.00	1.00
Gonadotropin releasing hormone	100.00	1.00	100.00	1.00
Thyrotropin releasing hormone	100.00	1.00	85.70	0.93
Melatonin	100.00	1.00	100.00	1.00
Viral	100.00	1.00	33.30	0.58
Lysospingolipids	100.00	1.00	58.80	0.76
Overall	99.93	0.99	97.30	0.97

c. The results were obtained from [2].

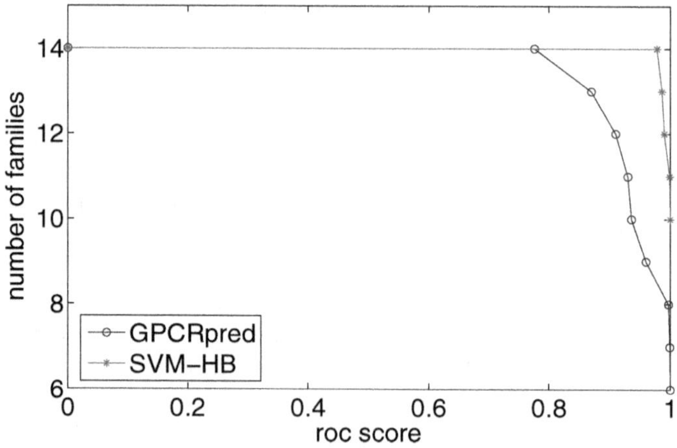

Fig. 1. Relative performance of the SVM-HB method and the GPCRpred method, where the curves show the number of families vs a ROC score threshold for the two methods

be seen from Fig.1 that using the hydropathy blocks, the SVM classifiers can better discriminate GPCR subfamilies than using the features generated from the dipeptide compositions. The results in Table 4 and Fig.1 show again that using the hydropathy blocks, we can extract more biological information than using the dipeptide compositions.

4 Conclusions

Hydropathy is one of important physicochemical properties of amino acids, and is better conserved than protein sequences in evolution. In this paper, we introduce a new method for identifying families or subfamilies of GPCR in a set of proteins using information extracted from protein sequences based on hydropathy blocks. Our proposed method has been applied to differentiating GPCR proteins from non-GPCR proteins, identifying the five major GPCR classes, and discriminating subfamilies of rhodopsin-like family of GPCR. Experimental results show that our proposed method using hydropathy blocks is indeed efficient and outperforms the method based on dipeptide composition in classifying GPCRs.

Acknowledgments

This work was partly supported by the NSF of China (Nos.60472111 and 60405002), Faculty Research Grant of Hong Kong Baptist University with Project Number: FRG/02-03/II-40, and the Research Grant Council of Hong Kong SAR under Project HKBU 2156/04E.

References

1. Bouvier, M.: Structural and functional aspects of g protein coupled receptors oligomerization. Biochem. Cell Biol., vol. 76. (1998) 1–11
2. Bhasin, M. and Raghava, G. P. S.: Gpcrpred: an svm-based method for prediction of families and subfamilies of g-protein coupled receptors. Nucleic Acids Research, vol. 32. (2004) W383–389
3. Karchin, R., Karplus, K., Haussler, D.: Classifying g-protein coupled receptors with support vector machines. Bioinformatics, vol. 18. (2002) 147–159
4. Altschul, S., Madden, T., Schafer, A., Zhang, J., Zhang,Z., Miller, W., D. Lipman: Gapped blast and psi-blast: A new generation of protein data. Nucleic Acids Research, vol. 25. (1997) 3389–3402
5. Krogh, A., Brown, M., Mian, I., Sjolander, K., Haussler, D.: Hidden markov models in computational biology: Applications to protein modeling. Journal of Molecular Biology, vol. 235. (1994) 1501–1531
6. Vapnik, V.: The Nature of Statistical Learning Theory. 2nd edn. Springer-Verlag, Berlin Heidelberg New York (1996)
7. Zhao, X., Huang, D., Cheung, Y., Wang, H., Huang, X.: A novel hybrid ga/svm system for protein sequences classification. Lecture Notes in Computer Science, vol. 3177. Springer-Verlag, (2004) 11–16
8. Lolkema, J., Slotboom, D.: Hydropathy profile alignment: a tool to search for structural homologues of membrane proteins. FEMS Microbiol Rev., vol. 22. (1998) 305–322

9. Esposti, M., Crimi, M., Venturoli, G.: A critical evaluation of the hydropathy profile of membrane proteins. European Journal of Biochemistry, vol. 190. (1990) 207–219
10. Lolkema, J., Slotboom, D.: Estimation of structural similarity of membrane proteins by hydropathy profile alignment. Mol Membr Biol, vol. 15. (1998) 33–42
11. Klingler, T., Brutlag, D.: Discovering structural correlations in a-helices. Protein Science, vol. 3. (1994) 1847–1857
12. Schmidler, S. C., Liu, J. S., Brutlag, D. L.: Bayesian segmentation of protein secondary structure. Journal of Computational Biology, vol. 7. (2000) 233–248
13. Panek, J., Eidhammer, I., Aasland, R.: A new method for identification of protein (sub)families in a set of proteins based on hydropathy distribution in proteins. Proteins, vol. 58. (2005) 923–934
14. Taylor, W.: Identification of protein sequence homology by consensus template alignment. Journal of Molecular Biology, vol. 188. (1986) 233–258
15. Vinga, S., Almeida, J.: Alignment-free sequence comparison-a review. Bioinformatics, vol. 19. (2003) 513–523
16. Cristianini, N., Shawe-Taylor, J.: An Introduction to Support Vector Machines. Cambridge University Press, Cambridge, United Kingdom (1986)

HSPPIP: An Online Tool for Prediction of Protein–Protein Interactions in Humans

Yu Xue[1], Changjiang Jin[1], and Xuebiao Yao[1,2]

[1] School of Life Science, University of Science and Technology of China, Hefei, Anhui 230027, P.R. China
[2] Department of Physiology, Morehouse School of Medicine, Atlanta, GA 30310, USA
{yxue, jcjin, yaoxb}@ustc.edu.cn

Abstract. Recently, protein-protein interaction prediction (PPIP) has been emerging as an appealing question. Although several in silico approaches have been developed to delineate the potential protein-protein interaction (PPI), there are few online tools of human PPIP for further experimental design. Here we present an online service, hsPPIP (Protein-Protein Interaction Predicting of Homo Sapiens), to predict or evaluate the potential PPIs in human. The annotations of functional domain (Interpro) and GO (Gene Ontology) for proteins are employed as two informative features, and are integrated by the naïve Bayesian approach. The prediction accuracy is comparable to the existing tools. Based on the hypothesis that the features correlated with PPIs are conserved in different organisms, the web server hsPPIP is established and could predict the PPIs of human dynamically. hsPPIP is implemented in PHP+MySQL and can be freely accessed at: http://973-proteinweb.ustc.edu.cn/hsppip/.

1 Introduction

Dissecting the Protein-Protein Interaction (PPI) network in a cell is the foundation for elucidating all kinds of cellular mechanisms. In vivo or in vitro PPI identifications are usually time-consuming and labour intensive. Although high-throughput approaches have generated many data sets of PPIs, they are fairly noisy and need further confirmation by in vivo or in vitro experiments. So for its statistical accuracy and convenience, in silico PPI prediction (PPIP) might be a great help and insightful for further experimental consideration.

In this work, we have developed an easy-to-use online tool, hsPPIP (Protein-Protein Interaction Predicting of Homo Sapiens), to predict or evaluate the potential PPIs of human. We choose the annotations of functional domain (Interpro) [1] and GO (Gene Ontology) [2] for proteins as two features associated with PPIs. The naïve Bayesian approach is adopted to integrate the features together to improve the prediction accuracy. Since there is increasing evidence that interacting protein pairs are usually co-conserved through evolution and the PPI network structures are also conserved in different organisms [3, 4], we hypothesize that the features correlated with PPIs are also conserved in different organisms. So we might predict human PPIs

D.-S. Huang, K. Li, and G.W. Irwin (Eds.): ICIC 2006, LNBI 4115, pp. 603 – 611, 2006.

based on the associated features from budding yeast. hsPPIP is especially fast and convenient for the construction of small scale human PPI networks. Although there have been several other web servers or databases for human PPIP, we propose that hsPPIP could be a useful tool for the experimentalists.

2 Materials and Methods

2.1 Training Data Set

The PPI data set for training was collected from six public PPI database, the MIPS CYGD (Comprehensive Yeast Genome Database) PPI (Mar. 2004) [11], BIND (Biomolecular Interaction Network Database) (Apr. 2004) [12], GRID (General Repository for Interaction Database) (Mar. 2004) [13], SGD (Saccharomyces Genome Database) (Apr. 2004) [14], MINT (Molecular INTeraction database) (Apr. 2004) [15], and Database of Interacting Proteins (DIP) (Mar. 2004) [16]. We only used budding yeast PPI data in the current analysis. The PPI information of each PPI database is listed in table 1. We combined the six PPI data sets into a non-redundant PPI data set including 21, 295 unique protein pairs. We also manually validated the data set.

2.2 Gold Standards and Testing Data

To test our method, we adopt both positive and negative control data sets reported previously [5]. The positive control contains protein pairs from the same MIPS complex (pos_MIPS), and the negative control is protein pairs with different subcellular localization and couldn't be found in verified PPI list (L_neg). As in the literature [5], the positive control is regarded as the set of positive PPIs (*P*), and the negative control is regarded as the set of negative PPIs (*N*).

For comparison, we also randomly generate 200, 000 protein pairs (Random200K), in which the real PPIs are expected to be only a small fraction. The gold standard and testing data sets are listed in table 1.

2.3 Collection of Human and Yeast Proteins

The protein sequences were downloaded from ExPASy (ftp://cn.expasy.org). The hsPPIP used the protein accession numbers as standard entries for prediction. All accession numbers of human and yeast proteins are retrieved and stored in the local MySQL database for further processing.

2.4 Protein Annotation by Functional Domain (Interpro) and Gene Ontology (GO)

Interpro is an integrated resource of annotation for protein families, functional domains and motifs [1]. Gene Ontology Consortium [2] is to produce a controlled vocabulary as the knowledge of gene and protein roles in cells. We downloaded the files containing all Interpro and GO annotations for the proteins from the EBI repository (ftp://ftp.ebi.ac.uk/pub/databases/), and mapped the Interpro and GO annotations to human and yeast proteins, respectively. The information is also stored in our local MySQL database and could be searched and browsed. Users could search or browse our local database for each protein entry in human or yeast.

We excluded three GO annotations from the analysis: GO:0000004 (biological_process unknown), GO:0008372 (cellular_component unknown), and GO:0005554 (molecular_function unknown), since such unknown and ambiguous assignments might introduce noises into the analysis.

Table 1. The total data sets we used for this work. Training data set is retrieved and combined from six public PPI databases into a non-redundant PPI data set, including 21, 295 protein pairs. The gold standard is taken from the previous work [5] with both positive and negative control. We also randomly generate 200, 000 protein pairs as an additional testing data set.

Data type	Data set		# Protein pairs	Used for …
Training data	CYGD (MIPS) (2004-03-21)		11,862	
	BIND (2004-04-01)		5,875	
	GRID (2004-03-18)		19,038	
	SGD (2004-04-01)		4,624	Training
	MINT (2004-04-01)		14,014	
	DIP (2004-03-07)		15,393	
	TOTAL (non-redundant PPI data set)		21,295	
Gold standards	Positive	Protein pair in the same MIPS complex	8,617	
	Negative	Protein pair separated by localization	2,705,844	Testing
Testing data	Random	Random PPI (200K)	200,000	

2.5 Algorithm

PPIP (protein-protein interaction prediction) with functional domain-based strategies is widely used [17, 18, 19, 20]. In this work we adopt the Naïve Bayesian approach to integrate two features, functional domains and GO annotations, for PPIP. We considered if two proteins P_i, P_j have *a, b* Interpro annotations and *c, d* GO annotations respectively, then the probability of P_i and P_j to be interacting pair is shown below:

$$P(PPI_{ij}=1)=1-\prod_{(I_k,I_l)\in(P_i\times P_j)}(1-P(I_{ab}=1))\times\prod_{(G_m,G_n)\in(P_i\times P_j)}(1-P(G_{cd}=1))\cdot \quad (1)$$

PPI_{ij}=1: Protein P_i interacts with Protein P_j.
I_{ab}= 1: Interpro annotation I_a and I_b are interacting functional domain.
G_{cd}= 1: GO annotation G_c and G_d are functional associated.
a, b: Numbers of Interpro annotations in P_i and P_j, respectively.
c, d: Numbers of GO annotations in P_i and P_j, respectively.
$(I_{a'}, I_{b'}) \in (P_i \times P_j)$: Interpro pair (I_a, I_b) is included in protein pair $Pi\times Pj$.

$(G_{c'}, G_{d'}) \in (P_i \times P_j)$: GO pair ($G_c$, G_d) is included in protein pair $Pi \times Pj$.

The $P(I_{ab}=1)$ and $P(G_{cd}=1)$ could be obtained from training our non-redundant PPI data set. And the equation for calculating the two *probabilities* could be proposed as:

$$P(I_{ab}=1)=\frac{Int_{ab}}{N_{ab}}. \quad (2)$$

$$P(G_{cd}=1)=\frac{Int_{cd}}{N_{cd}}. \quad (3)$$

Int_{ab} : Number of PPIs that include (I_a, I_b);
N_{ab}: Number of all potential PPIs that include (I_a, I_b).
Int_{cd} : Number of PPIs that include (G_c, G_d);
N_{cd}: Number of all potential PPIs that include (G_c, G_d).

3 Results and Discussion

3.1 Testing the Accuracy of hsPPIP

Table 1 gives the basic information about the PPI data sets used in this work. We test our approach with four data sets: our training data set (nr_ppi), random PPI (Random200K), positive control (pos_MIPS) and negative control (L_neg). For random PPI, we randomly pick out two proteins of budding yeast to form 200,000 pairs. Obviously, this data set contains some real PPI, but its overall characteristics should be similar to the negative control. The *cut-off value* of hsPPIP is changed from 0-0.9 with per 0.1 per step. The result is diagramed in figure 1.

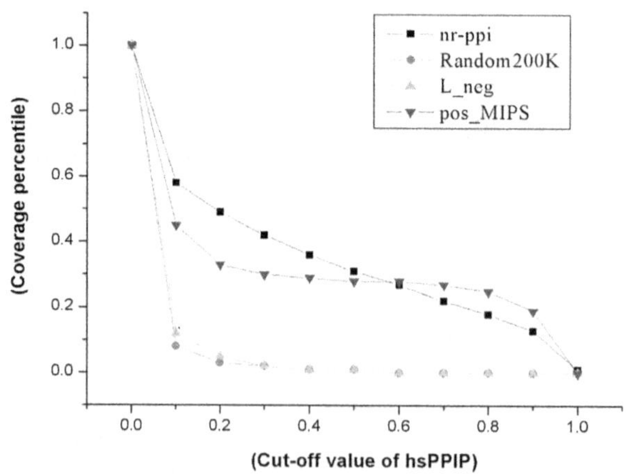

Fig. 1. Prediction accuracy of hsPPIP. We use four data sets to test the prediction accuracy: the training data set (nr-ppi), positive control (pos_MIPS), random PPIs (Random200K) and negative control (L_neg). It's clear that the positive control is much similar with the training data. And both the negative and random PPIs get little hits. The default cut-off value of hsPPIP is taken as 0.3.

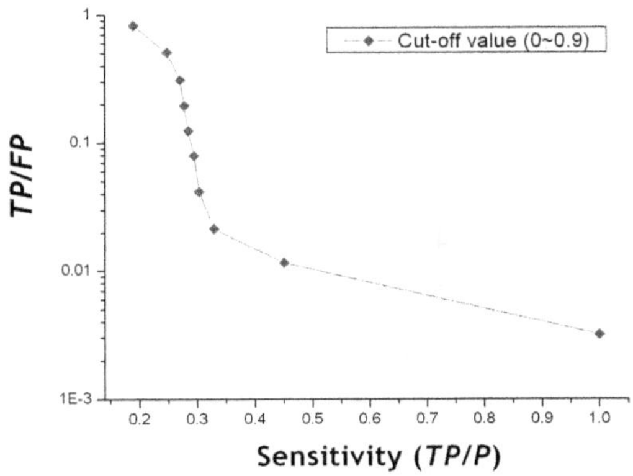

Fig. 2. The curve of TP/FP vs. Sensitivity (TP/P) under certain cut-off value of hsPPIP. We change the cut-off value of hsPPIP from 0-0.9 with per 0.1 per step. With the cut-off value of hsPPIP as 0.3, the TP/FP is 0.041 and Sensitivity (TP/P) is about 30.32%. If the cut-off value is taken as 0.7, the TP/FP is 0.309 and Sensitivity (TP/P) is about 27.04%, which could be comparable with the previous work.

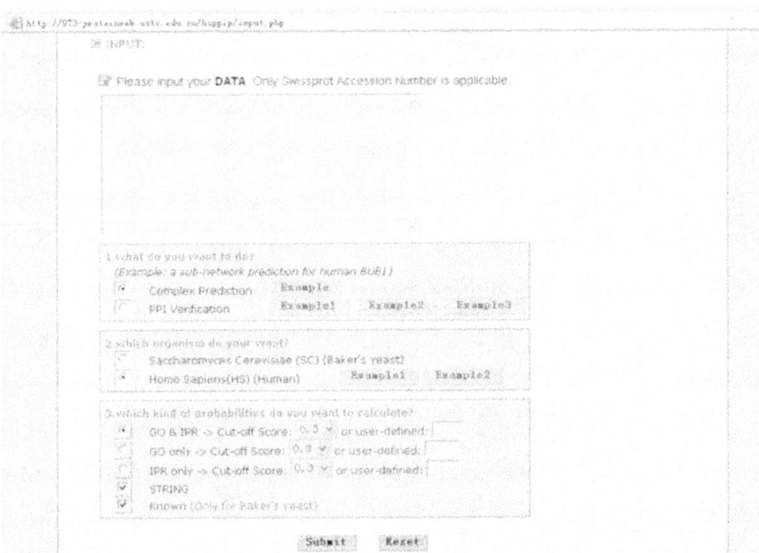

Fig. 3. The prediction page of hsPPIP WWW server. For convenience, hsPPIP supports PPIP in both budding yeast and human. The default cut-off value is taken as 0.3. Two accessing methods are provided. The first is Complex Prediction, which gives out all the potential PPIs among the given list of proteins with the predicted probabilities higher than the cut-off. The second is PPI Verification, which verifies the given list of PPIs with the predicted probabilities higher than the cut-off. Users could either choose Interpro & GO or one of the two features for PPIP.

It's obvious that the training data set gets the best accuracy. And the curve of positive control is much similar with the training data set. The curves of both the negative control and random PPI are similar and very different to our positive control and training data set. So we propose that hsPPIP could distinguish true positive PPIs from random PPIs accurately. The curve of the positive inclines to smooth when *cut-off value* is greater than 0.3. And even when *cut-off value* is great than 0.9, our approach could still predict ~20% PPIs in positive control properly. For convenience, we set the default *cut-off value* of hsPPIP as 0.3. Users could choose their favorite *cut-off value*.

We also perform an additional test with *TP/FP* vs. Sensitivity (*TP/P*) and compare the prediction results with previous work [5]. The *P* is total positive PPIs and *TP* is true positive PPIs which hsPPIP could predict properly. The *FP* is false positive predictions from hsPPIP. We change the *cut-off value* of hsPPIP from 0-0.9 with per 0.1 per step. The result is diagramed in figure 2.

With the *cut-off value* of hsPPIP as 0.3, the *TP/FP* is ~0.041 and Sensitivity (*TP/P*) about 30.32%. If the *cut-off value* is taken as 0.7, the *TP/FP* is ~0.309 and Sensitivity (*TP/P*) about 27.04%, which could be comparable with the previous work (Ref. 5, Figure 2C, *TP/FP*: *~0.3-0.4,* Sensitivity (*TP/P*): ~27%).

3.2 The Implementation of Web Server

The web server of hsPPIP has been implemented in PHP+MySQL. The web server is developed mainly for human PPIP. But for convenience, we also include the PPIP of budding yeast. In our system, only protein's Swissprot accession numbers are available. In addition, the information for proteins included in hsPPIP could be visited by either searching or browsing. The correlated probabilities of two features associated with PPIs, Interpro and GO annotations, are pre-computed and stored in a MySQL database. There are two approaches to access hsPPIP. Firstly, for complex prediction, users could submit a list of their required proteins with the Swissprot accession numbers. hsPPIP will find all the potential PPIs among the given list of proteins with predicted probabilities higher than the cut-off. Secondly, for PPI verification, users could submit a list of protein pairs, and hsPPIP will find all the potential PPIs whose probabilities are higher than the cut-off. The prediction page of hsPPIP is shown in Figure 3.

The hsPPIP could export the prediction results into multiple file formats. The plain text format could be downloaded and viewed locally. The .dot format file could be imported into the professional graphic software Graphviz (http://www.research.att.com/sw/tools/graphviz/) to get a publication-quality figure. Users could modify the color setting for nodes (proteins) and edges (interactions) of the .dot file. Moreover, the .sif format file could be generated for Cytoscape [6]. We also adopt a Java applet program [7] (a modified version for hsPPIP) to visualize the prediction results directly in the web browser.

3.3 An Example of Usage

Here we use human Bub1 protein (Swissprot accession number: O43683) and its putative interacting partners as an instance to diagram the usage of hsPPIP. Human Bub1 protein is one of the spindle checkpoint components localized to kinetochore during mitosis, blocking the onset of anaphase until all chromosomes are attached to microtubules properly [8]. Aberrant organization of spindle checkpoint complex will contribute the genomic instability and aneuploidy, which could induce a variety of cancers with high rate [9]. Although many experimental studies were performed to depict the PPIs implicated in spindle checkpoints, it's still unclear how many proteins will interact with Bub1 and Bub1-related sub-network.

Firstly, we used STRING web server [10] to get the putative interacting partners of human Bub1 with default parameters (see in figure 4A). The sub-network contains eleven human proteins that interact with each other. Then we retrieve the Swissprot accession number of the eleven proteins and input them into hsPPIP for Complex Prediction. The default *cut-off value* for hsPPIP is 0.3. And the prediction results are visualized by Java applet in figure 4B. Only protein with at least one interaction with other proteins is shown. Moreover, the .dot file of prediction results from hsPPIP is modified directly and imported into Graphviz with the output format .png (see in figure 4C). In addition, .sif file format could be visualized by Cytoscape (see in figure 4D). By comparing the prediction results of hsPPIP to STRING, we find the prediction results of hsPPIP are similar with of STRING. Some putative PPIs are existed in STRING while missed by hsPPIP, such as Mxd3 protein (Q9BW11) with its binding partners. But there are several putative PPIs not existed in STRING but

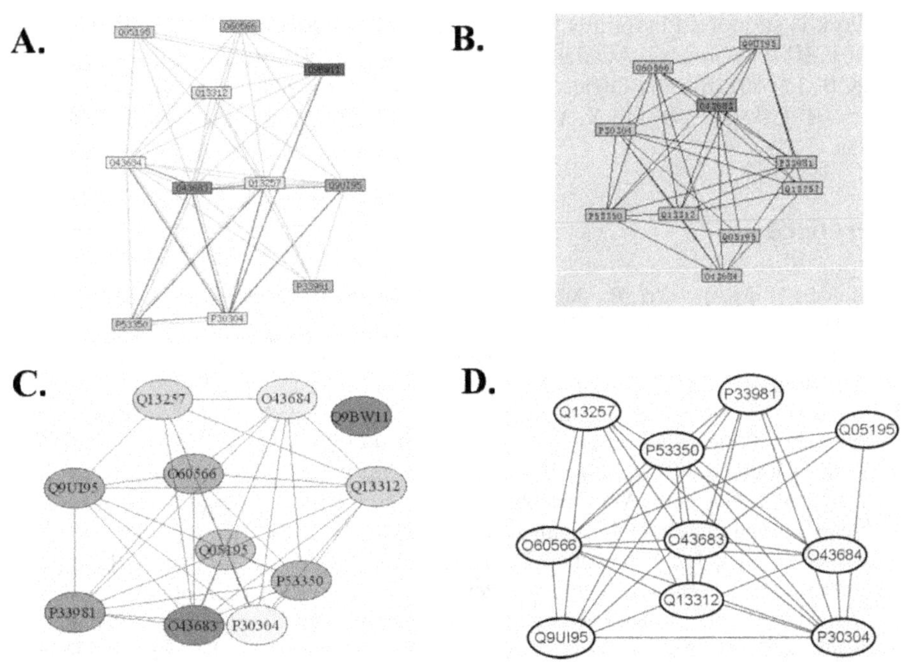

Fig. 4. Predicted sub-network for human Bub1 protein (O43683). (A) The prediction results of STRING, one of the most popular web servers for PPIP. (B) The prediction results of hsPPIP is visualized by java applet online. (C) The prediction results of hsPPIP is modified directly and then imported to Graphviz for visualization. (D) The prediction results of hsPPIP is viewed in Cytoscape.

predicted by hsPPIP. For instance, another spindle checkpoint protein human BubR1 (O60566) is predicted to interact with MAD (Q05195) by hsPPIP, but STRING doesn't find this potential interaction. And the predicted PPIs for human Bub1 may be insightful and needs the further experimental verification.

4 Conclusions

In this work, we provide a web server hsPPIP for protein-protein interaction prediction in human. For convenience, the hsPPIP could also predict the PPIs in budding yeast. The prediction results could be downloaded, modified locally, or visualized online directly. We also use human Bub1 protein to predict its potential PPIs and construct its sub-network by hsPPIP. Compared to another popular web server STRING, the prediction results are similar. Users could submit their favorite proteins in a list to predict the PPIs among them. The prediction results from hsPPIP may be helpful and insightful for further experimental design.

Acknowledgements

The work is supported by grants from Chinese Natural Science Foundation (39925018 and 30121001), Chinese Academy of Science (KSCX2-2-01), Chinese 973 project (2002CB713700), Beijing Office for Science (H020220020220) and American Cancer Society (RPG-99-173-01) to X. Yao. X. Yao is a GCC Distinguished Cancer Research Scholar.

References

1. Mulder, N. J., Apweiler, R., Attwood, T. K., Bairoch, A.: InterPro, Progress and Status in 2005. Nucleic Acids Res. 33 (2005) 201-205
2. Harris, M. A.: The Gene Ontology (GO) Database and Informatics Resource. Nucleic Acids Res., 32 (2004) 258-261
3. Hahn, M. W., Kern, A. D.: Comparative Genomics of Centrality and Essentiality in Three Eukaryotic Protein-Interaction Networks. Mol Biol Evol. 22 (2004) 803-806
4. Yu H., Luscombe, N. M., Lu ,H. X., Zhu, X., Xia, Y., Han, J. D., Bertin, N., Chung, S., Vidal, M., Gerstein, M.: Annotation Transfer Between Genomes: Protein-protein Interologs and Protein-DNA regulogs. Genome Res. 14 (2004) 1107-1118
5. Jansen, R., Yu, H., Greenbaum, D., Kluger, Y., Krogan, N. J., Chung, S., Emili, A., Snyder, M., Greenblatt, J. F., Gerstein, M.: A Bayesian Networks Aapproach for Predicting Protein-Protein Interactions from Genomic Data. Science, 302 (2003) 449-453
6. Shannon, P., Markiel, A., Ozier, O., Baliga, N. S., Wang, J. T., Ramage, D., Amin, N., Schwikowski, B., Ideker, T.: Cytoscape: A Software Environment for Integrated Models of Biomolecular Interaction Networks. Genome Res. 13 (2003) 2498-2504
7. Mrowka, R.: A Java Applet for Visualizing Protein-protein Interaction. Bioinformatics, 17 (2001) 669-671
8. Shah, J. V., Botvinick, E., Bonday, Z., Furnari,,F., Berns, M., Cleveland, D. W.: Dynamics of Centromere and Kinetochore Proteins, Implications for Checkpoint Signaling and Silencing. Curr. Biol. 14 (2004) 942-952

9. Cahill, D. P., Lengauer, C., Yu, J., Riggins, G. J., Willson, J. K., Markowitz, S. D., Kinzler, K. W., Vogelstein, B.: Mutations of Mitotic Checkpoint Genes in Human Cancers. Nature, 392 (1998) 300-303
10. Von, M. C., Jensen, L. J., Snel, B., Hooper, S. D., Krupp, M., Foglierini, M., Jouffre, N., Huynen, M. A., Bork, P.: String: Known and Predicted Protein-protein Associations, Integrated and Transferred Across Organisms. Nucleic Acids Res. 33 (2005) 433-437
11. Mewes, H. W., Amid, C., Arnold, R.: MIPS: Analysis and Annotation of Proteins from Whole Genomes. Nucleic Acids Res. 32 (2004) 41-44
12. Alfarano, C., Andrade, C. E., Anthony, K., Bahroos, N., Bajec, M.: The Biomolecular Interaction Network Database and Related Tools 2005 Update. Nucleic Acids Res.33 (2005) 418-424
13. Breitkreutz, B. J., Stark, C., Tyers, M.: The GRID: The General Repository for Interaction Datasets. Genome Biol. 4 (3) (2003) 233-245
14. Cherry, J. M., Adler, C., Ball, C., Chervitz, S. A.: SGD: Saccharomyces Genome Database. Nucleic Acids Res. 26 (1998) 73-79
15. Zanzoni, A., Montecchi-Palazzi, L., Quondam, M., Ausiello, G., Helmer-Citterich, M., Cesareni, G.: MINT: a Molecular INTeraction database. FEBS Lett. 513 (2002) 135-140
16. Salwinski, L., Miller, C. S., Smith, A. J., Pettit, F. K., Bowie, J. U., Eisenberg, D.: The Database of Interacting Proteins: 2004 Update. Nucleic Acids Res. 32 (2004) 449-451
17. Sprinzak, E., Margalit, H.: Correlated Sequence-signatures as Markers of Protein-protein Interaction. J. Mol Biol. 311 (2001) 681-692
18. Kim, W. K., Park, J., Suh, J. K.: Large Scale Statistical Prediction of Protein-protein Interaction by Potentially Interacting Domain (PID) Pair. Genome Inform Ser Workshop Genome Inform. 13 (2002) 42-50
19. Obenauer, J. C., Yaffe, M. B.: Computational Prediction of Protein-protein Interactions. Methods Mol Biol., 261 (2004) 445-468
20. Han, D. S., Kim, H. S., Jang, W. H., Lee, S. D., Suh, J. K.: PreSPI: A Domain Combination Based Prediction System for Protein-protein Interaction. Nucleic Acids Res. 32 (2004) 6312-6320

Prediction of Ribosomal -1 Frameshifts in the *Escherichia coli* K12 Genome

Sanghoon Moon, Yanga Byun, and Kyungsook Han*

School of Computer Science and Engineering, Inha University,
Inchon 402-751, Korea
jiap@inhaian.net, quska@inhaian.net, khan@inha.ac.kr

Abstract. Ribosomal frameshifting at a particular site can yield two protein products from one coding sequence or one protein product from two overlapping open reading frames. Many organisms are known to utilize ribosomal frameshifting to express a minority of genes. However, finding ribosomal frameshift sites by a computational method is difficult because frameshift signals are diverse and dependent on the organisms and environments. There are few computer programs available for public use to identify frameshift sites from genomic sequences. We have developed a web-based application program called FSFinder2 for predicting frameshift sites of general type. We tested FSFinder2 on the *Escherichia coli* K12 genome to detect potential -1 frameshifting genes. From the genome sequence, we identified 18,401 frameshift sites following the X XXY YYZ motif. 11,530 frameshift sites out of the 18,401 sites include secondary structures. Comparison with the GenBank annotation produced 11 potential frameshift sites, including 3 known frameshift sites. The program is useful for analyzing frameshifts of various types and for discovering new genes expressed by frameshifts.

1 Introduction

Ribosomes in general terminate translation at three kinds of stop codons (UAG, UGA and UAA), but some ribosomes continue to decode after the stop codons. This alternative translational event is called ‘recoding’. Recoding events include frameshifting, read-through and bypassing [1-3]. In frameshifting, ribosomes shift reading frame by one or more nucleotides at a specific mRNA signal between overlapping genes [4]. Frameshifts are classified into different types depending on the number of nucleotides shifted and the shifting direction. The most common type is a -1 frameshift, in which the ribosome slips a single nucleotide in the upstream direction. -1 frameshifting requires a frameshift cassette that consists of a slippery site, a stimulatory RNA structure and a spacer. +1 frameshifts are much less common than -1 frameshifts, but have been observed in diverse organisms [1].

No program exists to predict general types of frameshift. In addition, existing computational models predict too many false positives. In previous work we developed a program called FSFinder (Frameshift Signal Finder) for predicting -1 and +1

* Correspondence author.

D.-S. Huang, K. Li, and G.W. Irwin (Eds.): ICIC 2006, LNBI 4115, pp. 612 – 621, 2006.

frameshift sites [5]. That program is written in Microsoft C# and is executable on Windows systems only. To remove these limitations and to handle frameshifts of general type, we developed a new web-based application called FSFinder2. Users can predict frameshift sites of any type online from any web browser and operating system.

In previous experimental results of testing FSFinder2 on ~190 genomic and partial DNA sequences showed that it predicted frameshift sites efficiently and with greater sensitivity and specificity than other programs, because it focused on the overlapping regions of ORFs and prioritized candidate signals (For -1 frameshifts, sensitivity was 0.88 and specificity 0.97; for +1 frameshifts, sensitivity was 0.91 and specificity 0.94) [5-7].

Using the web service of the FSFinder2, we analyzed the *Escherichia coli* (*E. coli*) K12 genome sequence to find the -1 frameshifting genes with high probability. From the *E. coli* K12 genome sequence, we found 18,401 frameshift sites after the X XXY YYZ motif. Among these sequences, 11,530 frameshift sites included secondary structure such as pseudoknots or stem-loops. Using the GenBank description we found 312 overlapping regions of two genes with more than 1 base. Using FSFinder we found 309 overlapping regions with more than 30 bases. After removing redundant ones, we obtained 195 overlapping regions and found 66 potential frameshift sites in the 195 overlapping regions. Among these sites, 11 sites including 3 known frameshift sites were considered significant based on the gene length, shape and the length of overlapping region. We believe that at least 4 new frameshift sites are highly likely to be frameshift sites.

2 Analyzing Method

2.1 Finding Frameshift Motifs

The cassettes of -1 frameshift consist of three parts: slippery site, spacer and secondary structure. The slippery site is usually a heptameric sequence in the form X XXY YYZ (in the incoming 0-frame), where X, Y and Z can be same nucleotides [5-7]. The spacer is a short sequence of 5 to 11 nucleotides separating slippery site and the downstream secondary structure. The downstream structure is usually a pseudoknot or simple stem-loop, as shown in Fig. 1.

For analyzing frameshift sites in the *E. coli* genome sequence, we detected heptameric sequence with a secondary structure. In previous work by others [5-7], frameshift sites have two constraints. In slippery site X XXY YYZ, X is any nucleotide, Y is A or U, and Z is either A, U or C. However, in our work, any kind of nucleotide with secondary structure can be located in the slippery sequence.

Fig. 2 shows the parameters for the FSFinder web service. The web service and web application of FSFinder2 were implemented using XML, XSLT and JavaScript. If the user sends a query to the FSFinder2 server after setting parameters or defining a new model, all the computations are done on the server side. After computation, the server sends the results to the user.

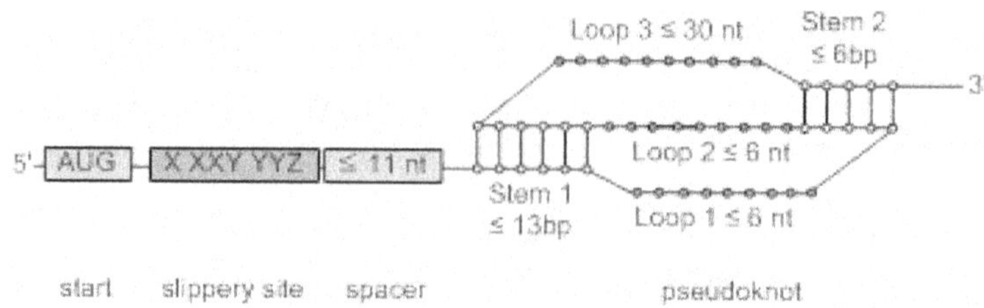

Fig. 1. A programmed -1 ribosomal frameshift signal with H-type pseudoknot. Stem 1 has 11 base pairs, stem 2 has 6 base pairs, and both loops of the pseudoknot have 6 nucleotides. In particular, any nucleotide can be located in the slippery site not the same as previous work.

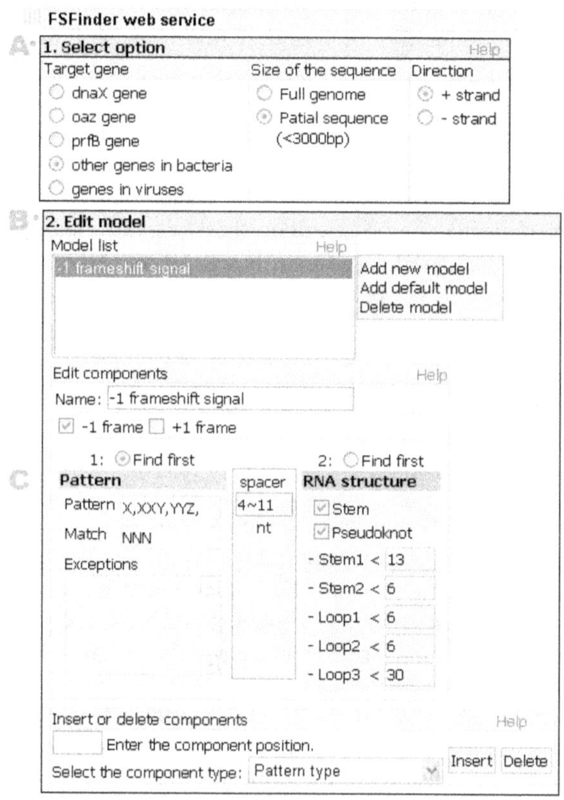

Fig. 2. The input page of the FSFinder web service. (A) The select option lets the user choose the type of genes expressed via frameshifts, the size of the sequence and its direction. (B, C) The user can define a new model by specifying its components and their locations. To analyze *E. coli* genome, we set the selection option as other genes in bacteria, partial sequence and +1 strand. Because we focused on -1 frameshift model, we made a model which is the general motif X XXY YYZ of the -1 frameshift. There was no limitation of frameshift site, thus any kind of nucleotide can be located in the frameshift site. For downstream secondary structure, the length of the stems and loops set by default.

```
<FSFinder_run xmlns="http://wilab.inha.ac.kr/WSFSFinder/">
  <FSFinder_Input xmlns:xsd=http://www.w3.org/2001/XMLSchema
  xmlns:xsi="http://www.w3.org/2001/XMLSchema-instance">
    <Sequence_information>
      <Target_gene>other genes in bacteria</Target_gene>
      <Sequence_size>Patial</Sequence_size>
      <Sequence_direction>Plus strand</Sequence_direction>
      <Sequence>input sequence</Sequence>
      <Sequence_name>E. coli K-12</Sequence_name>
    </Sequence_information>
    <FSFinder_Model_List>
      <FSFinder_Model Find_first="0" Overlap_with="-1">
        <Model_name>-1 frameshift signal</Model_name>
        <Components>
          <Pattern_type Spacer="0">
            <Pattern>X,XXY,YYZ,</Pattern>
            <Pattern_match>NNN</Pattern_match>
          </Pattern_type>
          <RNA_structure_type Spacer="4~11" Structure=
            "StemLoop;Pseudoknot">
            <Stem1_size_max>13</Stem1_size_max>
            <Stem2_size_max>6</Stem2_size_max>
            <Loop1_size_max>6</Loop1_size_max>
            <Loop2_size_max>6</Loop2_size_max>
            <Loop3_size_max>30</Loop3_size_max>
          </RNA_structure_type>
        </Components>
      </FSFinder_Model>
    </FSFinder_Model_List>
  </FSFinder_Input>
</FSFinder_run>
```

Fig. 3. XML schema with the parameters shown in Fig. 2. If the user sets the parameters of a model in the web page of the FSFinder web service, the parameters are converted to XML. After the request of a web service is sent to the FSFinder server, all the computations are done on the server side.

To analyze *E. coli* genome, we set the selection option as other genes in bacteria, partial sequence and + strand (Fig. 2A). Because we focused on -1 frameshift sites (Fig. 2B), we defined a signal that fits all kinds of motif X XXY YYZ of the -1 frameshift. There was no limitation of frameshift site, thus any kind of nucleotide can be located in the frameshift site. Thus we set the match type as NNN (N is any nucleotide) and no exception of arrangement. That is, frameshift site can occur even at A AAA AAA or U UUU UUU. Default values were used for the lengths of the stems and loops in the downstream secondary structure (Fig. 2C). Fig. 3 shows the XML schema for the parameters of FSFinder2. When the user sets the parameters of a model in the web page of the FSFinder web service, the parameters are converted to XML. After the request of a web service is sent to the FSFinder server, all the computations are done on the server side.

2.2 Finding Overlapping Regions of Genes

Fig. 4 shows the frameshift site of the genome of SARS corona virus (NC_004718). ORF1a (265..13398) and ORF1b (13398..21485) partially overlap each other. U UUA AAC is the slippery sequence and the secondary structure is a pseudoknot [8]. In the GenBank description, the overlapping region of ORF1a and ORF1b has only one 1 nucleotide. ORF1b starts from codon AAA instead of a regular start codon. The start codon of ORF2 (red triangle) exists outside the overlapping region. Frameshifting can occur even in overlapping regions with 1 nucleotide. To solve these problems, finding an overlapping region is divided into two processes: finding motifs and finding overlapping regions.

2.2.1 Finding Overlapping Regions Using FSFinder

The shape of an overlapping region is considered to find an overlapping region. As shown in Fig. 5, the overlapping region is extended from stop codon 1 of open reading frame 2 to stop codon 2 of open reading frame 1 (green color). The minimum length of both genes should be longer than 100 nucleotides, and the length of the overlapping region should be longer than 30 nucleotides. In addition, the location of the start codon of ORF2 does not matter. As in the SARS corona virus, the start codon is not always located in the overlapping region of frameshifting genes.

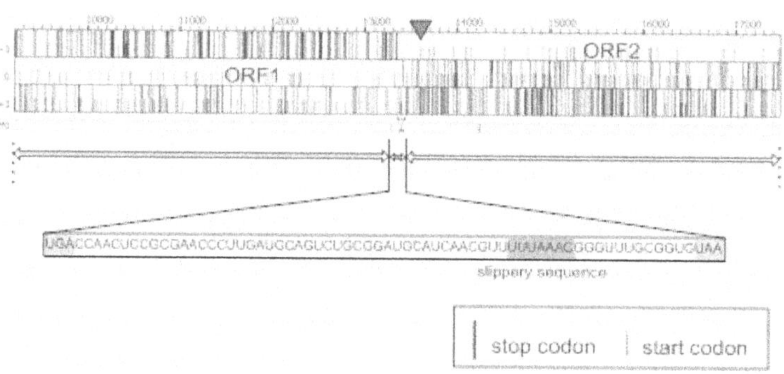

Fig. 4. The frameshift site of the SARS corona virus (the slippery sequence UUUAAAC shown in green background)

2.2.2 Finding Overlapping Regions Using the GenBank Description

Additional method to find overlapping regions is to use the description of the GenBank file. If more than 1 nucleotide is overlapped, we consider two open reading frames as a candidate of partially overlapping genes. Finally, the overlapping regions found both by FSFinder and by the GenBank description are used for further analysis.

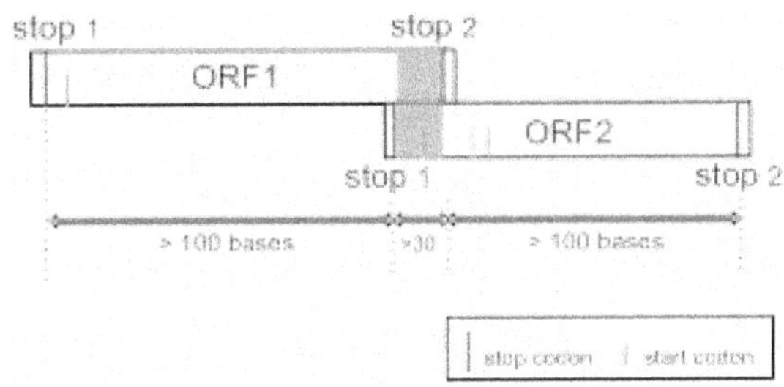

Fig. 5. Finding overlapping regions by FSFinder. Unlike general overlapping regions, FSFinder extends the overlapping region from stop codon 1 of open reading frame 2 to stop codon 2 of open reading frame 1 (green color). The length of both ORFs should be longer than 100 nucleotides and the length of the overlapping region should be longer than 30 nucleotides.

3 Results and Discussion

The *E. coli* K12 genome sequence (NC_000913) was obtained from GenBank. As shown in Fig. 6, all the heptameric sequences that follow X XXY YYZ motif were

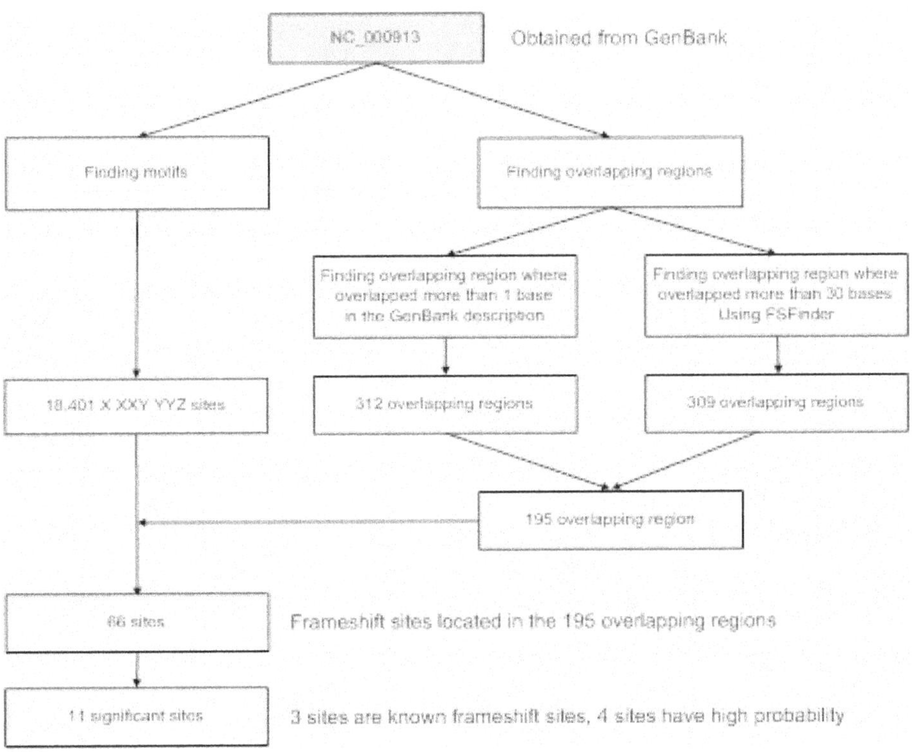

Fig. 6. The prediction process of frameshift sites from the *Escherichia coli* K 12 genome sequence

examined. As a result of this process 18,401 sites were found. Whether these sites are located in the overlapping region or not was not considered when finding these sites.

For overlapping regions of ORFs, we found 312 candidate regions that were partially overlapped more than 1 base according to the GenBank description and 309 candidate regions that were overlapped more than 30 bases using the FSFinder web service. After removing redundant sites, we obtained 195 overlapping regions. From these regions, heptameric sequences that were not located in the overlapping region were filtered out. Just 66 sites remained in the overlapping regions. From the gene length, shape and the length of the overlapping region, 11 sites including 3 known frameshift sites were identified as significant candidates.

This process consists of two sub-processes: finding motifs and finding overlapping regions. The *E. coli* K 12 genome sequence (NC_000913) was obtained from the GenBank. To find motifs, all the heptameric sequences that follow X XXY YYZ motif were found. 18,401 sites were detected. According to the GenBank description, there were 312 candidate overlapping regions that were partially overlapped more than 1 base in. The FSFinder web service found 309 candidate overlapping regions that were overlapped more than 30 bases. After removing redundant sites, we obtained 195 overlapping regions. After further removing heptameric sequences that were not located in the overlapping region, 66 sites remained in the overlapping regions. Considering the gene length, shape and length of the overlapping region, 11 frameshift sites including 3 known sites were considered significant candidates.

Table 1 shows the names of overlapping genes, the locations of the slippery sites, slippery sequences and number of overlapped nucleotides from our analysis. The three genes, yi21_1-yi22_1, yi21_5-yi22_5, yi21_6-yi22_6, are known genes expressed via -1

Table 1. The predicted frameshift sites in the E. coli K 12 genome sequence. The first threegenes marked with * symbol are those with known frameshift sites. PK represents a pseudoknot.

Gene name	Slippery site	Slippery sequence + secondary structure	# bases in the overlapping region found by FSFinder	# bases in the overlapping region based on GenBank
i21_1..yi22_1*	380892	A AAA AAG + PK	61	43
i21_5..yi22_5*	3184526	A AAA AAG + PK	61	43
i21_6..yi22_6*	4496612	A AAA AAG + PK	61	43
insA_6..insB_6	3581757	A AAA AAC +PK	103	82
entB..entA	627723	A AAC CCG + PK	76	1
tehA..tehB	1499529	G GGA AAA + PK	154	4
xdhB..xdhC	3001496	G GGG GGA + stem	37	4
mhpD..mhpF	372127	C CCA AAA only	61	4
fliM..fliN	2019082	U UUA AAU only	40	4
atoS..atoC	2319873	G GGA AAU only	85	4
yijC..yijD	4159773	G GGA AAU only	46	1

frameshifts. These three genes and insA_6-insB_6 are insertion sequences. We believe that entB-entA, tehA-tehB, and xdhB-xdhC have a high probability using frameshift events. All these seven genes have either pseudoknot or stem loop as a downstream secondary structure. The rest four genes, mhpD-mhpF, fliM-fliN, atoS-atoC and yijC-yijD, have no downstream secondary structures and have a lower probability of frameshifting than the other 7 genes.

There exist previous works similar to our approach. Hammell *et al.* [7] studied -1 ribosomal frameshift signals in the large databases. Using their well-established model, they found that -1 frameshifts occur with frequencies from two- to six-fold greater than random. They considered the nucleotides of a frameshift site, spacer, and pseudoknot. However, they focused on the -1 frameshifts only. Bekaert *et al.* [6] performed a computational method similar to Hammell's approach. From their model, they designed a model for -1 eukaryotic frameshifting. But these two programs are not available for public use.

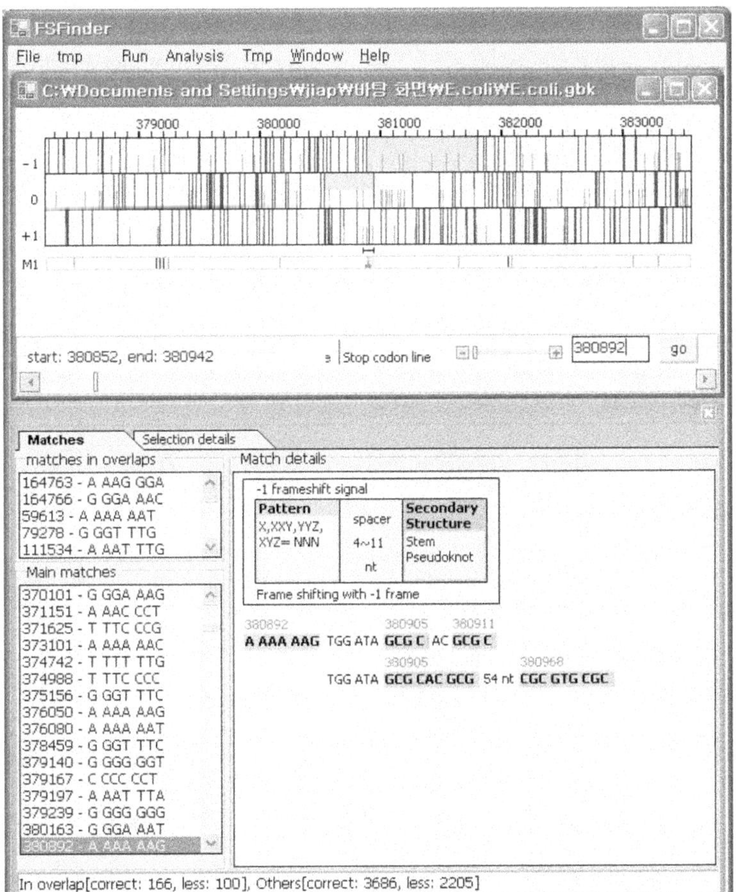

Fig. 7. Frameshift site of overlapping genes i21_1..yi22. FSFinder2 found the sequences of the frameshift cassettes and locations of the slippery site.

The FreqAnalysis [9] is an application program available for public use. FreqAnalysis was implemented in the Java language and can find putative translational frameshift from probabilistic calculation. In fact, Shah *et al.* [9] found putative frameshift sites from the *Sacharomyces cerevisiae* ORFs. But FreqAnalysis does not consider the X XXY YYZ frameshift motif. Thus it is hard to compare the results of FreqAnalysis with ours directly.

Fig. 7 shows the result of the FSFinder2. Two rectangles filled with sky blue color in the upper window represent i21_1 and yi22_1, respectively. The i21_1 gene located in 0 frame and the yi22_1 gene located in -1 frame are partially overlapped. When translational frameshift occurs, translation continues past the stop codon at 380,938 until the stop codon at 381,801.

4 Conclusion

Ribosomal frameshifting is unusual event which is known to affect producing heterogeneous proteins, auto-regulation. Prediction of frameshift sites is very difficult. On the other hand if prediction is possible, this is very useful to understand of biological phenomenon. And unveiling of unknown protein production mechanisms can be realized. For this valuable advantage, we develop a web application and serve web service for prediction of frameshift sites.

Using the FSFinder, we analyzed *Escherichia coli* K 12 genome sequence to detect potential -1 frameshifting genes. From the E. coli K 12 genome sequence, we have got 18,401 frameshift sites followed X XXY YYZ motif. Among these sequences, 11,530 frameshift sites included secondary structure. Comparing with GenBank description, we have got 11 sequences including 3 known frameshift sites. Among these sequences, we believe that at least 4 sequences have a high probability to use frameshift event and all these 4 sites have a downstream secondary structure such as pseudoknot and stem loops. Other 4 gene do not have downstream structure, but we believe that these sequences have less probabilities than above 7 genes but significant.

Acknowledgements

This work was supported by the Korea Science and Engineering Foundation (KOSEF) under grant R01-2003-000-10461-0 and in part by the Brain Korea 21 Project.

References

1. Farabaugh, P.J.: Programmed Translational Frameshifting. Ann. Rev. Genetics 30 (1996) 507-528
2. Gesteland, R.F., Atkins, J.F.: Recoding: Dynamic Reprogramming of Translation. Annu. Rev. Biochem. 65 (1996) 741-768
3. Herr. A.J., Gesteland, R.F., Atkins, J.F.: One Protein From Two Open Reading Frames: Mechanism of a 50 ntTranslational Bypass. EMBO J. 19 (2000) 2671-2680
4. Baranov, P.V., Gesteland, R.F., Atkins, J. F.: Recoding: Translational Bifurcations in Gene Expression. Gene 286 (2002) 187-201

5. Moon, S., Byun, Y., Kim, H.-J., Jeong, S., Han, K.: Predicting Genes Expressed via -1 and +1 Frameshifts. Nucleic Acids Research 32 (2004) 4884-4892
6. Bekaert, M., Bidou, L., Denise, A., Duchateau-Nguyen, G., Forest, J., Froidevaux, C., Hatin, Rousset, J., Termier, M.: Towards a Computational Model for -1 Eukaryotic Frameshifting Sites. Bioinformatics 19 (2003) 327-335
7. Hammell, A.B., Taylor, R.C., Peltz, S.W., Dinman, J.D.: Identification of Putative Programmed -1 Ribosomal Frameshift Signals in Large DNA Databases. Genome Research 9 (1999) 417-427
8. Ramos, F.D., Carrasco, M., Doyle, T., Brierley, I.: Programmed -1 Ribosomal Frameshifting in the SARS Coronavirus. Biochemical Society Transactions 32 (2004) 1081-1083
9. Shah, A.A., Giddings, M.C., Parvaz, J.B., Gesteland, R.F., Atkins, J.F., Ivanov, I.P.: Computational Identification of Putative Programmed Translational Frameshift Sites. Bioinformatics 18 (2002) 1046-1053

Using a Stochastic AdaBoost Algorithm to Discover Interactome Motif Pairs from Sequences

Huan Yu[1], Minping Qian[1], and Minghua Deng[1,2]

[1] LMAM, School of Mathematical Sciences and Center for Theoretical Biology, Peking University, Beijing 100871, P.R. China

[2] dengmh@pku.edu.cn

Abstract. Protein interactome is an important research focus in the post-genomic era. The identification of interacting motif pairs is essential for exploring the mechanism of protein interactions. We describe a stochastic AdaBoost approach for discovering motif pairs from known interactions and pairs of proteins that are putatively not to interact. Our interacting motif pairs are validated by multiple-chain PDB structures and show more significant than those selected by traditional statistical method. Furthermore, in a cross-validated comparison, our model can be used to predict interactions between proteins with higher sensitivity (66.42%) and specificity (87.38%) comparing with the Naive Bayes model and the dominating model.

1 Introduction

With the finishing of genome sequencing projects, the protein level functional annotation of the entire genome has been a major goal in the post-genomic era. Furthermore, the relationships between proteins play more important role than the function of individual proteins since proteins interact and regulate each other in most biological process. Many experimental and computational methods are developed to identify protein-protein interactions (PPIs).

The traditional genetic, biochemical or biophysical techniques such as protein affinity chromatography, immunoprecipitation, sedimentation and gel-filtration [1] are not suitable for genome-wide interaction detection. Several high-throughput methods have being developed to identify protein-protein interactions in parallel [2–7]. However, the high throughput experiment methods suffer from both high false positives and false negatives[8]. As a result, computational (*in silico*) methods are required to complete the interactome and explore the interacting mechanisms.

The computational approaches for protein interaction discovery have been developed over the years. Earlier methods focus on predicting genome-wide functional linkages by phylogenetic profiles [9, 10], gene fusion information[11, 12] or conservation of gene neighborhoods and gene orders[13, 14]. With the high-throughput interaction data, several machine learning approaches have been developed to predict protein interactions in the domain level [15–29] or motif level

D.-S. Huang, K. Li, and G.W. Irwin (Eds.): ICIC 2006, LNBI 4115, pp. 622–630, 2006.

[20]. The central point of those works is interaction prediction. Most of them considered protein domain interactions as the mediate of protein interaction. Gomez [20] tried to show predicting ability of short tuple pairs. Actually, the classifier used by Gomez was a modified Bayes model. To avoid the independent assumption, only the most informative feature was included in the joint probability computing, which is equivalent to dominating Bayes model with uniform prior distribution. In fact, protein interaction motifs themselves are very important for drug and protein design, often referred as protein contact interfaces or binding sites. Interactome motif finding would be of considerable value. Protein-protein interaction data provide more information of motif pairs [30, 31] than one-side interaction motifs [32]. But both Sprinzak's [30] and Wang's [31] works are based on known one-side motifs from database such as PROSITE [33]. We want to discover hidden motif pairs behind the interaction data. In our previous work [34], statistical significant motif pairs set was selected and successfully validated in several ways. But in this paper we will show that statistical significant is not equivalent to biological functional.

In this study, we use an AdaBoost [35] strategy to study the protein-protein interactions in the motif level. Starting from protein sequences and known protein interactions only, the predicting effective motif pairs are identified as the signatures of the protein-protein interactions. By validating them on PDB structure, our interacting motif pairs are more significant than those selected by traditional statistical method. Furthermore, we predict genome-wide protein-protein interactions with higher sensitivity and specificity comparing with Gomez's dominating Bayes model [20] and the Naive Bayes model as a benchmark baseline.

2 Methods

2.1 Data

Protein-protein interaction data and corresponding protein sequences for *Saccharomyces cerevisiae*(yeast) were collected from the DIP (Database of Interacting Proteins) [36]. The data was obtained on July 31, 2005 containing 15529 interactions among 4786 proteins. Since negative interaction data are not available, several methods have been proposed to construct negative interaction data set. The two most popular approaches are described as follows. In the first approach, proteins in separate subcellular compartments are considered as negatives[37]. The shortcoming of this approach is that there is no information showing whether proteins can bind with each other if they are being put together. Different subcellular locations only indicates that the proteins have no chance to bind. The second approach assumes that two proteins do not interact if there is no positive evidence of interaction[20]. The huge false negative data set is not acceptable for computing. We defined a negative PPI pair as the one proposed in our previous work [34]. A negative PPI pair is a pair of proteins that have chance to interact with each other but fail to be observed in the interaction map. To be precise, we first built a protein interaction graph with proteins as nodes and positive interactions as edges with equally weights. Then all pairs among nodes in connected graph with shortest path length greater than a cutoff were treated as putative negative protein-protein interactions. The cutoff was selected to balance the sample sizes of positive and negative training data. For example, when cutoff = 7, we obtained 26605 negative protein-protein interactions.

To estimate the reliability of our results, 135 protein complex structures were obtained from PDB (RCSB Protein Data Bank) [38] on October 31, 2005 with following conditions:

1. Contains RNA Chain(s): No;
2. Contains Carbohydrate Chain(s): No;
3. Number of Chains (min.): 2;
4. Contains Enzyme Chain(s): Yes
5. Contains DNA/RNA Hybrid Chain(s): No
6. Contains Glycoprotein Chain(s): Yes
7. Contains DNA Chain(s): No

The PDB data contain the 3-dimension coordinates of all amino acid atoms, which are direct evidences for protein-protein interactions. The protein complex structures allow us to find out the protein contact interface by simply calculating the Euclidean distances between atoms. The shortcoming of PDB structures is lack of data compared to sequences due to difficulty of biological wet experiments.

2.2 Feature Vectors Extraction

Feature vectors extraction used in this work is similar to the one discussed in Gomez's study [20]. 20 amino acids were divided into six categories of biochemical similarity [{IVLM}, {FYW}, {HKR}, {DE}, {QNTP} and {ACGS} [39]]. After the reduction, there are $M = \binom{6^4}{2} + 6^4 = 840456$ possible 4-tuple pairs in total if the two 4-tuples are exchangeable. Each pair of interaction (positive or negative) could be represented by a binary vector of length M, in which each bit indicated whether the corresponding 4-tuple pair occurred in the interaction. These M 4-tuple pairs are considered as the interacting motif pair candidates.

2.3 Algorithm

Our method is a boosting approach on regulated stochastic linear-threshold classifier. AdaBoost is not a classifier but a procedure to learn a voting ensemble of many weak classifier. The procedure is shown in Table 1. The only missing things in AdaBoost strategy is a system complexity parameter T and a weak classifier with sample weights. In each step, a error adjustment method was performed on the sample weights. Usually, the larger T leads to more computation and risk of overfitting. In our experiment, the factors of weak classifiers were convergence to zero after four rounds, so T was fixed to 4.

Given the sample weights calculated by each boosting round, our goal was mining the different effects of each vector component. Furthermore, we used a regulated stochastic WINNOW2 (in Table 2, with $R = 200,000$) as the weak

Table 1. The AdaBoost procedure. T weak classifier are constructed according to its accuracy in previous round. The final classifier is a weighted linear combination of the T weak classifier.

- Given training data: $(x_1, y_1), (x_2, y_2), ..., (x_n, y_n)$ where $y_i = 1, -1$ for positive and negative samples respectively.
- Initialize weight $D_1(i) = 1/n$ for $i = 1, 2, ..., n$.
- For $t = 1, ..., T$:
  1. Train weak learner using distribution D_t and get back a weak classifier h_t : $\{0,1\}^M -> \{1, -1\}$.
  2. Calculate the error rate with respect to D_t: $r_t = \sum_i D_t(i) h_t(x_i) y_i$.
  3. Update the weights: $D_{t+1}(i) = D_t(i) exp(-\alpha_t h_t(x_i) y_i)$, where $\alpha_t = \frac{1}{2} log(\frac{1+r_t}{1-r_t})$.
  4. Normalize the weights D_{t+1} into a probability distribution.
- The final strong classifier is: $H(x) = \text{sign}(\sum_{t=1}^{T} \alpha_t h_t(x))$.

Table 2. The regulated stochastic WINNOW2. The algorithm is applied according to the sample weights. Feature factors are regulated to capture the effective components.

- Given training data $(x_1, y_1), (x_2, y_2), ..., (x_n, y_n)$ and its corresponding weights $D_t(1), D_t(2), ..., D_t(n)$.
- Initialize learner factor $\omega_i = 1$ for $i = 1, 2, ..., M$, threshold $\theta = M/2$
- For $r = 1, ..., R$:
  1. Draw a sample (x_k, y_k) according to the sample weights, we denotes vector x_k as $(x_{k1}, x_{k2}, ..., x_{kM})$.
  2. The learner responds as follows:
     - $h = 1$, if $\sum_{i=1}^{M} \omega_i x_{ki} > \theta$
     - $h = -1$, if $\sum_{i=1}^{M} \omega_i x_{ki} \leq \theta$
  3. Update learner factors $\omega_i := \omega_i 2^{(1-x_{ki})(y-h)/2}$
- We define regulated classifier at level c, $h(c)$, as follows:
  - $h(c) = 1$, if $\sum_{i=1}^{M} \omega_{i,c} x_{ki} > \theta$,
  - $h(c) = -1$, if $\sum_{i=1}^{M} \omega_{i,c} x_{ki} \leq \theta$,

  where $\omega_{i,c} = 0$, if $\omega_i \geq c$ and $\omega_{i,c} = \omega_i$, if $\omega_i < c$.
- let N_c denotes the number of positive prediction on the training data with classifier $h(c)$. Output the classifier $h(C)$ where $C = \text{argmax}(c : N_c = N_0)$, where N_0 is the number of positive prediction at the cutoff level c=0.
- The features with non-zero factor $w_{i,c}$ are identified as effective components.

classifier, which was somewhat a error adjustment method on the feature weights. The original version of WINNOW2 was introduced by Littlestone [40]. The major difference of our version from the original one is that our sample weights are not equal. The sample for next learning round was drawn according to the sample weights. Finally, an additional regulation step was applied to discover the effective components. The features with large enough factor were identified as effective components. The union of effective components in these T rounds were considered as the motif pairs of the protein-protein interactions.

2.4 Cross-Validation and Structural Validation

The performance of classifier was measured using five-fold cross-validation. The data were divided into five subsets. The classifier was trained on four subsets and tested on the remaining one. Then the average sensitivity and specificity across all five trails were computed and compared with Naive Bayes classifier (described in appendix) and dominating Bayes model [20].

The reliability of identified motif pairs sets was estimated by bootstrap of PDB protein complex structures validation. A motif pair is verifiable is defined as follows:

Definition 1. *Motif pairs verifiable: Given two protein chains with 3-D structural coordinates and a motif pair ($Motif_A$,$Motif_B$), $Motif_A$ occurs in one protein chain and $Motif_B$ occurs in the other and at least one atom from $Motif_A$ and one from $Motif_B$ contact.*

Atoms contacting if and only if $dist(Atom_A, Atom_B) < \varepsilon$, where $dist(\cdot,\cdot)$ is the Euclidean distance, and ε is an empirical threshold 5Å.

For a given motif pairs set S, the bootstrap method for p-value estimation is applied as follows. We repeatly drawn the same number of 4-tuple pairs randomly as set $R_1, R_2, \cdots, R_m$. If m is large enough, the p-value of the set S can be estimated as $P(S) = \#(V(R_i) \geq V(S))/m$, where $V(S)$ denotes the number of verifiable motif pairs in set S.

The results were compared with χ^2 feature selection method (described in appendix).

3 Results

At the motif level, we combined the five-fold information together as follows: M_i = {motif pairs identified in at least i folds}, where $i = 2, 3, 4, 5$. Let M_{all} denote the motif pairs set identified with all training data. Table 3 shows the results of 100,000 times bootstrap of PDB structure validation comparing with χ^2 feature selection.

It is a shock that more statistical conserved motif pairs are less structural verifiable. Noting the fact that the totally random selection should have average p-value of 0.5, the number of verifiable motif pairs in the most conserved set selected by χ^2 method is only a little more than that of random pairs. Whereas all motif pairs sets identified by our AdaBoost algorithm are significantly verifiable.

At the protein interaction level, with threshold at zero, our AdaBoost model achieves 87.38% in specificity and 66.42% in sensitivity in average. The sensitivities comparison with comparable fixed specificity at approximately 87.38% are shown in Table 4. It is clearly shown that AdaBoost outperforms the others in prediction accuracy.

Table 3. Motif validation. the V(Set) column and the V(Stat) column are the verifiable number of the Set and the same size set selected by χ^2 method, respectively. p-values are estimated by 100,000 times bootstrap method described above.

Set	number of motif pairs in Set	V(Set)	p-value	V(Stat)	p-value
M_{all}	660	77	6.43e-3	73	2.49e-2
M_2	775	87	1.13e-2	84	2.76e-2
M_3	490	58	1.16e-2	50	0.147
M_4	307	45	2.67e-4	28	0.424
M_5	157	27	1.93e-4	15	0.374

Table 4. Accuracy Comparison. the Sensitivity and Specificity of three methods with comparable fixed specificity.

	AdaBoost	Dom-Bayes	Naive Bayes
Sensitivity	66.42%	54.85%	32.08%
Specificity	87.38%	87.14%	87.38%

4 Conclusion and Discussion

This work describes a stochastic AdaBoost method for discovering the interactome motif pairs and predicting protein-protein interactions from sequences. To apply the machine learning procedure, the putative negative protein-protein interactions were generated by known positive interactions. By resampling validation, our identified motif pairs fit known protein complex structures very well while traditional statistical method is not. We notice a phenomenon that conservation is not a guarantee of functional sites. Moreover, the motif pairs are validated at protein interaction level. In contrast using all features in the dominating Bayes model, only those effective features are included in our classifier. Our method achieves higher specificity and sensitivity in protein-protein interaction prediction.

In the real world, the whole protein interaction network is more complex since protein may have different behavior with others depending on the many conditions, e.g., protein location, expression time, temperature. In the future, We expect to expand our work with more types of data such as the phylogenetic information, expression data and so on.

Acknowledgments

This research is supported by the grants from National Natural Science Foundation of China (No. 30570425, No. 90208022) and the National Key Basic Research Project of China (No. 2003CB715903), and supported in part by Microsoft Research Asia (MSRA).

References

1. Phizicky, E.M., Fields, S.: Protein-Protein Interactions: Methods for Detection and Analysis. Microbiol Rev **59**(1) (1995) 94–123
2. MacBeath, G., Schreiber, S.L.: Printing Proteins as Microarrays for High-Throughput Function Determination. Science **289**(5485) (2000) 1760–1763
3. Uetz, P., Giot, L., Cagney, G., et al.: A Comprehensive Analysis of Protein-Protein Interactions in Saccharomyces Cerevisiae. Nature **403**(6770) (2000) 623–627
4. Ito, T., Chiba, T., Ozawa, R., et al.: A Comprehensive Two-Hybrid Analysis to Explore the Yeast Protein Interactome. Proc Natl Acad Sci U S A **98**(8) (2001) 4569–4574
5. Zhu, H., Bilgin, M., Bangham, R., et al.: Global Analysis of Protein Activities Using Proteome Chips. Science **293**(5537) (2001) 2101–2105
6. Gavin, A.C., Bosche, M., Krause, R., et al.: Functional Organization of the Yeast Proteome by Systematic Analysis of Protein Complexes. Nature **415**(6868) (2002) 141–147
7. Ho, Y., Gruhler, A., Heilbut, A., et al.: Systematic Identification of Protein Complexes in Saccharomyces Cerevisiae by Mass Spectrometry. Nature **415**(6868) (2002) 180–183
8. Mrowka, R., Patzak, A., Herzel, H.: Is There a Bias in Proteome Research? Genome Res **11**(12) (2001) 1971–1973
9. Huynen, M.A., Bork, P.: Measuring Genome Evolution. Proc Natl Acad Sci U S A **95**(11) (1998) 5849–5856
10. Pellegrini, M., Marcotte, E.M., Thompson, M.J., et al.: Assigning Protein Functions by Comparative Genome Analysis: Protein Phylogenetic Profiles. Proc Natl Acad Sci U S A **96**(8) (1999) 4285–4288
11. Enright, A.J., Iliopoulos, I., Kyrpides, N.C., et al.: Protein Interaction Maps for Complete Genomes Based on Gene Fusion Events. Nature **402**(6757) (1999) 86–90
12. Marcotte, E.M., Pellegrini, M., Ng, H.L., et al.: Detecting Protein Function and Protein-Protein Interactions from Genome Sequences. Science **285**(5428) (1999) 751–753
13. Dandekar, T., Snel, B., Huynen, M., et al.: Conservation of Gene Order: A Fingerprint of Proteins that Physically Interact. Trends Biochem Sci **23**(9) (1998) 324–328
14. Overbeek, R., Fonstein, M., D'Souza, M., et al.: The Use of Gene Clusters to Infer Functional Coupling. Proc Natl Acad Sci U S A **96**(6) (1999) 2896–2901
15. Wojcik, J., Schachter, V.: Protein-Protein Interaction Map Inference Using Interacting Domain Profile Pairs. Bioinformatics **17 Suppl 1** (2001) S296–S305
16. Deng, M., Mehta, S., Sun, F., et al.: Inferring Domain-Domain Interactions from Protein-Protein Interactions. Genome Res **12**(10) (2002) 1540–1548
17. Kim, W.K., Park, J., Suh, J.K.: Large Scale Statistical Prediction of Protein-Protein Interaction by Potentially Interacting Domain (pid) Pair. Genome Inform Ser Workshop Genome Inform **13** (2002) 42–50
18. Bock, J.R., Gough, D.A.: Whole-Proteome Interaction Mining. Bioinformatics **19**(1) (2003) 125–134
19. Gomez, S.M., Rzhetsky, A.: Towards the Prediction of Complete Protein-Protein Interaction Networks. Pac Symp Biocomput (2002) 413–424
20. Gomez, S.M., Noble, W.S., Rzhetsky, A.: Learning to Predict Protein-Protein Interactions from Protein Sequences. Bioinformatics **19**(15) (2003) 1875–1881

21. Han, D.S., Kim, H.S., Jang, W.H., Lee, S.D., Suh, J.K.: PreSPI: A Domain Combination Based Prediction System for Protein-Protein Interaction. Nucleic Acids Res **32**(21) (2004) 6312–6320
22. Hayashida, M., Ueda, N., Akutsu, T.: Inferring Strengths of Protein-Protein Interactions from Experimental Data Using Linear Programming. Bioinformatics **19 Suppl 2** (2003) II58–II65
23. Ng, S.K., Zhang, Z., Tan, S.H.: Integrative Approach for Computationally Inferring Protein Domain Interactions. Bioinformatics **19**(8) (2003) 923–929
24. Chen, X.W., Liu, M.: Prediction of Protein-Protein Interactions Using Random Decision Forest Framework. Bioinformatics **21**(24) (2005) 4394–4400
25. Espadaler, J., Romero-Isart, O., Jackson, R.M., Oliva, B.: Prediction of Protein-Protein Interactions Using Distant Conservation of Sequence Patterns and Structure Relationships. Bioinformatics **21**(16) (2005) 3360–3368
26. Liu, Y., Liu, N., Zhao, H.: Inferring Protein-Protein Interactions through High-Throughput Interaction Data from Diverse Organisms. Bioinformatics **21**(15) (2005) 3279–3285
27. Nye, T.M., Berzuini, C., Gilks, W.R., Babu, M.M., Teichmann, S.A.: Statistical Analysis of Domains in Interacting Protein Pairs. Bioinformatics **21**(7) (2005) 993–1001
28. Riley, R., Lee, C., Sabatti, C., Eisenberg, D.: Inferring Protein Domain Interactions from Databases of Interacting Proteins. Genome Biol **6**(10) (2005) R89
29. Lehrach, W.P., Husmeier, D., Williams, C.K.: A Regularized Discriminative Model for the Prediction of Protein-Peptide Interactions. Bioinformatics **22**(5) (2006) 532–540
30. Sprinzak, E., Margalit, H.: Correlated Sequence-Signatures as Markers of Protein-Protein Interaction. J Mol Biol **311**(4) (2001) 681–692
31. Wang, H., Segal, E., Ben-Hur, A., et al.: Identifying Protein-Protein Interaction Sites on a Genome-Wide Scale. Advances in Neural Information Processing Systems 17, MIT Press (2005) 1465–1472
32. Fang, J., Haasl, R.J., Dong, Y., Lushington, G.H.: Discover Protein Sequence Signatures from Protein-Protein Interaction Data. BMC Bioinformatics **6**(1) (2005) 277
33. Falquet, L., Pagni, M., Bucher, P., et al.: The PROSITE Database, its Status in 2002. Nucleic Acids Res **30**(1) (2002) 235–238
34. Yu, H., Qian, M., Deng, M.: Understanding Protein-Protein Interactions: From Domain Level to Motif Level. Proceeding of Sino-Germany Conference: Network, From Biology to Theory, Springer-Verlag (2005) in press
35. Freund, Y., Schapire, R.E.: A Decision-Theoretic Generalization of On-Line Learning and an Application to Boosting. Journal of Computer and System Sciences **55**(1) (1997) 119–139
36. Salwinski, L., Miller, C.S., Smith, A.J., et al.: The Database of Interacting Proteins: 2004 Update. Nucleic Acids Res **32**(Database issue) (2004) D449–D451
37. Jansen, R., Gerstein, M.: Analyzing Protein Function on a Genomic Scale: The Importance of Gold-Standard Positives and Negatives for Network Prediction. Curr Opin Microbiol **7**(5) (2004) 535–545
38. Deshpande, N., Addess, K.J., Bluhm, W.F., et al.: The RCSB Protein Data Bank: A Redesigned Query System and Relational Database Based on the mmCIF Schema. Nucleic Acids Res **33**(Database issue) (2005) D233–D237

39. Taylor, W.R., Jones, D.T.: Deriving an Amino Acid Distance Matrix. J Theor Biol **164**(1) (1993) 65–83
40. Littlestone, N.: Learning Quickly when Irrelevant Attributes Abound: A New Linear-Threshold Algorithm. Machine Learning **2**(4) (1988) 285–318

Appendix: Methods for Comparison

In the Appendix, we described two methods for comparison. The first is Naive Bayes Model, to be a benchmark for protein-protein interaction prediction accuracy. The second is χ^2 statistical feature selection, which is widely used based on the philosophy of that statistical significance leads to biological significance.

4.1 Naive Bayes Model

Naive Bayes model is based on so-called Bayesian theorem and the independent assumption of features. Given the same training data described in the text. The posterior probability of a new sample w in class z is

$$P(z|w) \propto p(w|z)p(z) = p(z)\prod_{i=1}^{M} w_i P(w_i|z), \tag{1}$$

where $P(w_i|z)$ can be easily estimated by simply counting the feature in training data and $p(z)$ is the prior probability estimated by proportion of class z in training data.

w is classified as positive sample if and only if $P(1|w)/P(0|w) > c$, where c is a given constant threshold.

4.2 χ^2 Feature Selection

For each feature F we calculate the statistic with Yates' correction of the 2×2 contingency table test as follows:

	Positive sample	Negative sample
#(F occurrence)	a	c
#(All other feature occurrences)	b	d

$$K = \frac{N(|ad - bc| - N/2)^2}{(a+b)(c+d)(a+c)(b+d))} \sim \chi^2 \tag{2}$$

where N is the sum of a, b, c, and d; a and b are the occurrences of a given feature and other features than the given feature (non-feature) in the foreground set, respectively; c and d are the same in the background set. K approximately follows χ^2 distribution.

Larger K indicates more significant systematic association between the feature occurrence and the sample label.

The most significant positive occurrence features are selected as statistical significant features.

Web Service for Predicting Interacting Proteins and Application to Human and HIV-1 Proteins

Byungkyu Park and Kyungsook Han*

School of Computer Science and Engineering, Inha University, Inchon 402-751, Korea
bpark@inhaian.net, khan@inha.ac.kr

Abstract. Finding all the proteins that potentially interact with a query protein is useful when studying a biological mechanism, but it is much more difficult than finding interactions between a set of given proteins. This is because it involves intensive computation to search the relevant data in databases, and different databases have different accession numbers and names for the same protein. Recently a few servers have been developed for finding interactions between user-specified proteins, but none of these can find all the potentially interacting proteins for a given protein, including the former version of our prediction server, HPID version 1.0. This paper describes a new online prediction system, HPID 2.0 (http://www.hpid.org), for finding partners interacting with a query protein as well as for finding interactions between query proteins. We applied the new system to predicting the interactions between the entire human proteins, and to identifying human proteins interacting with HIV-1 proteins (http://hiv1.hpid.org). We believe that this is the first online server for predicting proteins interacting with a given protein, and that it will be a useful resource for studying protein-protein interactions.

1 Introduction

Studying a biological mechanism or function often requires a small-scale network of protein-protein interactions related to the mechanism or function rather than a genome-wide network of protein-protein interactions. However, small-scale subnetworks of the protein-protein interactions of interest are not readily available from databases. This is partly because computing all possible subnetworks involves prohibitive computation of searching data in relevant databases, and partly because different databases have different accession numbers and names for a same protein.

We have previously developed a database called the Human Protein Interaction Database (HPID; http://www.hpid.org) for predicting potential interactions between proteins submitted by users, as well as for providing human protein interaction information pre-computed from existing structural and experimental data [1]. However, the previous version of HPID, HPID version 1.0, is not capable of finding potentially interacting partners for a given protein. Finding all the proteins potentially interacting with a protein of interest is useful when studying a biological mechanism,

* Correspondence author.

D.-S. Huang, K. Li, and G.W. Irwin (Eds.): ICIC 2006, LNBI 4115, pp. 631 – 640, 2006.

but much more difficult than finding interactions between given proteins. Recently a few servers have been developed for finding interactions between a set of given proteins [2, 3], but none of these can search relevant databases and find potentially interacting proteins with a protein given by a user.

In an attempt to improve the capability of the online prediction system as well as the reliability of the predicted interactions, we developed a new online prediction system (HPID 2.0; http://www.hpid.org) using several types of information: protein domain, protein function, and sub-cellular localization. We used the new online prediction system to predict human proteins potentially interacting with a human immunodeficiency virus type 1 (HIV-1) protein. The results were built into a web-based system (http://hiv1.hpid.org) for analyzing and visualizing the large-scale interactions between HIV-1 and human proteins and the comparative analysis of the interactions. HPID 2.0 is the first online system for predicting proteins interacting with a given protein and will be a useful resource for studying protein-protein interactions.

2 Methods of Predicting Interacting Proteins

To predict interacting partners of a protein of interest, the user should first find the superfamilies of the protein using any of the Superfamily (http://supfam.mrc-lmb.cam.ac.uk/SUPERFAMILY), InterPro (http://www.ebi.ac.uk/InterProScan) or Pfam (http://www.sanger.ac.uk) databases. The HPID online prediction system then takes one or more superfamilies of the protein and its protein name as input. The reason for taking the superfamilies as input instead of the amino acid sequence is partly because the assignment of superfamilies by these databases is reliable and partly because some proteins have no superfamily or assigning them a superfamily takes too long. The HPID online prediction system cannot predict interacting partners for a protein with no superfamily.

Protein interactions at the superfamily level were predicted from the Protein Structural Interactome MAP (PSIMAP) [4]. PSIMAP was constructed by extracting all the structural domain–domain interactions from PDB. It contains 37,387 interacting domain pairs that have five or more contacts within 5Å and these domain pairs were grouped into 2,171 protein family–family pairs using the SCOP family definition [4]. Unlike HPID 1.0, HPID 2.0 is not dependent on protein IDs to find interacting proteins. Interacting partners for a query protein are found by homology search of protein sequences in relevant databases, which currently include Ensembl (http://www.ensembl.org), BIND (http://bind.ca), DIP (http://dip.doe-mbi.ucla.edu), HPRD (http://www.hprd.org), and NCBI (http://www.ncbi.nlm.nih.gov).

We used this method to find potential interaction partners with a human protein or human immunodeficiency virus type 1 (HIV-1) protein. Currently the search for potential interaction partners is limited to human proteins, but should be extendable to other organisms.

We developed a web service using Microsoft C# language and .NET framework. We used XML (eXtensible Markup Language) schema and the Simple Object Access

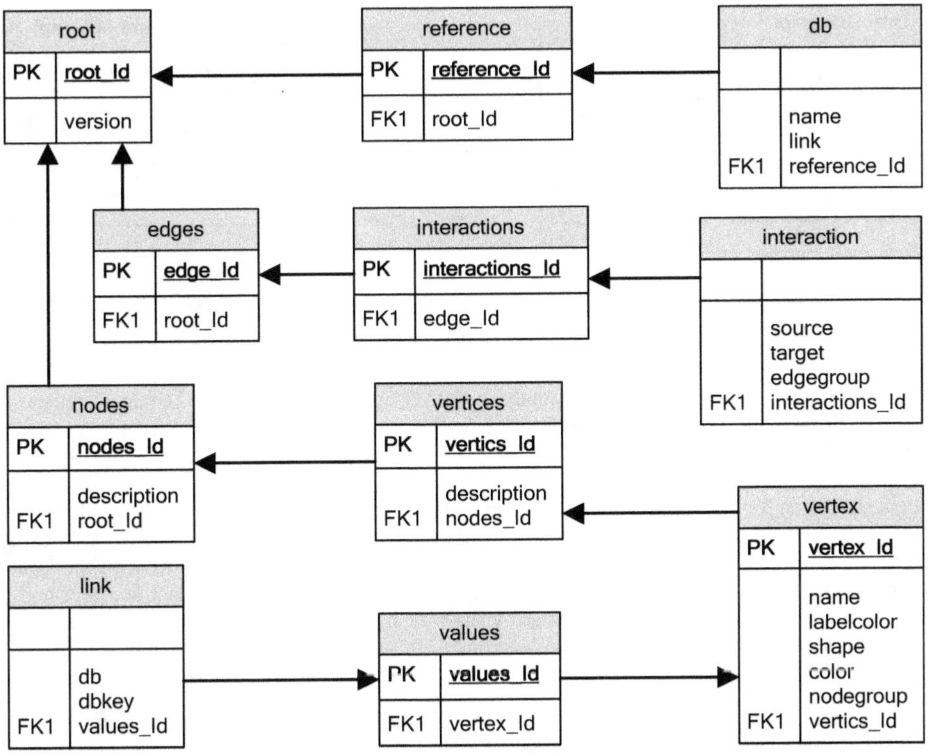

Fig. 1. Structure of XML schema. The 'root' element of the XML schema is an identification of a protein-protein interaction network. The 'reference' element has the addresses of the databases that have information about the proteins in the network. The user can access the information through the addresses. The 'nodes' element represents the node type such as protein, protein structure, protein function, and gene. Different node types can be displayed in different shapes and colors. The 'link' and 'db' elements make the user access protein information from many databases. The 'edges' element displays protein-protein interactions.

Protocol (SOAP) version 1.1 of W3C [5] for a data structure and a communication protocol. XML Schemas express shared vocabularies and allow machines to carry out rules made by people [6]. Fig. 1 shows the high-level structure of our XML schema (for the full XML schema see http://hiv1.hpid.org/GraphDataSet.xsd). SOAP is a lightweight protocol for exchange of information in a decentralized, distributed environment [7]. We used the Web Service Description Language (WSDL) for XML Remote Procedure Call (XML-RPC). WSDL is an XML format for describing network services as a set of endpoints operating on messages containing either document-oriented or procedure-oriented information [8]. Fig. 2 shows how to access the interactions between HIV-1 and human proteins from a local computer of a user. An application in the user computer can be either any web browser, WebInterViewer [9, 10], or any application program developed by the user.

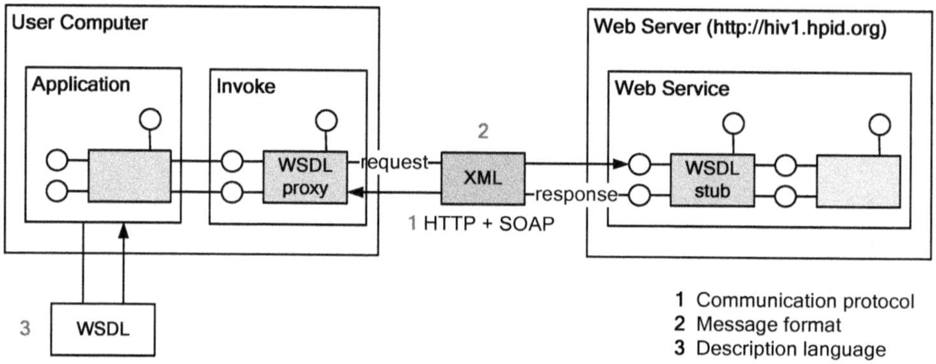

Fig. 2. The web service of the HIV1 database (http:// hiv1.hpid.org). When the user requests to explore protein interactions, a web browser or the standalone version of WebInterViewer is executed in the user computer. When the user clicks a hyperlink of an interesting HIV-1 gene, a request of a web service is sent to the web server. There are two methods of responding to the request. One is for the application of the user computer to get XML data directly, and the other is to instantiate the WDSL proxy object in the user computer and then to let the instance get XML data. WebInterViewer uses the first method. For the interacting proteins with the HIV-1 gene specified by the user, WebInterViewer gets results in the XML schema, as described in Fig. 1. Finally, the interacting proteins are visualized by WebInterViewer. The other method is for an application program that is to analyze protein interactions.

The user can generate a network of publishable quality from the analysis results using WebInterViewer. If the user wants to embed the Active-X control of WebInterViewer in the user's web server, he or she can include the following code in the web page at http://hiv1.hpid.org/usage.htm.

```
<script language="javascript">
function View(xml)    // xml: URL or Web Service
{
   document.IVAx1.Options("VL,NS2");
   document.IVAx1.ConnectPIEx2("HIV-1", xml);
}
</script>
<a
href="JavaScript:View('http://hiv1.hpid.org/data/hiv1.xml')">tes
t1</a>
<OBJECT id="IVAx1"
codeBase="http://interviewer.inha.ac.kr/download/IVF_DiF_Ax.cab#
version=1,1,6,3"
classid="clsid:D8D13FEA-2AD6-42AC-B3D5-17AFC6C55907"
name="IVAx1"></OBJECT>
```

If the user wants to run the standard alone version of WebInterViewer, he or she can use the following code.

```
<a
href="javascript:document.wivDownloader.runIV();javascript:docum
ent.wivDownloader.ConnectPIEx2('HIV-1',
'http://hiv1.hpid.org/data/hiv1.xml')">test2</a>
```

```
<OBJECT
codeBase="http://interviewer.inha.ac.kr/Download/IVF_DiF_Down.ca
b#version=1,2,1,4" height="0" width="0" classid="clsid:4BB61BDA-
DADE-4F94-ADB6-7EA1E8962613"
name="wivDownloader" VIEWASTEXT ID="wivDownloader"></OBJECT>
```

3 Results and Discussion

We built a web-based server for predicting interacting proteins and for visualizing protein interaction networks (Fig. 3). The web services provided by the server and their methods and parameters are summarized in Tables 1 and 2, respectively. The basic process of the web service is exchanging SOAP messages described in XML. When the server of web services receives the SOAP request with the parameters, the server returns the SOAP message in response to the method. Different application programs developed by various languages can use the web services, whose results can be a request message for another web service. Example XML files for the responses to the HTTP GET request can be found at the following URL addresses.

1. `http://hiv1.hpid.org/WebServices/WebInterViewer.asmx/GetInteractions?nodeID=capsid&gene=gag`
2. `http://onlineprediction.hpid.org/WebService/psimap.asmx/DoPredict?Protein=test&Superfamily=a.1.1`
3. `http://hiv1.hpid.org/WebServices/pubmed.asmx/GetPubMed?uids=10074203`

Table 1. Web services provided by our system

Web Services	Web Service Description Language (WSDL)
WebInterViewer	http://HIV1.HPID.org/WebServices/WebInterViewer.asmx?WSDL
HPID Online prediction	http://Onlineprediction.hpid.org/WebService/psimap.asmx?WSDL
NCBI PubMed service	http://HIV1.HPID.org/WebServices/PubMed.asmx?WSDL

Table 2. Methods and parameters of the web services

Web Services	Methods	Parameters
WebInterViewer	GetInteractions	nodeID: HIV-1 protein name gene: HIV-1 gene name
HPID Online prediction	DoPredict	Protein: user's protein name Superfamily: SCOP superfamily
NCBI PubMed service	GetPubMed	uids: PubMed ID

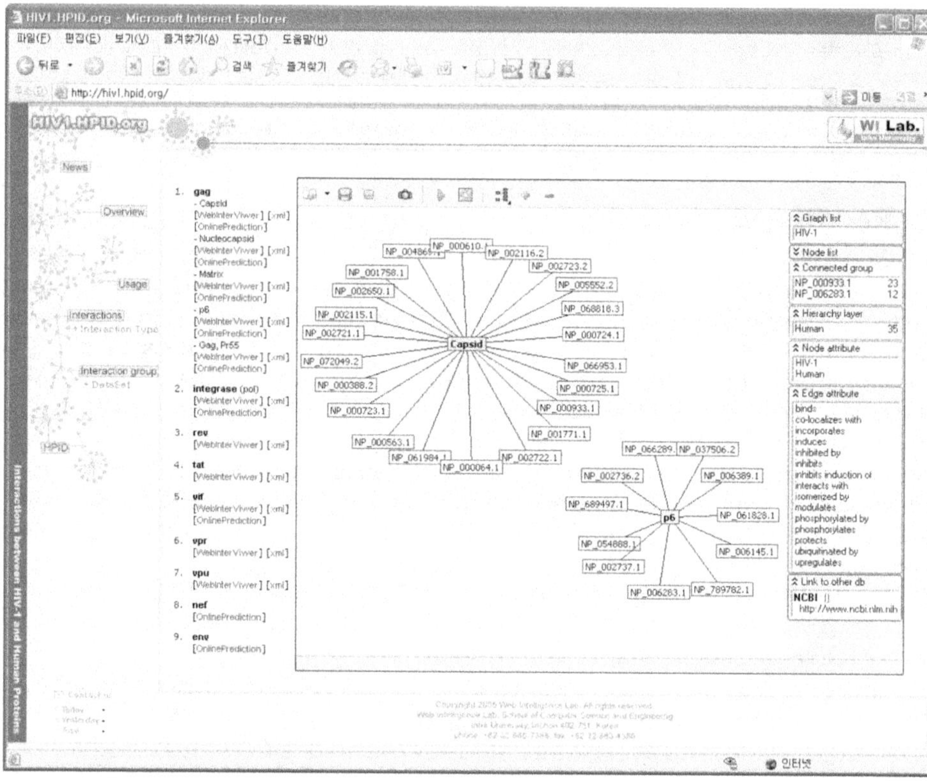

Fig. 3. The user interface of the HIV-1 database. The 'OnlinePrediction' button each HIV-1 protein lets the user predict potential human proteins interacting with the HIV-1 protein. When the user clicks the 'WebInterViewer' button below an HIV-1 protein, all human proteins interacting with it are listed. Several interactions networks can be compared and analyzed in the same window.

Our HIV1 database (http://hiv1.hpid.org) can be used for analyzing and visualizing the large-scale interactions between HIV-1 and human proteins and for the comparative analysis of the interactions. Relevant articles to any human or HIV-1 protein in the interaction networks, which are archived in PubMed (http://www.pubmed.gov), are also linked from the NCBI_Accession in the 'Link to other db' menu of our system (Fig. 4). Detailed explanation is available in the usage section of the web page.

The rest of this section gives an example of predicting human proteins potentially interacting with the HIV-1 *capsid* protein. The online prediction server finds 208 human proteins from Ensembl and 147 human proteins from NCBI as the potential interaction partners with the HIV-1 *capsid* protein. As a supporting basis of the prediction, the prediction report shown in Fig. 5A lists two superfamilies, a.28.3 and a.73.1, assigned to the HIV-1 *capsid* protein. The superfamily a.28.3 has two interacting superfamilies, a.73.1 and b.1.1, and the superfamily a.73.1 has five interacting superfamilies, a.28.3, a.61.1, a.73.1, b.1.1, and b.62.1. If the user clicks on one of these interacting superfamilies, he or she can see the interacting human proteins in the superfamily together with their reliability scores. Fig. 5B shows the network of human proteins interacting with HIV-1 *capsid* protein, visualized by WebInterViewer. It is a star-shaped network, centered on the HIV-1 *capsid* protein (yellow node in the network). The interaction network can be analyzed interactively using WebInterViewer and saved in the XML format for later analysis.

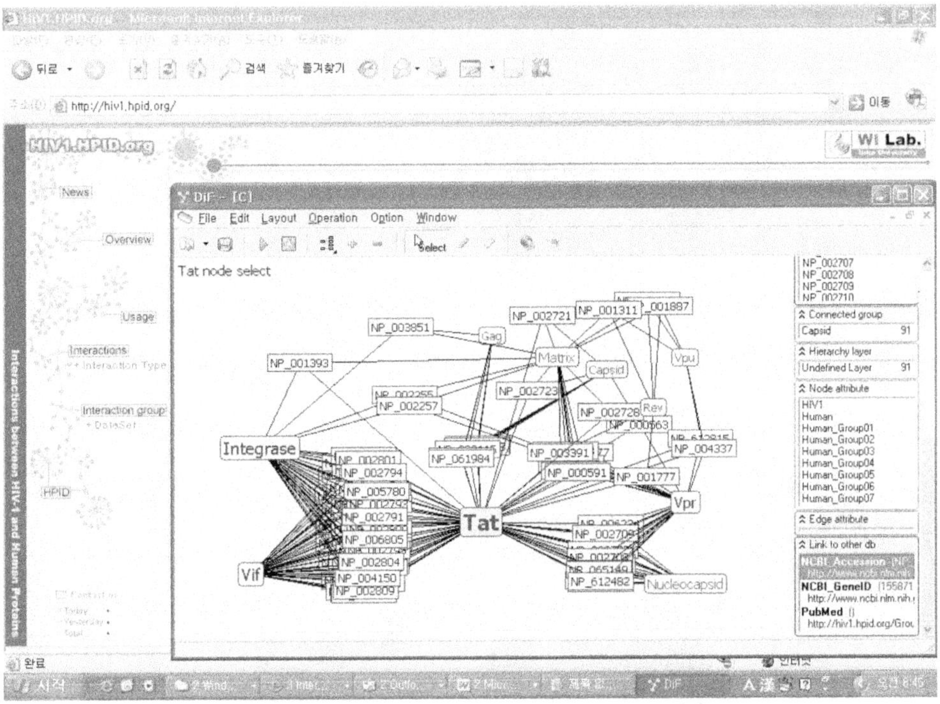

Fig. 4. Example of accessing relevant articles to a protein in the interaction network. When the user clicks a protein (yellow node) in the network and double clicks the NCBI_Accession of the protein in the 'Link to other db' menu (lower right pane), all relevant articles are displayed.

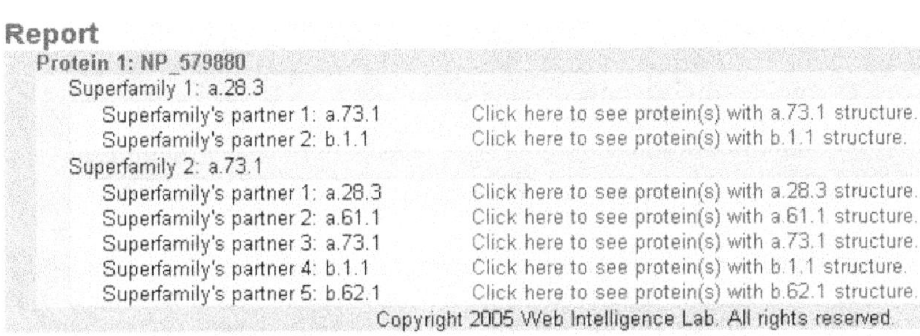

(A)

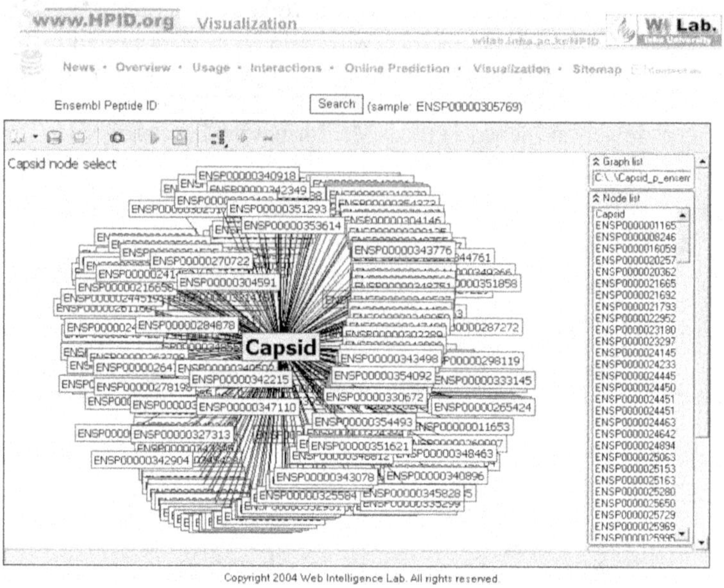

(B)

Fig. 5. (A) The prediction report for the HIV-1 *capsid* protein (NCBI accession number: NP_579880). (B) A network of 208 human proteins potentially interacting with the HIV-1 *capsid* protein, visualized by WebInterViewer.

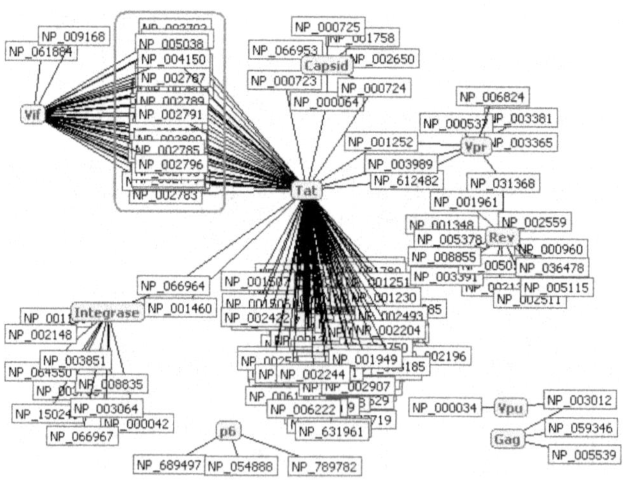

Fig. 6. Interaction network of HIV-1 and human proteins, consisting of interactions of "interact with" type only. The network shows 98 interactions between 11 HIV-1 proteins and 49 human proteins. The human proteins shared by the *vif* and *tat* genes (enclosed in a red box in the network) are proteasome (prosome, macropain) subunits of a highly ordered ring-shaped 20S core structure.

We also used the HIV-1 database for the comparative analysis of the experimental data on the interactions between HIV-1 and human proteins. The experimental data was extracted from NCBI at http://www.ncbi.nlm.nih.gov/RefSeq/HIVInteractions. The whole interaction network contains 1,768 interactions of 65 different types with 810 human proteins. The comparative analysis identified several interesting interaction patterns from the topological analysis of the interaction networks. For example, the network shown in Fig. 6 shows an interaction network of HIV-1 and human proteins, consisting of interactions of "interact with" type only. The network has 98 interactions between 11 HIV-1 proteins and 49 human proteins. The human proteins interacting with both the *vif* and *tat* genes (enclosed in a red box in the network) have only the *vif* and *tat* genes as their interacting HIV-1 partners. The human proteins are proteasome (prosome, macropain) subunits of a highly ordered ring-shaped 20S core structure.

4 Conclusion

Small-scale interaction networks relevant to a protein of interest are not readily available from databases partly because computing all possible subnetworks involves prohibitive computation of searching data in the databases and because different databases have different accession numbers and names for a same protein. We constructed an online prediction system called HPID 2.0 (http://www.hpid.org) for searching for interaction partners of a protein given by a user. The online prediction system was also used to predict human proteins potentially interacting with HIV-1 proteins. The prediction results and experimental data were built into a web-based system (http://hiv1.hpid.org) for analyzing and visualizing the large-scale interactions between HIV-1 and human proteins and the comparative analysis of the interactions. We believe that this is the first online system for the comparative analysis of the interactions networks of HIV-1 and human proteins and that it is a valuable tool for scientists in the field of protein-protein interactions and HIV/AIDS research.

Acknowledgements

This work was supported by LG Yonam Foundation and KOSEF through the Systems Bio-Dynamics Research Center.

References

1. Han, K., Park, B., Kim, H., Hong, J., Park, J.: HPID: The Human Protein Interaction Database. Bioinformatics 20 (2004) 2466 – 2470
2. Goffard, N., Garcia, V., Iragne, F., Groppi, A., de Daruvar, A.: IPPRED: server for proteins interactions inference. Bioinformatics 19 (2003) 903-904
3. Brown, K.R., Jurisica, I.: Online Predicted Human Interaction Database. Bioinformatics 21 (2005) 2076 – 2082

4. Kim, W.K., Bolser, D.M., Park, J.H.: Large-scale Co-Evolution Analysis of Protein Structural Interlogues Using the Global Protein Structural Interactome Map (PSIMAP). Bioinformatics 20 (2004) 1138-1150
5. http://www.w3.org
6. http://www.w3.org/XML/Schema
7. http://www.w3.org/TR/2000/NOTE-SOAP-20000508
8. http://www.w3.org/TR/wsdl
9. Han, K., Ju, B., Jung, H.: WebInterViewer: Integrated Framework for Visualizing and Analyzing Molecular Interaction Networks. Nucl. Acids Res. 32 (2004) W89-W95
10. Ju, B.-H., Han, K.: Complexity Management in Visualizing Protein Interaction Networks. Bioinformatics 19 (2003) i177–i179

An Immunity-Based Dynamic Multilayer Intrusion Detection System

Gang Liang, Tao Li, Jiancheng Ni, Yaping Jiang, Jin Yang, and Xun Gong

Department of Computer Science, Sichuan University, Chengdu, China
Gangliang56@163.com

Abstract. A real computer network produces new network traffic continuously in real time, thus the normal behaviors of network traffic are different in different time, but the self set of current network detection systems based on immunity are static. If the network environments change, the false negative rates and false positive rates will increase rapidly. So the traditional method can not adapt to changing network environments. In order to get over the limitation of the traditional means, an immunity-based dynamic intrusion detection system is proposed in this paper. In the proposed model, a dynamic renewal process of self set is described in detail. In addition, we establish a new set to improve the detection precision and shorten the training phase by adding the characters of the current known attacks to memory cell set. The experimental results show that the new model not only reduces the false negative rates and false positive rates effectively but also has the feature to adapt to continuous changing network environments.

1 Introduction

The growing numbers of machines connect to Internet and the ever increasing dependency of our world on network-based computer systems has raised the network security concerns to a crucial stake. Meanwhile, the internet revolution has been accompanied with its share of crimes and malicious activities. The continuous increase in computer crimes suggests that the traditional security systems are no longer effective to protect today's computer infrastructure [1]. Intrusion detection systems (IDSs) have been developed for that matter and are widely deployed.

An IDS is an automated system for the detection of computer system intrusions using audit trails provided by operating systems or network monitoring tools. The goal of an IDS is not only helps the administrators to detect intrusions and limit damages, but also helps to identify the source of attacks, which sometimes acts as a deterrent especially in case of insider attacks [2]. Intrusion detection functions include:

- Monitoring and analyzing both user and system activities
- Analyzing system configurations and vulnerabilities
- Assessing system and file integrity
- Ability to recognize patterns typical of attacks
- Analysis of abnormal activity patterns
- Tracking user policy violations

D.-S. Huang, K. Li, and G.W. Irwin (Eds.): ICIC 2006, LNBI 4115, pp. 641 – 650, 2006.

IDSs mainly employ two detection techniques: anomaly detection and misuse detection. The anomaly detection approach establishes the profiles of normal actives of users, systems or system resources, network traffic and services using the audit trails generated by a host operating system or a network-scanning program [3]. The misuse detection approach defines suspicious misuse signatures based on known system vulnerabilities and security policy [4]. The IDSs base on anomaly detection detect intrusion by identifying signification deviation from the normal behavior patterns of profiles, and the IDSs based on misuse detection probe whether these misuse signatures present or not in the auditing trails. The advantage of misuse detection approach is its low rate of false errors, since signatures are specific to suspicious activities. However, misuse detection approach fail to detect any attack for which they were not provided with signature, so new breeds of attacks, can go undetected in this way [5]. The strength of the anomaly detection approach is that a prior knowledge of the security flaws of the target system is not required. Thus it is able to detect not only known intrusions but also unknown intrusions, but the false negative error of this approach is very high. Because each technique has different strengths and drawbacks, the intrusion detection system should employ both anomaly diction system and misuse detection system in parallel. These two components should be reciprocal in an IDS. But currently, most of IDSs only carry out misuse detection.

Human immune systems have been successful at protecting the human body against a vast variety of foreign pathogens or organisms. One most interesting feature of the human immune system is the discriminative power to detect harmful pathogens without attacking human body self cells. The human immune system distinguishes previously known and unknown pathogens from human body self cells via its own evolutionary mechanism, which is similar to evolution of organisms in nature. The human immune system has a multi-layered architecture. It consists of passive layers such as the skin,mucus membranes, pH,temperature and generalised inflammatory responses, and adaptive layers including both the humoral (B cell) and cellular (T cell) mechanisms [4] [5] [6]. The self recognition of immune system (IS), alone with the dis4tributed, self-organized and lightweight nature of the mechanisms by which it achieves this protection, has in recent years made it the focus of increased interest within the computer science and intrusion detection communities. A number of artificial immune systems (AIS's) have been built for intrusion detection. In 1997, Forrest proposed the application of computer immune systems to computer security for the first time. In 1999, Kim and Bentley have identified the features of human immune system that can contribute to the development of effective network-based IDS. In 2000, Hofmeyer and Forrest develop an AIS for network intrusion detection, called LYSIS [6]. Different work inspired by LYSIS has recently been report in [7]. But there are two problems in most of these works. The first problem is that most systems only adopt to anomaly detection and the FP error rates of these systems are very high. Another problem is that these systems carry out static detection, and the self and detector libraries become very large and can not adapt to dynamic changing network environment [7].

In order to overcome the limitations, an immune-based multilayer model for the dynamic intrusion detection was proposed in this paper. A dynamic renewal process of self et is described in detail. In addition, we establish a new set to improve the detection precision and shorten the training phase by adding signature detection layer.

The remainder of the paper is organized as follows. In Section 2, we establish an immune-based multilayer model for the dynamic intrusion detection. Then in Section3, the simulations and experimental results are provided. Finally, Sections 4 gives the summary and conclusions.

2 Proposed Model

The IS makes use of both misuse detection and anomaly detection. It has mechanisms for detecting deviations from a set of normal patterns, and it has way of storing and recalling specific patterns associated with previous pathogenic attacks. The traditional immune IDSs only adopt anomaly detection and ignore misuse detection, so the false negative error and false positive error all are very high of these systems. In addition, the traditional systems carry out static detection, and the self and detector libraries become very large and can not adapt to dynamic changing network environment

The aim of the framework presented in this section is to develop a dynamic multilayered detection system that can be used to create a population of individuals which are able to identify a variety of known and unknown anomalous behavior to overcome the limitations of the traditional systems.

2.1 Model Overview

The immune-based dynamic multilayer IDS model consists of signature detection layer and adaptive detection layer. When the system starts .the antigens (IP packets captured from the monitored network) are fed into system. If one antigen is recognized as malicious behavior by either detection layer, alarming messages and some messages about this malicious behavior will be send to network administrators to help them take corresponding measures. There is a subtle difference between the signature detection layer and adaptive detection layer in running mechanism. In adaptive detection layer, if one antigen is recognized as malicious behavior by the elements of memory detector set, it is send alarming messages soon like the signature detection layer, else if an attack is recognized by the elements of the mature detector set , it just increase the value of the count filed of the corresponding mature detectors. When a mature detector's count field is over predefined threshold, model will send a signal to network administrator. If administrators give an affirmative signal which we call it co-simulation signal in fix period, the mature detector will be become memory detector and the character field will be added into the gene library to update the gene library. Otherwise, system will consider this mature detector as an invalid detector and will delete it from mature detector set. During the detection course, if the monitored network adds or uninstall some network services, the corresponding signature of these services will also be added or deleted from the self library and all the memory detectors must be computed the affinity to the elements of the self library. If the affinity of any memory detectors is over threshold, this memory detector will be removed.

The overall architecture of our proposed model is presented in Fig.1.

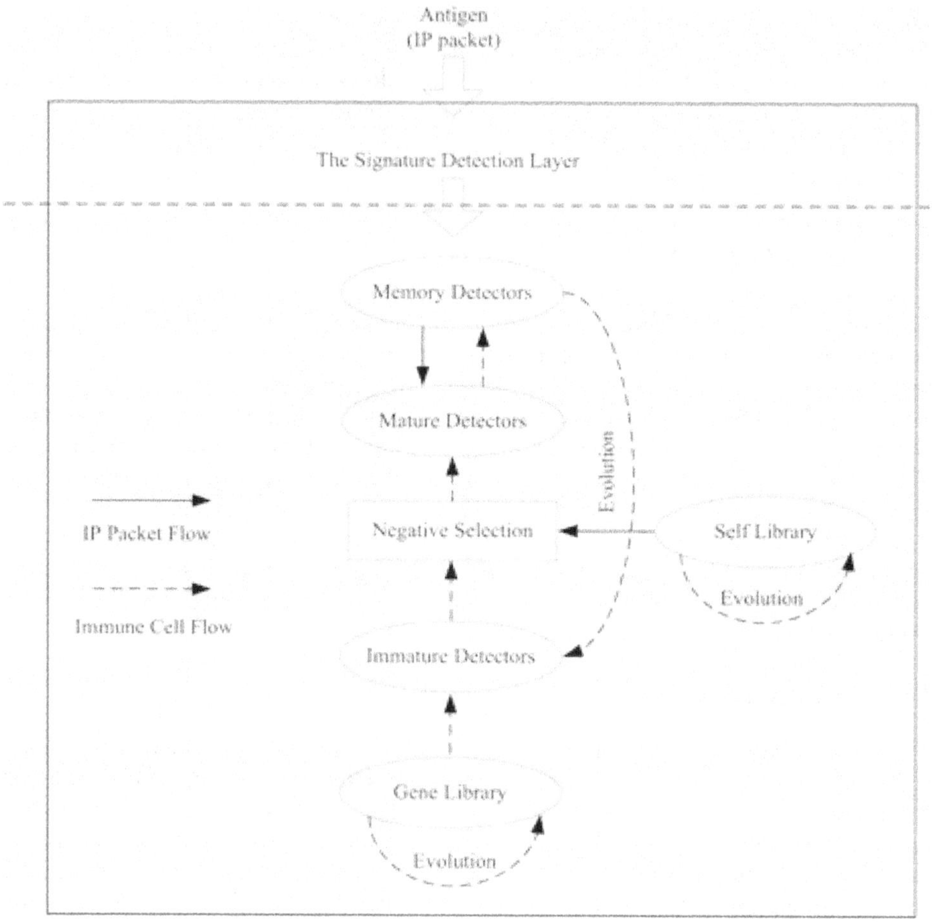

Fig. 1. Architecture of Immunity-Based Dynamic Multilayer Intrusion Detection Model

2.2 Signature Defense Layer

Definition 1 Antigen Self and Nonself : Antigens (Ag, $Ag \subset D$, $D = \{0,1\}^l$) in our approach are fixed-length (l=56) binary strings extracted from the Internet Protocol (IP) packets transferred in the network. An antigen consists of the source and destination IP addresses, port number, protocol type, IP flags, IP overall packet length, TCP/UDP/ICMP fields, etc. Self is defined as normal sanctioned network service transactions and harmless background clutter, while Nonself represent IP packets from a computer network attack, and Self ∪ Nonself = Ag.

In the course of foetus forming, the immunoglobulin G can enter into the body of foetus by placenta or colostrum to construct the passive immune ability and this passive immune ability can make foetus get the homologous immune ability to his matrix's. Similar to innate immune system of IS, signature defense layer adopts to misuse detection.

Similar to innate immune system of IS, signature defense layer adopts to misuse detection. There is only existed one type detector---memory cell I in this layer. The memory cell I is represented as a set of binary string feature masks which are highly specific to known type malicious behavior. The antigens firstly compare to the cells of memory cell I. If an antigen match to any cells of memory cell I, alarming messages and assistant messages will be send to network administrators. Or else the antigens will be fed to adaptive detection layer.

The network administrators can add new known attack's signatures, arranged follow the format of memory cell (described in section 2.3) to the memory cell I set to update memory cell I.

The aim we introduce signature defense layer is to decrease the false negative error and false positive error and detect familiar attacks. Experimental results show we can not only decrease the false errors but also increase system efficiency by introducing signature defense layer

2.3 Adaptive Defense Layer

As opposed to the innate immune system which is nonspecific in its defence against harmful pathogens, the adaptive immune system initiates a response specific to the pathogen that has entered the body. The adaptive immune system also provides memory capabilities to the immune system [9].We can make use of adaptive detection layer to detect novel and complex unknown intrusions.

Definition 2 Mature Detector and Memory Detector : mature detector, which is not match self antigen, start monitoring new antigens. Each mature detector attempts to bind to new antigen when a mature detector matches any new antigen and it does not immediately regard the detected antigen as nonself. Memory detector is activated mature detector, in contrast to mature detectors, memory detectors immediately regard the detected antigen as nonself when they match any single antigen and they have longer life span than that of mature detectors.

There are two mainly phases in the adaptive detection layer. The first is training phase. This involves the generation of self and detector sets. Before the system start to work, we must study the monitored network behaviors and extract the characters from these behaviors to form self set. Unlike the traditional model [10-11], the self elements changes dynamically. This is because the network behaviors are changes continuously. Some network activities, which were regarded as normal behavior before, are forbidden after the bugs are fixed. Furthermore, the network may provide more services or uninstall some services. So, the elements also need to change to keep up with the changing of the network behavior. Equation (1) depicts the dynamic evolution of self, where $Self_{new}$ is the set of newly defined self elements at time t, and $Self_{outdated}$ is the set of outdated self elements representing current abnormal behaviors.

$$Self(t) = \begin{cases} initial \quad Self \quad set & t = 0 \\ \\ Self(t-1) \cup Self_{new} - Self_{\text{outdated}} & t >= 1 \end{cases} . \tag{1}$$

Equation (1) depicts the dynamic evolution of self, where $Self_{new}$ is the set of newly defined self elements at time t, and $Self_{outdated}$ is the set of outdated self elements representing current abnormal behaviors.

Immature detector is created with a randomly generated receptor string that has the same number of fields as the self string, and remains immature for a certain period of time. During this maturation period, the detector is compared to every element of self set, and if the detector matches any elements, it "dies" and is replaced by a detector with a new, randomly generated receptor string. This is analogous to negative selection of thymocytes in the thymus. If the detector survives, the maturation period without matching anything, it becomes mature, and future matches may indicate that a potentially dangerous network connection pattern was detected. However, if a mature detector stays in a mature detector population for a long period, it is highly likely that the match count of a mature detector will eventually meet the activation threshold and it becomes a memory detector. In contrast to mature detectors, memory detectors activate immediately when they match any single antigen, that is to say that the activation threshold of memory detectors is one generation. This is because memory detectors have matched true non-self antigens in the past, and thus any antigen detected by these detectors is considered as a definite non-self antigen without extra checking [12].

The match rule considered is based on hamming distance and is termed the hamming match. Two string a and b match under the hamming match rule if they have same bits in at least r positions.

$$D(a,b) = \sum_{i=1}^{L} \delta, \begin{cases} \delta = 1, & a_i \neq b_i \\ \delta = 0, & a_i = b_i \end{cases} . \quad (2)$$

Equation (2) depicts how to measure the hamming distance, where a, b are two matching strings, and i is the cursor which indicate the matching position.

$$F(ab, ag, r) = \begin{cases} 1, & D(ab,ag) \geq r \\ 0, & D(ab,ag) < r \end{cases} . \quad (3)$$

Equation (3) gives the hamming matching function.

When self changes, mature detectors or memory detectors should be compared to every element of $self_{new}$. If a detector matches any elements of $self_{new}$, it will be removed immediately.

The detectors in the detectors set have finite lifetimes and they are continually changing. This mechanism ensures that an attack can not repeatedly exploit the same gaps in the coverage and it also prevents the scaling problem [8]. In contrast to mature detectors, memory detectors have much extend life spans and have lower thresholds of activation.

The second phase is detection phase. In this course, memory detectors monitor antigen first when antigens are presented to the adaptive detection layer from signature detection layer. Because memory detectors remember nonself antigen patterns that are proven to be intrusions, it is expected that memory detectors detect real intrusions from presented antigen. Then, the antigen set is presented to mature detectors after being filtered by memory detectors. Although memory detectors can detect some intrusions, there are still exited some further intrusions that have never occurred before. Mature detectors were expected to detect these new intrusions. When a mature detector binds antigen, system sends a requirement signal to system administrator to ask for co-simulation signal. In fixed period, if system receives the affirmative co-stimulation signal, it will send alarm signal to user, other system only increases the match counts of the mature detector. If the accumulated mature detector match counts meet the activation threshold and they receives the affirmative co-stimulation signal from system administrator, this mature detector is active and become a memory detector. Meanwhile, system will clone and mutate this detector.

$$Num = \xi(1 - conc)\,. \tag{4}$$

Equation (4) indicates that the clone number when a mature detector become a memory detector. Where ξ is a constant, and *conc* denotes the concentration of the new memory detector. The reason that we adopt to this measure is to control clone number is that we want to the generated detectors can cover more wide shape space in order to reduce the false negative error.

3 Experiment

The experiment has been carried out in the Laboratory of Computer Network and Information Security at Sichuan University. A total of 40 computers in a network are under surveillance. According to the model we propose in this paper, an antigen is defined as a 56 bits length binary string composed of the source/destination IP address, port number, protocol type, IP flags, IP overall packet length, TCP/ UDP/ICMP fields, and etc. And we can use true positive rate (TP) and false positive rate (FP) to evaluate the performance of the model.

Define 6 TP and FP : TP refers to the situation where the classification predicts an intrusion correctly. FP refers to the situation where a normal record is (falsely) identified as an intrusion [15] [16].

$$TP = \frac{The \;\; number \;\; of \;\; attacks \;\; which \;\; are \;\; judged \;\; correctly \;\; by \;\; system}{The \;\; total \;\; number \;\; of \;\; the \;\; real \;\; attacks} \tag{5}$$

$$FP = \frac{The \;\; number \;\; of \;\; attacks \;\; which \;\; are \;\; not \;\; judged \;\; by \;\; system}{The \;\; total \;\; number \;\; of \;\; the \;\; real \;\; attacks}\,. \tag{6}$$

There are several important parameters in the model such as tolerisation period , life span of mature detectors, and the size of initial self-set and etc. In order to evaluate every parameter's impact on model's performance, we have carried out many group experiments. The parameters that are used in the experiments are summarized in

Table 1. Parameters used for experiments

Parameters	Values
Tolerisation Period	50
Life span of Mature Detectors	40
The Size of Initial Self-Set	40
The Number of New Generated Immature Cells	400
The Number of Immature Cells Generated in the Process of Mutation	100
Generation	10000
Activation Threshold of Mature Detectors	5

table 1, and these values of the parameters can get relative satisfied results (FP=0.048, TP=0.962) in our experiments, which have been verified by hundreds of tests in our laboratory.

In order to verify the adaptability, and test the efficiency of our model, experiments were undertaken with regards to the LISYS [9].

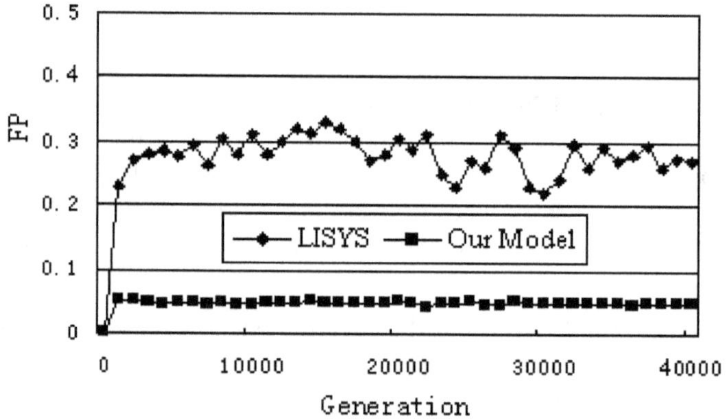

Fig. 2. FP rates for LISYS and our model

As the Fig.2 shows that the FP rates of our model is much lower than that of LISYS and the FP rates are not fluctuant with changing of the time. So we can safely conclude that the dynamic model is more adaptable to real network environment than LISYS.

From the following Fig.3, we can see that the TP rates of LISYS are much lower than that of our model. There are two main reasons for that. One is that we add signature detection layer to the model and system can detect known intrusion rapidly and reliable. The second is that the elements of self set and detector set can vary with the changing of the network environment. So the TP rates of our model are higher than that of LISYS.

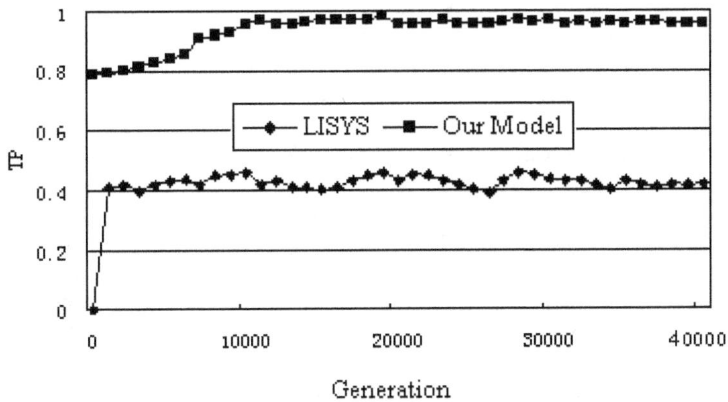

Fig. 3. TP rates for LISYS and our model

4 Conclusion

Inspired by the biological immune system, an immunity-based dynamic multilayer intrusion detection model was proposed in this paper. In this model, we employ misuse detection and anomaly detection in parallel, and these two detection component are reciprocal in the system. Meanwhile, in order to adapt to dynamic changing network environment, the elements of the self set and detector set can vary according to the changes of the network environment. The practical experimental results prove that this model can efficiently reduce both the false-positive error rates and false-negative error rates, and enhance the ability of self-adaptation and diversity for the network detection systems.

Acknowledgments

This work is supported by the National Natural Science Foundation of China under Grant No. 60373110, 60573130 and 60502011, the National Research Foundation for the Doctoral Program of Higher Education of China under Grant No.20030610003, the New Century Excellent Expert Program of Ministry of Education of China under Grant No. NCET-04-0870, and the Innovation Foundation of Sichuan University under Grant No.2004CF10.

References

1. Steven A. H.: An Immunological Model of Distributed Detection and Its Application to Computer Security. USA:University of New Mexico (1999)
2. Li T.: Computer Immunology. Publishing House of Electronics Industry, Beijing (2004)
3. Jonathan I. T.: Artificial Immune System:A Ovel Data Analysis Technique Inspired by The Immune Network Theory. University of Wales.Wales (2001)

4. Kim, J. W., J.Bentley P.: Towards An Artificial Immune System for Network Intrusion Detection. The Congress on Evolutionary Computation, Honolulu (2002) 1015 – 1020
5. Kim, J. W., J.Bentley P.: Immune Memory in the Dynamic Clonal Selection Algorithm. Proceedings of the First International Conference on Artificial Immune Systems (ICARIS), Canterbury (2002) 57-65
6. Inoue, H., Forrest, S.: Anomaly Intrusion Detection in Dynamic Execution Environments. New Security Paradigms Workshops (2002)
7. Li T.: Idid: An Immunity-Based Dynamic Intrusion Detection Model. Science in china (2005) 1912-1919
8. Esponda, F., Forrest, S., Helman, P.: A Formal Framework for Positive and Negative Detection. IEEE Transactions on Systems, Man, and Cybernetics 34: (2004) 357-373
9. Esponda, F.,Forrest, S., Helman, P.: Positive and Negative Detection. IEEE Transactions on Systems, Man and Cybernetics (2002)
10. Hofmeyr, S., Forrest, S.: Architecture for an Artificial Immune System. Evolutionary Computation 7(1), Morgan-Kaufmann, San Francisco, CA (2000) 1289-1296
11. Kim, J. W.: Integrating Artificial Immune Algorithms for Intrusion Detection , PhD Thesis, Department of Computer Science, University College London (2002)
12. Chao, D. L., Davenport, M. P., Forrest, S., Perelson, A. S.□Stochastic Stage-Structured.: Modeling of the Adaptive Immune System. Proceedings of the IEEE Computer Society Bioinformatics Conference (2003) 124-131
13. Li, T.: A New Model for Dynamic Intrusion Detection. Lecture Notes in Computer Science. Springer-Verlag, Berlin Heidelberg New York (2005) 72 - 84
14. Li, T.: An Immune Based Dynamic Intrusion Detection Model. Chinese Science Bulletin (2005) 2650 - 2657
15. Li, T.: An Immunity Based Network Security Risk Estimation. Science in China Ser. F Information Sciences. (2005) 557 - 578
16. Li, T..: An Immune-Based Model for Computer Virus Detection. Lecture Notes in Computer Science. Springer-Verlag, Berlin Heidelberg New York (2005) 59 - 71

Immunity and Mobile Agent Based Grid Intrusion Detection

Xun Gong, Tao Li, Gang Liang, Tiefang Wang, Jin Yang, and Xiaoqin Hu

School of Computer Science, Sichuan Univ., Chengdu 610065, China
orchid_xun@tom.com

Abstract. This paper analyzes the unique characteristics of a grid environment and the deficiencies of current grid intrusion detection systems, and proposes a novel immunity and mobile agent based grid intrusion detection (*IMGID*) model. Then, the concepts and formal definitions of *self*, *nonself*, *antibody*, *antigen* and *agent* in the grid security domain are given. Furthermore, the mathematical models of *self*, *mature MoA* (mature monitoring agent) and *dynamic memory MoA* (memory monitoring agent) *survival* are established. Our theoretical analysis and experimental results show that the model is a good solution to grid intrusion detection.

1 Introduction

Grid Computing shows a nice future, but also presents tough security challenges. A grid environment has many distinctive characteristics that are different from those of common network environments:

- The user population and the resource pool are large and dynamic.
- A computation (or processes created by a computation) may acquire, start processes on, and release resources dynamically during its execution.
- Resources and users may be located in different countries [1].

So the mechanisms and policies for securing the gird are more complicated.

There has been abundant research on intrusion detection for common network environments, but current research on grid intrusion detection is still in its infancy, and a few references can be used [2] [3]. M. F. Tolba, M. S. Abdel-Wahab proposed a grid intrusion detection architecture (GIDA) [4] in 2005. In the architecture, the anomaly detection is implemented using LVQ neural network. Increasing the number of IDSs (intrusion detection servers) will reduce the training time of LVQ, but increasing the number of grid users will increase the training time. Obviously, if we increase IDSs to reduce the training time, the cost will be higher and the system will not be self-adaptive to the grid environment where resources and users increase dynamically and may be large. So the detection performance of GIDA is not good.

Artificial immune system has the features of dynamicity, self-adaptation and diversity [5] [6] [7] [8] [9] [10] [11], and mobile agent has many same appealing properties as that of artificial immune system [12] [13] [14]. Thus, we apply mobile agent

D.-S. Huang, K. Li, and G.W. Irwin (Eds.): ICIC 2006, LNBI 4115, pp. 651 – 659, 2006.

technology as support for intrusion detection, and propose a novel immunity and mobile agent based grid intrusion detection (*IMGID*) model. The model that has the features of artificial immune system just meets the constraints derived from the characteristics of the grid environment and addresses the above mentioned performance problem.

In *IMGID*, the concepts and formal definitions of *self*, *nonself*, *antibody*, *antigen* and *agent* (that simulates the lymphocyte and is used as a detector to recognize nonself antigens, i.e. intrusions,) in the grid security domain are given. We define three kinds of agents: monitoring agents (*MoA*), communicator agents (*CoA*) and beating off agents (*BoA*). MoAs simulate B-lymphocytes in the immune system and patrol grid nodes. They are responsible for monitoring parameters simultaneously at four levels (user level, system level, process level, and packet level) and detecting intrusion. As B-lymphocytes consist of mature and memory lymphocytes, MoAs are divided into mature and memory MoAs that simulate, respectively, mature and memory cells. CoAs serve as communicators and are responsible for message transmission among agents. They simulate lymphokines secreted from T cells to stimulate B cells to clone themselves or eliminate cloned cells. BoAs simulate T killer cells that deal with intrusive activities. Once MoAs detect intrusions, they will stimulate CoAs and present the features of intrusions to BoAs. CoAs will activate BoAs. BoAs will move to the suspected place to counterattack intrusions. The counterattack strategies consist of disconnecting a node, discarding dubitable packets, killing a process, and so on. MoAs can clone themselves if an intrusion is detected. Some cloned MoAs are left here to detect more intrusions, and others are moved to other grid nodes to detect the similar intrusions. Then, the mathematical models of *self*, *mature MoA* and *dynamic memory MoA survival* are established.

In the present immunity-based intrusion detection systems [15], the memory lymphocytes have an unlimited lifecycle except they match the newly added selfs. Obviously, all the mature lymphocytes will become memory ones and the system will not have the feature of dynamicity and diversity in the end.

However, *IMGID* introduces a new idea to overcome this problem: the least recently used memory MoAs will degrade into mature MoAs, if the number of memory MoAs reaches or exceeds a given maximum.

2 The Proposed Grid Intrusion Detection Model (*IMGID*)

IMGID has two sub-models (see Fig. 1). The first is the model of the generation of new MoAs. The second is the intrusion detection model.

In the first model, we define antigens (*Ag*) to be the features of grid services and accesses, and given by:

$$Ag \in D \,. \tag{1}$$

$$D = \{ d \mid d = < ueser\_level, system\_level, process\_level, packet\_level > \} \,. \tag{2}$$

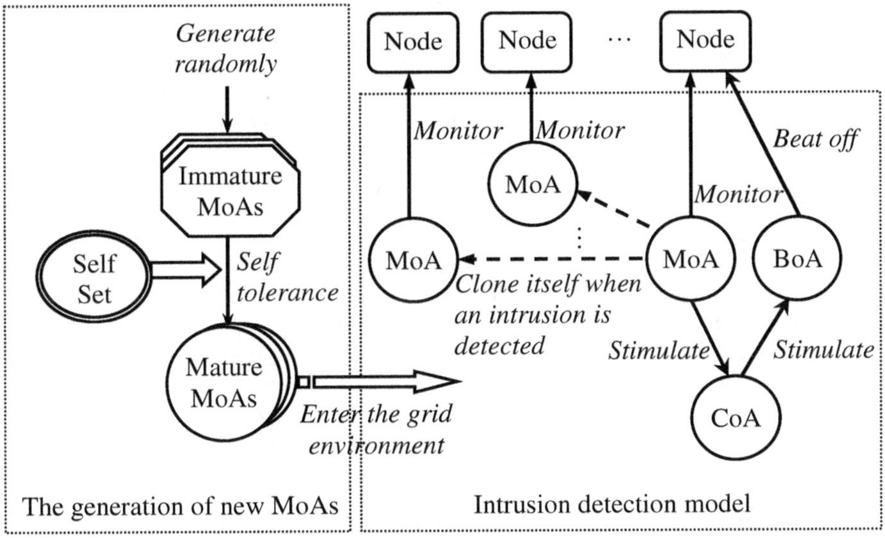

Fig. 1. The proposed grid intrusion detection model

The evolvement of MoAs in grid environment is shown in Fig. 2.

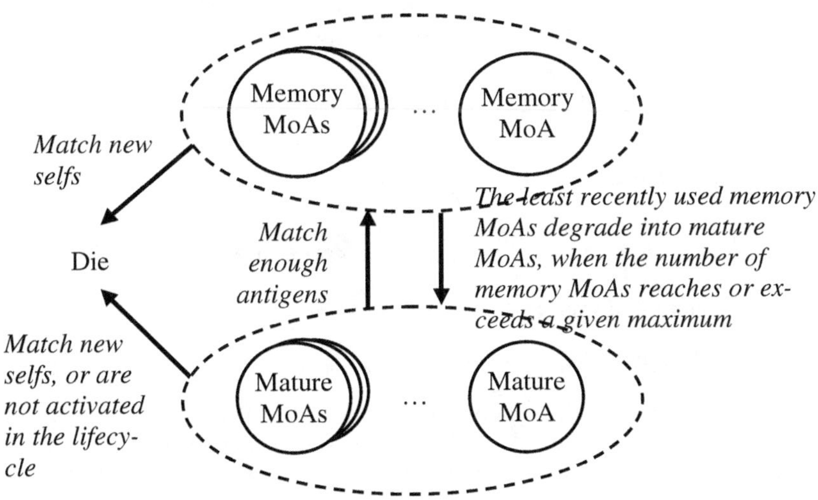

Fig. 2. The evolvement of MoAs in grid environment

user_level, *system_level*, *process_level* and *packet_level* are, respectively, the parameter monitored by monitoring agents at user level, system level, process level, and packet level. We define *Self* to be the set of normal grid services and accesses. Similarly, *Nonself* is also a set of abnormal services and accesses. *Ag* contains two subsets, *Self* and *Nonself*, where $Self \subset Ag$ and $Nonself \subset Ag$ such that

$$Self \cup Nonself = Ag, \quad Self \cap Nonself = \Phi \,. \tag{3}$$

In the intrusion detection model, all the agents form a set (*Agent*). *Agent* contains three elements, *MoA*, *CoA* and *BoA*. Monitoring agents, communicator agents and beating off agents form, respectively, the set *MoA*, *CoA* and *BoA*. Thus, we have:

$$Agent = \{MoA, CoA, BoA\}\,. \tag{4}$$

$$MoA = \{< d, age, count > \mid d \in D, age \in N, count \in N\}\,. \tag{5}$$

d is the monitoring agent that is used to detect an intrusion, *age* is the agent age, *count* (affinity) is the intrusion number detected by agent *d*, and N is the set of natural numbers. *d*, *age* and *count* are also called, respectively, field *d*, *age* and *count* of an agent. For the convenience using the fields of a agent *x*, a subscript operator "." is used to extract a specified field of *x*, where

$$x.fieldname = \text{the value of field } fieldname \text{ of } x\,. \tag{6}$$

MoAs consist of Mature MoAs and Memory MoAs (see Fig. 2). A mature MoA is a MoA that is tolerant to self but is not activated by antigens. A memory MoA evolves from a mature one that matches enough antigens in its lifecycle. Mature MoAs form a set (MA_{MoA}). Mature MoAs detect novel intrusions that have not previously been identified. Memory MoAs form a set (ME_{MoA}). Memory MoAs greatly enhance detection of previously seen intrusions. Therefore, we have:

$$MoA = MA_{MoA} \cup ME_{MoA}\,. \tag{7}$$

$$MA_{MoA} \cap ME_{MoA} = \Phi\,. \tag{8}$$

$$MA_{MoA} = \{x \mid x \in MoA, \forall y \in Self \ \ (< x.d, y > \notin Match \wedge x.count < \beta)\}. \tag{9}$$

$$ME_{MoA} = \{x \mid x \in MoA, \forall y \in Self \ \ (< x.d, y > \notin Match \wedge x.count \geq \beta)\}. \tag{10}$$

Match is a match relation in *D* defined by

$$Match = \{< x, y > \mid (x, y \in D), \ f_{match}(x, y) = 1\}\,. \tag{11}$$

$f_{match}(x, y)$ is based on the affinity between *x* and *y*: if the affinity greater than a specified threshold, then 1 is returned, otherwise, 0 is returned. In *IMGID*, the affinity function can be *r-contiguous-bits matching rule*, *Hamming distance*, *Landscape-Affinity Matching*, etc [16].

2.1 The Definition of Self

In a dynamic grid environment, as time goes on, some grid services and accesses, which were normal in the past, are forbidden now. Equation (12) depicts the evolution of self, where *Self*(*t*) evolves from *Self*(*t*-1), the mutated self antigens ($Self_{variation}$) at time *t* are eliminated, and the newly defined self antigens ($Self_{new}$) at time *t* are added.

$$Self(t)=\begin{cases}\{x \mid x \text{ is the initial self antigen}\}, & t=0\\ Self(t-1)-Self_{variation}(t)\cup Self_{new}(t), & t\geq 1\end{cases}. \tag{12}$$

$$Self_{variation}(t)=\{x \mid x \text{ is the self antigen mutated at time } t\}. \tag{13}$$

$$Self_{new}(t)=\{y \mid y \text{ is the self antigen added at time } t\}. \tag{14}$$

The model of self assures that the number of selfs is small and remains steady.

2.2 Mature MoAs Model

$$MA_{MoA}(t)=\begin{cases}\Phi, & t=0\\ MA_{retain}(t)\cup MA_{new}(t)\cup MA_{cycle}(t)-MA_{activation}(t), & t\geq 1\end{cases}. \tag{15}$$

$$MA_{retain}(t)=MA'_{MoA}(t)-A(t)\cup A'(t). \tag{16}$$

$$\begin{aligned}MA'_{MoA}(t)=\{y \mid y\in MoA, x\in MA_{MoA}(t-1), x.age<\lambda, \forall s\in Self_{new}(t) <x.d.s>\notin Match,\\ y.d=x.d, y.age=x.age+1, y.count=x.count\}.\end{aligned} \tag{17}$$

$$A(t)=\{x \mid x\in MA'_{MoA}(t), \exists y\in Ag(t-1) <x.d,y>\in Match\}. \tag{18}$$

$$\begin{aligned}A'(t)=\{y \mid y\in MoA, x\in A(t), y.d=x.d, y.age=x.age,\\ y.count=x.count+1\}\end{aligned}. \tag{19}$$

$$MA_{cycle}(t)=\{y \mid y\in MoA, y.d=x.d, y.age=T, y.count=\beta-1, x\in ME_{degradation}(t), T<\lambda\}. \tag{20}$$

$$MA_{activation}(t)=\{x \mid x\in A'(t),\ x.count\geq\beta\}. \tag{21}$$

$$\begin{aligned}MA_{new}(t)=\{y \mid y\in MoA, y.d=x.d, y.count=0, y.age=0,\\ x\in (I_{new}(t)-\{z \mid z\in I_{new}(t), \exists s\in Self(t) <z,s>\in Match\})\}.\end{aligned} \tag{22}$$

$$I_{new}(t)=\begin{cases}\Phi, & |MA_{MoA}(t-1)|\geq\Gamma\\ \{y_1,y_2,\ldots,y_{\Gamma-|MA_{MoA}(t-1)|}\}, y_i\in D, 1\leq i\leq\Gamma-|MA_{MoA}(t-1)|, & otherwise\end{cases}. \tag{23}$$

$MA_{retain}(t)$ simulates the process that the mature cells evolve into the next generation ones, where the cells do not tolerate to those newly added self elements or have not match enough antigens ($\beta>0$) in lifecycle λ, will be eliminated. $MA_{new}(t)$ depicts the generation of new mature MoAs. Γ (>0) is the max number of mature MoAs in *IMGID*. $MA_{activation}(t)$ is the set of mature MoAs which match enough antigens and will be activated and evolved into memory MoAs at time t. $MA_{cycle}(t)$ is the set of the least recently used memory MoAs which degrade from memory MoAs into mature

MoAs and be given a new age T (>0) and count $\beta-1$ ($\beta \geq 1$). Once a mature MoA with count $\beta-1$ matches an antigen in lifecycle λ , the count will be set to β and the mature MoA will be activated again. The method assures degraded memory MoAs priority in evolvement. Because memory MoAs have better detection capability than mature MoAs, the method enhances detection efficiency.

2.3 Dynamic Memory MoAs Survival Model

$$ME_{MoA}(t) = \begin{cases} \Phi, & t = 0 \\ ME_{retain}(t) \cup ME_{new}(t) - ME_{\deg radation}(t), & t \geq 1 \end{cases}. \tag{24}$$

$$ME_{retain}(t) = ME'_{MoA}(t) - E(t) \cup E'(t). \tag{25}$$

$$\begin{aligned} ME'_{MoA}(t) = \{ y \mid y \in MoA, x \in ME_{MoA}(t-1), \forall s \in Self_{new}(t) < x.d.s > \notin Match, y.d = x.d, \\ y.age = x.age + 1, y.count = x.count \}. \end{aligned} \tag{26}$$

$$E(t) = \{ x \mid x \in ME'_{MoA}(t), \exists y \in Ag(t-1) < x.d, y > \in Match \}. \tag{27}$$

$$\begin{aligned} E'(t) = \{ y \mid y \in MoA, x \in E(t), y.d = x.d, \\ y.age = 0, y.count = x.count + 1 \}. \end{aligned} \tag{28}$$

$$\begin{aligned} ME_{new}(t) = \{ x \mid x \in ME_{MoA}, y \in MA_{activation}(t), \\ x.d = y.d, x.age = 0, x.count = y.count \}. \end{aligned} \tag{29}$$

$$ME_{\deg radation}(t) = \begin{cases} \{ x \mid x \in ME_{retain}(t), x.age > \gamma \}, & |ME_{MoA}(t-1)| \geq \kappa \\ \Phi, & otherwise \end{cases}. \tag{30}$$

Equation (24) depicts the dynamic evolvement of memory MoAs, where the memory MoAs that match the newly added selfs will die and are matched by an antigen will be activated immediately. $ME_{retain}(t)$ simulates the process that the memory MoAs evolve into the next generation ones. $ME_{new}(t)$ is the set of new memory MoAs evolved from mature ones. $ME_{degradation}(t)$ is the set of memory MoAs that are not activated by antigens lately and degrade from memory MoAs into mature MoAs when the number of memory MoAs reaches or exceeds a given maximum κ . γ (>0, natural number) is the lifecycle of memory MoAs.

3 Simulations and Experiment Results

The experiments were carried out in the Laboratory of Computer Network and Information Security at Sichuan University. We developed a grid simulation toolkit based on GridSim [17] to satisfy our needs. The simulation environment simulates users with different behaviors, resources and agents. A total of 40 computers in a grid were under surveillance. 10 immature MoAs was given.

To prove the steady performance and high efficiency of *IMGID*, we developed the first set of experiments to show the effect of selfs number on training time in *IMGID*. The initial number of selfs was 40. The training time increased due to the increasing of selfs number. The number of selfs was small and the increase rate of selfs was low. So the training time for mature MoAs was not long and the increase rate of the training time was low. The experiment results are shown in Fig.3. The short training time illustrates the high efficiency and the low increase rate of the training time illustrates the steady performance in *IMGID*.

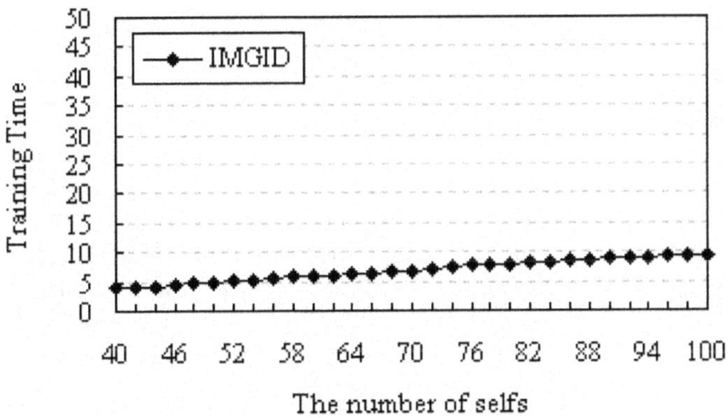

Fig. 3. Effect of the number of selfs on the training time in *IMGID*

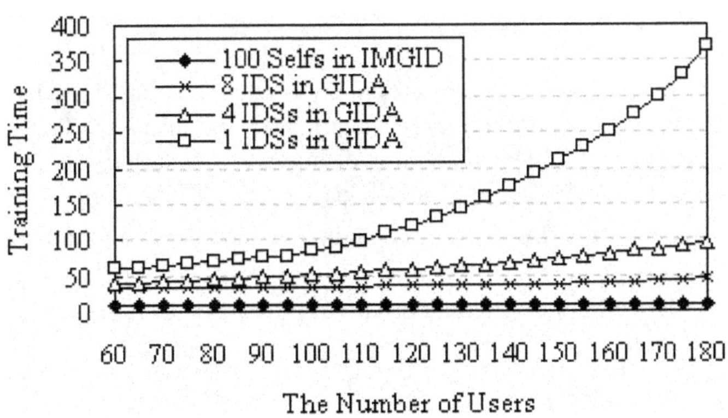

Fig. 4. Effect of the number of IDSs and users on the training time in *IMGID* and GIDA

To prove the self-adaptive capacity and steady performance of *IMGID*, we developed the second set of experiments to compare *IMGID* with LVQ. In GIDA, the number of IDSs was set to 1, 4, 8, respectively. The training time for detectors in GIDA increased due to the increasing of users or reducing of IDSs. Specially, the training time increased rapidly when only one IDS was used. In *IMGID*, the number of tolerated selfs was 100. We increased gird users and made experiments on the training time of *IMGID* when the number of selfs kept 100. The training time depended on the number of selfs and was hardly affected by dynamically varying IDSs and users. The results in Fig.4 illuminate that the number of IDSs and users has a great effect on the training time in GIDA, while no effect on that in *IMGID*.

Comparison between the curve trend of the training time in *IMGID* (Fig.3) and those three in GIDA (Fig.4) shows the increase rate of the training time in *IMGID* was lower than that in GIDA.

Therefore, the experiment results show that *IMGID* has high efficiency and steady performance to adapt the dynamically varying grid environment.

4 Conclusions

The proposed immunity and mobile agent based grid intrusion detection (*IMGID*) model is an active grid security technique.

In *IMGID*, the concepts of *self*, *nonself*, *antigen*, *antibody* and *agent* have been abstracted and extended. Besides, *IMGID* introduces a new idea of *dynamic memory MoAs survival*, which assures dynamicity and diversity in immunity-based IDS. Furthermore, the mathematical models of *self*, *mature MoA* and *dynamic memory MoA survival* are presented.

The theoretical analysis and the experiment results show that the proposed method is an effective solution to intrusion detection for grid security.

5 Future Researches

The effects of several import parameters on system performance are very important in the models of mature MoA and dynamic memory MoA survival. The set of appropriate parameters could enhance detection efficiency and assure steady performance. These parameters include the new age T and count $\beta-1$ of degraded memory cells in the models of mature MoA, κ and γ in the models of dynamic memory MoA survival.

The *match algorithm* and the mathematical models of *CoA* and *BoA* need to be abstracted and extended.

Acknowledgment

This work is partially supported by the National Natural Science Foundation of China under Grant No. 60373110, 60573130 and 60502011, the National Research Foundation for the Doctoral Program of Higher Education of China under Grant No.20030610003, the New Century Excellent Expert Program of Ministry of Education of China under Grant No.NCET-04-0870, and the Innovation Foundation of Sichuan University under Grant No.2004CF10.

References

1. Foster, Carl Kesselman, Gene Tsudik, Steven Tuecke: A Security Architecture for Computational Grids. In Proc. of Computer and Communication Security (1998)
2. Leu, Fang-Yie, Lin, Jia-Chun, Li, Ming-Chang, Yang, Chao-Tung: A Performance-Based Grid Intrusion Detection System. In Proc. of International Computer Software and Applications, Vol.1 (2005) 525–530
3. De Castro.L. N, Timmis.J. I: Artificial Immune Systems as a Novel Soft Computing Paradigm. Soft Computing journal, Vol. 7 (2003) 526–544
4. Tolba, M., Abdel-Wahab, M., Taha, I., Al-Shishtawy, A.: GIDA: Toward Enabling Grid Intrusion Detection Systems. In Proc. of the Second International Intelligent Computing and Information Systems Conference (2005)
5. Forrest, S., Hofmeyr, S., Somayaji, A.: Computer Immunology. In Proc. of Communications of the ACM, Vol. 40 (1997) 88–96
6. Somayaji, A., Hofmeyr, S. A., Forrest, S.: Principles of a Computer Immune System. In Proc. of the New Security Paradigms'97 (1997) 75–82
7. Li, T.: Compute immunology. Publishing House of Electronics Industry, Beijing (2004)
8. Li, T.: A New Model for Dynamic Intrusion Detection. Lecture Notes in Computer Science. Springer-Verlag, Berlin Heidelberg New York (2005) 72–84
9. Li, T.: An Immune based Dynamic Intrusion Detection Model. Chinese Science Bulletin (2005) 2650–2657
10. Li, T.: An Immunity based Network Security Risk Estimation. Science in China Ser. F Information Sciences (2005) 557–578
11. Li, T.: An Immune-Based Model for Computer Virus Detection. Lecture Notes in Computer Science. Springer-Verlag, Berlin Heidelberg New York (2005) 59–71
12. Dasgupta, D.: Immunity-based Intrusion Detection System: A general framework. In Proc. of 22nd National Information Systems Security Conference (1999) 147–160
13. Machado, R. B., Boukerche, A., Sobral, J. B. M. , Juca, K. R. L., Notare, M. S. M. A.: A Hybrid Artificial Immune and Mobile Agent Intrusion Detection Base Model for Computer Network Operations. In Proc. of the 19th IEEE International Parallel and Distributed Processing (2005)
14. Lin, Jie, Wang, Cong, Gou, Yanhui: Agent-based Access Control Security in Grid computing Environment. In Proc. of Networking, Sensing and Control (2005)
15. Paul, K. Harmer, Paul, D. Williams, Gregg, H. Gunsch, Gary, B. Lamont: An Artificial Immune System Architecture for Computer Security Applications. In Proc. of IEEE Transactions on Evolutionary Computation, Vol. 6 (2002)
16. Paul, K. Harmer, Paul, D. Williams, Gregg, H. Gunsch, Gary, B. Lamont: An Artificial Immune System Architecture for Computer Security Applications. In Proc. of IEEE Transaction on Evolutionary Computation, Vol. 6 (2002) 252–280
17. M. Murshed, R. Buyya, D. Abramson: GridSim: A Grid Simulation Toolkit for Resource Management and Scheduling in Large-Scale Grid Computing Environments. In Proc. of the 17th IEEE International Symposium on Parallel and Distributed (2002)

Immune-Based Peer-to-Peer Model for Anti-spam

Feng Wang, Zhisheng You, and Lichun Man

Computer College, Sichuan University, Chengdu 610065, P.R. China
wangfeng_scu@yahoo.com.cn, zsyou@mail.sc.cninfo.net, manlichun@sohu.com

Abstract. Spam (or junk email) has been a major problem on the Internet. A lot of solutions have been proposed to deal with it. However, with the evolvement of spammers' techniques and the diversification of email content, the traditional anti-spam approaches alone are no longer efficient. In this paper, a new anti-spam Peer-to-Peer (P2P) model based on immunity was presented. *Self*, *Non-self*, Antibody, Antigen and immune cells in email system were defined. The model architecture, the process of Antigen presenting, clone selection and mutation, immune tolerance, immune response, life cycle of immune cells and some other immune principles were described respectively. The analyses of theory and experiment results demonstrate that this model enjoys better adaptability and provides a new attractive solution to cope with junk emails in P2P environment.

1 Introduction

Nowadays, the overrun of spam brings people a lot of troubles. In order to deal with spam, a lot of approaches have been proposed: such as Blacklisting and Whitelisting method, rule-based method [1], statistics-based method [2][3][4], and secure authentication-based method [5]. All these approaches have achieved some effects.

However, there are still some limitations to these approaches. As the source address of an email can be forged, the Blacklisting and Whitelisting method is not so effective; As rule-based method is static and statistics-based method needs a lot of training examples, they are often got through by spammers using various tricks, for example text obfuscation, random space or word insertion, HTML layout, and text embedded in images; Though the secure authentication-based method enjoys high security, the easiness of usage is affected. On the other hand, with the continuous growth of Internet, the type and content of emails are becoming more and more diverse. And users' interests are changing continuously too. In addition, there are great differences between Chinese and English [6]. Finally, filters based on these methods work independently, so they are lack of knowledge about new forms of spam.

With the development of distributed system, some anti-spam approaches based on Peer-to-Peer (P2P) network have been proposed [7][8][9] to collaboratively share knowledge about spam. But some of the limitations mentioned above still exist and most of these approaches are designed for structured network environment. Considering the Biological Immune System (BIS) has the characteristics of high distribution

D.-S. Huang, K. Li, and G.W. Irwin (Eds.): ICIC 2006, LNBI 4115, pp. 660 – 671, 2006.

and self-adaptation [10], immunity is introduced in this paper to build an anti-spam model in P2P environment. The analysis of theory and experimental results demonstrate that it enjoys better adaptability and provides a new attractive solution to cope with junk emails in P2P environment.

2 Anti-spam P2P Model Architecture

Due to the collaborative computation advantage of P2P technology, it is a good choice to propose an anti-spam solution in P2P environment, based on the following considerations: (1) Users who joined the same peer group share approximately the same interests. Therefore, the spam information can be shared among them. (2) In central control system, as mail server is the bottleneck, it is not desirable to implement complicated filtering strategies on mail server, or the quality of service will be affected. (3) It is a most dynamic environment that all the peer nodes and junk emails keep changing all the time, so a model that is adaptive to changes is needed.

The architecture of the proposed model is shown in figure 1.

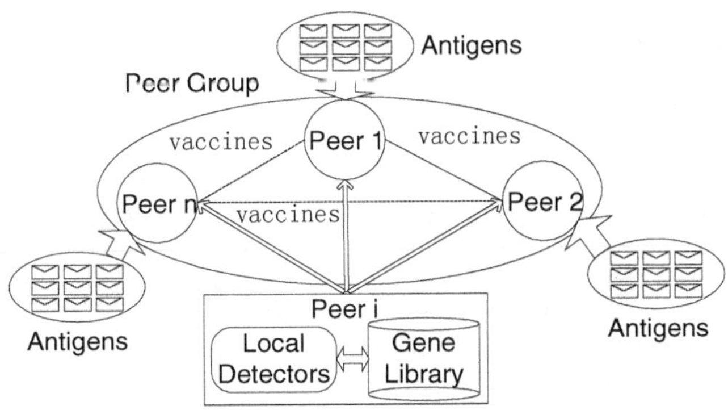

Fig. 1. Architecture of the proposed anti-spam P2P model

Once a user joins the peer group, the peer will keep communication with several nearest peers in the peer group and exchange information about spam with each other. Every time a peer receives an email (*Antigen*), its local detectors that are used to simulate immune cells in the model will be activated (immune response) to execute filtering function. While some other immune mechanisms (e.g. clone selection and mutation, life cycle of immune cells) occur too.

As each peer only faces part of the Antigen space, it is necessary to share the spam information in the peer group. This is done by dynamically handing around spam vaccines, which can be implemented by mobile agent or some other approaches. Through this mechanism, the global spam information can be collected with the collaborative work of all the peers in the peer group.

3 Immune Principles

For convenience, the following notations are established:

N: the set of natural numbers;
Set_String: the set of string;
Set_C: the set of immune cells;
Set_MC: the set of memory cells;
Set_IC: the set of immature cells;
Th_Deg: the threshold of degeneration for memory cells;
Th_aff_Rec: the threshold of affinity for recognition;
Th_aff_Mutate: the threshold of affinity for mutation;
Th_Spam: the threshold of score for spam;
Num_init_Mcount: the initial matching number of memory cells;
Num_init_Icount: the initial matching number of immature cells;
Num_const_Clone: clone constant;
Num_const_Mutate: mutation constant;
m: the number of Antigen fields;
n: the number of vector fields;
.: the operation of citing a field from a vector.

3.1 Antigen Presenting

Let Antigens be the emails received by users. Let E be the set of string contained in Antigens and $E \subset Set\_String$.

In order to simulate the Antigen presenting of BIS, the features of an email are extracted to get the features of an Antigen: *Antigenic determinant* [10]. First of all, the procedure of preprocess [11] is executed to reduce the influence of noise (e.g. removing HTML tags and tokens in the stop-list) after an email is decoded. And some of the common features found in most junk emails are extracted. If it is Chinese, the process of words segmentation is done subsequently [6]. Then keywords and the associate weights are extracted and the email is processed into the form of Antibody. Actually, it is a data mining process [11] and content-based filtering strategy can be brought into effect.

$$Ag = \{< F_1, ..., F_m > | F_i \in E, i \in [1, m], m \in N\} \tag{1}$$

Equation (1) defines the set of string that are extracted from the header and body of an email, standing for the transaction features of an email. F_i is the ith transaction feature vector, i.e. the ith gene segment of Ag, which is described by (2).

$$F_i =< Str_1, ..., Str_n >, Str_i \in Set\_String, i \in [1, n], n \in N\} \tag{2}$$

L is the set of Lymphocytes as defined by (3), where d is the Antigenic Determinant and *count* is the matching number of d. Actually, L is the detectors (*Antibodies*) used in the model to determine whether an email is spam or not.

$$L = \{< d, count > | d \in Ag, count \in N\} \tag{3}$$

Self is defined to mean the set of legitimate emails and *Nonself* represents the set of junk emails. Equation (4) expresses the relationship between *self* and *Nonself.*

$$Self \cup Nonself = Ag \wedge Self \cap Nonself = \varnothing, Self \subset Ag \wedge Nonself \subset Ag \quad (4)$$

The set of memory cells *Set_MC* denotes cells that don't match *Self* and whose count value is high above the activation threshold of memory cells. And *Set_IC* is the set of immature cells. Obviously, there is the following relationship as (5).

$$Set\_C = Set\_MC \cup Set\_IC \wedge Set\_MC \cap Set\_IC = \varnothing \quad (5)$$

3.2 Affinity Measure

$$f_{match}(c,v) = \begin{cases} 1 & if\ f_{affinity}(c,v) \ge Th\_aff\_Rec \\ 0 & otherwise \end{cases}, c \in L, v \in Ag \quad (6)$$

Equation (6) defines the matching relationship in E, where $f_{affinity}(c,v)$ is used to compute affinity between an Antigen and a detector and 1 represents Antigen v is matched by a specific detector c.

The function of computing affinity between an Antigen and a detector is described as follows, representing the matching degree between feature sets.

```
function f_affinity(c,v)
  begin
    f_affinity(c,v)  := 0;
    for i := 1 to m do
    begin
      cs := the shorter vector in c.d.F_i and v.F_i;
      cl := the longer vector in c.d.F_i and v.F_i;
      num := the number of string in cs present in cl;
      len := the number of string in cs;
      f_affinity(c,v)  :=  f_affinity(c,v)  + ( num / len ) * w_i;
    end;
end.
```

In the above pseudocode, w_i is the weight that is corresponded to the transaction feature F_i of an Antigen.

3.3 Clone Selection and Mutation

$$Set\_C_{clone\_mutate} = \bigcup_{j=1}^{\lceil Num\_const\_Clone * f_{affinity}(c,v) \rceil} f_{clone}(c) \bigcup_{i=1}^{\lceil Num\_const\_Mutate * (1 - f_{affinity}(c,v)) * Len \rceil} f_{mutate}(f_{clone}(c)), c \in Set\_C \quad (7)$$

Equation (7) describes clone selection and mutation, where *Len=max* (*Length* (*c.d*, *v*)), $c \in Set\_C$, $v \in Ag$, which means the longer length of vector between *c.d* and *v*. *c* belongs to the immune cells set where every detector matches current Antigen *v* and it is affirmed by user, i.e. the set of candidate immune cells to be cloned and mutated. Function $f_{clone}(c)$ is the clone process for detector *c*. And function $f_{mutate}(c')$ denotes the mutation operation for the cloned cell *c'*. From (7), it is shown that the relationship between clone number and affinity is direct ratio while the relationship between mutation degree and affinity is inverse ratio. At the same time, in order to avoid producing too random or too similar Antibodies, the mutation range is limited during this operation (i.e. immune tolerance, referring to section 3.4). Through the process of clone selection and mutation, the diversity of detector is assured. And it is possible to detect the new mutated junk emails.

Equation (7) also shows the dynamic evolvement of the immune model. The procedure of clone selection and mutation is detailed by the following pseudocode.

```
procedure clone_selection_mutation(Ag: v, float: aff)
  begin
    Set_Candidate := Φ;
    //get the candidate set of cells to be cloned
    for each(c ∈ Set_C) do
    begin
      if ( f_affinity(c,v) > aff) then
        Set_Candidate := Set_Candidate ∪ {c};
    end;
    for each( c ∈ Set_Candidate) do
    begin
      //clone operation
      c'.d := c.d;
      c'.count := Num_init_Icount;
      aff := f_affinity(c',v);
      num_clones := round(Num_const_Clone * aff);
      Len := max(Length(c'.d,v));
      num_mutate := round(Num_const_Mutate *
                            (1 - aff) * Len);
      for i := 0 to num_clones do
      begin
        b := c';
        //mutation operation
        for j := 0 to num_mutate do
        begin
          pos_mutate := get the position of random
                            mutation for b;
          replace the string at pos_mutate of b with
            the string from appropriate gene library
        end;
        //immune tolerance
        if ( f_tolerance(b) = 1) then
          Set_IC := Set_IC ∪ {b}
      end;
    end;
end.
```

3.4 Immune Tolerance

In BIS, new immature cells are produced in the Bone Model [10]. Before a new produced immature cell joins immune circle, it must pass through the process of immune tolerance, avoiding *Auto-immune Reaction*. Equation (8) defines the procedure of immune tolerance, where 0 means tolerance failure and 1 means success.

$$f_{tolerance}(b)=\begin{cases}1, & if\ \forall c\in Set\_C, f_{affinity}(c,b)\in(Th\_aff\_Rec, Th\_aff\_Mutate)\\0, & otherwise\end{cases}, b\in Ag \tag{8}$$

The immune tolerance is realized by *Negative Selection Algorithm* [10], which simulates the process of immature cells' maturation. If a new produced detector is too similar to one of Antibodies (the existing detectors) or too random, it means tolerance failure and will be deleted.

3.5 Immune Response

$$f_{classfy}(c,v)=\begin{cases}1, & if\ \exists(f_{score}(c,v))\geq Th\_Spam\\0, & otherwise\end{cases}, c\in Set\_C, v\in Ag \tag{9}$$

$$f_{score}(c,v)=f_{score\_innate}(c,v)+f_{score\_adaptive}(c,v), c\in Set\_C, v\in Ag \tag{10}$$

$$f_{score\_x}(c,v)=f_{affinity}(c,v)\times 100,\ x=innate\vee adaptive, c\in Set\_C, v\in Ag \tag{11}$$

Equation (9) defines the classification function, which determines whether an email is classified as spam or not. For convenience, a scoring function is defined by (10), which consists of innate immune response and adaptive immune response. Equation (11) shows the scoring function of the two parts. The affinity of innate immune response and adaptive immune response can be calculated according to the pseudocode in section 3.2.

The immune response consists of three layers. Layer 1: innate immune response. Layer 2: the immune response of memory cells. Layer 3: the immune response of immature cells. Obviously, with the increase of search scale, the response speed of the above three layers decreases in turn. The procedure is detailed by the following pseudocode.

```
procedure immune_response(Ag: v)
  begin
    aff := 0;
    flag_spam := FALSE;
    //innate immune response
    f_score := score from innate immune response;
    if (f_score >= Th_Spam) then
    begin
      move the email to spam box;
      flag_spam := TRUE;
    end
    else //adaptive immune response
    //the second and the first immune response in turn
    for (each(c ∈ (Set_MC ∪ Set_IC))) do
    begin
      if ((f_affinity(c,v) >= Th_aff_Rec) and
          ((f_score + f_score_adaptive(c,v)) >= Th_Spam)) then
```

```
      begin
        move the email to spam box;
        flag_spam := TRUE;
        aff := f_affinity(c,v);
        break;
      end
    end;
    // costimulation from end users
    if ( f_costimulation(v) = 1)
    begin
      add Antigenic determinant of v to the
        appropriate Antibody gene library;
      if (aff > 0) then
      begin
        clone_selection_mutation(v, aff);
        //evolvement of immune cells
        Set_Match := all the detectors that matched v;
        for each( c ∈ Set_Match) do
        begin
          if (c ∈ Set_MC) then
            c.count++;
          if (c ∈ Set_IC) then
          begin
            c.count := Num_init_Mcount;
            Set_MC := Set_MC ∪ {c};
            Set_IC := Set_IC - {c};
          end
        end;
      end;
    end;
    //False Positive
    if (flag_spam = TRUE and f_costimulation(v) = 0)
    begin
      Set_DC := all the detectors that matched v;
      for (each(c ∈ Set_DC)) do
      begin
        Delete Antigenic determinant of detector c from
          the appropriate Antibody gene library;
        Set_C := Set_C - {c}
      end
    end
end.
```

Because whether an email is spam or not is determined by end users eventually, it is inevitable to classify a legitimate email as a spam (*False Positive*) or classify a junk email as a legitimate one (*False Negative*). So the final step of immune response is user's costimulation, as described by (12). According to user's feedback, an Antigen's Antigenic determinant is added into or deleted from the appropriate Antibody gene library. While the operation of clone selection and mutation is executed.

$$f_{costimulation}(v) = \begin{cases} 1 \text{ ,outside signal treats } v \text{ as } NonSelf \\ 0 \text{ ,} otherwise \end{cases} ,v \in Ag \tag{12}$$

In P2P environment, once a new spam is detected in one peer, the new produced detector will be delivered to its nearest peers in the same peer group, simulating the process of vaccination in BIS [10]. Therefore, other peers in the group can detect the new junk email rapidly without further learning. Through the mechanism of vaccine distribution, the information about spam can be shared in the peer group and the efficiency of spam filtering can be improved.

Considering security and privacy factors in P2P network, it is necessary to encrypt the vaccine with *Secure Hash Algorithm* [9]. At the same time, an enhanced authentication mechanism is needed in the model too.

3.6 Life Cycle of Immune Cells

In the course of immune tolerance, if a new produced cell interacts with an element of immune cell set (*Auto-immune Reaction*) or is too random, it will be deleted. Otherwise it will be saved in *Set_IC* and join immune circle. During immune circle, once a memory cell matches an Antigen correctly, its matching number will be increased by one and its life cycle will be lengthened. Once an immature cell matches an Antigen (the first immune response) correctly, it will be evolved to be a memory cell. A memory cell has a much longer life cycle and can response rapidly to an Antigen which has been recognized before (the second immune response).

In order to simulate the senescence of immune cells, all the immune cells' count value is decreased regularly. If a memory cell's count value is less than the threshold *Th_Deg*, it will be degenerated to be an immature cell. And if an immature cell hasn't matched any junk emails and its count value is decreased to zero, it will die naturally. Meanwhile, any detectors that don't pass through user's costimulation will be deleted too. This mechanism assures the diversity of immune cells, the continuous search ability towards Antigen space and the survival of the best immune cells.

The evolvement of immune cells has been described by the pseudocode in section 3.5 and the degeneration of immune cells is shown in the following.

```
procedure degeneration
  begin
    for each (c ∈ Set_C) do
      c.count--;
    //degeneration of memory cell
    if (c ∈ Set_MC and c.count < Th_Deg) then
    begin
      Set_MC = Set_MC - {c};
      c.count = Num_init_Icount;
      Set_IC = Set_IC ∪ {c};
    end;
    //death of immature cell
    if (c ∈ Set_IC and c.count = 0) then
    begin
      Set_IC = Set_IC - {c};
```

```
        Delete Antigenic determinant of detector c from
            the appropriate Antibody gene library;
    end;
end.
```

4 Results

4.1 Experimental Setup

For lack of common Chinese email corpus, the experimental data was collected by authors and authors' friends. Emails were extracted to be Lymphocytes (detectors), which were consisted of six gene segments, e.g. IP addresses, mail addresses, subject, hyperlinks, etc. Some of the main parameters in the experiments are shown in table 1.

In order to simulate the real environment, the average frequency of receiving emails was 10 emails per day, including 3 junk emails averagely. And the emails in the test set were arranged in the receiving time. The final results were calculated by taking the mean of these data. And six peers consisted of the peer group in order to simulate the distribution of vaccine in P2P environment. Recall, precision, and accuracy [12] were taken to evaluate performances of the model.

Table 1. Experimental parameters

Parameter	Value	Range
Num_init_Mcount	20	>0
Num_init_Icount	10	>0
Th_Deg	15	>0
Num_const_Clone	7	>0
Num_const_Mutate	0.8	0-1
Th_Spam	50	>0
Th_aff_Rec	0.2	0-1
Th_aff_Mutate	0.5	0-1

For convenience, the proposed model in the paper is named "ASID" (Anti-spam system based on immunity and data mining). ASID is a self-learning and self-adaptability system, it can learn knowledge about spam from the continually received emails. And this accords with the realistic situation, because there is no or less emails that can reflect his or her personality to train when he or she begins to use email system. So there is no any training process for ASID in the following experiments.

In our comparison experiment between ASID and rule-based method, we found that rule-based method could achieve good performances only when the features of spam were very common, because it was lack of learning ability. So ASID achieved better performances than rule-based method. And Bayesian method, one of the most popular used anti-spam methods, was taken to compare with the proposed approach. The security level in the two approaches was set at the same level: middle.

4.2 Experiment 1: Comparison Between ASID and Bayesian Method

This experiment was performed with 750 emails where 519 were manually classified as legitimate and the rest 231 were assumed of spam. While 140 junk emails and 300 legitimate emails were chosen to be the training set for Bayesian method. And 66,480 non-spam words and 58,885 spam words were collected respectively after the training. Detectors were then tested against a collection of 219 legitimate emails and 91 junk emails. Table 2 shows the experimental results.

Table 2. The comparison results between ASID and Bayesian method

Performance	ASID	Bayesian method
Recall	88.17%	77.42%
Precision	74.41%	82.15%
Accuracy	85.48%	88.06%

From table 2, we can find that the performances on precision and accuracy of Bayesian method are superior to those of ASID. It is reasonable because Bayesian filter learns lots of knowledge about junk and legitimate emails from training set. While ASID is lack of such pre-knowledge and the number of test emails in the experiment is not big enough. However, considering they are achieved without any training, it is inspiring to get the results whose differences are not so big.

With the increase of Antigen space, the knowledge learned from receiving emails is increased too. So the performances of ASID are improved gradually. This tendency can be found from figure 2. It can be anticipated that ASID will achieve better performances than Bayesian method with more test emails. And our later experiment verified this anticipation. When the test emails were increased to 450 (77 new junk emails and 63 new legitimate emails were added into the above test set), the performances of ASID on recall, precision and accuracy were 84.79%, 77.57% and 84.22%

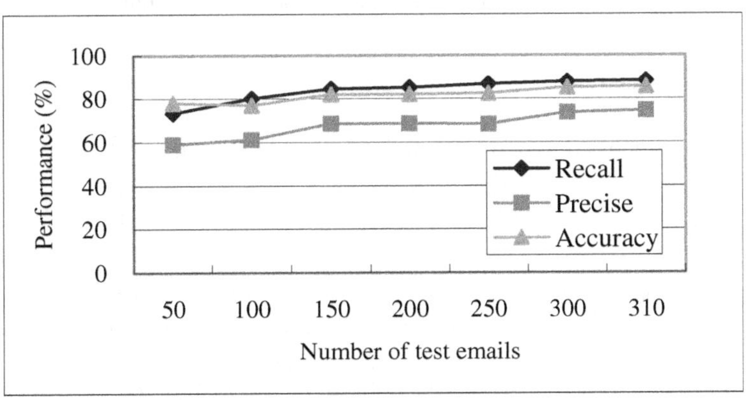

Fig. 2. Performances of ASID

respectively, while the data of Bayesian method were 69.04%, 81.07% and 81.56%. Because more new junk emails were contained in the new added test emails, there was a little decrease on recall and accuracy of ASID compared with those in table 2.

During the experiment, we found that if the scale of training set used for Bayesian method was small, the performances were very bad. We also found those junk emails whose features were not contained in the scope of statistics escaped the filtering of Bayesian method. Conversely, such junk emails could be learned in the first immune response and be recognized quickly in the second immune response using ASID.

From the above experiments, we found another feature of ASID was that the number of detectors produced during immune circle was not so big. The reason is that innate immune response is introduced to deal with normal junk emails using traditional methods (e.g. list-based method and rule-based method). And adaptive immune response is used to cope with junk emails with various tricks.

4.3 Experiment 2: Distribution of Spam Vaccine in P2P Environment

In the simulated P2P environment, we firstly separated the test emails into 6 portions and use ASID to filter junk emails in the six peers respectively. Then we exchanged the test emails with each other and found that the exchanged emails could be classified correctly in other peers. And the idea of vaccine distribution is validated.

5 Conclusion

This paper presents a new model to fight against junk emails (especially Chinese junk emails) in P2P environment. Many filtering strategies are integrated in this model. And the implementations of immune principles are described in detail. The experimental results show that better self-adaptability is achieved. Using fewer detectors, it achieves similar or better performances to some other anti-spam methods. Therefore, it provides a new attractive solution to anti-spam problem.

Further research will be focused on an improved affinity function, ways to simplify detector and integrate other feature extraction methods, e.g. visual features [13].

Acknowledgment

This work was supported in part by Tianjin Hailiang Co. Ltd.

References

1. X. Carreras and L. Mrquez: Boosting Trees for Anti-Spam Email Filtering. In: Proceedings of the 4th International Conference on Recent Advances in Natural Language Processing (RANLP 2001) (Tzigov Chark, Bulgaria, September 2001) 58–64.
2. M. Sahami, S. Dumais, D. Heckermany, and E. Horvitz: A Bayesian Approach to Filtering Junk E-mail. In: Proceedings of the AAAI Workshop on Learning for Text Categorization (Madison, Wisconsin, July 1998) 55–62.

3. I. Androutsopoulos, G. Paliouras, V. Karkaletsis, G. Sakkis, C.D. Spyropoulos, and P. Stamatopoulos: Learning to Filter Spam E-mail: A Comparison of a Naive Bayesian and a Memory-Based Approach. In: Proceedings of the Workshop on Machine Learning and Textual Information Access, 4th European Conference on Principles and Practice of Knowledge Discovery in Databases (PKDD 2000) (Lyon, France, September 2000) 1–13.
4. A. Kolcz and J. Alspector: SVM-based Filtering of E-mail Spam with Content-Specific Misclassification Costs. In: Proceedings of the TextDM 2001 Workshop on Text Mining, IEEE International Conference on Data Mining (San Jose, USA, November 2001) 1048–1054.
5. Hall R. J.: Channels: Avoiding Unwanted Electronic Mail. Communications of ACM (March 1998) 85–102.
6. Huaping Zhang and Qun Liu: Model of Chinese Words Rough Segmentation Based on N-Shortest-Paths Method. Journal of Chinese Information Processing, vol. 16 (2002) 1–7.
7. Feng Zhou, Li Zhuang, Ben Y. Zhao, Ling Huang, Anthony D. Joseph and John Kubiatowics: Approximate Object Location and Spam Filtering on Peer-to-peer Systems. In: Proceeding of ACM/IFIP/USENIX International Middleware Conference (Rio De Janeiro, Brazil, June 2003) 1–20.
8. E. Damiani, S. De Capitani di Vimercati, S. Paraboschi, and P. Samarati: P2P-Based Collaborative Spam Detection and Filtering. In: Proceedings of the Fourth International Conference on Peer-to-Peer Computing (P2P'04) (Zurich, Switzerland, August 2004) 176–183.
9. J. Metzger, M. Schillo, and K. Fischer: A Multiagent-Based Peer-to-Peer Network in Java for Distributed Spam Filtering. In: Proceedings of the International Central and Eastern European Conference on Multi-Agent Systems (CEEMAS 2003) (Prague, Czech Republic, June 2003) 616–625.
10. LiTao: Computer Immunology. Publishing House of Electronics Industry, Beijing (2004).
11. HAN Jia-wei and KAMBER M.: Data Mining Concepts and Techniques. China Machine Press, Beijing (2002).
12. Ion Androutsopoulos, John Koutsias, Konstantinos V. Chandrinos and Constantine D. Spyropoulos: An Experimental Comparison of Naïve Bayesian and Keyword-based Anti-Spam Filtering with Personal E-mail Messages. In: Proceedings of the Annual International ACM SIGIR Conference on Research and Development in Information Retrieval (Athens, Greece, July 2000) 160-167.
13. Ching-Tung Wu, Kwang-Ting Cheng, Qiang Zhu, and Yi-Leh Wu: Using Visual Features for Anti-Spam Filtering. In: Proceedings of the International Conference on Image Processing (ICIP 2005) (Genoa, Italy, September 2005) 509-512.

NASC: A Novel Approach for Spam Classification

Gang Liang, Tao Li, Xun Gong, Yaping Jiang, Jin Yang, and Jiancheng Ni

Department of Computer Science & Technology, Sichuan University,
610065 Chengdu, China
gangliang56@163.com

Abstract. The technology of spam filters has recently received considerable attention as a powerful approach to Internet security management. The traditional spam filters almost adopted static measure, and filters need be updated and maintained frequently, so they can not adapt to dynamic spam. In order to get over the limitation of the traditional means, an immunity-based spam classification was proposed in this paper. A brief review about traditional technology of spam filter is given first, particularly, the Bayes probability model. Then the NASC is described in detail by introducing self, non-self, detector and detect algorithm. Finally, a simulation experiment was performed, and the important parameters of the model were analyzed. The experimental result shows that the new model increases the recall of filter greatly in condition that precision also increasing, and demonstrate that the model has the features of self-learning and self-adaptation.

1 Introduction

The effective development and full utilization of Internet have propelled the social and economic development, and have changed the means of people's work, life and communications. Following the enlarging of Internet scale and business, the amount of netizen is increasing rapidly. Among the services of Internet, the e-mail service is the mostly welcomed one with the most users. However, a few users domestic and abroad send enormous Unsolicited Commercial email (or UCE or "spam") that not only occupy network resources, but also affect normal communications, which makes the spam e-mail the serious international problem [1], [2], [3], [4].

By the late 1990s, spam had become a major problem for Internet Service Providers (ISPs) and businesses. ISPs and enterprises were forced to take steps to stop spam from overwhelming their email servers and flooding computer networks.

True anti-spam technologies, referred to below as second and third generation technologies [5], are a relatively new development that followed the advent of spam in the mid 1990s. Anti-spam technologies are entirely new and are sharply differentiated from basic MTA controls. In particular, Bayesian algorithms and more recent advances in text analysis are a radical departure from pre-existing mechanisms characterized or repurposed as anti-spam tools [5].

Although the anti-spam technologies have get rapid development, but most of these technologies adopt to static methods and they can not adapt to the change of spam.

D.-S. Huang, K. Li, and G.W. Irwin (Eds.): ICIC 2006, LNBI 4115, pp. 672 – 681, 2006.

Spam changes over time as new products become available or popular, but it also changes because the problem is co-evolutionary spammers adapt to filters, and filters adapt to spam.

The immune system model lends itself reasonably well to creation of adaptive system [5] and artificial immune systems (AIS) have been used for a diverse set of things including spam detection. AIS produced contains a unique set of detectors, making it harder for spam senders to create messages that will go through many such systems. It can detect not only repeat messages, but also materials that have not already been seen by the system. AIS has many of the advantages, and in addition it can use fairly complex heuristics as detectors, that should lead to more accurate matching.

The aim of this paper is to establish an immune-based model which called NASC for dynamic spam detection. The experimental results show it is a good solution for spam detection.

The rest of this paper is organized as follows. In Section 2, we review the technologies that have used to combat spam. In Section 3, a novel approach for spam classification was proposed and described in detail. Section 4 the simulations and experimental results are provided. The conclusions are given in Section 5.

2 Anti-spam Technologies

There are many solutions that can be used to combat the various types of email-based spam. Filtering messages, blocking message senders, ensuring senders are authentic, and Challenge/Response Systems are all methods used to combat spam. In this section, we briefly review the technologies that have used to combat spam.

2.1 Message Filtering

In general, implementing message filtering is straightforward and does not require any modifications to existing email protocols. A good filter design will minimize the number of false positives (filtering of a message that is not spam) and maximize the efficiency of the filter. Filters simply prevent spam from entering inboxes, but do not stop the generation of spam.

Message filtering often rely on humans to create detectors based on the spam they've received. Humans can create good heuristic detectors, but with spam senders intelligently adapting to anti-spam solutions, it is hard to keep up. A dedicated spam sender can use the frequently publicly available information about such heuristics and their weightings to evade detection [1], [3], [5].

2.2 Black List Sender Email Addresses

This is a simple spam blocking technique that is often used. Users create a black list of from addresses that should be prevented from entering the network and reaching the user's inbox. There are a few disadvantages with using this technique. Because spammers can create many false from email addresses, it is difficult to maintain a black list that is always updated with the correct emails to block. Also, some spammers do not even use a from address so a black list would not be able to catch these cases. Even a rule to block emails without a from address would not be sufficient because some legitimate emails, such as newsletters to which a user may subscribe, may also not include a from address. Black list sender email addresses is effective in temporarily blocking a small amount of spam but ineffective as an overall anti-spam solution. As an alternative to black lists, some users set up an e-mail white list consisting of acceptable e-mail addresses or domains. Users only accept e-mail from users that are listed on their white list, while all other e-mail is blocked. However, it does not solve the problems associated when people who legitimately want to send you e-mail have not previously corresponded with you via e-mail, have multiple e-mail addresses, or have a new e-mail address.

2.3 Challenge/Response Systems

Challenge/response systems are used to counter spammers who use automated mailing programs to generate millions of spam emails per day. These systems are designed to slow down spammers by putting roadblocks up for the incoming spam.

Challenge/response systems work under the assumption that spammers using fake sender email addresses would never receive the challenge, and spammers using real email addresses would not be able to reply to all of the challenges. But Challenge/response systems also have a number of limitations [1], [2], [3], [4].

3 Immunity-Based Spam Classification Model

The aim of this paper is to establish an immune-based model which called NASC for dynamic spam detection. The model is composed of three parts: mail character presenting module (MCPM), training module and filtering module. MCPM use vector space Model (VSM) and present the received mail in discrete words. Training module generates various immature detectors from gene library. However, some of these new immature detectors are false detectors and they will be removed by the negative selection process [5], which matches them to the training mails. If the match strength between an immature detector and one of the training mails is over the pre-defined threshold ,this new immature detector is consider as a false detector. Filtering module matches the received mails to the mature and detectors. If the match strength between a received mail and one of detectors, the mail will be consider as the spam.

3.1 Mail Character Presenting Module (MCPM)

Computer is not in possession of human being's intelligence and it only can recognize 0 and 1. So, the mails must be converted the form which computer can catch on at first, if we want to implement the spam filtering function by using computer. According to the Bayes-hypothesis, we can represent the document by means of keywords and index terms. The vector space model (VSM) has been widely used in the traditional information retrieval (IR) field [4], [6].Most search engines also use similarity measures based on this model to rank Web documents. Documents are mapped into term vector space. A document is conceptually represented by a vector of keywords extracted from the document, with associated weights representing the importance of the keywords in the document and within the whole document. The form of the vector is listed below:

$$V = (t_1, w_1, t_2, w_2 \cdots t_n, w_n) \quad (1)$$

Where, t_i represent keyword and w_i is the associated weights representing the importance of the keywords in the document.

MCPM contains two subsets: Chinese dividing module and keywords retrieving module. The task of Chinese dividing module is to separate the document into words and phrase. The algorithm used in dividing model proposed in this paper is maximum matching method [6], [7].

Once the Chinese dividing module has finished its job, the keywords retrieving model is to pick up suited keywords according to the weight of keywords and form the term vector space. The weight of a term in a document vector can be determined in many ways. A common approach uses the so called *tf x idf* method [6], in which the weight of a term is determined by two factors: how often the term *j* occurs in the document *i* (the term frequency $tf_{i,j}$) and how often it occurs in the whole document collection (the document frequency df_j).Precisely, the weight of a term *j* in document *i* is $w_{i,j} = tf_{i,j} \times idf_j = tf_{i,j} \times \log N \div df_j$, where *N* is the number of documents in the document collection and *idf* stands for the inverse document frequency. This method assigns high weights to terms that appear frequently in a small number of documents in the document set.

3.2 Training Module

Define 1 Gene: the keywords which are include in the uninteresting e-mail.

According to the document's structure, the gene library is divided into two subsets: mail list gene library and content gene library. Mail list gene library consists of suspicious email addresses or domains which are issued by the authoritative institution. User can update the library not only by the internet but also by adding the unacceptable email addresses by himself. Content gene library consists of keywords which are picked up from spasm. The initial genes of content gene library used in this paper are all from China anti-spam alliance. We can also update content gene library by making use of filtering module after the system works.

Define 2 Immature Detector: the term vector which is organized in pre-defined structure <sender mail address, subject, keyword1,keyword2···keywordn> and every elements of the term vector are all from gene library.

Define 3 Affinity: the match strength between two documents, when documents are mapped into term vector spaces, common similarity measure, known as the cosine measure, determines the angle between the document vectors and the query vector when they are represented in a n-dimensional Euclidean space, where n is the vocabulary size. Precisely, the similarity between a document *D1* and a query *D2* is defined as

$$Sim(D_1, D_2) = \frac{\sum_{k=1}^{n} w_{1k} \times w_{2k}}{\sqrt{\left((\sum_{k=1}^{n} w_{1k}^2)(\sum_{k=1}^{n} w_{2k}^2) \right)}} \tag{2}$$

Where $w_{2,j}$ is the weight of term *j* in the query, and is defined in a similar way as $w_{i,j}$. The denominator in this equation, called the normalization factor, discards the effect of document lengths on document scores.

If the match strength between an immature detector and one of the training mails is below the pre-defined threshold , the detector is consider as a mature detector. The matching function *f-aff* is defined below:

$$f\_aff(D_1, D_2) = \begin{cases} 1, \; iff\,(sim(D_1, D_2)) \geq \delta \\ \\ 0, \; other \end{cases} \tag{3}$$

Where δ is the pre-defined threshold.

Define 4 Mature Detector: the detectors whose affinity between nonself is over the pre-defined threshold. The structure of detector is list below:

MC=<term vectors, count, age>. (4)

Where count represent the detected spam number, and age is the survival period and the initial value of age is 0.

During the training stage, the goal is to populate the gene libraries, produce an initial set of immature cells from training examples. The training module generates immature detectors by choosing random sections from gene library. Negative selection [5] is used to precensor the generated immature detectors against training example. This guarantees that false-positive errors [5] do not occur against a self as any detector matching self is removed before fielding. An overview of this algorithm is described by the Pseudocode list below

```
program Train (var l, m:integer)
  begin
    repeat
       generate l immature detectors
       foreach(detectors ∈ the set of generated immature
                 detectors)
       compute the affinity between detector and self
       if the affinity of a detector is over threshold
          remove the detector
       else
          add the detector to the set of mature detectors
        end foreach
     until the number of the mature detectors is over m
end.
```

3.3 Filtering Module

Define 5 Memory Detector: the detector which is considered as reliable detector. A mature can become a memory detector if the matching threshold is over active threshold. The structure of memory detector is list below:

MMC=<term vectors, time, count> (5)

Where count represent the detected spam number, and time indicate the time when the detector detect the spam and this parameter is used in the memory detector update procedure.

Define 6 Co-stimulation: the signal which comes from administrator to system. This signal is used in the following two situations:

1. When mature detectors detect a suspicious mail, system need this signal to judge the suspicious mail whether is junk.
2. When the count field of one mature detector is over active threshold, system will need the co-stimulation signal. If an affirmative co-stimulation signal writes back to system from network administrator in fixed period, this mature detector will become memory detector. Otherwise, it will be recognized as invalid detector and deleted from mature detector set.

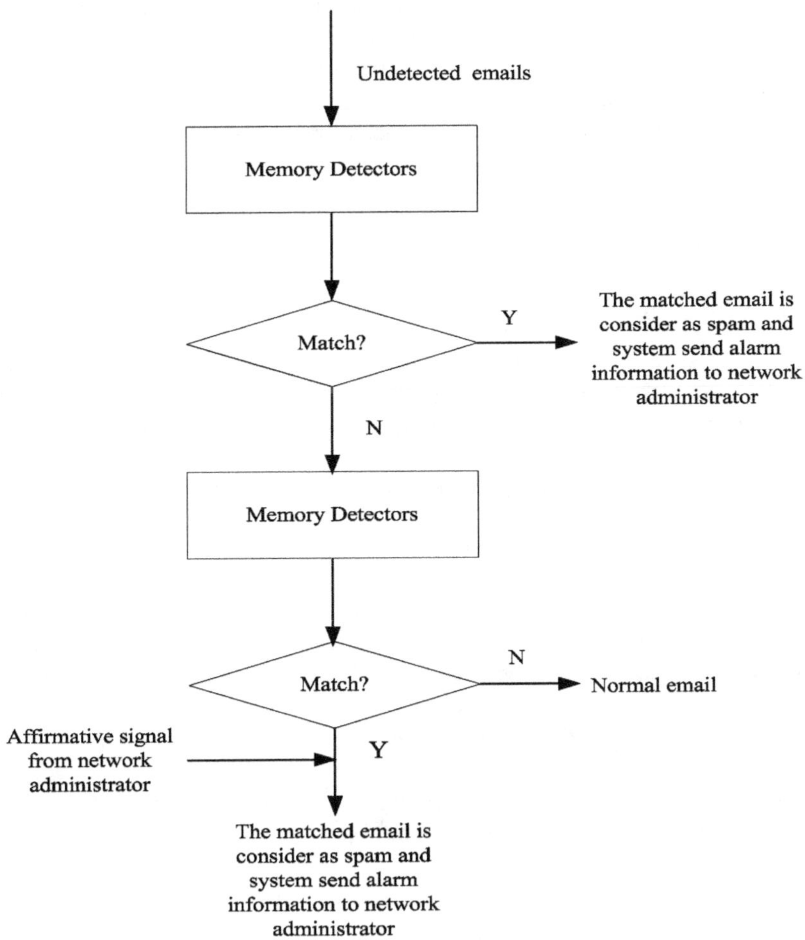

Fig. 1. The system working flow of filtering model

As the Fig.1 shows, the undetected e-mails are compared with memory detectors firstly. If one e-mail match any elements of memory detector set, this e-mail is classified as spam and send alarming information to user. In NASC, E-mail classified as junk is not deleted but removed to a temporary store. Mature Detectors must have become stimulated to classify an e-mail as junk, and therefore it is assumed the first stimulatory signal has already occurred. Feedback from administrator is then interpreted to provide a co-stimulation signal. If an e-mail is simply deleted from this store we assume the algorithm has performed a correct classification as the user really was not interested in that e-mail and so a co-stimulation signal has occurred. The cell is rewarded by being allowed to reproduce. If, on the other hand, the user does not delete the e-mail the algorithm has performed a misclassification, signal two does not occur and mature detector or memory detector is removed appropriately.

To both counteract the increase in population size brought about by reproduction and keep the algorithm dynamic, all detectors in the model are all given an expiry date. Expiry allows the system to remove old detectors that have never matched any messages so that they do not waste processing time of the system. In addition, this expiry is what allows the system to forget detectors that are no longer in active use. Once old detectors have been removed, new detectors can be generated to replace them, starting the cycle again.

4 Simulations and Experimental Results

4.1 The Evaluation Method of the Model

The essence of spam classification is a special text classification. So the evaluation standard of spam classification is same to text classification's, we can use recall ratio and precision ratio to evaluate the performance of the model [6], [7].

4.2 Experimental Environment and Procedure

The experiment has been carried out in the Laboratory of Computer Network and Information Security at Sichuan University. The whole experiment is divided into two phase: training phase and application phase. The different between the training phase and application phase is that training phase does not use filtering module. At the beginning of the system working, the set of detectors is empty. Experiments were performed with 2000 e-mails of which 750 were manually classified as spam, the remaining 1250 were assumed of normal email. We partition these 2000 emails into 10 equal parts and choose one part randomly as a training example, then the remain parts are used for test and we can get 9 group recall and precision ratios. The average value of these 9 group values is considered as the model's recall and precision ratio.

4.3 Experimental Result and Result Analysis

There are several important parameters in the model such as initial detectors scale M, active threshold δ, matching threshold β,mature detector life cycle$\lambda$and the vector's dimension k. In order to evaluate every parameter's impact on model's performance, we have carried out 9 group experiments. The experimental result and result analysis are described below:

Table 1. The effect on the experimental Result by modifying the parameter

M	δ	β	Λ	k	recall (%)	precision (%)
50	10	0.5	120	10	77.65	84.8
100	10	0.5	120	10	79.46	85.9
150	10	0.5	120	10	78.54	85.7
100	5	0.5	120	10	83.98	85.11
100	10	0.5	120	10	79.46	85.9
100	15	0.5	120	10	56.96	87.2
100	10	0.3	120	10	84.62	82.3
100	10	0.5	120	10	78.54	85.7
100	10	0.8	120	10	63.47	86.1
100	10	0.5	60	10	74.63	84.8
100	10	0.5	120	10	79.46	85.7
100	10	0.5	360	10	82.39	85.3
100	10	0.5	120	5	78.14	85.11
100	10	0.5	120	15	75.86	86.6
100	10	0.5	120	20	60.68	87.9

The effect of initial detectors scale M: As the table 1 shown, with the increasing of M, the recall and precision rises rapidly at beginning, but the recall and precision all hold the line when system has worked for a time whatever the change of the value of M. This is because M controls the scale of the detectors. There are more detectors if m increases, so the recall and precision also increase. But after the system running for some time, the recall and precision change little because the match algorithm adopt to imprecise match and the detectors of low M model also can recognize familiar spam. So the initial detectors scale is not decisive factor to the model's performance. But the value of M should not be too low.

The effect of active threshold δ: According to table 1, the recall is in positive proportion toδ.There are more memory detectors ifδ is low and model can detect more spasm. Otherwise, cause of the mature detectors has expiry date, many mature detectors has been invalid before it become memory detectors and model can not afford to enough detectors to detect spasm, So the recall is decline. We can also get a conclusion from table 1 thatδ has little affection on the precision.

The effect of life cycle λ: The life time of mature detectors is more long, the number of the mature detectors is more big and the number of the memory detectors also more big. So the recall increase with the increasing λ.As the table 1 shown , λ has no impaction on precision.

The effect of match threshold β: As table 1 shown, theβ is more high ,the recall is more high and the precision is more low. This is because system can correctly detect really spam in strict match condition, but due to the strict match condition, some of spam can not be detected by model and the precision decreases.

The effect of the vector's dimension k: According to table 1, with the increasing of k, the recall can increase gradually and it will reach max when k=15, then it will decrease gradually. This is because the dimension is low and vector space can not stand for document, but if dimension is high, it will introduce classification noises.

Experiment was repeated 10 times when turning the parameters. A satisfied result (recall=75.84% ,precision=85.7%)is obtain with $M=100$, $\delta=10$, $\beta=0.5$, λ=120, k=15.In the same condition, the recall and precision of Bayes method [9], [10] are 67.76% and 83.98%.This is immune method can adapt to dynamic spam and has the features of self-learning and self-adaptation.

5 Conclusion

The previous models or methods almost adopted static measure, and filters need be updated and maintained frequently, so they can not adapt to dynamic spam and lack of self-adaptation.

In this paper, an immune-based dynamic model called NASC for spam detection has been built. The experiment shows that this model can efficiently raise both the recall ratio and precision ratio, and enhance the ability of self-adaptation and diversity for the spam.

Acknowledgments

This work is supported by the National Natural Science Foundation of China under Grant No.60373110, 60573130 and 60502011, the National Research Foundation for the Doctoral Program of Higher Education of China under Grant No.20030610003, the New Century Excellent Expert Program of Ministry of Education of China under Grant No.NCET-04-0870, and the Innovation Foundation of Sichuan University under Grant No.2004CF10.

Reference

1. Crawford, E., Kay, J., McCreath, E.: IEMS – The Intelligent Email Sorter. In Proc. of the Nineteenth International Conference on Machine Learning, Sydney, Australia (2002)
2. Brutlag, J.D., Meek, C: Challenge of the E-mail Domain for Text Classification. International Conference on Machine Learning (2000) 103-110
3. Liu, B., Ma, Y., Yu, P. S.: Discovering unexpected information from your competitors' web sites. In Proc. of the Seventh International Conference on Knowledge Discovery and Data Mining, San Francisco, USA (2001)
4. Diao,Y., Lu, H., Wu, D.: A Comparative Study of Classification Based Personal E-mail Filtering. PAKDD (2000)

5. Li, Tao: Compute immunology, Publishing House of Electronics Industry, Beijing (2004)
6. Salton, G., Wong, A., Yang, C. S.: A vector space model for automatic indexing. Communications of the ACM (1975) 613-620
7. LEE, DIK L., CHUANG, HUEI: Document Ranking and the Vector-space model. IEEE SOFTWARE (1997) 67-75
8. Fabrizio, Sebastiani: Machine learning in automated text categorization. ACM Computing Surveys (2002) 11-12, 32-33
9. Zhang, Le, Zhu, Jingbo, Yao, Tianshun: An Evaluation of Statistical Spam Filtering Techinques. ACM Transactions on Asian Language Information Processing (TALIP) Vol. 3, No.4 (2004) 243-269
10. Schneider, K. M.: A Comparison of Event Models for Naive Bayes Anti-Spam E-Mail Filtering. In Pmc. 11th Conf. Euro, Chop. ACL, Budapest, Hungary (2003)
11. Li, T.: A New Model for Dynamic Intrusion Detection. Lecture Notes in Computer Science, Springer-Verlag, Berlin Heidelberg New York (2005) 72 - 84
12. Li, T.: An immune based dynamic intrusion detection model. Chinese Science Bulletin (2005) 2650 - 2657
13. Li, T.: An immunity based network security risk estimation. Science in China Ser. F Information Sciences (2005) 557 - 578
14. Li, T.: An Immune-Based Model for Computer Virus Detection. Lecture Notes in Computer Science, Springer-Verlag, Berlin Heidelberg New York (2005) 59 - 71

Prediction Algorithms in Large Scale VOD Network Collaborations

Bo Li[1], Hualin Wan[2], and Depei Qian[1]

[1] School of Computer Science and Engineering,
Beijing University of Aeronautics & Astronautics,
Beijing, China
{libo, depeiq}@buaa.edu.cn
[2] Sino-German Joint Software Institute,
Floor 16, Shining Building, No.35 Xueyuan Lu,
Beijing, China
hualinwan@jsi.cn

Abstract. VOD (Video on Demand) is one of significant services for next generation networks. VOD networks mean local networks to provide VOD services to communities in a scale about 500 to 1000 users simultaneously. Many VOD networks could cooperate through external networks link internet. The cooperation requires not only huge local bandwidth but great backbone bandwidth. This paper presents prediction algorithms to reduce the cost in external communications which is very important for VOD operators. Basic algorithms can reduce overall costs about 30% and well trained ANN can provide extra 10% performance.

1 Introduction

VOD systems have been implemented in market for nearly ten years. In early years, it started from KTV discs based system, which was operated by a PC to provide single video clip through local Ethernet networks. Video clips were compressed in RM or MPEG1 standards with a CIF resolution, the quality was very close to VCD. The services could be delivered to tens of customers if there were enough PCs available. The second generation VOD services were implemented in hotels for customers to enjoy personal TV programs. Some of them were provided in 2M TS stream with MPEG2 standards similar to DVD quality. Single service cost 2Mbps bandwidth and it was difficult to provide the service out of the hotel. Now, VOD systems enter the third generation, which targets provide VOD services to everyone through local networks. And the rapid popularizing is expected when IPTV and HDTV enter families. From now on, VOD not only is an engineering term, but raises some academic interests. How to provide good service with current infrastructure is very important to promote VOD services in market.

It is very clear that for each local area, and local networks, to provide VOD, a server or a server group is required and it transfer video clips from storage to customers. There are lots works done on the single VOD networks[1][2][3][4][5].

D.-S. Huang, K. Li, and G.W. Irwin (Eds.): ICIC 2006, LNBI 4115, pp. 682 – 692, 2006.

Most of them concentrated on the efficiency of local server group or local bandwidth and tried to reduce the latency for a single requirement. Now VOD system can be implemented in community scale and it is not difficult and not expensive to run a VOD system to serve 1,000 users simultaneously. Recent works concern more on mobile devices and large scale VOD problems [6][7][8][9].

Single VOD networks, especially ones serve local communities can not be equipped with all video clips available to deliver and customers always want more to watch and most of video clips not available in where they get the services. In these cases, VOD networks can seek to cooperate to provide more programs and because there are many video clips transferred through internet, it costs great internet bandwidth and even backbone bandwidth. Bandwidth consuming is core problem for VOD operators to consider, and if there is no good methods to reduce the costs, operators can hardly tend to cooperative mode. Another problem is to build a complete new video warehouse; served clips should be in good format like MP4 or H.264, where input video should be in different format as DVD, VCD, or even VHS. For huge transcending works, it is better to distribute works on many local VOD systems and share all clips through internet.

For collaborations, there are two kinds of basic sharing strategies. The first is all sharing are not cached and each customer can require video clips from external accessing. Obviously, it costs bandwidth a lot and it avoids any problems invoked by copying from one networks to another. The second methods assume two or more networks belong to the same group and there is nothing wrong to cache programs from one. For any single requirement, it makes a copy of required video clip and transfers it to customers locally. This method avoids some unnecessary transferring but gets another kind of unnecessary transferring when the user just wants to skim those clips. In most cases, it is difficult to say which method is better.

This paper presents a few prediction algorithms for VOD networks to cooperate and it also presents predicting methods based on the working mode. With prediction, systems could know if the transferring is necessary or in which cases it is necessary. The prediction also could show the probabilities for each transferred copy to be used in short future, which is very important for server groups to optimize the performance of whole system. Following sections will presents cooperative architecture of VOD, predicting approaches, and cooperative training to improve prediction result. At the end of this paper, some simulation results will be presented to indicate the performance of prediction.

2 Cooperative VOD Networks Architecture

For a single VOD networks, it works in following way:

The video clips are stored in a SAN or similar device. A server or server group get clips from storage and transferred to terminal devices. In the testbed for this paper, the video clips follow H.264 and enhanced AAC+ standard, packed into TS stream format, which can be implemented in general broadcasting devices. For video stream with a resolution of 640 by 360 at 30 FPS, it can provide a DVD quality video with 512Kbps bandwidth. In this case, the overall performance is much better than most of VOD systems.

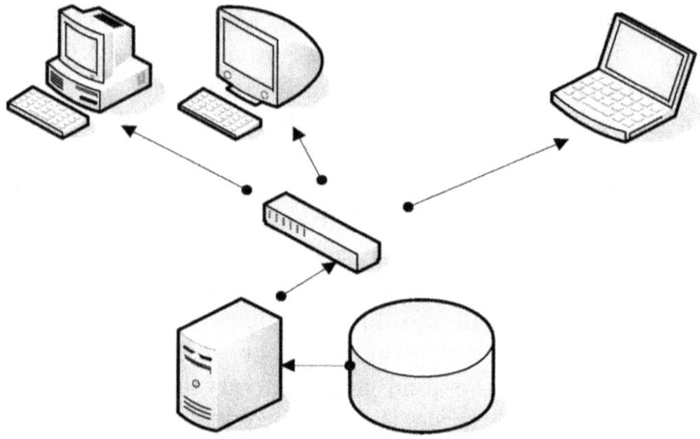

Fig. 1. VOD networks

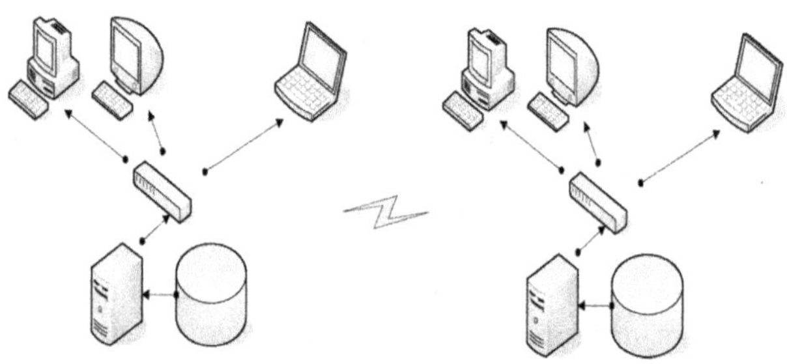

Fig. 2. Cooperative VOD networks

Commonly, two VOD networks are connected through internet, and for special reason, they can be connected through a dedicated cable. For the first case, the internet bandwidth is also shared by other purposes and costs considerable expensive. This is why to optimize cooperative methods for VOD networks.

The testbed for this paper can serve about 500 VOD request simultaneously. The main obstacle to serve more customers is the effective local bandwidth and server groups output capability. Assuming there are about 2,000 users in a communities use this kind of VOD services, 3% programs are not available in local VOD networks, each programs lasts about half an hour, getting all clips from external system needs about 54G bits each day. For some new and hot programs, the costs can be great. The person in Charge of those services can manage it in this way: for some clips they think hot and popular, get them from external and serve them locally, and this solution leads to the prediction methods described in following section.

In most cases, two VOD networks work in four modes as following:

- Mirror mode, one networks works as another one's mirror, when there are any new clips available in main networks, mirror can pull it to local storage;
- Peer to peer mode, two networks work independently, they will exchange clips only when users require, when transcoding works are distributed in VOD

networks, this mode can be basic mode;

- Two networks are managed by a system, which manage all clips like distributed storage; this mode involves different sharing strategies.

In this paper, first two kinds of modes are discussed and prediction algorithms are implemented in systems to reduce external bandwidth consuming.

3 Prediction Algorithms for VOD Networks

It is not difficult for human to decide which clip is hot or not, but it is not easy for computer system to know about it. A clip is required by a user, maybe in different conditions:

- User want to watch it from beginning to end
- User want to know about the story of it, he will skip most of details
- User want to know if he had watched or not
- User want to know if there is any player he likes
- User want to continue watching from last time position
- User make a mistake to click on it
- ……

Generally, different conditions show different signal to system, and system could analyze what the original intention of the user.

If each user's intention is clear, system could get an overall trends for users to a video clips. Operators can provide preview clips and detailed information for each clip to help users to find what to watch and how to watch. It is very important for systems to know well about users' intentions.

For each video program, there are also many conditions different from each other:

- New video program never being required
- New video program required many times
- Old video program required many time recently
- Old video program required many time, but last requirement was made a year ago
- Old video program seldom being required, but until now there still some requirements made for it
- ……

Generally, system could make a rank list for each video program to be evaluated, if the rank higher than a threshold, it could be a hot program and of course, for any external requirement, it is better to get the full video and put it in local storage.

The statistics of video requirements and user access log can help external visitor know about the programs and ranked list can help it decide what to do next. For prediction methods is to find what will happen next, especially for video program without a clear future. And all the predictions are based on statistics and rules produced by the system.

In a VOD network, for a user's single requirement, assume A represents all operations for this requirement, each A_i represents a single operation and its timestamp like watching from position a to position b. Then a user operation model $U(A)$ can be implemented to decide the type of this requirement.

For a video program, R represents all requirements since it put into the storage. R_i represents a single requirement for this video program. Then a video program usage model $V(R,U(R_1),...U(R_n))$ can be implemented to decide the type of this video program.

When $U(A)$ and $V(R,U(R_1),...U(R_n))$ are available, a prediction can be made through $P(V(R,U(R_0),...,U(R_n),U_P))$, where U_p means current requirement type.

The main problem for the prediction is model is not easy to build, for different user group, user models is much different from each other. A young people model can not be used on old people. Another problem is how to estimate time factor in the models. Time factors will vary from different. Two basic factors can not be ignored: one is strategy of operator, and another one is internet tariff standard, both of them can affect models above greatly. If tariff is higher enough, the cooperative transferring can be ignored. If the system can provide detailed introduction, the users can decide if to require the program before send requirement. Any way, it is possible to estimate user behavior and program accessing pattern, which lead to a prediction to show how much probabilities this program can be required by other users in a short time.

For stochastic estimation, it is not easy to tell how close from a model to reality, but the estimation should be limited with a border, not to introduce too many errors to later prediction.

An ANN was implemented to estimate video program model as figure 3. The input is video program properties and a single requirement, with parameters as time, operation, duration, time to previous requirement and time to next requirement. The ANN is trained with real data from system and could output a popular rate.

For a few cooperative VOD networks, it is not difficult to evaluate a program's popular rate in each network. Overall result can be used to calculate the related probabilities.

There are three prediction models for program popular rate compared in the testbed. The first one is simple fuzzy logic model. Preset the rules and formulas to calculate the probabilities. For example, rules can be in this way:

- If the same type of program is hot, this program with a middle above popular rate, the probability is high
- If the program is new and the type is popular and the topic is OK, the probability is middle
- If the program is old, and recent usage is not bad, the probability is middle
- ……

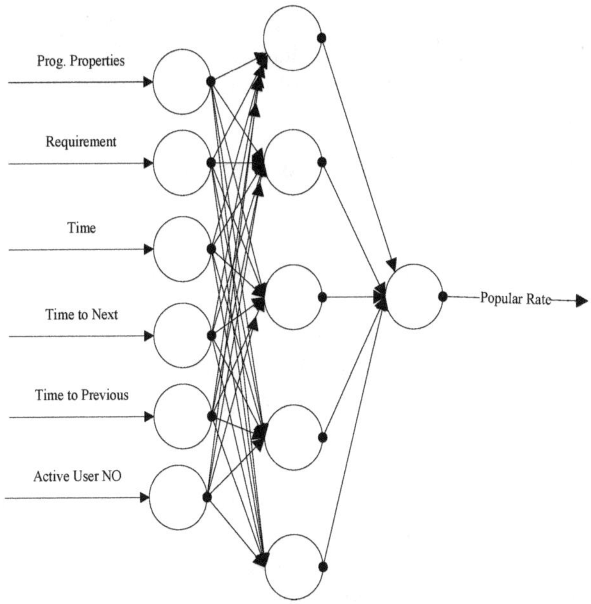

Fig. 3. ANN to produce program model

The performance of this prediction depends on the rules very much. For different system, rules should be adjusted to adapt new system, boring and low efficient.

The second prediction method is by Monte-Carlo methods and Bayesian methods. The main problem is the observed result is not probabilities what the prediction produced. In this case, another parameter was used to evaluate usage of supposed to be transferred video programs, the time span to next requirement in local networks. The prediction algorithm needs to use requirement recodes in required networks to estimate the time span and it is very difficult to make a right estimation for new programs while there are not enough data to be used in perdition.

The third method is ANN. ANN methods can also be implemented in predicting the time span.

ANN performance bases on the number of samples. For a small network, it is difficult to get a converged ANN for general usage.

It is recommended to combine different methods at same time to achieve better performance.

User behaviors should be also predicted based on the user profiles and user operation. The basic ideas are the same with program popular rate predictions. And the affective issues are more than popular rate predictions. For instance, the time of a requirement affects not only the number of simultaneous requirements, but also accessing patterns of users. And a user's accessing operations are in his own pattern should be analyzed individually.

Two VOD networks cooperate is not enough to provide highly cooperative environment. For more than three networks, sharing their model and prediction ANN will be also very efficient to improve the overall performance.

4 Experimental Results and Explanation

It is still difficult for our testbed to simulate a real VOD system provides 500 programs simultaneously. So a simulation program developed to evaluate the performance of prediction algorithms.

The simulation was set in following way:

- Video clips are compressed with H.264 and Enhance AAC+, resolution 640 by 360, 30 fps, stereo audio, packed in TS stream, 512Kbps
  Storage is an 30T raid5 SAN, connected with Fiber Channel Card to VOD server
- VOD server is about 32 CPU and 64GB RAM, maximal capability to serve 1,000 users simultaneously
- Local connection is 1000M Ethernet to desktop
- Users are classified into six group, by age and sex
- Video program are classified into ten type, some subtypes are stressed like cartoons and hot movies

Figure 4 compare external bandwidth usage for three strategies. Here, prediction strategy is combination of fuzzy logic and ANN.

The data in Figure 4 stresses more hot programs to share and the share rate is about 3%.

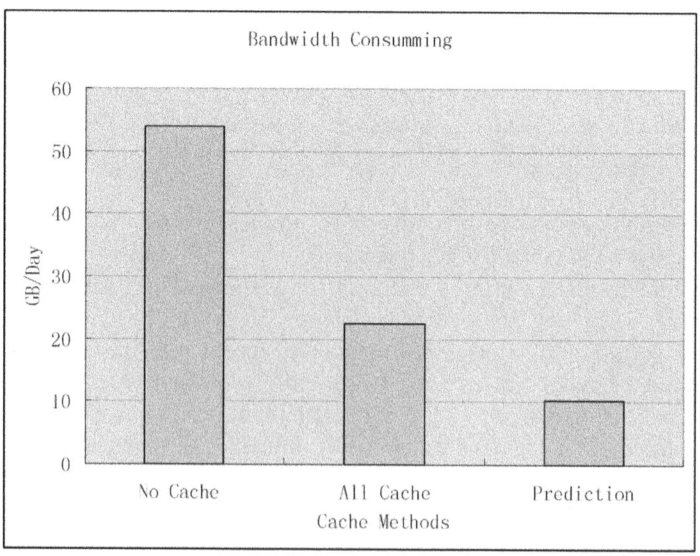

Fig. 4. External bandwidth consuming

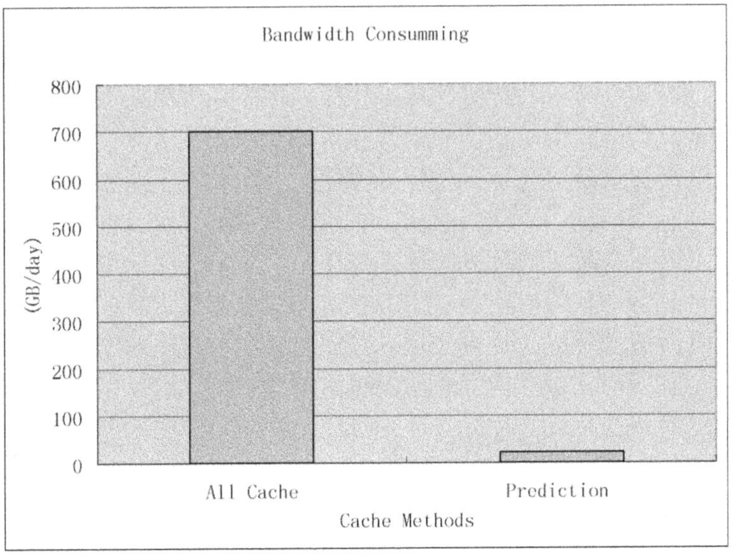

Fig. 5. External bandwidth consuming for mirror mode

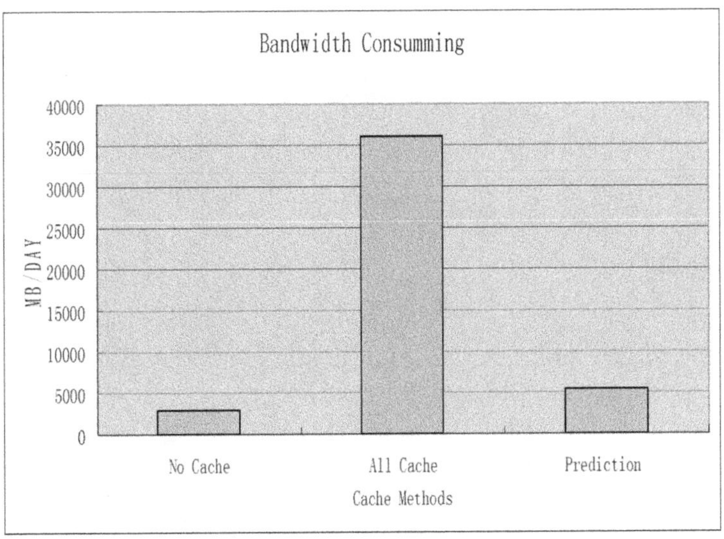

Fig. 6. Flowing bytes of external networks

Figure 5 shows in mirror mode, the performance of the prediction methods.

In a short time the prediction methods allow only very popular programs to be transferred, in the experiment, it ranged from 3% to 5%. There may be some in future, about 10% to 15% to be transferred. For a huge program warehouse, most of them are ignored by most of users.

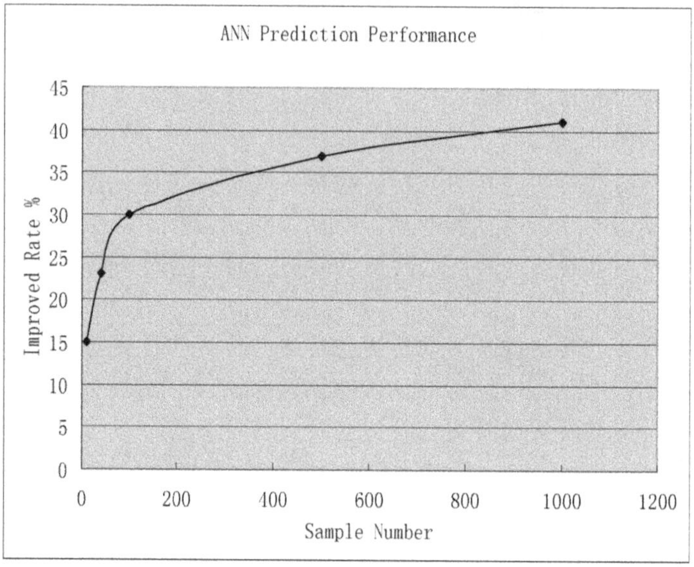

Fig. 7. Trained ANN prediction performance

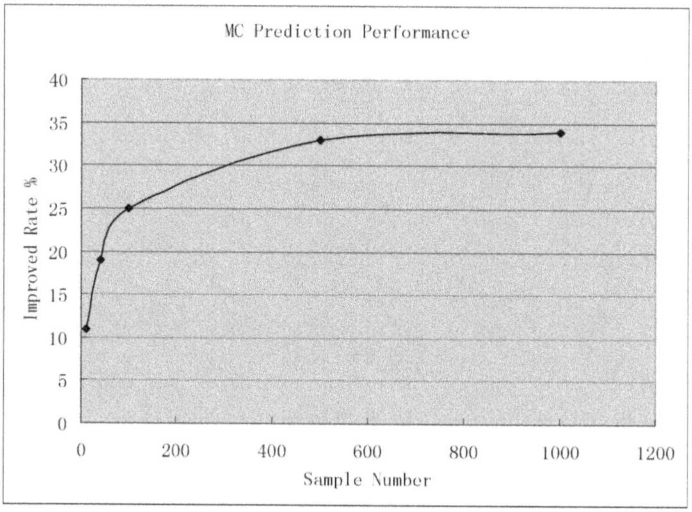

Fig. 8. Monte-Carlo prediction performance

For some requirements just to watch no more than 10 minutes. Figure 7 shows the external bandwidth consuming in a simulation process.

The performance is very bad when the system caches all things to play. In this case, prediction can reduce the transferring costs by user model.

It is not difficult to find, prediction methods can help VOD networks to reduce costs invoked in cooperative works. The question is what the extent and where is the limitation.

Figure 7 shows the performance curve for ANN prediction trained to achieve better performance. It was improved with extra fuzzy based rules and it can be improved with more training samples and more rules.

ANN can be trained to close the reality, but it can't get too close. It affected the performance of prediction algorithms. Before a good model to describe prediction and user behaviors in VOD area, ANN could be used for a good balance system performance.

Figure 8 shows Monte-Carlo prediction performance. Fixed PDFs were implemented in these experiments. Because the PDFs is preset and can't describe system very well, the performance is a bit poor compared to ANN results.

5 Conclusion and Future Work

Some starting works in prediction algorithm for VOD networks to cooperative presented in this paper. A testbed to VOD service and VOD networks to cooperate is build. During the development of VOD system, we found it very important to do some preparation for future cooperative works. Multi VOD networks and even other kinds of video providers can work together to serve better and save resources.

For VOD networks cooperation, cooperative learning and remote learning will be our future work to do. We also hope to push these research works to other area.

Acknowledgement

This paper based on the research of Chinese 863 project, Service Oriented Supporting Platform.

References

1. Allen, R.J., Heltai, B. L., Koenig, A. H., Snow, D. F., Watson, J. R.: VCTV: A Video-Ondemand Market Test. AT&T Technical Journal, 72(1) (1993) 7–14
2. Chou, C. F., Golubchik, L., Lui ,J. C. S.: Striping Doesn't Scale: How to Achieve Scalability for Continuous Media Servers with Replication. In Proc. IEEE ICDCS'00 (2000) 64–71
3. Lee, Y. B.,Wong, P. C.: Performance Analysis of a Pullbased Parallel Video Server. IEEE Trans. on Parallel andDistributed Systems, 11(12) (2000) 1217–1231
4. Zhou, X., L¨uling, R., Xie, L.: Solving a Media Mapping Problem in a Hierarchical Server Network with Parallel Simulatedannealing. In Proc. 29th ICPP (2000) 115–124
5. Zink, M., Griwodz, C., Steinmetz, R., KOM Player-A Platform for Experimental VoD Research, In Proc. ISCC 2001 (2001)
6. Sato K., Katsumoto, M., Miki, T.,: Fragmented Patching: New VOD Technique That Supports Client Mobility, in Proc. Of AINA 2005 (2005) 527-532

7. Bellavista, P., Corradi , A., Foschini, L.: MUMOC: An Active Infrastructure for Open Video Caching, In Proc. Of DFMA2005 (2005) 64-71
8. Chen, Y., Dai, Q.: Research on Model and Scheduling Algorithm for Double-Mode VoD Servers, In Proc. Of ICCNMC 2003 (2003) 297-304
9. Bruneo, D., Villari, M., Zaia, A., Puliafito, A.: VOD Services for Mobile Wireless Devices, in Proc. Of Eighth IEEE International Symposium on Computers and Communications (2003) 602-609

A Computational Korean Lexical Access Model Using Artificial Neural Network*

Heui Seok Lim[1], Kichun Nam[2,**], Kinam Park[3], and Sungho Cho[4]

[1] Division of Computer, Information, and Software, Hanshin University, Korea
limhs@hs.ac.kr
http://nlp.hs.ac.kr
[2] Department of Psychology, Korea University, Korea
kichun@korea.ac.kr
http://coglab.korea.ac.kr
[3] Department of Computer Education, Korea University, Korea
superkn@nate.com
http://comedu.korea.ac.kr
[4] Department of Information Communication, Hanshin University, Korea
zoch@hs.ac.kr
http://nlp.hs.ac.kr

Abstract. In this paper, we propose a computational Korean lexical access model based on connectionist approach. The model is designed to simulate the behaviors observed in human lexical decision task. The proposed model adopts a simple recurrent neural network architecture which takes a Korean string of 2-syllable length as an input and makes an output as a semantic vector representing semantic of the input. As experimental results, the model shows similar behaviors of human lexical decision task such as frequency effect, lexical status effect, word similarity effect, semantic priming effect, and visual degradation effect.

1 Introduction

Researches on the process of mental lexicon access and the representation of mental lexicon are very important to find out the internal mechanism of lexical processing in human brain. In order to measure the processes that are involved in mental lexical access, we need a task that guarantees that a lexical entry is accessed. A naming task, stroop task, and lexical decision task(henceforth **LDT**) are frequently used in the study of lexical access. Among them, LDT is a most frequently adopted methods. The LDT has been proven to be a fruitful method for exploring the nature of the mental lexicon. In LDT, subjects are asked to discriminate words from nonwords by pressing YES or NO button if the stimulus is a word or a nonword.

There are several basic findings that are consistently obtained using the visual LDT and these findings are observed independently on languages. The major

* This study has been supported by the Korea Research Foundation(2004-074-HM0004).

** Corresponding author.

D.-S. Huang, K. Li, and G.W. Irwin (Eds.): ICIC 2006, LNBI 4115, pp. 693–701, 2006.

findings are frequency effect, lexical status effect, word similarity effect, semantic priming effect, and visual degradation effect. Many researchers agree with that a persuasive lexical access model must be able to explain the major effects.

Many different lexical access models have been proposed to explain internal mechanism involved in mental lexical process in human brain. The proposed models are such as the search model[2], the logogen model[6], the interactive-activation model[5] and the verification model[1]. These models successfully account for most findings but for exceptional effects. However, as all of the models are theoretical models, it is very hard to perform computational experiments and simulations without human subjects.

Computational experiments and simulations are very helpful to elaborate or to make the model complete by handling experimental environment. In addition, computational model may suggest fruitful areas for experimentalists to investigate. Also, partial lesions of neurological areas and pathways can be modeled in a straightforward if connectionist modeling is used. The intensional lesion study using the connectionist computational model is never possible with human subjects.

In spite of the advantages of computational model in the study of cognitive neuro psychology, researches on developing computational mental lexicon model have been rarely tried[3], [7]. Furthermore, all the previous works have been focused on the models of English and researches on developing computational Korean mental lexicon have been never performed except for the symbolic computational model of [4].

Lim and Nam propose a computational model for Korean mental lexicon which simulate length effect and frequency effect using FB-trie[4]. FB-trie is a modified trie data structure to enable more frequent words are accessed faster by visiting minimal elements and nodes. Lim and Nam aimed to model language ability or proficiency with the size of training corpus. Correlations of frequency of Trie1 and Trie2 are very small while those of FB-trie1 and FB-trie2 are relatively large and very similar to those of human recognition. The reason of high correlations of length of Trie1 and Trie2 is due to an inherent characteristic of the trie indexing structure requiring more time for retrieving a longer string. The experiment results were very promising in that correlations of frequency and length with human reaction times are higher with the larger training corpus than with the smaller. But, the proposed model tried to simulate only length effect and frequency effect. It did not show the extendability to simulate other language phenomena.

In this paper, we propose a computational Korean lexical access model which can simulate major language phenomena in LDT by using connectionist approach. The organization of this paper is as follows: We briefly introduce the major language phenomena which our model must be able to explain to be a persuasive lexical access model. We will describe input/output pairs which is used to train Korean lexical access model, the network architecture, training procedure. Then, we will show some experimental results whether the proposed model can account for the basic findings aimed to mimic. Finally, we will present some conclusion and future works.

2 Major Language Phenomena in LDT

There are several basic findings that are consistently obtained using the visual lexical decision task[8]. Among them, the major findings are frequency effect, lexical status effect, word similarity effect, semantic priming effect, and visual degradation effect. These effects must be explained by any model which purports to be an account of lexical access.

Frequency effect. The time taken to make a lexical decision reponse to a highly frequently word is less than that to a word of low frequency. For example, subjects will recognize the common word HOUSE more quickly than the rarer word ROUSE.

Lexical status effect. The time taken required to classfy a letter string as a nonword is longer than a word. For example, it takes longer for the subjects to categorize the word BLINK into word than the word FLINK into nonword.

Word similarity effect. When a letter string which is sufficiently similar to a word, it is hard to classify as a nonword

Semantic priming effect. When a word is preceded by a semantically related word, it becomes easier to classify as a word. For example a nonword like TRIAN, which is a word with two letters transposed, is associated with longer response times and a higher error rate than a nonword like TRUAN which is not generated from a word in this way.

visual degradation effect. When a letter string is visually degraded, it becomes more difficult to classify as a word or nonword

3 Overview of the Model

The proposed model adopts a simple recurrent neural network architecture which takes a Korean string of 2-syllable length as an input and makes an output as a semantic vector representing semantic of the input. The model consists of three layers: input layer, hidden layer, and output layer. The input layer consists of input units which represents orthography and context units which represent the history of the early activations of hidden units. When the network makes the output, it is regarded that the network make a lexical decision over the input data. But, as the time to spend in making an output in the network is always identical for every input data, response time to an input is calculated indirectly by using the error rate of the output.

Developing the model involves making four sets of design decisions that apply to the development of connectionist model. The following four sections describe the characteristics of the model, which are the main sections of this paper.

Input/Output Design. What input/out pairs is the network trained on to deal with Korean?

The Network Architecture. What type of units is used, how are the units organized into groups, and in what manner are the groups connected?

First Consonants(19)

ㄱ	1000000000000000000
ㄲ	0100000000000000000
ㄴ	0010000000000000000
:	:
ㅍ	0000000000000000010
ㅎ	0000000000000000001

Middle Vowels(21)

ㅏ	100000000000000000000
ㅑ	010000000000000000000
ㅐ	001000000000000000000
:	:
ㅡ	000000000000000000010
ㅣ	000000000000000000001

Last Consonants(27)

ㄱ	100000000000000000000000000
ㄲ	010000000000000000000000000
ㄳ	001000000000000000000000000
:	:
ㅍ	000000000000000000000000010
ㅎ	000000000000000000000000001

Fig. 1. Representation of Orthography

The Training Procedure. How are examples presented to the network, what procedure is used to adjust the weights to accomplish the task, and what is the criterion for halting training?

Evaluation of the Model. How is performance of the network evaluated-specifically, how is time of lexical decision task measured in the network? What are the experimental results?

4 Input/Output Design

Some representation methods of English orthography have been proposed such as in [5], [3], and [7]. But, as Korean is very different from English, such representation methods can not be used in representation of Korean orthography. We propose a method for representing Korean orthography for connectionist models. A typical form of a Korean syllable is CVC(first Consonant-middle Vowel-last Consonant) or CV(first Consonant-middle Vowel). The number of the first consonant and the last consonant are 19 and 27 respectively. The number of vowel is 21. We may represent a Korean syllable with 15 bits: 5 bits are used to represent each consonant and vowel. But, with this scheme, two different syllables can be represented with the very similar bit stream. This may be problematic in visual LDT experiment. Thus, we propose a orthogonal representation in which different consonants or vowels are represented with different bit streams.

The proposed orthography, a Korean syllable is represented as a 67-bit vector; 19 bits are used to represent first consonant, 21 bits for middle vowel, and 21 bits for last consonant. Every possible Korean syllable can be represented with this representation scheme. Figure 1 shows some example of vector representation of Korean syllables.

We represent the semantics of the input words in terms of 210 pseudo semantic features. The assignment of semantic features to words has the property that words in the same category tend to be more similar than words in different categories. The number of the semantic categories is seven and thirty words are assigned to each category.

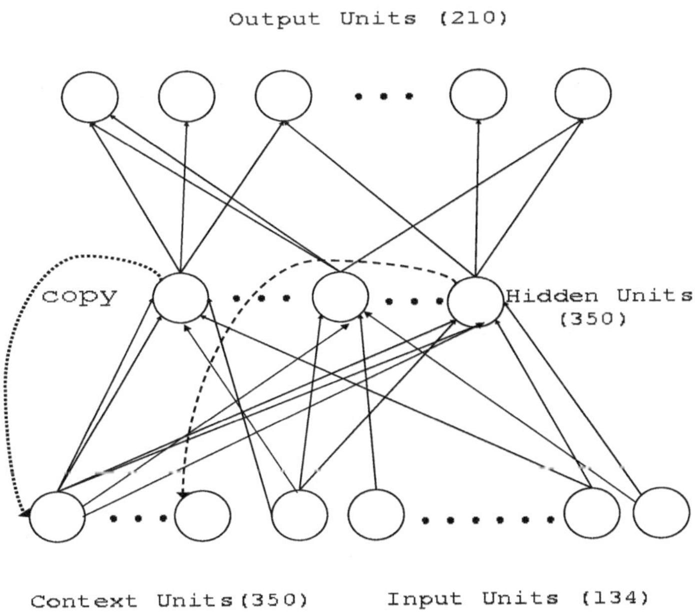

Fig. 2. The Network Architecture

5 The Network Architecture

Figure 2 depicts the network architecture of the proposed model. The network is a simple recurrent network which contains connections from the hidden units to a set of context units. The network consists of three layers: input layer, hidden layer, and output layer. The input layer includes 134 input units(grapheme units), and 350 context units. The 134 grapheme units and contexts units are fully connected to a group of 350 hidden units. The hidden units are also fully connected to 210 output units and context units.

Each unit except context unit has a real-valued activity level ranging between 0 and 1, computed by a smooth nonlinear function of the summed input received from other units. The activities of the context units are 0s or direct copies of the hidden unit activities on the previous time step whether the experimental task is priming task or not. If the task is not priming task, all of context units are set to 0s. If the task is priming task such as semantic priming task, the units are set by copying from hidden units.

6 Training and Evaluation of the Model

The network was initialized to have small random weights, so that at the beginning of training the pattern of semantic activity produced by the word was quite different from its correct semantics. An modified version of backpropagation algorithm was used to computed the way the each weight in the network should change so as to reduce this difference. The error of output unit was calculated by sum of cross entropy of each output unit. Response time in lexical decision time was measured in terms of semantic stress measure which was used in [7]. More formally, the semantic stress S_j of $unit_j$ is a measure of the information content (entropy) of its activation a_j, corresponding to the degree to which it differs from the "neutral" output of 0.5. The higher value of semantic stresses means that the network makes more correct semantic output pattern. If the network makes higher average value of semantic stresses of all output nodes in a fixed time, the network is taken to response more quickly on the input data.

7 Experimental Results

Figure 3-figure 7 show our experimental results. In the figures, x-axis and y-axis represent the number of training iterations and value of the semantic stress respectively. Theses figures show differences of semantic stress between two sets

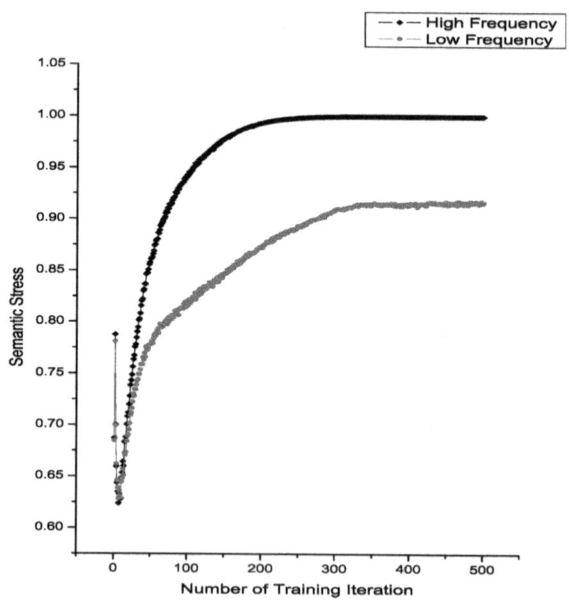

Fig. 3. Frequency Effect

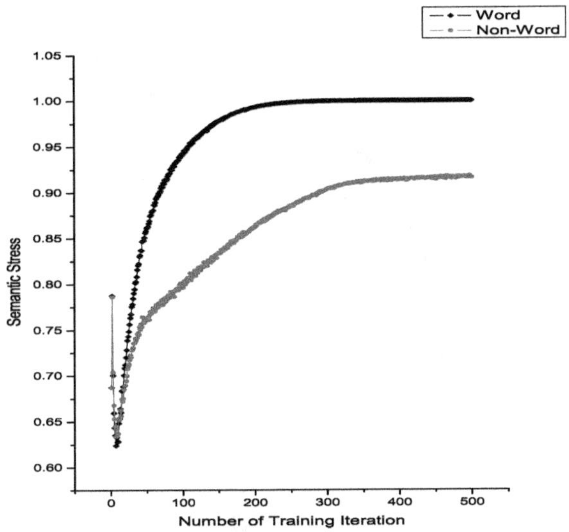

Fig. 4. Lexical Status Effect

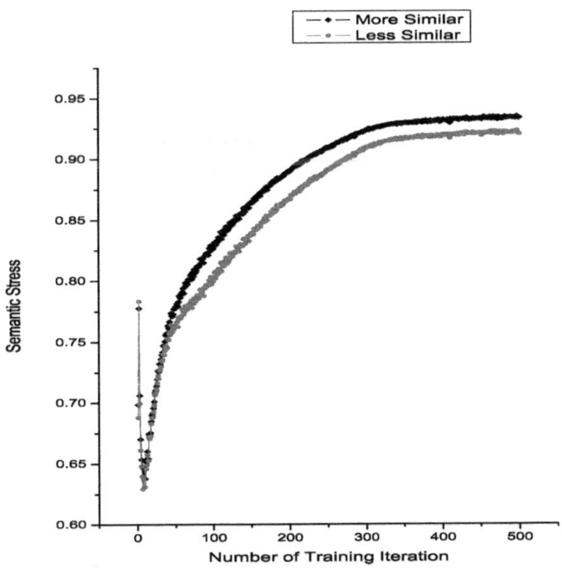

Fig. 5. Word Similarity Effect

of each material. In every experiment, the difference of the average semantic stress was not shown while it is very definite after about 100 training iterations. Although it seems that the difference of the average semantic stress is not obvious as shown in figure 5, every aimed language phenomena including similarity effect

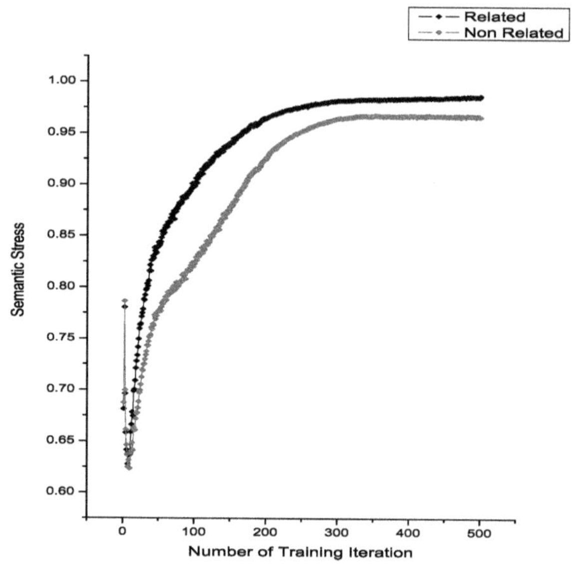

Fig. 6. Semantic Priming Effect

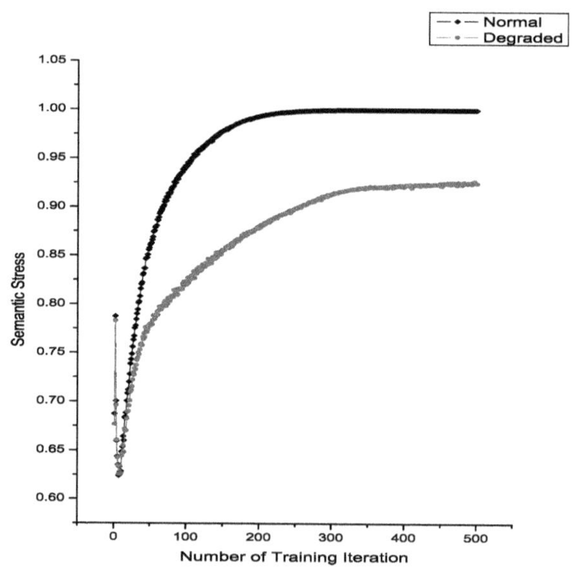

Fig. 7. Visual Degradation Effect

can be simulated statistically significantly after about 100 training iterations. The simulations reveal that the model is reasonably successful in replicating the human performance in LDT.

8 Conclusion

This paper proposes a computational model which can simulate Korean LDT. The model is designed to simulate the behaviors observed in human lexical decision task. The proposed model adopts a simple recurrent neural network architecture which takes a Korean string of 2-syllable length as an input and makes an output as a semantic vector representing semantic of the input. As experimental results, we acquired quite encouraging results; the model successfully simulate basic findings observed in human lexical decision task. To our knowledge, the proposed model is only Korean computational model that attempts to simulate human performance in LDT. However, the model is incomplete in that it still has room to be improved to simulate many other symptoms of language disorders such as aphasia and dyslexia. In the future, we will try to perform lesion study to reproduce characteristics of dyslexia by deleting links or modifying weights of links of the propose model.

References

1. Becker, C. A.: Semantic Context and Word Frequency Effects in Visual Word Recognition. Jounal of Experimental Psychology: Human Preception and Performance, **5** (1979) 252-259
2. Forster, K. I.: Accessing The Mentafl Lexicon. In E. C. J. Walker & R. J. Wales (Eds.), New approaches to language mechni, Amsterdam, North-Holland (1976)
3. Hinton, G. E., Shallice, T.: Lesioning An Attractor Network: Investigatins of acquired dyslexia. Psychological Review, **98** (1) (1991) 74-95
4. Lim, H.S., Nam, K., Hwang, Y.: A Computational Model of Korean Mental Lexicon. ICCSA 2005, Lecture Notes in Computer Science, Springer-Verlag Berlin Heidelberg **3480** (2005) 1129-1134
5. McClelland, J. L., Rumelhart, D. E: An Interactive Activation Model of Context Effects in Letter Perception: Part 1. An Account of Basic Findings. Psychological Review, **88** (1981) 375-407
6. Morton, J.: Interaction of Information in Word Recognition, Psychological Review, **76** (1969) 165–178
7. Plaut, D. C., Shallice, T.: Deep Dyslexia: A Case Study of Connectionist Neuropsychology. Cognitive Neuropsychology, **10** (5) (1993) 377-500
8. Taft, M.: Reading and The Mental Lexicon. Hove (etc.) : Lawrence Erlbaum Associates (1993)

An Artificial Retina Chip Using Switch-Selective Resistive Network for Intelligent Sensor Systems

Jae-Sung Kong, Sang-Heon Kim, Jang-Kyoo Shin, and Minho Lee

Department of Electronics, Kyungpook National University, Sankyuk-dong, Buk-gu, Daegu 702-701, South Korea
{kongjs, myobfdc, jkshin}@ee.knu.ac.kr, mholee@knu.ac.kr

Abstract. We designed and fabricated a bio-inspired CMOS vision chip for edge detection using a switch-selective resistive network. A CMOS buffer circuit, which is embodied in the structure of a common drain amplifier, is commonly used for both raw and smoothed images by using additional switches. By using the switch-selective resistive network, the total number of MOSFETs in the unit pixel and fixed-pattern noise (FPN) was reduced. In addition, by applying the saturating resistive network, the chip outputs a uniform edge signal under various lighting conditions. From the experimental results of a fabricated one-dimensional array with 20 pixels, we could confirm the operation of the vision chip.

1 Introduction

Vision systems, which have image sensors and subsequent processing units for particular purposes, do not use raw images from image sensors such as charge-coupled devices (CCD) and CMOS image sensors (CIS) [1-2]. Indeed, they use filtered images to improve performance and reduce error rate. In particular, Laplacian filtering, by which the edge part of an image is emphasized, have been used in many image processing field such as pattern recognition, and the enhancement of noisy images (e.g. medical images, silhouettes, infrared red images) [3-5]. In order to get edge information, computer vision systems, which consist of CCD and digital computers, have been conventionally used [1-2]. The systems, however, are limited in terms of size, power consumption, and speed regarding real applications, because they consist of two separate modules, which do not interact. Recently, bio-inspired vision chips have been proposed to overcome these problems [6-12]. The vision chips, which mimic the functions of image capturing and processing in the human retina, offer several advantages including compact size, high speed, low power dissipation, and dense system integration.

Two challenges of bio-inspired vision chips include low resolution and noise problems [7-12]. Comparing with conventional CISs, the unit pixel of a bio-inspired vision chip requires additional analog/digital circuits (such as an in-pixel noise reduction circuit, an analog buffer, and some other functional circuits), therefore the size is about several hundred times larger than that of conventional CIS. Noise also occurs from additional circuits, due to the mismatches of the MOSFETs; the process variation; and the temporal fluctuation of power, control signals and biases. In particular,

D.-S. Huang, K. Li, and G.W. Irwin (Eds.): ICIC 2006, LNBI 4115, pp. 702 – 710, 2006.

the vision chips for edge detection require a resistive network, which functions as Gaussian filtering in digital image processing and specialized circuits for the acquisition of both raw and smoothed images. These additional circuits can cause additional area consumption, power dissipation and noise. Therefore, the design of the structure, as well as circuits, is very important for the optimization of their operation. Previously, two structures of bio-inspired vision chips have been proposed [7-8, 10-12]. One type is a voltage division type. The other uses two analog buffers to achieve raw and smoothed images. Otherwise, these structures have their own disadvantages in power and area consumption, signal aliasing and additional noise. To overcome these problems, a new structure, a so-called switch-selective resistive network, is proposed.

In addition, a saturating resistive network is applied to the vision chip to get uniform edge information under various lighting conditions, so-called light adaptation [11-12]. It can help enhance the quality of a final edge image.

The vision chip, which is based on these ideas, has been designed and fabricated with a 20 pixel array using 0.35 μm 2-poly 4-metal standard CMOS technology. The operation of the vision chip was tested via experiments.

2 Theory

2.1 The Principle of Edge Detection

It is well known that the photoreceptors, horizontal cells, and bipolar cells of the retina are concerned with edge detection [13]. Photoreceptors transform the input light into electrical signals and yield voltage in proportion to the input light intensity. Horizontal cells perform spatial smoothing under the influence of the neighboring photoreceptor output. Bipolar cells respond to differences between the photoreceptors and horizontal cells i.e. a signal is yielded which corresponds to the edges of an input pattern. All operations for edge detection are done in a parallel manner. Therefore, we can obtain the edge information of an input image in real-time.

In Figure 1, the structure of a biological retina and the principle of edge detection in a retina are shown. The x-axis represents the position of each cell and the y-axis represents the normalized output curves of photoreceptors, horizontal cells, and bipolar cells, respectively. If a bright light is projected only on the right-hand side of the photoreceptor array, the output of the photoreceptor is a high-level signal. On the other hand, when a dim light is projected on the left-hand side of the photoreceptor array, the output of the photoreceptor is a low-level signal. Horizontal cells receive signal of the photoreceptors and spatially smooth them. Edge signals, the differences between the outputs of the photoreceptor and the horizontal cell, are yielded through bipolar cells.

2.2 The Modeling of a Retinal Structure in Electrical Devices

Photoreceptors, horizontal cells, and bipolar cells in the retina are three key elements that embody vision chip for edge detection [7-13]. First, a raw image is necessary to extract edge information. CCDs are used to sense incident image in high quality. They require a special process for fabrication; thus, it is impossible to embed other circuits for image processing. The problem of on-chip integration can be solved by using CIS technology. The function of horizontal cell can be embodied by using a resistive network, which has been proposed by Mead's research group in which all photo-sensors in unit pixels are connected horizontally and vertically through resistive circuits [5]. The current flow, from the higher-potential area to the lower-potential area, contributes to image smoothing. Image smoothing is done in a spatially parallel manner, therefore, the mechanism is proper for real-time applications. The function of bipolar cells can be embodied by using differential circuits. In addition, addressing circuits and noise suppression circuits are necessary for practical design.

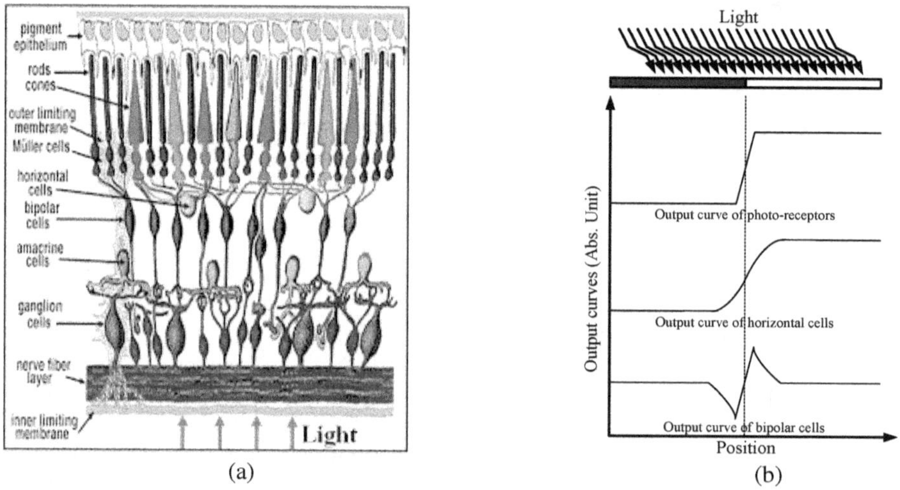

Fig. 1. Edge detection mechanism: (a) Structure of a biological retina; (b) Principle of edge detection

3 Circuit Design

3.1 Previous Structures of a Resistive Network

Figure 2 shows previously proposed structures of a resistive network. Actually, the commercial linear resistor requires a large area in order to materialize it in CMOS technology. The resistors in the figure are special circuits, so-called resistive circuits, which control current the flow. Fig. 2 (a) shows a voltage division type [7-8]. This type requires a vertical resistive circuit and several horizontal resistive circuits for a unit pixel. The number of horizontal resistive circuits depends upon the designer's choice. If a designer wants to create a hexagonal resistive network, the number of horizontal resistive circuits should be 6. Vertical and horizontal resistive circuits are different in obtaining optimum output. By using this type of resistive network, raw and smoothed images can be obtained at the upside and downside nodes of a photodetector, respectively. This resistive network has several weak points. It consumes a great amount of power. Bias circuits become the main source of power dissipation because of the continuously flowing current which keeps the operating point of analog circuits. To operate resistive circuits under optimum conditions, vertical and horizontal resistive circuits individually require a different bias circuit. It means that this type of resistive network requires at least two bias circuits (when all horizontal resistive circuits commonly use a bias circuit) in order to operate each unit pixel. For an entire chip, the total number of bias circuits increases linearly according to the resolution. The complex biases make it difficult to control the chip under optimum condition. The final edge information of a vision chip delicately depends on the characteristics of a resistive network. This control difficulty is another weak point. The third weak point is the distortion of a raw image. The node for a raw image is also connected to a neighboring unit pixel via resistive circuits, therefore, the node cannot keep the original information, the output of a photodetector.

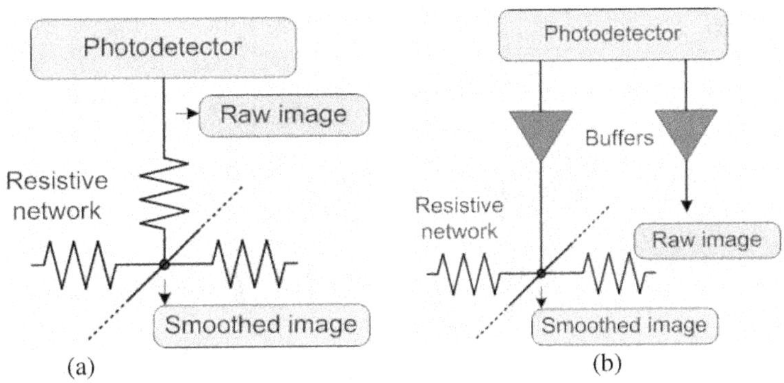

Fig. 2. Previous structures of a resistive network: (a) Using voltage division (b) Using two parallel buffers

Figure 2 (b) shows a resistive network using two parallel buffers [10-12]. This type has some advantages, such as a distortion-free raw image and convenience in optimizing the resistive network because of its simple structure; nevertheless, it also has many problems such as large power dissipation, large area consumption, and additional noise. First, the circuit requires additional power dissipation in order to operate the two additional buffers. The total power dissipation from the buffers increases in proportion to the resolution. These buffers also require area consumption as well as power dissipation. This is the second weak point. The third weak point is that the mismatch between these two separated buffer causes fixed pattern noise (FPN). The third weak point is a critical problem regarding realization.

3.2 Proposed Structure and Embodiment

To overcome these problems, a novel structure is required. Figure 3 shows a block diagram of the proposed circuit. Switches were embedded in the unit pixel in order to connect neighboring unit pixels via a resistive circuit; i.e. the node at the downside of a photodetector outputs a raw image when the switch is open. The output node, otherwise, outputs a smoothed image when the switch is closed. The first advantage of this structure is low power dissipation, because the switches do not require any additional current. Raw and smoothed images are achieved at the same node; thus there is no FPN from the structure. This is the second advantage. The third advantage is that the resistive network requires less area. The proposed circuit requires additional switches; it does not require a vertical resistive circuit or buffers comparing with previous types. Switches only require very small area for fabrication in CMOS technology. The fourth advantage is that there is no aliasing problem between the raw and smoothed images, because the resistive network is physically disconnected when the switches are open. The other advantage is that the characteristic of resistive network is easily controlled, due to their simple structure. By using this advantage, we could apply a saturating resistive network to the vision chip in order to enhance the final edge image. Structural disadvantages include additional switching noise and additional control complexity, but they are not critical problems.

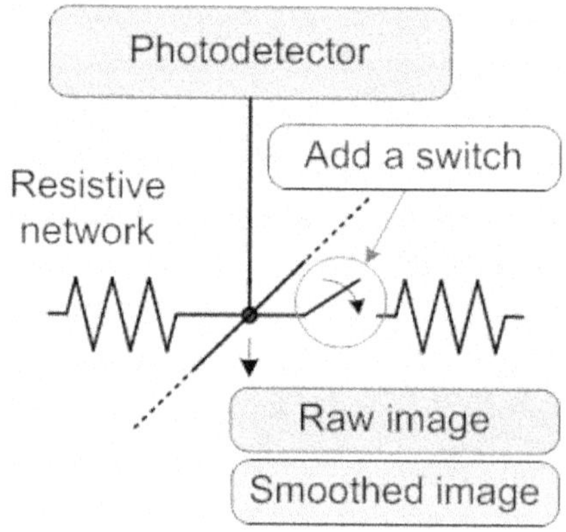

Fig. 3. A block diagram of a switch-selective resistive network

Figure 4 shows a unit pixel's circuit schematic of the proposed vision chip. The circuit contains a active pixel sensor(APS) for image capturing, two simplified correlated-double sampling (SCDS) circuits for reducing FPN and yielding the difference between raw and smoothed image data, and a saturating resistive element (SRE) which consists of four MOSFETs (MDM, ML1, ML2, and MNB). 'RE' in the Figure 4 means resistive element.

Tree-transistor APS and SCDS were used for image capturing and noise reduction, respectively. APS is an imager, where by every pixel includes at least one active transistor. Transistors in the APS may operate as both an amplifier and a buffer in order to isolate the photogenerated charge from the large capacitance of the common output line. The photodiode capacitance is charged to a reset voltage by turning on the RST. During the integration time, electrons are accumulated in the photodiode capacitance according to the light intensity. After the integration time, the light signal is sampled when the SW1 is turned on. The sampled signal has FPN due to the process variation for fabrication. In order to reduce the FPN, a correlated-double sampling (CDS) circuit is required.

For the vision chip, the area consumption of the noise reduction circuit should be considered, as well as the performance of noise reduction. By this reason, a SCDS, which was proposed by Kavadias, was applied to our vision chip [14]. A dominant cause of area consumption of the CDS is its capacitors for analog memory. From this point of view, SCDS is very effective method for noise reduction. The key advantage of SCDS is that the circuit requires one capacitor. Therefore, we can easily minimize the area consumption. Compared with SCDS, conventional CDSs use two capacitors in order to memorize both image and reference signals. Detailed operation of SCDS can be found in the reference [8, 14].

We applied a switch-selective resistive network. The structure used a common drain amplifier for both raw and smoothed images. Each image could be selected by using the switch, DMSW. FPN was reduced, because the output node for the raw and smoothed images is the same. Area consumption was also reduced.

To yield the difference between the raw and smoothed images, a modified SCDS was used. By using the modified SCDS, we could achieve the difference in unit gain without extra noise. When SW3 is in a high state, the noise-reduced raw image is sampled in C2. After sampling the raw image, the DMSW is turned on, then the smoothing operation starts. For a sufficient smoothing operation, about 100 μsec is required. After the smoothing operation, the smoothed image is sampled by turning on SW4.

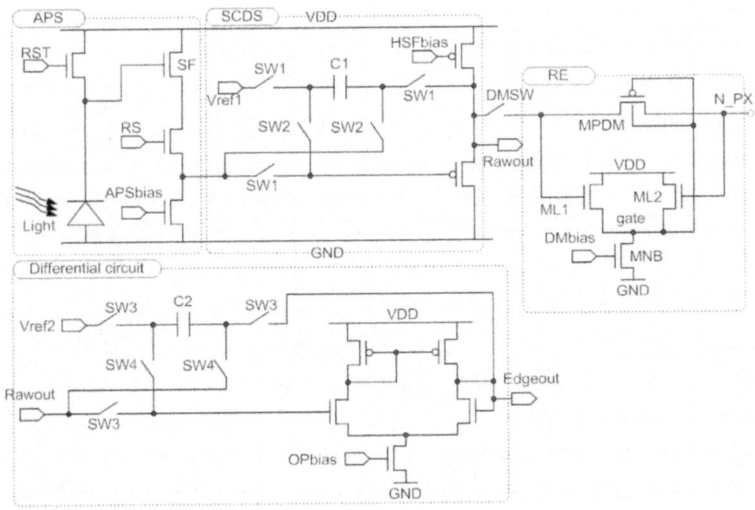

Fig. 4. A circuit schematic of the proposed unit pixel

In order to achieve the enhanced edge image, a so-called light adaptation, a saturating resistive network was applied to the vision chip [11-12]. SRE limits current flow to a constant level when the pixel-to-pixel potential difference is above the threshold voltage. Otherwise, when it is below the threshold voltage, there is little current flow through the resistive element. This is the operating principle of the light adaptation in the edge detection circuit. The circuit sends uniform signal under various light conditions. Therefore, improved edge image could be achieved.

Figure 5 shows the timing diagram that controls the proposed circuit.

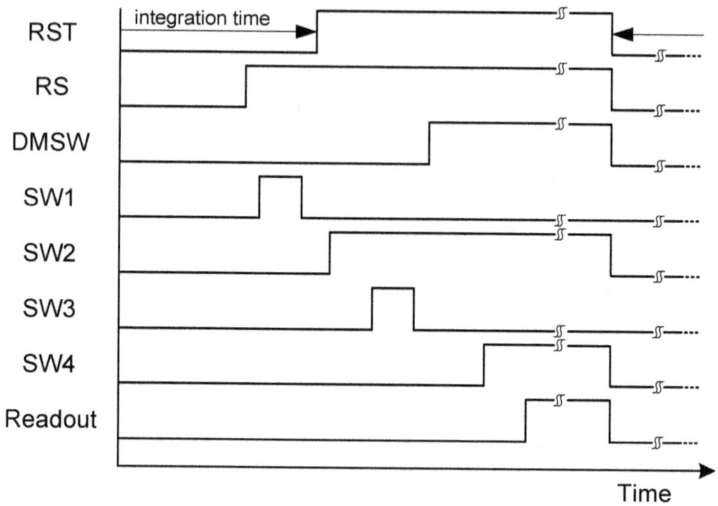

Fig. 5. Timing diagram

4 Experimental Results and Discussion

The chip was materialized by using 0.35 ㎛ 2-poly 4-metal standard CMOS process. We fabricated a 20-pixel array in order to investigate the characteristic of the circuit. A data acquisition card (DAQ) and LabVIEW of the National Instrument Company was used to generate control signals and achieve output signals. A lens with a focal length of 16 mm was mounted in a C-mount format, for projecting an input image onto the chip surface. Figure 6 shows the measurement setup and the fabricated vision chip on a test PCB. The test PCB contains a interconnection component for digital I/O, several digital signal buffers for shifting voltage-level (DAQ outputs 5 V signal through digital I/Os, but the operating power of the vision chip is 3.3 V), several variable resistors for biasing, an analog signal buffer and several interconnection components for transferring output signals.

Figure 7 shows the edge output of the chip. We investigated the light adaptation capability by using two different light conditions of 1200 Lux and 2400 Lux. These light conditions are quite different, because the photo-current in the APS exponentially increases by an increase in the light intensity. From the results of the two light conditions, we confirmed the light adaptation of the vision chip.

Our chip spent several hundred msec for charge integration under the illumination condition of 2000 Lux. It is a rather long time, as compared with the conventional CCD camera. The reason of the problem was that the CMOS process for fabrication is not optimum for image sensors; i.e. the photo-efficiency was relatively poor because of the layers for insulation. Although the photodetector did not perform well, we could confirm the operation of the proposed circuit.

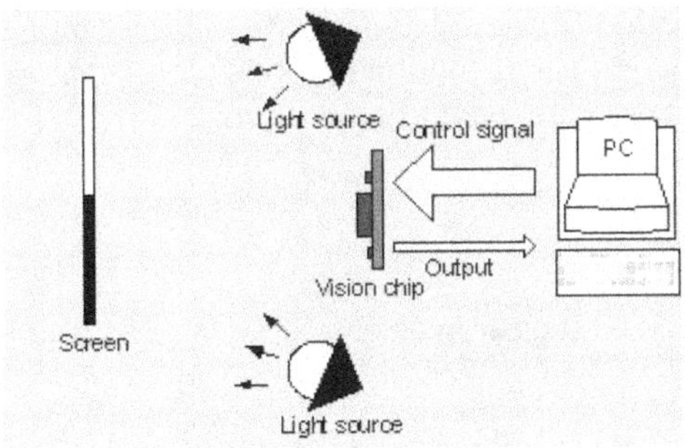

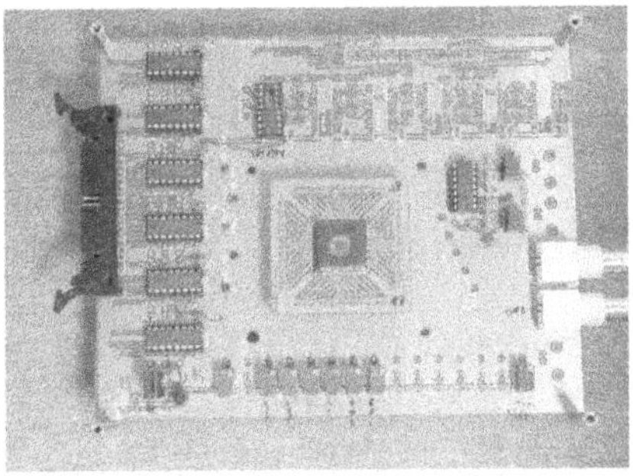

(a) (b)

Fig. 6. Measurement: (a) Measurement setup; (b) Fabricated vision chip on a test PCB

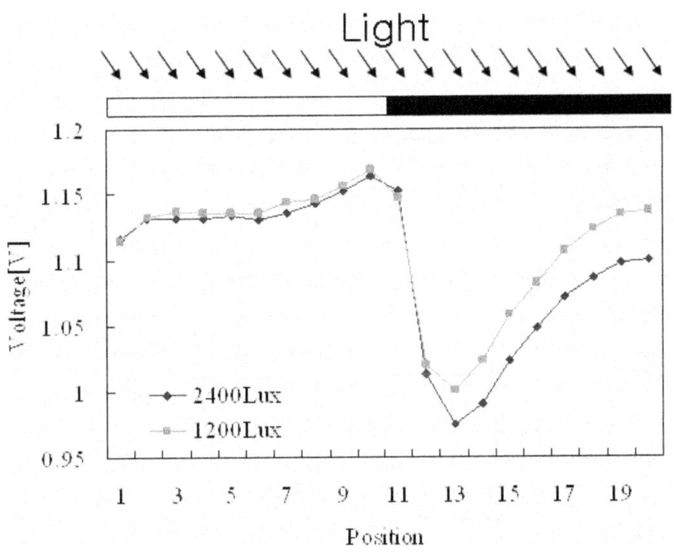

Fig. 7. Experimental results

5 Conclusion

We designed and fabricated a bio-inspired vision chip for edge detection. By using a switch-selective resistive network, several advantages, such as low power dissipation, low area consumption, less noise, and convenience of control, were achieved. A SCDS to reduce FPN and a saturating resistive network to improve the quality of a final edge image were applied to the vision chip. The fabricated chip, with a 20 pixel array, was working well although the process for fabrication was not optimum for image sensors. Light adaptation of the vision chip was also investigated. From the results, we can expect that the edge image of an incident image can be effectively extracted by applying the proposed structure to the edge-detection vision chip. By employing the proposed vision chip in the pattern recognition applications, real time and robust computations, with a compact hardware, could be achieved.

Acknowledgement

This work was supported by the Brain Science Research Center at the KAIST and the BK21 Program in Korea.

References

1. Moini A.: Vision Chips or Seeing Silicon. CHiPTec (1997)
2. Mead, C. A.: Analog VLSI and Neural Systems. Addison-Wesley (1989)
3. Zhao, J., Tow, J., Katupitiya, J.: On-tree Fruit Recognition Using Texture Properties and Color Data. Intelligent Robots and Systems 2005 (IROS 2005), (2005).
4. R.A. Lotufo, A.D. Morgan, A. S. Johnson.: Automatic Number-Plate Recognition. Proc. IEE Colloquium on Image Analysis for Transport Applications, 31 London (1990).
5. Kim, W.-C., Kim, J.-H., Lee, Minho, Shin, J.-K., Yang, H.-S., Yonezu, Hiroo.: Smooth Pursuit Eye Movement System Using Artificial Retina Chip and Shape Memory Alloy Actuator. IEEE Sensors Journal, 5 (2005) 501-509.
6. Wu, C.-Y., Chiu, C.-F.: A New Structure of the 2-D Silicon Retina. IEEE J. Solid-State Circuit, 30 (1995) 890-897.
7. Kameda, S., Honda, A., Yagi, T.: Real Time Image Processing with An Analog Vision Chip System, International Journal of Neural Systems, 9 (1999) 423-428.
8. Suh, S.-H., Kim, J.-H., Kong, J.-S., Shin, J.-K.: Vision Chip for Edge Detection with A Function of Pixel FPN reduction. J. of the Korean Sensors Society, 14 (2005) 191-197
9. Park, J.-H., Kim, J.-H., Suh, S.-H., Shin, J.-K., Lee, Minho, Choi, P., Yagi, T.: A Complementary Metal-Oxide-Semiconductor Vision Chip for Edge Detection and Motion Detection with A Function for Output Offset Cancellation. Optical Review, 12 (2005) 15-19.
10. Kim, J.-H., Kong, J.-S., Suh, S.-H., Lee, Minho, Shin, J.-K., Park, H. B., Choi, C. A.: A Low Power Analog CMOS Vision Chip for Edge detection Using Electronic switches. ETRI Journal, 27 (2005) 539-544.
11. Kong, J.-S., Kim, J.-H., Shin, J.-K., Lee Minho,: Image Quality Improvement of An Edge Detection Vision Chip Based on A Vertebrate Outer Retina Using Saturating Resistive Network. International Conference on Neural Information Processing(ICONIP2005), Taipei (2005) 292-295.
12. Kong, J.-S., Sawa, S., Kim, J.-H., Shin, J.-K., Lee, M., Yonezu, H.: A Light-Adaptive CMOS Vision Chip for Edge Detection Using Saturating Resistive Network. Proceeding of ITC-CSCC2004, Sendai (2004) 8C2L-2-1 - 8C2L-2-4.
13. Kandel, E. R., Schwartz, J. H., Jessell, T. M.: Principles of Neural Science. 3nd Edition. Appleton & Lange Norwalk, CT.
14. Kavadias, S.: Offset-Free Column Readout Circuit for CMOS Image Sensors. Electronics Letters, 35 (1999) 2112-2113.

An Efficient Feedback Canceler for Hearing Aids Based on Approximated Affine Projection*

Sangmin Lee[1], Inyoung Kim[2], and Youngcheol Park[3]

[1] Division of Bionics & Bioinfomatics, Chonbuk National University, Jeonju, Korea
[2] Dept. of Biomedical Eng., Hanyang University, Seoul, Korea
[3] Computer & Telecomm. Eng. Division, Yonsei University, Wonju, Korea
young00@yonsei.ac.kr

Abstract. In the feedback cancelation systems in hearing aids, signal cancelation and coloration artifacts can occur for a narrow-band input. In this paper, we propose a new adaptive feedback cancelation algorithm that can achieve fast convergence by approximating the affine projection (AP) algorithm and prevent signal cancelation by controlling the step-size. A Gram-Schmidt prediction error filter (GS-PEF) is used for a stable approximation of the AP algorithm, and the step-size is varied using the prediction factor of the input signal. Simulations results are presented to verify efficiency of the proposed algorithm.

1 Introduction

Howling, caused by acoustic feedback, is a persistent and annoying problem for hearing aid users. It also limits the maximum usable gain of hearing aids and, in turn, degrades overall performance. To alleviate this well-known problem, many adaptive feedback cancelation systems have been developed, in which the entire effect of the feedback is eliminated by estimating the acoustic feedback path.

In the adaptive feedback cancelation system, fast adaptation to the variations of feedback path and input spectrum is required. However, fast adaptation often generates ringing and coloration artifacts. Thus, when the feedback environment is time-varying, the desired adaptation behavior is a compromise between fast adaptation required to track changes in the feedback path and slow adaptation required to improve the sound quality of the system.

Another problem associated with the hearing aids is the correlation between the input and output signals, which leads to a bias in the estimate of the feedback path, unless adequate delay is provided [1]. For a broadband input, such delays enable an adaptive system to converge to an accurate estimate of the feedback path. But for a narrowband input, even with a delay, the feedback cancelation system tends to minimize the error signal by canceling the input instead of modeling the feedback path. Adaptation with a sinusoidal input, in general,

* This paper was supported (in part) by research funds of Chonbuk National University in 2005, and by the grant from the Korea Health 21 R&D Project, Ministry of Health & Welfare, Korea (02-PJ3-PG6-EV10-0001).

D.-S. Huang, K. Li, and G.W. Irwin (Eds.): ICIC 2006, LNBI 4115, pp. 711–720, 2006.

causes system instability and ringing or coloration of the output signal [2]. One approach to maintaining system stability is to constrain adaptations [3]. However, the constrained adaptation will not distinguish between deviation caused by error in the noise model and deviation caused by a change in the external feedback path.

In this paper, we propose an efficient adaptive feedback cancelation system for hearing aids. The proposed system employs a new adaptive algorithm that approximates the affine projection (AP) algorithm with much less computational cost.

Due to its simplicity and efficiency, the LMS adaptive filter has been widely used for the feedback cancelation in hearing aids [1-3]. The convergence performance of the LMS algorithm, however, is often deteriorated by colored input signals. To overcome this problem, the affine projection (AP) algorithm [4] that updates the weight vector based on a number of recent input vectors can be used. The AP algorithm allows a higher convergence speed than the LMS algorithm, especially for the colored input signal. But it is computationally complex. Many fast versions of the AP algorithm have been suggested to provide significant simplifications [5], but they require a process of matrix inversion, which is not only computationally expensive but also a source of numerical instability.

The proposed algorithm approximates the AP algorithm by orthogonalizing the input vectors using a Gram-Schmidt prediction error filter (GS-PEF) [6]. To attain a stable approximation, the proposed algorithm applies a weighted error vector to the GS-PEF for the weight vector update. In addition, the proposed algorithm uses a new step-size control scheme that can prevent the signal cancelation and coloration artifacts caused by adaptation with narrowband inputs. The proposed algorithm performs the AP approximation and step-size control in a unified filter structure with a minimum structural complexity.

2 Approximation of the Affine Projection (AP) Algorithm

2.1 AP Algorithm with Orthogonalized Input Vectors

The AP algorithm updates the weight vector based on K most recent input vectors [4]. Similar to the well-known normalized LMS (NLMS) algorithm, a step-size is used to control the rate of convergence and the steady-state excess mean-square error. Let $\mathbf{w}(n)$ be an $(N \times 1)$ estimate of an unknown weight vector, the affine projection algorithm with a projection order M is summarized as

$$\mathbf{w}(n) = \mathbf{w}(n-1) + \mu\boldsymbol{\Delta}(\mathbf{n}), \tag{1}$$

$$\boldsymbol{\Delta}(n) = \mathbf{X}(n)\left(\mathbf{X}^T(n)\mathbf{X}(n)\right)^{-1}\mathbf{e}(n), \tag{2}$$

$$\mathbf{e}(n) = \mathbf{d}(n) - \mathbf{X}^T(n)\mathbf{w}(n-1), \tag{3}$$

where $\mathbf{X}(n) = [\mathbf{x}(n)\mathbf{x}(n-1)\cdots\mathbf{x}(n-M+1)]$ denotes an $(N \times M)$ input data matrix, $\mathbf{e}(n)$ and $\mathbf{d}(n)$ represent $(M \times 1)$ primary input and error output vectors, respectively. μ is the step-size.

Consider a Gram-Schmidt (GS) orthogonalization of the reference input vectors $\mathbf{x}(n-i), 0 \leq i \leq M-1$. The output vectors of GS orthogonalization, denoted by $\mathbf{u}_i(n), 0 \leq i \leq M-1$, can be written in a compact matrix form [6]:

$$\mathbf{U}(n) = \mathbf{X}(n)\mathbf{L} \tag{4}$$

where $\mathbf{U}(n) = [\mathbf{u}_{M-1}(n)\mathbf{u}_{M-2}(n) \cdots \mathbf{u}_0(n)]$ and $\mathbf{L}$ is an $(M \times M)$ unit lower-triangular matrix. Transformation matrix $\mathbf{L}$ in Eq. (4) is obtained by solving M linear prediction problems of orders of 0 through $M-1$, which are described as

$$\min_{\mathbf{a}_{M-i-1}} ||\mathbf{x}(n-i) - \bar{\mathbf{X}}(n-i-1)\mathbf{a}_{M-i-1}||_2, \; i = 0, 1, \cdots, M-1, \tag{5}$$

where $\mathbf{a}_i$ is a $(M-i-1) \times 1$ weight vector. Bar marks that the related matrix contains only their $M-i-1$ rightmost vectors of $\mathbf{X}(n-i-1)$, and $||\cdot||_2$ denotes $l2$ norm. The output vectors of the GS orthogonalization are given as the residue of the prediction error filter:

$$\mathbf{u}_{M-i-1}(n) = \mathbf{x}(n-i) - \bar{\mathbf{X}}(n-i-1)\mathbf{a}_{M-i-1}, \; i = 0, 1, \cdots, M-1. \tag{6}$$

Let $\mathbf{D}$ denote the correlation matrix of the output vectors $\mathbf{u}_i(n)$. We can write $\mathbf{D}$ as

$$\begin{aligned} \mathbf{D} &= \mathbf{U}^T(n)\mathbf{U}(n) \\ &= \mathbf{L}^T(\mathbf{X}^T(n)\mathbf{X}(n))\mathbf{L} \\ &= diag\{||\mathbf{u}_{M-1}(n)||_2, ||\mathbf{u}_{M-2}(n)||_2, \cdots, ||\mathbf{u}_0(n)||_2\}. \end{aligned} \tag{7}$$

Due to a one-to-one correspondence between $\mathbf{x}(n-i)$ and $\mathbf{u}_i(n)$, we have

$$\mathbf{X}(n) = \mathbf{U}(n)\mathbf{L}^{-1}. \tag{8}$$

Thus, we can rewrite Eq. (2) as the following:

$$\mathbf{\Delta}(n) = \mathbf{U}(n)\mathbf{D}^{-1}\epsilon(n) = \sum_{i=0}^{M-1} \frac{\varepsilon_i(n)}{||\mathbf{u}_i(n)||_2}\mathbf{u}_i(n) \tag{9}$$

where $\epsilon(n) = [\varepsilon_{M-1}(n), \varepsilon_{M-2}(n), ..., \varepsilon_0(n)]^T$ denotes an $(M \times 1)$ transformed error vector given by $\epsilon(n) = \mathbf{L}\mathbf{e}(n)$.

On the other hand, it is straightforward to show that the error vector can be written as [9]

$$\mathbf{e}(n) = \begin{bmatrix} e(n) \\ (1-\mu)\bar{\mathbf{e}}(n-1) \end{bmatrix}, \tag{10}$$

where $\bar{\mathbf{e}}(n-1)$ is a vector consisting of the uppermost $M-1$ elements of $\mathbf{e}(n-1)$. Thus, for $\mu = 1$, the error vector is simplified to $\mathbf{e}^T(n) = [e(n) \; \bar{\mathbf{0}}_{M-1}^T]$, where $\bar{\mathbf{0}}_{M-1}$ is an $(M-1) \times 1$ zero vector. In this case, $\mathbf{L}\mathbf{e}(n) = e(n)$, and we have

$$\mathbf{\Delta}(n) = \frac{e(n)}{||\mathbf{u}_{M-1}(n)||_2}\mathbf{u}_{M-1}(n). \tag{11}$$

Eq. (11) indicates that the process of matrix inversion can be implemented by solving the problem: $min_{\mathbf{a}_{M-1}}||\mathbf{x}(n) - \bar{\mathbf{X}}(n-1)\mathbf{a}_{M-1}||_2$. Thus, with a linear predictive pre-processor, the AP update is simplified to an NLMS-like equation. Similar approximation of the AP algorithm was presented in [8].

2.2 Approximation of the AP Algorithm Using Gram-Schmidt Predictors

As shown in Eqs. (10) and (11), significant approximation is possible for the case that $\mu = 1$. However, μ in practical cases is arbitrary, since it governs the convergence speed and the excessive mean-square error in the steady-state. For arbitrary μ, we generally need to solve M systems of M equations.

In this paper, a new adaptive algorithm approximating the AP algorithm with much less computational cost is proposed and applied to the problem of feedback cancelation in hearing aids. The proposed algorithm is based on a Gram-Schmidt prediction error filter (GS-PEF) [6] that iteratively performs Gram-Schmidt orthogonalization.

Given $\hat{\mathbf{h}}_{M-1}(n)$, the equivalent weight vector of an Mth-order GS-PEF, the proposed algorithm is summarized as

$$\hat{\mathbf{w}}(n) = \hat{\mathbf{w}}(n-1) + \mu \frac{\hat{\varepsilon}_{M-1}(n)}{||\tilde{\mathbf{u}}(n)||_2 + \delta} \tilde{\mathbf{u}}(n), \tag{12}$$

$$\tilde{\mathbf{u}}(n) = \begin{bmatrix} \hat{u}_{M-1}(n) \\ \bar{\tilde{\mathbf{u}}}(n-1) \end{bmatrix}, \ \hat{u}_{M-1}(n) = \hat{\mathbf{h}}_{M-1}(n)\mathbf{x}(n), \tag{13}$$

$$\hat{\varepsilon}_{M-1}(n) = \hat{\mathbf{h}}_{M-1}(n)\mathbf{e}_a(n), \ \mathbf{e}_a(n) = \begin{bmatrix} e(n) \\ (1-\mu)^{\eta}\bar{\mathbf{e}}_a(n-1) \end{bmatrix}, \tag{14}$$

where $\hat{u}_{M-1}(n)$ is the prediction error of the Mth-order GS-PEF using $\mathbf{x}(n)$ as an input vector, and $\hat{\varepsilon}_{M-1}(n)$ is a prediction residue of the weighted error vector obtained using the same GS-PEF. Again, bars in the equations mark that the related vectors with dimension i contain only their $(i-1)$ uppermost elements. The input vector $\tilde{\mathbf{u}}(n)$ for the weight update is obtained by appending $\hat{u}_{M-1}(n)$

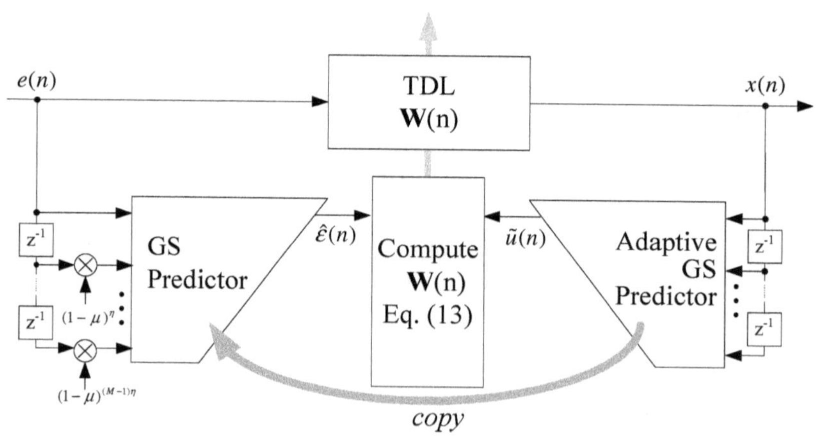

Fig. 1. Schematic diagram of the proposed algorithm approximating the AP algorithm using Gram-Schmidt predictors

to the beginning of the previous input vector $\tilde{\mathbf{u}}(n-1)$, which will reduce the computational complexity required to obtain the input vector.

If $\mu \neq 1$ and $\eta = 1$, $\mathbf{e}_a(n)$ is identical to $\mathbf{e}(n)$ in Eq. (10) and the algorithm approximates Eq. (9) using only an $(M-1)$th order linear prediction filter implemented through the GS-PEF. As η increases, less weight on $\bar{\mathbf{e}}_a(n-1)$ is used. Thus, η is a factor of compromising between the cases that $\mu = 1$ and $\mu \neq 1$ in the approximation of the AP algorithm by using an $(M-1)$th order prediction filter. Fig. 1 shows a schematic diagram of the proposed algorithm approximating the AP algorithm using Gram-Schmidt prediction error filters.

Coefficients of the GS-PEF can be adaptively adjusted, and any adaptive algorithm can work for the proposed algorithm. However, as a low-complexity system is required, the normalized LMS algorithm is suitable to this application. Fig. 2 shows a schematic diagram of the GS-PEF for the projection order (M) of 4.

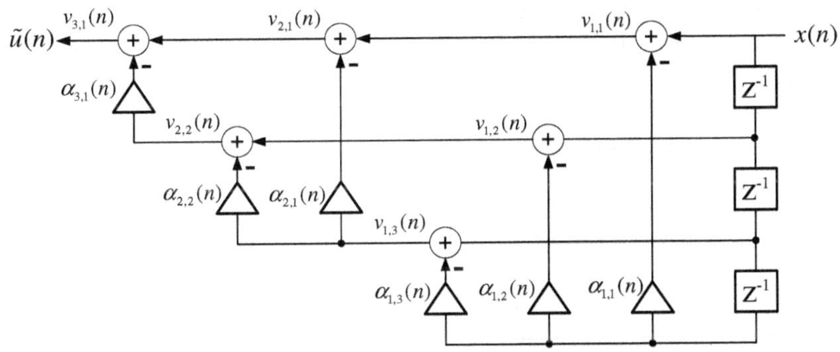

Fig. 2. The NLMS Gram-Schmidt predictor for M=4

The NLMS updates for the GS-PEF can be summarized as the following pseudo code [6]:

$$
\begin{aligned}
& v_{0,i} = x(n-i+1),\ 1 \le i \le M \\
& \text{for } l = 1 : M \\
& \quad \chi_l(n) = (1-\hat{\beta})\chi_l(n-1) + \hat{\beta}\,|v_{l-1,M-l+1}(n)|^2 \\
& \quad \text{for } m = 1 : M-l \\
& \qquad v_{l,m}(n) = v_{l-1,m}(n) - \alpha_{l,m}(n) v_{l-1,M-l+1}(n) \\
& \qquad \alpha_{l,m}(n+1) = \alpha_{l,m}(n) + \frac{\hat{\beta}}{\chi_l(n)} v_{l-1,M-l+1}(n) v_{l,m}(n) \\
& \quad \text{end of } m \\
& \text{end of } l \\
& \tilde{u}(n) = v_{M-1,1}(n)
\end{aligned}
\tag{15}
$$

where $0 < \hat{\beta} < 1$ is a convergence parameter controlling time span of the adaptation.

The proposed algorithm is similar to the approximated AP algorithm proposed in [7]. However, the algorithm in [7] updates the weight vector based on the assumption that $\mu = 1$, i.e., the error sample $e(n)$ is used instead of the residue of the weighted error vector as in Eqs. (12)∼(14). Thus, when μ is close to 1, the algorithm in [7] works quite well. But, in practice, the weight update is performed with an arbitrary μ. Moreover, μ should be small to control the steady-state performance of the adaptive filter. When small μ is used, the algorithm suffers from the convergence problem, and even becomes unstable.

3 Step-Size Variation

For the variation of the step-size, we define the prediction factor defined as a prediction error power normalized by the input power: $\rho = \frac{||\mathbf{u}_{M-1}(n)||_2}{||\mathbf{x}(n)||_2}$. In case that the order of the predictor is at least of the order of the input AR process $x(n)$, the prediction error will be statistically white. Then, the minimum prediction error power becomes $J_{min} = ||\mathbf{u}_{M-1}(n)||_2$. Furthermore, since $\bar{\mathbf{X}}^T(n-1)\mathbf{u}_{M-1}(n) = \mathbf{0}_{M\times 1}$, J_{min} can be written as $J_{min} = ||\mathbf{x}(n)||_2 - ||\bar{\mathbf{X}}(n-1)\hat{\mathbf{a}}_{M-1}(n)||_2$. Thus, $\rho = J_{min}/||\mathbf{x}(n)||_2 = 1 - ||\bar{\mathbf{X}}(n-1)\hat{\mathbf{a}}_{M-1}(n)||_2/||\mathbf{x}(n)||_2$. Note that the ratio ρ can never be negative and the ratio $||\bar{\mathbf{X}}(n-1)\hat{\mathbf{a}}_{M-1}(n)||_2/||\mathbf{x}(n)||_2$ is always positive, we therefore have $0 \leq \rho \leq 1$. Thus, the prediction factor always lies between zero and one.

Given an instantaneous estimate of the prediction factor, denoted by $\hat{\rho}(n)$, the step-size of the proposed algorithm is determined as

$$\beta(n) = \lambda\beta(n-1) + (1-\lambda)\min\{T_{\max},\ \hat{\rho}(n)\} \tag{16}$$

$$\kappa(n) = \begin{cases} s_{\text{tone}}\beta(n), & \beta(n) < T_{\text{tone}} \\ \beta(n), & otherwise \end{cases} \tag{17}$$

$$\hat{\mu}(n) = \kappa(n)\mu_{\max} \tag{18}$$

where $\mu_{\max}$ controls the maximum magnitude of the step-size. The instantaneous estimate of prediction factor is computed using the GS-PEF as $\hat{\rho}(n) = \|\tilde{\mathbf{u}}(n)\|_2/\|\mathbf{x}(n)\|_2$. In the algorithm, $\hat{\rho}(n)$ is limited by $T_{\max}$ to disregard over-estimated ones, and it is averaged using a simple-pole IIR filter to have $\beta(n)$. $\beta(n)$ will be close to unity for wideband inputs, and it will become small for narrowband inputs. Therefore, the algorithm will use large step-sizes for wideband sections and small step-sizes for narrowband sections of the input, which will provide a more accurate estimate of the feedback path than using a fixed step-size. When tone signals such as DTMF or tone-like signals are added to the input, the prediction factor will decrease abruptly. The step-size then decreases too and adaptation will be inhibited. Thus, the system can avoid adaptation using tone-like inputs. To further inhibit adaptation for tone-like inputs, $\beta(n)$ is compared with a threshold T_{tone}. When the input is judged as a tone-like signal, a small constant s_{tone} less than 1 is multiplied to the smoothed prediction factor.

Fig. 3 shows a complete block diagram of the proposed adaptive feedback canceler, which uses the GS-PEF to approximate the AP algorithm using Eqs.

(12)~(14) and (15)~(17). The reference input signal $x(n)$ is fed to the adaptive GS-PEF and the residual is obtained. The coefficients of the GS-PEF are copied into another GS-PEF connected to the error signal, by which the residue of the weighted error vector is obtained. If the feedback path is rapidly changing, the

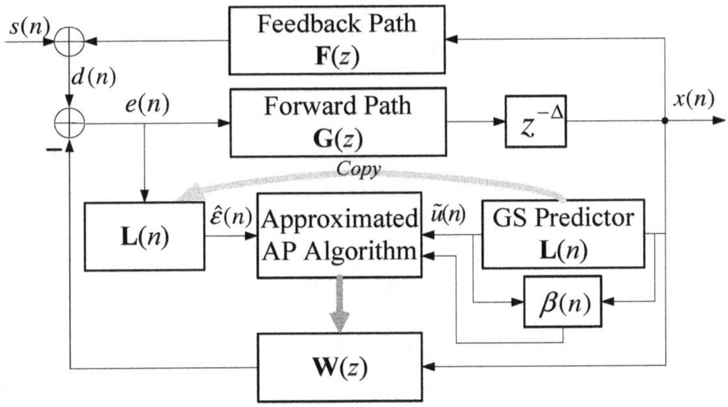

Fig. 3. Block diagram of the proposed adaptive feedback canceler

hearing aid output will also rapidly increase due to a mismatch between the current estimate and the new feedback path. The mismatch will emphasize the error at frequencies with small gain margin relative to other frequencies [2]. Consequently, the output spectrum will vary dramatically. The spectral variation, in turn, influences the prediction factor of the prediction filter. During the period of adaptation to the spectral variation, the GS-PEF will produce large prediction errors, which results in a big step-size. Thus, the feedback cancelation filter can quickly adapt to the new feedback path.

In a howling situation the feedback canceler must use a large step-size, but the prediction factor will rapidly decrease because howling is normally associated with strong tone signals at the output. However, since the power of howling tones is abnormally high, those tones can be easily discriminated from tones embedded in the input by monitoring the output power.

4 Computer Simulations

For the simulations, $20dB$ hearing aid gain was assumed, and an 80-sample decorrelation delay was inserted in the forward path together with a probe noise of $40dB$ SNR. The feedback path was modeled using a 128-tap FIR filter at the sampling rate of 16kHz, and a 32-tap adaptive filter was used for the feedback cancelation.

Firstly, the proposed algorithm was tested using a speech-shaped noise which was generated by passing the white noise through a 16th-order AR filter modeling the speech spectrum. For simulations we added a 1.2-second long DTMF tone corresponding to the digit '3' to the input at the time index 5*sec*, and the change of feedback path was simulated at the time index 8*sec*, at which the impulse response of the feedback path was modulated using a 10Hz sine with an amplitude 4. We also used a 3rd-order GS-PEF to approximate the AP algorithm of order 4. Parameter values used for the simulation were that $\mu_{\max} = 0.01$, $\lambda = 0.001$, $T_{\min} = 0.25$, $T_{\text{tone}} = 0.05$, and $s_{\text{tone}} = 0.25$, where T_{min} and T_{tone} were empirically determined to obtain the best performance. Step-sizes for the AP and NLMS were 0.002 and 0.02, respectively.

Misalignments are shown in Fig. 4, which shows that the proposed algorithm achieved close approximation to the AP algorithm in terms of convergence speed and steady-state misalignment. It always provided lower misalignments and faster convergence speed than the NLMS algorithm. Also, the new algorithm was not obviously affected by the temporary DTMF tone in the input. The NLMS and AP algorithms, on the other hand, quickly reacted to the input DTMF tone, which resulted in high misalignments and ringing sound during the presence of DTMF tone. Fig. 5 shows the output spectrograms extracted for the DTMF period. It can be seen that, unlike the NLMS and AP algorithms that canceled the input pure tone to minimize the modeling error, the proposed algorithm provides stable outputs.

Another simulation was performed using a true speech signal. 30dB SNR was assumed, and 1.2-second long DTMF tones corresponding to '3', '5', and '7' were added at the time indices of 4, 9 14 seconds, respectively. The feedback path was changed at 7 and 13 seconds, respectively, similarly to the previous case. Other parameters were the same as the previous simulation.

Misalignments are shown together with the input speech in Fig. 6. The results show that convergence rate of the proposed algorithm is comparable to the AP algorithm, but provides more stable outputs. The overall sound quality provided by the proposed algorithm was clean and stable. No ringing and coloration artifacts were heard except for a short click occurred in a transient period of adaptation to the new feedback path. But the NLMS and AP

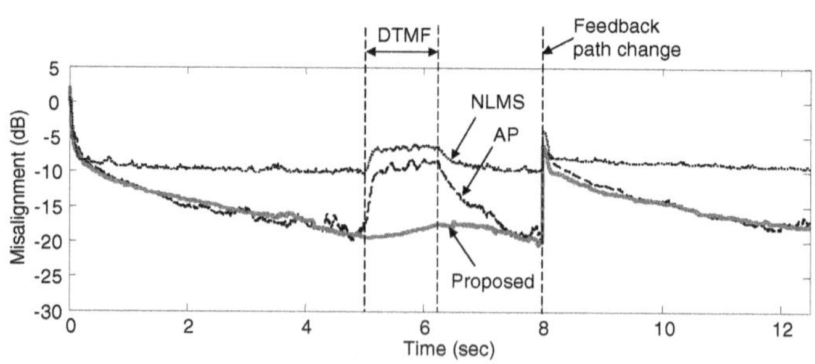

Fig. 4. Misalignments of feedback canceler for a speech-shaped noise: NLMS algorithm (dotted line), AP (dashed line), and proposed (solid line) algorithms

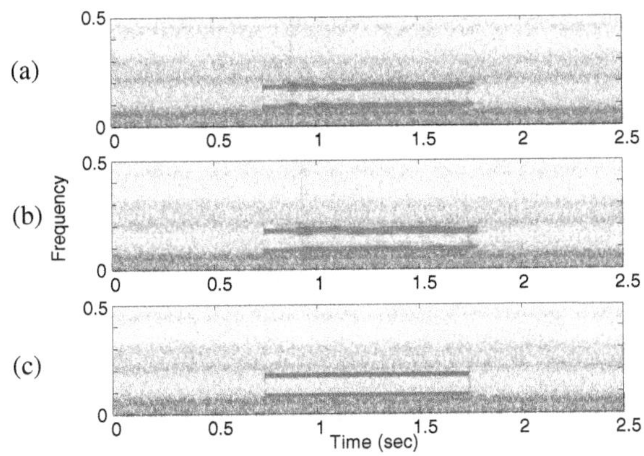

Fig. 5. Spectrograms of the hearing-aid output signals for the case in Fig. 4: (a) NLMS, (b) AP, and (c) proposed algorithms

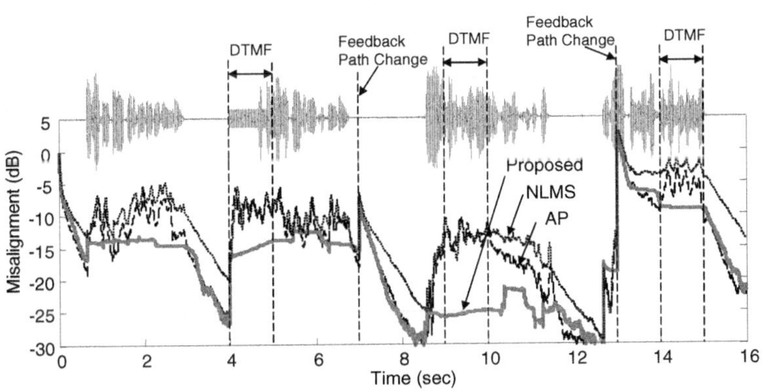

Fig. 6. Input speech and misalignments of feedback cancelers for a speech input: NLMS algorithm (dotted line), AP (dashed line), and proposed (solid line) algorithms

algorithms produced annoying artifacts during the periods of pure tones and some narrowband sections of the input speech.

5 Conclusions

A new adaptive algorithm for feedback cancelation in hearing aids has been presented. The proposed algorithm could achieve an approximation of AP algorithm with less computational cost, and by varying the step-size according to the spectral variation of the input, it could prevent system instability and coloration artifacts due to the narrowband inputs. Simulation results showed that the proposed algorithm was also capable of detecting and tracking the changes in the feedback path.

References

[1] Kates, J. M.: Feedback Cancellation in Hearing Aids: Results from A Computer Simulation. IEEE Trans. Signal Processing, **39** (9) (1991) 553-562

[2] Siqueira M. G., Alwan, A.: Steady-State Analysis of Continuous Adaptation in Acoustic Feedback Reduction Systems for Hearing-aids. IEEE Trans. Speech Audio Processing, **8** (4) (2000) 443-453

[3] Kates, J. M.: Constrained Aaptation for Feedback Cancellation in Hearing Aids. J. Acoust. Soc. Am., **106** (2) (1999) 1010-1019

[4] Ozeki, K., Umeda, T.: An Adaptive Filtering Algorithm Using An Orthogonal Projection to An Affine Subspace and Its Properties. Electron. Commun. Jpn., **67-A** (5) (1984) 19-27

[5] Gay, S. L., Tavathia, S.: The Fast Affine Projection Algorithm. In Proc. IEEE International Conference on Acoustics, Speech and Signal Processing (ICASSP), Detroit, MI (1995) 3023-3026.

[6] Jung, Y. W., Lee, J. H., Park, Y. C., Youn, D. H.: A New Adaptive Algorithm for Stereophonic Acoustic Echo Canceller. Proc. IEEE International Conference on Acoustics, Speech and Signal Processing (ICASSP), Istanbul, Turkey (2000) 801-804

[7] Albu, F., Kwan, H. K.: Combined Echo and Noise Cancellation Based on Gauss-Seidel Pseudo Affine Projection Algorithm. In Proc. IEEE International Symposium on Compound Semiconductors (ISCS), Vancouver, Canada (2004) 505-508.

[8] Rupp, M.: A Family of Adaptive Filter Algorithms with Decorrelating Properties. IEEE Trans. Signal Processing, **46** (3) 1998 771-775

[9] Benesty, J., Huang, Y.: Adaptive Signal Processing. Application to Real-world Problems, Springer-Verlag, Berlin Heidelberg (2003)

Robust Real-Time Face Detection Using Hybrid Neural Networks*

Ho-Joon Kim[1], Juho Lee[2], and Hyun-Seung Yang[2]

[1] School of Computer Science and Electronic Engineering
Handong University, Pohang, 791-708, Korea
hjkim@handong.edu
[2] Department of Computer Science, KAIST
Daejeon, 305-701, Korea
{jhlee, hsyang}@paradise.kaist.ac.kr

Abstract. In this paper, a multi-stage face detection method using hybrid neural networks is presented. The method consists of three stages: preprocessing, feature extraction and pattern classification. We introduce an adaptive filtering technique which is based on a skin-color analysis using fuzzy min-max(FMM) neural networks. A modified convolutional neural network(CNN) is used to extract translation invariant feature maps for face detection. We present an extended version of fuzzy min-max (FMM) neural network which can be used not only for feature analysis but also for pattern classification. Two kinds of relevance factors between features and pattern classes are defined to analyze the saliency of features. These measures can be utilized to select more relevant features for the skin-color filtering process as well as the face detection process.

1 Introduction

Convolutional neural networks (CNN) have been successfully applied to object recognition [1, 2]. The CNN model is a bio-inspired hierarchical multi-layered neural network that achieves some degree of shift and deformation invariance using three ideas: local receptive fields, shared weights, and spatial subsampling. Growing interest in computer vision has motivated a recent surge in research on problems such as face recognition, pose estimation, face tracking and gesture recognition. However, most methods assume human faces in their input images have been detected and localized [3, 4, 5]. Fuzzy min-max (FMM) neural network is a hyperbox-based pattern classification model [6, 7]. In this paper, we present an extended version of the WFMM model for face detection. Since the weight factor of the network can be adjusted by training process, the system can prevent undesirable performance degradation which may be caused by some environmental factors such as illumination

* This research was supported as a 21st Century Frontier R&D Program and Brain Neuroinformatics Research Program sponsored by Ministry of Information and Communication and Minister of Commerce, Industry and Energy in Korea.

D.-S. Huang, K. Li, and G.W. Irwin (Eds.): ICIC 2006, LNBI 4115, pp. 721 – 730, 2006.

changes in face detection system. We introduce a multi-stage face detection method using a hybrid neural network which combines a convolutional neural network and a fuzzy min-max neural network. The proposed method consists of three sub-processes: preprocessor, feature extractor and pattern classifier. We introduce an adaptive filtering technique which is based on a skin-color analysis using the proposed model. A modified convolutional neural network is used to extract translation invariant feature maps for face detection. We define two kinds of relevance factors between features and pattern classes. Through the feature analysis technique, we can select the most relevant features for the skin-color filter as well as the pattern classifier. Moreover, the training process makes it possible to adaptively adjust the feature ranges of the skin-color filter.

2 Multi-stage Face Detection

As shown in Fig. 1, the underlying face detection system consists of three modules: preprocessor, feature extractor and pattern classifier.

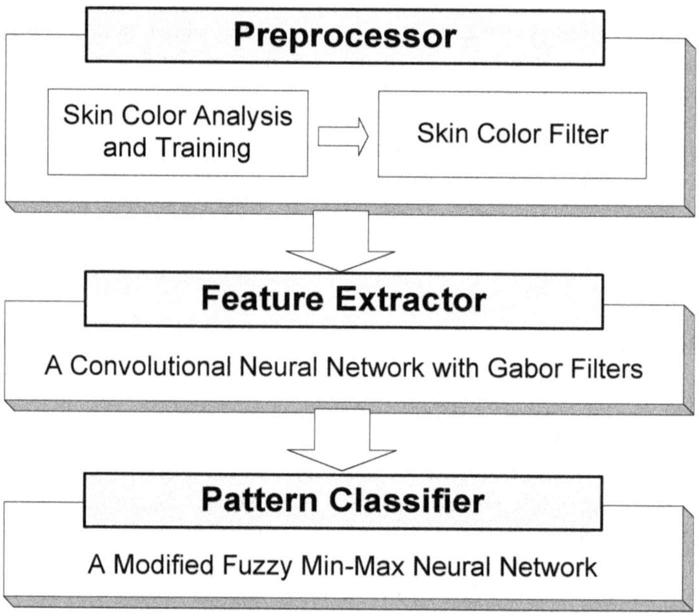

Fig. 1. The underlying face detection system

The feature extractor is implemented using a convolutional neural network in which a Gabor transform layer is added at the first layer. Through the skin color analysis and the training process, the system can generate an adaptive skin model and a relevant feature set for the given illumination condition. The feature extractor generates numerous features from the input image. The number of features affects the computation time and performance of the system. Therefore we propose a feature analysis technique to reduce the amount of features for the pattern classifier. More details of the method are described in Section 3 and Section 4.

3 Skin Color Filtering Using WFMM Neural Networks

In general, a skin color filtering problem is considered as a pixel classification problem. We have employed an extended version of FMM neural network for the adaptive skin color filter. In this paper we present a weighted fuzzy min-max(WFMM) neural network and its application to feature analysis technique. Through the learning capability of the model, the skin color filter can dynamically adjust the feature set and the feature ranges for the variance of environmental conditions such as illumination changes. Fig. 2 shows the structure of the WFMM model.

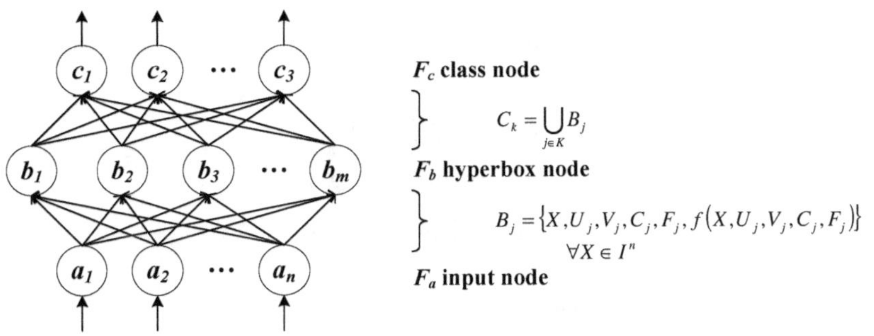

Fig. 2. The structure of the WFMM model

The membership function of a hyperbox is defined as Equation (1).

$$B_j = \{X, U_j, V_j, C_j, F_j, f(X, U_j, V_j, C_j, F_j)\} \ \forall X \in I^n \tag{1}$$

In the equation, U_j and V_j mean the vectors of the minimum and maximum values of hyperbox *j*, respectively. C_j is a set of the mean points for the feature values and F_j means the a set of frequency of feature occurrences within a hyperbox. As shown in Equation (2) and (3), the model employs a new activation function which has the factors of feature value distribution and the weight value for each feature in a hyperbox.

$$\begin{aligned} b_j(A_h) = \frac{1}{\sum_{i=1}^{n} w_{ji}} \bullet \sum_{i=1}^{n} w_{ji} [&\max(0, 1 - \max(0, \gamma_{jiv} \min(1, a_{hi} - v_{ji}))) \\ &+ \max(0, 1 - \max(0, \gamma_{jiu} \min(1, u_{ji} - a_{hi}))) - 1.0] \end{aligned} \tag{2}$$

$$\begin{cases} \gamma_{jiU} = \dfrac{\gamma}{R_U} & R_U = \max(s, u_{ji}^{new} - u_{ji}^{old}) \\ \gamma_{jiV} = \dfrac{\gamma}{R_V} & R_V = \max(s, v_{ji}^{old} - v_{ji}^{new}) \end{cases} \tag{3}$$

In the equation, w_{ij} is the connection weight which means the relevance factor between i-th feature and j-th hyperbox. The learning process of the model consists of three sub-processes: hyperbox creation, expansion, and contraction processes.

If the expansion criterion shown in Equation (4) has been met for hyperbox B_j, f_{ji}, u_{ji}, v_{ji} and c_{ij} are adjusted using Equation (5) and (6).

$$n\theta \geq \sum_{i=1}^{n} (\max(v_{ji}, x_{hi}) - \min(u_{ji}, x_{hi})) \tag{4}$$

$$\begin{cases} f_{ji}^{new} = f_{ji}^{old} + 1 \\ u_{ji}^{new} = \min(u_{ji}^{old}, x_{ki}) \\ v_{ji}^{new} = \min(v_{ji}^{old}, x_{ki}) \end{cases} \tag{5}$$

$$c_{ji}^{new} = (c_{ji} * f_{ji}^{old} + x_{hi}) / f_{ji}^{new} \tag{6}$$

As shown in the equations, the frequency value is increased by 1 at every expansion and the min and max point expansion operations are similar to the fuzzy intersection and fuzzy union operations [6]. The mean point value, c_{ji}, is updated by Equation (6). During the learning process the weight values are determined by Equation (7) and (8).

$$w_{ji} = \frac{\alpha f_{ji}}{R} \tag{7}$$

$$R = \max\left(s, v_{ji} - u_{ji}\right) \tag{8}$$

As shown in the equations, the weight value is increased in proportion to the frequency of the feature. In the equations, s is a positive constant to prevent the weight from having too high value when the feature range is too small. The value of f_{ji} is adjusted through the learning process. The contraction process is considered as an optional part of our model. The contraction process is to eliminate the possible overlappings between hyperboxes that represent different classes. We can expect that the weights concept of the model replace the role of overlapping handling because the weights reflect the relevance of feature values and hyperbox as different values. We define a new contraction method including the weight updating scheme. To determine whether or not the expansion has created any overlapping, a dimension by dimension comparison between hyperboxes is performed. If one of the following four cases is satisfied, then overlapping exists between the two hyperboxes.

For each of these cases, contraction process is performed. If $\delta^{old} - \delta^{new} > 0$, then $\Delta = i,\ \delta^{old} = \delta^{new}$, signifying that there was an overlap for Δ th dimension. Otherwise, the testing is terminated and the minimum overlap index variable is set to indicate that the next contraction step is not necessary, i.e. $\Delta = -1$. If $\Delta > 0$, then the Δ th dimension of the two hyperboxes are adjusted. Only one of the n dimensions is adjusted in each of the hyperboxes to keep the hyperbox size as large as possible and minimally impact the shape of the hyperboxes being formed.

$$
\begin{aligned}
&case1 : u_{ji} < u_{ki} < v_{ji} < v_{ki} \\
&\qquad \delta^{new} = \min(v_{ji} - u_{ki}, \delta^{old}) \\
&case2 : u_{ki} < u_{ji} < v_{ki} < v_{ji} \\
&\qquad \delta^{new} = \min(v_{ki} - u_{ji}, \delta^{old}) \\
&case3 : u_{ji} < u_{ki} < v_{ki} < v_{ji} \\
&\qquad \delta^{new} = \min(\min(v_{ki} - u_{ji}, v_{ji} - u_{ki}), \delta^{old}) \\
&case4 : u_{ki} < u_{ji} < v_{ji} < v_{ki} \\
&\qquad \delta^{new} = \min(\min(v_{ji} - u_{ki}, v_{ki} - u_{ji}), \delta^{old})
\end{aligned}
$$

We have defined new adjustment schemes from the new definition of hyperbox for the four cases. The frequency factor is increased in proportion to the relative size of the feature range, and the mean point value is adjusted by considering the expanded feature range.

4 Feature Analysis Technique Based on WFMM Model

This section describes a feature analysis technique for the skin-color filter and the classifier. We define two kinds of relevance factors using the proposed FMM model as follows:

$RF1(x_j, C_k)$: *the relevance factor between a feature value x_j and a class C_k*

$RF2(X_i, C_k)$: *the relevance factor between a feature type X_i and a class C_k*

The first measure *RF1* is defined as Equation (9). In the equation, constant N_B and N_k are the total number of hyperboxes and the number of hyperboxes that belong to class k, respectively. Therefore if the $RF1(x_i, k)$ has a positive value, it means an excitatory relationship between the feature x_i and the class k. But a negative value of $RF1(x_i, k)$ means an inhibitory relationship between them. A list of interesting features for a given class can be extracted using the *RF1* for each feature.

$$RF1(x_i, C_k) = \Big(\frac{1}{N_k}\sum_{B_j \in C_k} S(x_i, (u_{ji}, v_{ji})) \cdot w_{ij} - \frac{1}{(N_B - N_k)}\sum_{B_j \notin C_k} S(x_i, (u_{ji}, v_{ji})) \cdot w_{ij}\Big) / \sum_{B_j \in C_k} w_{ij} \tag{9}$$

In Equation (9), the feature value x_i can be defined as a fuzzy interval which consists of min and max values on the i-th dimension out of the n-dimension feature space. For an arbitrary feature X_i, let x_i^L and x_i^U be the min and max value, respectively, then the similarity measure S between two fuzzy intervals can be defined as Equation (10).

$$S(x_i, (u_i, v_i)) = S((x_i^L, x_i^U), (u_i, v_i)) = \frac{Overlap((x_i^L, x_i^U), (u_i, v_i))}{Max(x_i^U - x_i^L, v_i - u_i)} \tag{10}$$

In Equation (10), if two fuzzy intervals are all point data, then the denominator part of the equation, $Max(x_i^U - x_i^L, v_i - u_i)$ becomes zero. Therefore we define the similarity measure in this case as Equation (11). As shown in the equation, the similarity value is 1.0 when two intervals have an identical point, and 0 when they indicate two different points.

$$S((x_i^L, x_i^U), (u_i, v_i)) = \begin{cases} 1 & if \quad (x_i^L = x_i^U = u_i = v_i) \\ 0 & Otherwise \end{cases} \tag{11}$$

But if $Max(x_i^U - x_i^L, v_i - u_i)$ is greater than zero, the value is determined as described in Equation (16).

$$Overlap((x_i^L, x_i^U), (u_i, v_i)) = \begin{cases} x_i^U - u_i & if\,(x_i^L \le u_i \le x_i^U \le v_i) \\ v_i - u_i & if\,(x_i^L \le u_i \le v_i \le x_i^U) \\ x_i^U - x_i^L & if\,(u_i \le x_i^L \le x_i^U \le v_i) \\ v_i - x_i^L & if\,(u_i \le x_i^L \le v_i \le x_i^U) \\ 0 & Otherwise \end{cases} \tag{12}$$

The second measure *RF2* can be defined in terms of *RF1* as shown in Equation (13). In the equation, L_i is the number of feature values which belong to i-th feature.

$$RF2(X_i, C_k) = \frac{1}{L_i} \sum_{x_l \in X_i} RF1(x_l, C_k) \tag{13}$$

The *RF2* shown in Equation (13) represents the degree of importance of a feature in classifying a given class. Therefore it can be utilized to select a more relevance feature set for the skin color filter.

5 Feature Extraction and Face Classification

The most advantageous feature of convolutional neural network is invariant detection capability for distorted patterns in images [1, 2]. As shown in Fig. 3, the underlying system employs a convolutional neural network in which a Gabor transform layer is added at the first layer. The first layer of the network extracts local feature maps from the input image by using Gabor transform filters. The other layers of the feature extractor include two types of sub-layers called *convolution layer* and *sub-sampling layer*. Each layer of the network extracts successively larger features in a hierarchical set of layers. Finally a feature set is generated for the input of the pattern classifier. The pattern classifier has been implemented using the proposed WFMM neural network. For the feature extractor, a set of (38 x 42) candidate areas has been selected as input data. The Gabor filter layer generates eight (28 x 32) feature maps. Each unit in each feature map is connected to a (11 x 11) neighborhood into the input retina. In the sub-sampling layer, the feature map has half the number of rows and columns of the input data. Therefore the layer has eight feature maps of size (14 x 16).

The convolutional layer generates 44 feature maps. Each unit is connected to (3 x 3) neighborhood at identical locations in a subset of the feature maps of the Gabor transform layer. In the feature extraction layer, 44 kinds of feature values are generated. These data become the input data of the pattern classifier. The number of input features can be reduced through the feature analysis technique described in the previous section.

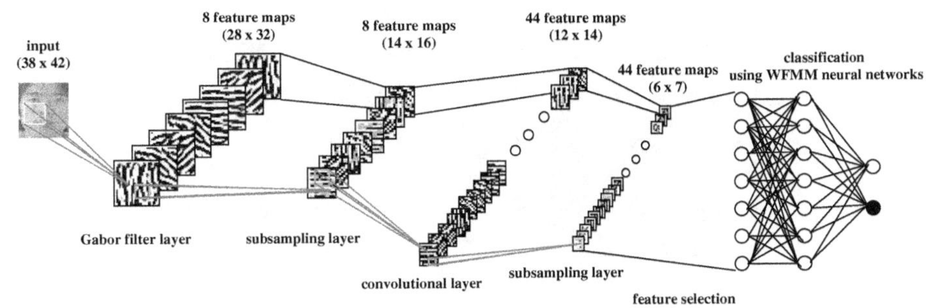

Fig. 3. Face detector using hybrid neural networks

6 Experimental Results

Two types of experiments have been conducted for a set of real images. For the training of skin-color filter, the system considers eleven color features: *Red, Green, Blue, Intensity, Cb, Cr, magenta, Cyan, Yellow, Hue,* and *Saturation*. Fig. 4 shows two input images captured under different illumination conditions and the face detection results.

Table 1 shows the skin-color analysis result and the feature range data derived from the training process. As shown in the table, different kinds of features can be adaptively selected for a given condition, and the feature ranges of skin-color can be also adjusted by the training process.

The table shows four features which have the highest value of the relevance factor *RF1*. As shown in the table, a number of hyperboxes for face and non-face patterns have been generated and the relevance factors are also adjusted through the training process. Therefore the system can select more effective features adaptively for the given environment.

We have selected face patterns from the real images and non-face patterns from the background images. 50 face patterns and 50 non-face patterns have been used for training process. 100 face patterns and 100 non-face patterns have been used for the test stage. Table 2 and Fig. 5 show the change of detection rate and false alarm rate by varying the number of training patterns, respectively. The result shows that the detection rate increases as more training patterns are used, and the false alarm rate decreases as more non-face counter examples are used for training.

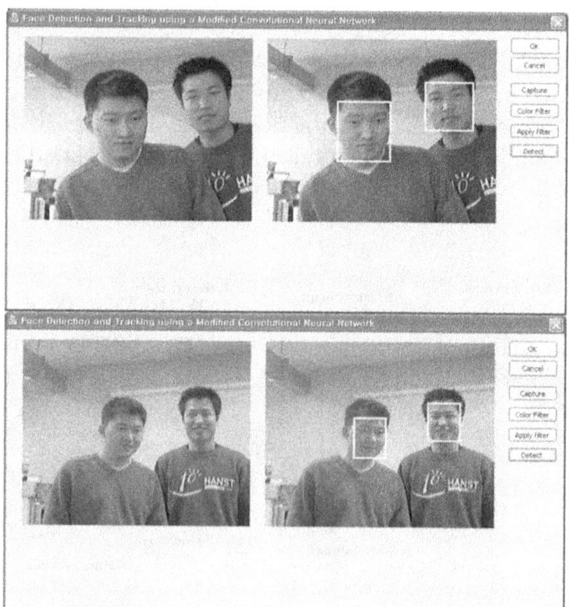

Fig. 4. Two training data captured under different illumination conditions

Table 1. Feature analysis results for the two images

image - 1			*image - 2*		
features	feature range	*RF1*	features	feature range	*RF1*
Saturation	0.072 ~ 0.322	6.734	*Cb*	0.538 ~ 0.742	9.274
Yellow	0.388 ~ 0.594	6.423	*Blue*	0.445 ~ 0.626	9.261
Cb	0.562 ~ 0.729	5.881	*Saturation*	0.074 ~ 0.283	8.716
Cr	0.114 ~ 0.355	5.217	*Cr*	0.123 ~ 0.368	7.662

Table 2. Detection rate and false alarm rate as varying the number of training patterns

number of training patterns		performance	
face patterns	non-face patterns	detection rate	false alarm rate
10	10	72	48
20	20	78	47
30	30	82	33
40	40	86	22
50	50	93	11

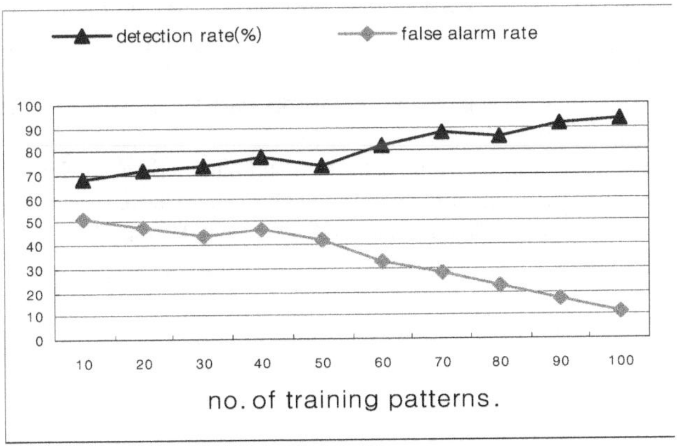

Fig. 5. Detection rate and false alarm rate as varying the number of training patterns

7 Conclusion

The weighted FMM neural network is capable of utilizing the feature distribution and frequency in learning process as well as in classification process. Since the weight factor effectively reflects the relationship between feature range and its distribution, the system can prevent undesirable performance degradation which may be caused by noisy patterns. Through the training process, the skin-color filter is adapted for the illumination condition under which the given images are captured. Two types of feature relevance measures have been defined for the feature selection technique. The first measure *RF1* makes it possible to eliminate the hyperbox contraction process since the measure represents different relevance values within overlapped hyperbox feature ranges. The learning algorithm of the weighted FMM neural network has been presented. The receptive field structure of CNN model provides invariant feature extraction capability and the use of shared weight also reduces the number of parameters in the system.

References

1. Garcia, Cristophe and Delakis, Manolis: Convolutional Face Finder: A Neural Architecture for Fast and Robust Face Detection, IEEE Transaction on Pattern Analysis and Machine Intelligence, Vol.26, No.11, (2004) 1408-1423
2. Lawrence, Steve, Giles, C. L., Tsoi, A. C. and Back, Andrew D.: Face Detection: A Convolutional Neural-Network Approach, IEEE Transaction n Neural Networks, Vol.8, No.1, (1997) 98-113
3. Feraud, R., Bernier, O. J., Viallet J. E. and Collobert, M.: A Fast and Accurate Face Detector Based on Neural Networks, IEEE Transaction on Pattern Analysis and Machine Intelligence, Vol.23, No.1.(2001) 42-53
4. Hsu, Rein-Lien, Mohamed Abdel-Mottaleb and Jain, Anil K.: Face Detection in Color Images, IEEE Transaction on Pattern Analysis and Machine Intelligence, Vol.24, No.5, (2002) 696-706
5. Zhu, Q, Cheng, K. T., Wu, C. T., Wu, Y. L.: Adaptive Learning of an Accurate Skin-Color Model, Proceeding of the Sixth IEEE International Conf. on Automatic Face and Gesture Recognitin, 1 (2004)
6. Simpson, P. K.: Fuzzy Min-Max Neural Networks Part 1: Classification. IEEE Transaction on Neural Networks, Vol.3. No.5. (1997) 776-786
7. Kim, H. J., Ryu, T. W., Nguyen, T. T., Lim, J. S. and Gupta, S.: A Weighted Fuzzy Min-Max Neural Network for Pattern Classification and Feature Extraction. Proceeding of International Conference on Computational Science and Its Application, Part.4 (2004) 791-798

The Novel Feature Selection Method Based on Emotion Recognition System*

Chang-Hyun Park and Kwee-Bo Sim

School of Electrical and Electronic Engineering, Chung-Ang University, 221, Heukseok-Dong, Dongjak-Gu, Seoul, 156-756, Korea
kbsim@cau.ac.kr

Abstract. This paper presents an original feature selection method for Emotion Recognition which includes many original elements. Feature selection has some merit regarding pattern recognition performance. Thus, we developed a method called an 'Interactive Feature Selection' and the results (selected features) of the IFS were applied to an emotion recognition system (ERS), which was also implemented in this research. Our innovative feature selection method was based on a Reinforcement Learning Algorithm and since it required responses from human users, it was denoted an 'Interactive Feature Selection'. By performing an IFS, we were able to obtain three top features and apply them to the ERS. Comparing those results from a random selection and Sequential Forward Selection(SFS) and Genetic Algorithm Feature Selection(GAFS), we verified that the top three features were better than the randomly selected feature set.

Keywords: Reinforcement Learning, Feature selection, Emotion Recognition, SFS, GAFS.

1 Introduction

Emotion recognition research has been typically attempted using four kinds of medium. They are speech, image, physiological signal, and gesture. In addition, our IEEE survey papers published from 1990 to 2005 show that papers using the speech medium have been published more often than others have. The reason for this result is probably due to feature set extraction from speech and image being easier than physiological signal or gesture and the possibility of classification is higher. In particular, EEG, ECG, and SC sensors are used to obtain a physiological signal but the signal from those sensors may be obstructed by electrical signals from fluorescent lamps or electric home appliances. This problem is the one obstacle in emotion recognition using a physiological signal. For an image, this means facial expression recognition and the main problem in this case is usually lighting conditions, which often change, or personal accessories like glasses which affect recognition performance. A problem of gesture recognition is similar to that of image recognition and the bigger problem is that it

* This research was supported by the brain Neuroinformatics research Program by Ministry of Commerce,Industry and Energy.

D.-S. Huang, K. Li, and G.W. Irwin (Eds.): ICIC 2006, LNBI 4115, pp. 731–740, 2006.

may not include much information regarding emotion. Apart from above the problems which these three media present, speech signal can send much more information regarding emotion. For example, talking over the telephone, one can recognize emotions and this shows the validity of speech signal for emotion recognition. Even a cheap microphone can be used sufficiently as a sensor for collecting speech signals and noise will not affect the extraction of the feature set unless it is too loud to be classified as a coming from the source of the signal. These reasons are why most researchers have focused on speech signal and why we have selected this medium for our paper. The commonly used feature set for emotion recognition from speech consists of pitch, energy, formant, and speech rate. Some researchers select all four of the feature sets, others select only one, and the features are generally extracted statistically from the four feature sets. In [1], 87 features was extrated from pitch, energy, and formant and they were classified into five emotions. In [2], 17 features were extracted from pitch, energy, speech rate and so on with sex also being classified. In addition, In [3], 11, 40, and 13 features were extracted. The fact that feature set selection is not fixed suggests that features may or may not be relevant to emotion recognition. This problem will plague researchers in this field until exceptional results are obtained. For this case, there is a GA based selection method, Floating search method and so on which can somewhat reduce difficulties for researchers [4]. Especially, the Sequential Forward Selection and Sequential Backward Selection methods of a Floating search method have been frequently used. In [2], a Forward Selection(FS) method was used and in [1],the 10 best features were selected out of 87 features by using a Sequential Floating Forward Selection algorithm(The extended version of SFS). In [5],a Sequential Forward Selection algorithm was also used and the best feature subset was selected out of 39 candidate feature sets and In [6], a good feature set was found using genetic programming for the music genre classification related problem. These feature selection methods provided a good solution for "The curse of dimensionality" and contributed to the performance of pattern recognitions. In addition, feature selection methods included supervised and unsupervised cases. Generally, a supervised case is employed more often than an unsupervised case. This is due to unsupervised feature selection methods having a high probability of incorrect results for corresponding patterns regarding perceived speech[7]. Although, there are many cases that cannot obtain an explicit supervised value, the unsupervised method has advantages. We propose a method using reinforcement learning taking advantage of both the supervised and unsupervised method, which can alleviate the shortcomings of both methods. Researches of the reinforcement learning have been proceeded using many methods, i. e. Dynamic programming, Monte Carlo method, TD method, Q learning etc. proposed by Sutton and Barto. Since there is such a variety of methods and the main elements such as "state","action" and "reward" may be freely defined and implemented by a developer, this method is thought to be a very important one for machine learning techniques[9]. In this study, rather than using a specific reinforcement learning method, we propose a method which selects feature sets by calculating rewards received when an action is performed in a state. In particular, this method does not only calculate the frequency of emotion transit but also the sum of the rewards for the evaluation of a feature selection. Therefore, this method has the advantage that the more frequently it contacts a user, the better its performance becomes. The outline of the paper is as follows, In Section II, it explains the emotion recognition method and Section III explains the proposed algorithm. The Section IV shows a simulation and result of using the proposed algorithm. Section V conclude and shows future works.

2 Emotion Recognition Method

This paper addresses emotion recognition by extracting features of speech. The emotion recognition with speech is largely divided into cases using acoustic information and language or discourse information. The former is a method that uses some feature sets such as pitch, formant, speech rate, timbre, etc. and the latter uses the meaning of a word. That is, whether the word is positive or negative to whether it represents a happy or sad state. The process of emotion recognition consists of collecting emotional speech, the acoustic analysis, implementing DB, feature set extraction and such features are trained and classified with emotions using a pattern classification method.

2.1 Database and Preparation

Emotional speeches were collected from 10 male graduate students. Their ages ranged from 24 to 31 years old and they were asked to say with 10 short sentences emotionally. The choice of the 10 sentences(scripts) was decided upon from the result of another survey or experiment. In the first stage, 30 sentences had been prepared and all the 30 sentences were asked to say by the subjects. After the recording, the speeches were listened to by other people and they were asked the question “What emotion do you feel when listening to the given

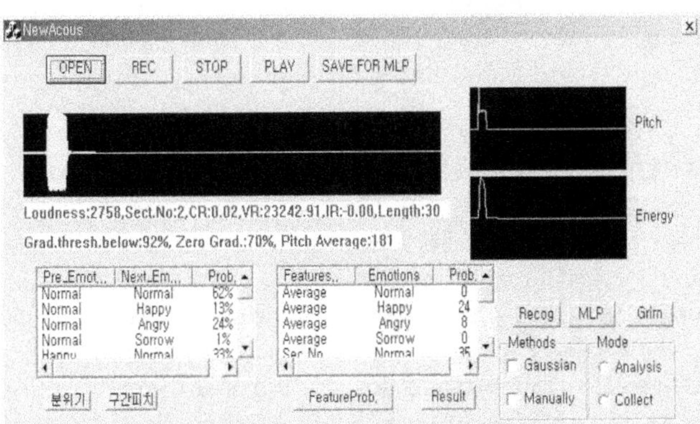

Fig. 1. Emotion recognition system

Table 1. Parameters of ANN

Parameter	Values
Input Units	3∼5
Hidden Units	11
Output Units	2
Learning Rate	0.003
Tolerance	0.25
Sigmoid function	$\frac{1}{1+e^{-3x}}$

recording?". The emotions conveyed in the 10 sentences that were read and the answers given by the subjects in this experiment were in agreement 90% of the time. In addition, the length of the prepared sentences was limited from 6 to 10 syllables. The recording format was 11Khz, 16bit, mono and the subjects were asked to keep a distance of 10 cm between themselves and the microphone. Since the distance of the microphone affected loudness or intensity, maintaining the required distance was very important. Recorded files were preprocessed and stored in DB(MS-ACCESS). In the preprocessing stage, there were several processes to signals such as FFT (extracting spectrum), Pitch extraction (by an autocorrelation method), IR(Increasing Rate) of pitch, CR(Crossing Rate), VR(Variance), and statistical values etc[10]. Fig. 1 shows emotion recognition system we implemented, which has the function of recording, preprocessing as uttered above and pattern recognition.

2.2 Pattern Classification Method

We used an artificial neural network for pattern classification, which commonly performs well and is robust to a signal with noise. It has been the most popular method to use in the pattern recognition field. This method commonly uses a Back Propagation Algorithm for tuning network parameters. In this study, we fixed the setting to ANN as follows, The number of Input Units and Hidden Units and Output Units and Learning rate and Tolerance and Sigmoid function are 3∼5 and 11 and 2 and 0.003 and 0.25 and $\frac{1}{1+e^{-3x}}$, respectively.

3 The Interactive Feature Selection Algorithm

Typically, researchers in the emotion recognition use various feature sets. Some researchers looked into the relation between acoustic analysis and emotion and used the feature sets based on that relation. However, because this method is subjective, it may easily lead to local minima. For this reason, recent studies consider a feature selection method for finding small superior features (4∼10) out of as many as 30 to 90 features. Most researchers do not use all features because they cannot confirm whether they are valid or not and noises with every features may deteriorate. Therefore, feature selection methods are popular in the pattern classification field[7]. According to John et al.[8], there are two main approaches to FS: filtering and wrapping. In the former, a feature subset is selected independently of the learning method that will use it. In the latter, a feature subset is selected using an evaluation function based on the same learning algorithm that will consider this subset of features. Although wrapper approaches have been shown to perform better, they can be rather time-consuming and it is sometimes infeasible to use them[7]. For this reason, the proposed algorithm tries to combine the characteristics of both the wrapper and filtering.

3.1 Reinforcement Learning Algorithm

Reinforcement learning consists of an agent and environment and is a learning method that leads the agent to perform a target action for a user. The process of learning is as follows,given an environment, an agent firstly performs an action and the agent receives a reward for the action from the environment. At that time, each time step is denoted as t, an environment state which the agent may be include is denoted as $s_t \in S$ (S is a set of possible environments) and an action is denoted as $a_t \in A(s_t)$ ($A(s_t)$ is a set of possible actions in a state). A reward for an action is denoted as r_t and when an episode has been completed, the r_t is expressed as the following equation.

$$R_t = \sum_{k=0}^{T} \gamma^k r_{t+k+1} \tag{1}$$

γ is a discount coefficient in the above equation and does not make the sum of rewards an infinity in the case of being defined as $t = \infty$. In addition, if the discount coefficient is zero, it means that only the current reward value is admitted. That is, we can give the weight to a future value differently according to the discount coefficient. Finally, reinforcement learning is a method that determines a policy to maximize the eq. 1.

3.2 Sequential Forward Selection(SFS) Algorithm

Sequential Forward Selection is the simplest greedy search algorithm. In this paper, we will briefly explain this algorithm. The fig2(a) shows the algorithm. Starting from the empty set, sequentially add the feature x^+that results in the highest objective function $J(Y_k + x^+)$ when combined with the feature $Y_k$that has already been selected.

3.3 Genetic Algorithm Feature Selection Algorithm (GAFS)

The Genetic Algorithm is popular method for finding an optimized solution. This algorithm has also good performance to the problems like nonlinear problems, which are hard to be solved by using the classic optimization techniques. The problem we are treating is also a nonlinear problem and thus, we think this problem may be solved by Genetic Algorithm. So, we tried to search good feature set using the Simple Genetic Algorithm. The fig.2(b) is the algorithm of the GAFS.

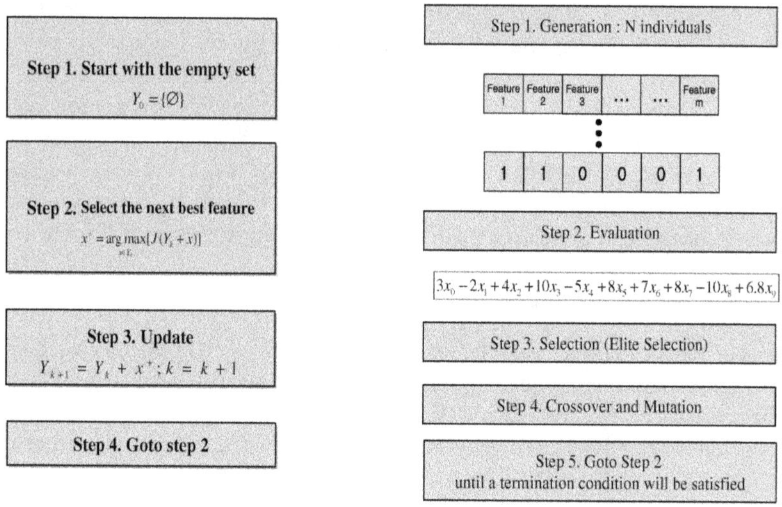

(a) SFS Algorithm (b) GAFS algorithm

Fig. 2. SFS and GAFS algorithm

The fig. 3(a), 3(b) is the performance graph of Genetic algorithm feature selection. Because this method is included in the part of wrapping, an evaluation function is needed. So, the function which is used for evaluation is as follows,

$$3x_0 - 2x_1 + 4x_2 + 10x_3 - 5x_4 + 8x_5 + 7x_6 + 8x_7 - 10x_8 + 6.8x_9 \qquad (2)$$

In the fig. 3(a), we can see the best fitness converged on the about 8 generations and the fig. 3(b) shows the average fitness is being increased. This simulation verifies that the GAFS is working well.

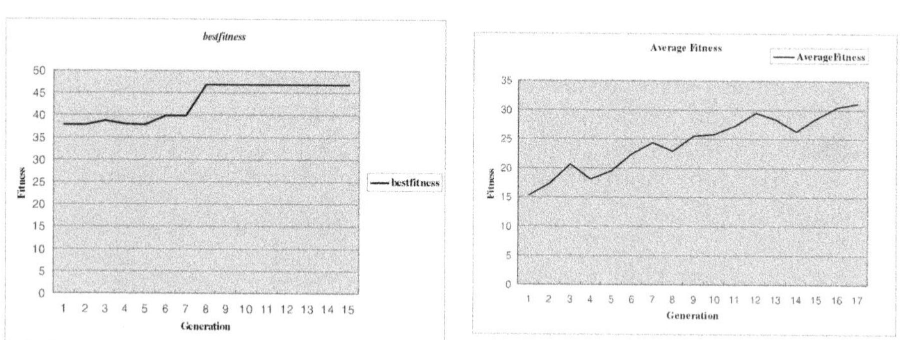

(a) GAFS Best fitness (b) GAFS Average fitness

Fig. 3. Performance graph of GAFS (Crossover rate: 0.5,Mutation rate: 0.02, Chromosome length: 10, Population No.: 100)

3.4 Interactive Feature Selection Algorithm (IFS)

The Interactive Feature Selection algorithm we are proposing is an algorithm based on reinforcement learning. Specially, popular algorithms such as SFS, SBS and so on, are deductive algorithms but our proposed algorithm is inductive. Also, these feature selection algorithms are based on the rationale of correlation and information-theoretic measures. Correlation is based on the raionale that good feature subsets contain features highly correlated with a class, yet are uncorrelated with each other. The IFS is also based on the correlation concept. Moreover, the feature selection algorithms consist of a search strategy and an evaluation by objective function but the conventional methods are incompetent in the search strategy part. Therefore, an IFS focuses on both the search strategy and evaluation by objective function. Fig 4(a) shows an IFS process. We assume that an emotion recognition system that includes this algorithm will be applied to a home robot or appliance. If such a situation is considered, users may be comfortable inputting emotional speech and a user's emotional state at that time(as a supervisor value). Due to this characteristic, this algorithm is a user adaptive system that can efficiently solve a problem and the more a user is in contact with this system, the better it will perform. The fig 4(b) shows an example of the IFS algorithm and is based on the Fig 4(a). First, this algorithm starts with a full feature set and when a new feature set and an emotion identifier is inputted, it assigns a +1 or -1 to the "return sign"(if an old emotion ID equals a new emotion ID then +1, Otherwise -1). Thereafter, the product of "return sign" and the difference of each feature is stored in an array "Pointstorage". This iteration is repeated for one

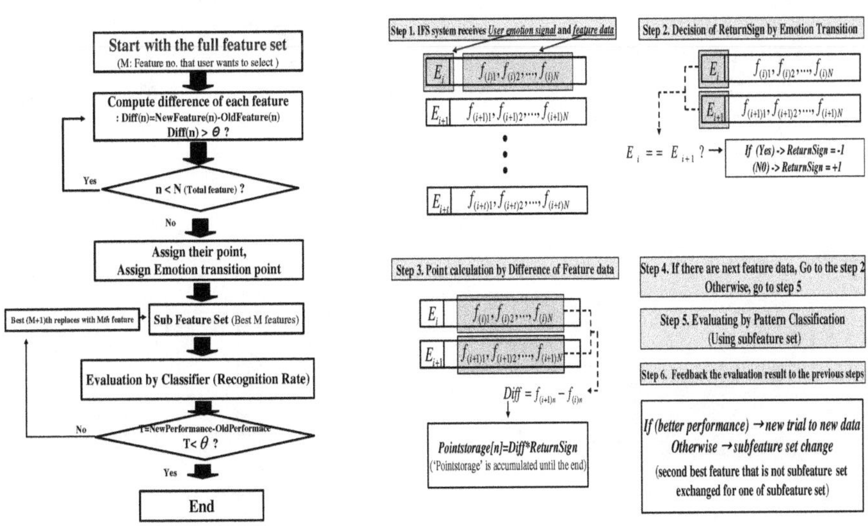

(a) IFS Algorithm (b) IFS example

Fig. 4. IFS algorithm and an example

episode(user can arbitrarily define an episode). After the episode, the feature set that was selected first is applied to an objective function(Pattern Classification System) and the evaluation result is stored. If the next evaluation result is worse than the previous, the worst feature of the selected feature set will be replaced with the best feature among those that were not selected(step 6 in Fig 4(b)).

4 Experimental Results

4.1 Simulation and Results

We applied 11 original features to the IFS simulator;Pitch features(max, min, mean, median, crossing rate, increasing rate), Loudness ;Intensity(max, min, mean),Sect.No and Speech rate. This program was made only for an IFS and the results from this program were applied to the Emotion Recognition System using ANN. That is, the feature set applied to ERS was the features previously outputted by IFS and then the emotion recognition experiment was performed. Classification was attempted using four methods. Results are shown in the fig 5(a) and 5(b). The fig 5(a) shows the fitness movement on every generation.

From the Fig 5(a),we can see that the IFS system searched for better results and improved gradually. In the algorithm,because the searching work was performed again when the new evaluation result was worse than the previous one, there was some range in the steady state(The range of 3~5 and 6~8 in the Time axis). In addition, from Fig 5(b), this graph compares four methods(IFS, SFS, GAFS and random selection) with the changed feature no. As expected, random selection performed poorly but IFS and SFS and GAFS similarly performed better. In the IFS and SFS and GAFS case, the results show a subtle distinction but the IFS with features 1,2 and 3 was better. However, with features 4 and 5, SFS showed better results.

Fig 6 shows the selected feature set from IFS system.

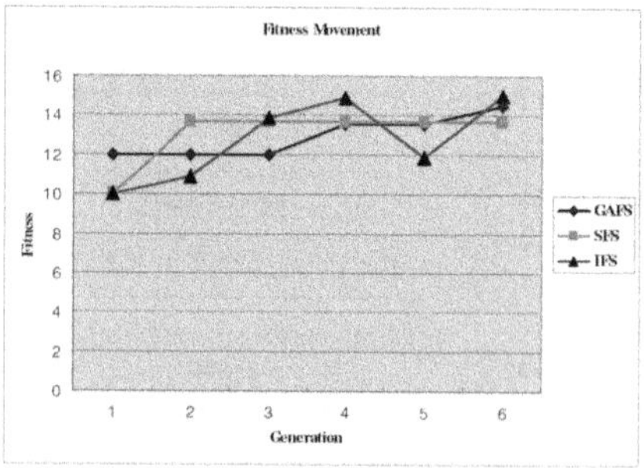

(a) Fitness movement for 3 methods

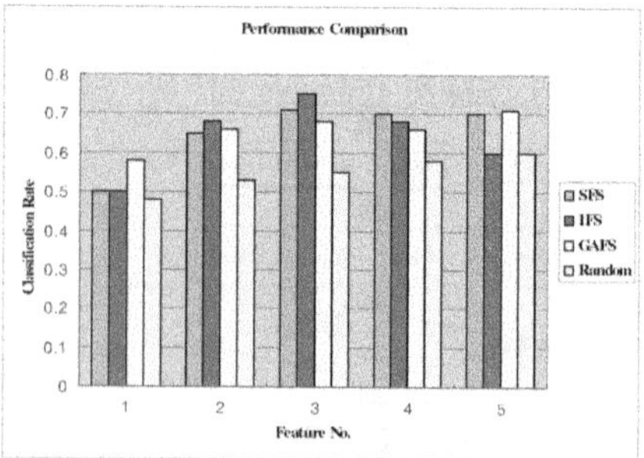

(b) Emotion Classification Rate Comparison

Fig. 5. Experiment Result

Feature No.	Features
1	Pitch Mean
2	Pitch Mean,Speech rate
3	Pitch Mean,Speech rate,Loudness
4	Pitch Mean,Speech rate,Loudness, Sect.No.

Fig. 6. Selected Feature Sets

5 Conclusions

This paper presents a solution to feature selection when there is an emotion recognition problem. The solution called an IFS performed as well as an SFS and GAFS. In particular, it is reinforcement based learning and supplements the role of search strategy in the feature selection process. Using the IFS simulator, we found some of the best features and used them in the emotion recognition experiment and results were compared to those of SFS and GAFS and Random selection. Performance was slightly better than SFS and GAFS. However, IFS has some disadvantages. If the amount of training data is too small, selection results may be not better than SFS. SFS does not require much training data. It is also sufficient that training data be only one set. If an objective function is clear, SFS will be adequate. However, in the case of emotion recognition, SFS may not perform as well as it had. In this case, the correlation-based method like the IFS will be better.

References

1. Ververidis, D., Kotropoulos, C.: Emotional Speech Classification Using Gaussian Mixture Models. IEEE International Symposium on Circuits and Systems (ISCAS2005), Kobe, Japan, (2005) 2871-2874
2. Lee, C. M., Narayanan, S. S: Toward Detecting Emotions in Spoken Dialogs. IEEE Transactions on Speech and Audio Processing. 13 293-303
3. Wagner, J., Kim, J. H., Andre, E.: From Physiological Signals to Emotions: Implementing and Comparing Selected Methods for Feature Extraction and Classification. Proceedings of IEEE International Conference on Multimedia and Expo (ICME-2005), Amsterdam, (2005) 940-943
4. Pudil, P., Novovicova, J.: Novel Methods for Subset Selection with Respect to Problem knowledge. IEEE Intelligent Systems, 13 (2) (1998) 66-74
5. Lin, Y.L., Gang, W.: Speech Emotion Recognition Based on HMM and SVM. Proceedings of Machine Learning and Cybernetics, 8 (2005) 4898-4901
6. Morchen, F., Ultsch, A., Thies, M., Lohken, I.: Modeling Timbre Distance With Temporal Statistics From Polyphonic Music. IEEE Transaction on Audio, Speech and Language Processing, 14 (1) (2006) 81-90
7. Combarro, E. F., Montanes, E., Diaz, I., Ranilla, J.,Mones, R.: Introducing a Family of Linear Measures for Feature Selection in Text Categorization. IEEE Transactions on Knowledge and Data Engineeringl, 17 (9)(2005) 1223-1232

8. John, G. H., Kohavi,R., Pfleger, K.: Irrelevant Features and The Subset Selection Problem. John, G. H., Kohavi, R. and Pfleger, K. (Eds.) In Proceedings of the Eleventh International Conference on Machine learning, New Brunswick, NJ, (1994) 121-129
9. Sutton, R. S., Barto, A. G.: Reinforcement Learning: An Introduction. A bradford book, London, (1998)
10. Park, C. H., Sim, K. B: The Implementation of The Emotion Recognition from Speech and Facial Expression System, Proc. of ICNC'05-FSKD'05, Changsha, China (2005) 85-88

Unsupervised Feature Extraction for the Representation and Recognition of Lip Motion Video

Michelle Jeungeun Lee[1], Kyungsuk David Lee[2], and Soo-Young Lee[1]

[1] Korea Advanced Institute of Science and Technology, Department of BioSystems, Daejeon 305-701, Korea (South)
{michelle_lee, sylee}@kaist.ac.kr

[2] University of Wisconsin at Madision, Department of Computer Science, Madison, Wisconsin, USA
ksdavidlee@gmail.com

Abstract. The lip-reading recognition is reported with lip-motion features extracted from multiple video frames by three unsupervised learning algorithms, i.e., Principle Component Analysis (PCA), Independent Component Analysis (ICA), and Non-negative Matrix Factorization (NMF). Since the human perception of facial motion goes through two different pathways, i.e., the lateral fusifom gyrus for the invariant aspects and the superior temporal sulcus for the changeable aspects of faces, we extracted the dynamic video features from multiple consecutive frames for the latter. The multiple-frame features require less number of coefficients for the same frame length than the single-frame static features. The ICA-based features are most sparse, while the corresponding coefficients for the video representation are the least sparse. PCA-based features have the opposite characteristics, while the characteristics of the NMF-based features are in the middle. Also the ICA-based features result in much better recognition performance than the others.

1 Introduction

Recently the extraction of basis features from motion video has attracted a lot attention. Especially extracted features from lip motion may be utilized to recognize noisy speeches in cooperation with lip-reading [1] as well as to graphically synthesize the lip motion for Text-To-Speech applications [2].

Different decomposition techniques are applied for extracting basis elements of video images. Principle Component Analysis (PCA) had been widely used for feature extraction from images [3], while Non-negative Matrix Factorization (NMF) [4] and Independent Component Analysis (ICA) [5] have been applied recently. The PCA-basis features are non-local, and ICA results in localized basis features [6]. The NMF comes with non-negative constraints, and usually results in regional features, i.e., between the global and local features. The non-negativity of NMF and the sparsity of ICA are motivated from the biological neural systems.

In previous studies the features are extracted from single-frame images and the sequential nature of the motion video is not utilized. However, it is commonly

D.-S. Huang, K. Li, and G.W. Irwin (Eds.): ICIC 2006, LNBI 4115, pp. 741 – 746, 2006.

understood that the human perception of facial motion goes through two different pathways, i.e., the lateral fusifom gyrus for the invariant aspects of faces and the superior temporal sulcus for the changeable aspects of faces such as lip movement. [7] Therefore, the features of lip motion may be different from the features for the face recognition, and require more representation from consecutive multiple frames.

In this paper, we applied the 3 decomposition techniques, i.e., PCA, NMF, and ICA, to multi-frame lip videos. The effects of the time frames and statistical characteristics are investigated for both the extracted features and corresponding representation coefficients. The extracted features are also utilized for the recognition of lip motion, and the multi-frame features demonstrate better recognition performance than the single-frame features.

2 Feature Extraction Algorithms

Let $\mathbf{s} = [s_1 \ldots s_N]^T$ be the unknown basis image (features), and $\mathbf{x} = [x_1 \ldots x_N]^T$ be the observed image which are considered as the linear mixture of basis images. Here, N is the number of pixels in the images, and **s** and **x** are 1-dimensional re-shaped versions of 2-dimensional images by concatenating all columns. The mixing model can be written as the combination of s and unknown mixing matrix **A**, i.e.,

$$\mathbf{x} = \mathbf{A}\mathbf{s} \text{ or } \mathbf{X}=\mathbf{A}\mathbf{S} \tag{1}$$

where $\mathbf{X} = [\mathbf{x}_1 \ldots \mathbf{x}_M]$ and $\mathbf{S} = [\mathbf{s}_1 \ldots \mathbf{s}_M]$ are matrices consisting of all the observed and basis images as the columns, respectively, and M is the number of images. The feature extraction algorithms find the unknown basis images (features) **S** and the mixing matrix **A** from the observed images **X** only. This is an underdetermined problem, and additional constraints are usually required. The popular choices are orthogonal basis images for PCA, linear independence for ICA, and non-negativity of NMF.

Principle Component Analysis (PCA) is probably the most commonly used technique for image processing tasks. PCA computes a set of orthogonal subspace basis vectors for an image database, and project the images into the compressed subspace. PCA generates compressed data with minimum mean-squared reproduction error, and is an unsupervised learning that produces global feature vectors. [3]

Non-negative Matrix Factorization (NMF) is a recently developed method for finding the representation of the data based on non-negative characteristics of biological neural signals. The non-negativity constraint of NMF only allows the addition of representation, and produces a part-based representation of the data which encodes much of the data using few active components [4].

ICA decomposes data image into linear combination of statistically independent mixtures. While PCA is based on the second-order statistics of the images, ICA decorrelates the high-order moments of the input in addition to the second-order moments. The ICA-based features are locally supported and usually very sparse [5].

3 Lip Motion Database

The lip motion videos employed for this research are the video part of the Tulips database [8], which consists of both video and audio data. The data set contains 96 digitized videos of 12 individuals (9 males and 3 females) saying the first four digits in English ("one", "two", "three" and "four") twice. The images are made for every 1/30 second, and the number of image frames for each video clip is between 6 and 16. The original dataset consists of 934 images, with 65 x 87 pixels each with 8-bit gray levels. Ten corners of the lips are marked by human hands for each of the images.

The images were first corrected for the rotation based on the manually-specified critical points (corners) on the lip boundary. Then, the lip images are segmented and normalized in fixed size, i.e., 32 x 18 pixels. To reduce the computational complexity, each image was vertically cut into the left-half and the right-half, and considered as two different images. The dataset now consists of 1868 images of which each one is re-sampled into 16 x 18 (288 in total) pixels. All the feature extractions were done on these half-images. However, in all the subsequent lip figures, the full images were restored with left-to-right symmetry for easy visual understanding.

The iterative ICA algorithm requires extensive computation for the large number of images. For the efficient computation of the ICA-based features it is commonly adopted to get the essence of the original images using the PCA. It this paper the dominant 60 principal components, which approximate the frame-difference images with about 95% accuracy, are used for the extraction of ICA-based features.

4 Results

The n-frame database was obtained by considering the n consecutive frame images as an element of the database. Therefore, the n-frame images consist of $288n$ pixels, while the number of the elements in the database becomes smaller. The PCA, NMF, and ICA algorithms are applied to these databases with the n from 1 to 4. The case of n=1 actually corresponds to the case of single-frame method.

In Table 1 the numbers of principle components are shown for given representation accuracy with PCA. Also, the compression ratio, i.e., C = (Number of PCAs) / (Number of Pixels), is shown. The number of principal components for a given accuracy does not change much on the number of frames. Therefore, the more frames are used, the more efficient features are extracted.

Table 1. Numbers of principle components and compression ratios for given representation accuracy

	Numbers of Principle Components				Compression Ratio			
Accuracy	Single Frame	2 Frames	3 Frames	4 Frames	Single Frame	2 Frames	3 Frames	4 Frames
90%	11	12	14	15	0.038	0.021	0.016	0.013
95%	19	23	27	30	0.066	0.040	0.031	0.026
99%	56	78	95	109	0.194	0.135	0.110	0.095

In Fig. 1 the extracted lip motion features are shown for the cases of n=1, 2, and 3 for comparison. Only top twelve features, sorted by the variance of the corresponding coefficients and their accumulated values are shown for simplicity. The delta (difference in imaging condition) has been plotted on the second row of each basis components for the 2-frame and 3-frame cases. The increased delta value denotes the features are dynamic, and the multiple-frame approaches may be advantageous over the popular single-frame approach.

The efficiency of data representation may be related to the sparseness of the features and corresponding coefficient values. With only a few non-zero values, the sparse representation results in efficient data coding, transmission, and storage. A qualitative measure of the sparseness is kurtosis, related to the 4th order moments. Positive kurtosis denotes super-Gaussian distributions, and the higher value of kurtosis means only a few is enough to represent the data with less error. In Table 2, the kurtosis values are summarized. Unlike the PCA-based and ICA-based features/coefficients, the NMF-based features/coefficients do not have zero mean and the kurtosis is calculated after adding negative data for symmetry.

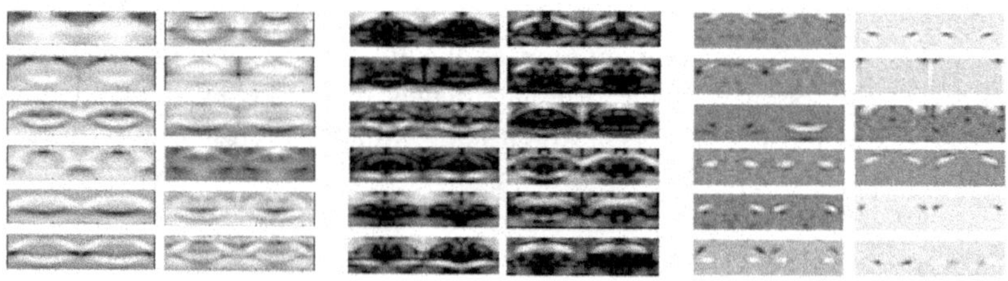

Fig. 1. Extracted lip motion features by PCA (*left figures*), NMF (*center figures*), and ICA (*right figures*) algorithms. Only features from 2-frames are shown for simplicity.

Table 2. Kurtosis values of extracted features and corresponding coefficients

Feature Extraction Method	Features				Coefficients			
	Single Frame	2 Frames	3 Frames	4 Frames	Single Frame	2 Frames	3 Frames	4 Frames
PCA	0.31	0.30	0.33	0.29	15.1	15.0	14.8	14.7
NMF	11.5	11.6	9.4	9.9	3.2	2.9	4.0	3.7
ICA	31.9	22.1	23.7	22.2	0.72	0.66	0.80	0.63

Using different number of frames to find the basis components, we found that the kurtosis values and the sparseness do not change much among the same decomposition method. The ICA algorithm has the greatest kurtosis for feature values. The PCA has the lowest value, and NMF is located between ICA and PCA. On the other hand, kurtosis values of coefficients are the other way around. Higher kurtosis value means there are more data concentrated near the zero. Since the efficiency of the data representation depends not only on the coefficient matrix but also the features matrix, it is difficult to predict the best technique based on the value of its kurtosis.

We also had conducted lip-motion recognition experiments based on the extracted features. Although Hidden Markov Model (HMM) is the popular choice for continuous speech recognition tasks, Dynamic Time Warping (DTW) is utilized here. The DTW shows similar performance for isolated-word recognition tasks [9], and also is more biologically-plausible with elastic neural networks. [10] To obtain better accuracy on the recognition rates with limited dataset, the *N-fold* hold-out methods with n=4, 6, and 12 divisions are incorporated. Actually 22 experiments are conducted for each algorithm and consecutive frame number, and the average recognition rates are summarized in Table 3. Also, to remove the effects of small number of frames for each word with 30 msec intervals, each video clip is interpolated to obtain simulated video clip with 15 msec frame intervals. Therefore, the (2N-1)-frame video clips are generated from the N-frame video clips, where N is between 6 and 16 in the Tulips1 database.

Table 3. Recognition rates (%) of lip motion video with different features and frame length

	Single Frame	2 Frames	3 Frames	4 Frames
PCA	57.8	53.8	51.2	57.6
NMF	57.3	57.5	63.4	58.9
ICA	69.6	66.5	65.8	64.8

As shown in Table 3, the ICA-based features result in better performance than the PCA and NMF-based features. It shows the advantage of sparse local representation of the ICA-based features. On the other hand, provided enough numbers of features were used for the recognition, the multi-frame features do not improve the performance. It may come from the fact that the recognition performance is closely related to the representation accuracy, not to the number of frames in the features.

5 Discussion

Comparison of the feature extraction methods based on PCA, NMF, and ICA is not straight-forward. The efficiency of the data representation depends not only on the coefficient matrix but also the feature matrix, and their relative significance is dependent upon the applications.

While ICA results in much sparse features, PCA results in much sparse representation coefficients. The characteristics of the NMF-based features are between those of the PCA and ICA-based features. For telecommunication applications the sparseness of the coefficients is very important, and the PCA-based features may be more advantageous. For the recognition of lip motion it is shown that the ICA-based features are superior.

Also the number of frames affects on the sparseness. Since only a small number of features are needed for the videos of the same frame length, the multiple-frame features have higher compression ratio and becomes more advantageous in efficiency.

The lip motion synthesis may be done by adding and subtracting these multi-frame features. Combining these images will create a video clip of lip motion. Also, the lip-motion features and phonetic speech features will be combined for the audio-visual speech recognition. Provided the lip-motion features were associated to the phonetic features, it will also be applied to the lip motion synthesis in close correlation to the human speeches.

Acknowledgment

S.Y. Lee was supported by the Brain Neuroinformatics Research Program sponsored by Korean Ministry of Science and Technology and Korean Ministry of Commerce and Industries.

References

1. Potamianos, G., Neti, C., Luettin, J., and Matthews, I.: Audio-Visual Automatic Speech recognition: An Overview. In: Bailly, G., Vatikiotis-Bateson, E., and Perrier, P. (eds.): Issues in Visual and Audio-Visual Speech Processing. MIT Press (2004)
2. Berthommier, F.: Direct Synthesis of Video from Speech Sounds for New Telecommunication Applications. Smart Object Conference (2003)
3. Turk, M., and Pentland, A.: Eigenfaces for recognition. Journal of Cognitive Neruoscience 3 (1991) 71-86
4. Lee, D.D., and Seung, H.S.: Learning the parts of objects by non-negative matrix factorization. Nature 401 (1999) 788-791
5. Barlett, M.S., Lades, H.M., and Sejnowski, T.J.: Independent component representations for face recognition. SPIE Symposium on Electronic Imaging: Science and Technology (1998)
6. Draper, B.A., Baek, K., Bartlett, M.S., and Ross Beveridge, J.: Recognizing faces with PCA and ICA. Computer Vision and Image Understanding 91 (2003) 115-137
7. Haxby, J.V., Hoffman, E.A., and Gobbini, M.I.: The distributed human neural system for face perception. Trens in Cognitive Sciences 4 (2000) 223-233
8. Movellan, J.R.: Visual speech recognition with stochastic networks. Advances in Neural Information Processing Systems 7 (1995) 851-858
9. Kim, D.S. Kim, Lee, S.Y., and Kil, R.M.: Auditory processing of speech signals for robust speech recognition in real-world noisy environments. IEEE Transactions on Speech and Audio Processing 7 (1999) 55 - 69
10. Lades, M., Vorbuggen, J.C., Buhmann, J., Lange, J., Malsburg, C., Wurtz, R., and Konen, W.: Distortion invariant object recognition in the dynamic link architecture. IEEE Trans. Computer 42 (1993) 300-311

A Novel Method for Expanding Current Annotations in Gene Ontology

Dapeng Hao[1], Xia Li[1,2,*], Lei Du[1], Liangde Xu[1], Jiankai Xu[1], and Shaoqi Rao[1,3,*]

[1] Department of Bioinformatics, Harbin Medical University, Harbin 150086, P.R. China
haorixin@hotmail.com, lixia6@yahoo.com, dulei@hrbmu.edu.cn, xuliangde@hotmail.com, xjk7@yeah.net, raos@ccf.org
[2] Biomedical Engineering Institute, Capital University of Medical Sciences, Beijing 100054, P.R. China
[3] Departments of Cardiovascular Medicine and Molecular Cardiology, The Cleveland Clinic Foundation, 9500 Euclid Avenue, Cleveland, Ohio 44195, USA

Abstract. Since the gap between the amount of protein sequence data and the reliable function annotations in public databases is growing, characterizing protein functions becomes a major task in the post genomic era. Some current ways to predict functions of a protein are based on the relationships between the protein and other proteins in databases. As a large fraction of annotated proteins are not fully characterized, annotating novel proteins is limited. Therefore, it is of high demand to develop efficient computation methods to push the current broad function annotations of the partially known proteins toward more detailed and specific knowledge. In this study, we explore the capability of a rule-based method for expanding the current annotations per some function categorization system such as Gene Ontology. Applications of the proposed method to predict human and yeast protein functions demonstrate its efficiency in expanding the knowledge space of the partially known proteins.

1 Introduction

Since the gap between the amount of sequence data and the reliable function annotation in public databases is growing, it becomes an immediate task to predict and to characterize the function facets for increasingly identified proteins. Some current methods to predict function(s) of a protein are based on the sequence homologies between the protein and other proteins in protein databases. Roughly 20%-40% of proteins can not be characterized in this way, which is also prone to making errors when the sequence identity is below 40% [10]. In recent years, several more involved computation methods are developed for inferring protein functions with help of other information sources, including analysis of phylogenetic profiling based on the hypothesis that proteins with a similar function(s) are more likely to be in the same evolutionarily conserved protein family or cluster across organisms [13]; analysis of

* Corresponding authors.

D.-S. Huang, K. Li, and G.W. Irwin (Eds.): ICIC 2006, LNBI 4115, pp. 747 – 756, 2006.

gene fusion events [6, 11]; analysis of gene expression data, by assuming the presence of a link between transcription similarity and function sharing [17]; and analysis of networks of protein-protein interactions which focus on neighboring nodes' functions in the network [10]. All these methods first identify the non-function relationships between a studied protein and known proteins, and followed by deriving (predicting) the function(s) of the protein based on the function knowledge of its associated proteins. Although many successes are achieved by using these methods, a large number of annotations remain incomplete and to be further characterized, perhaps via newly developed computation algorithms.

Most of current function prediction methods are limited in expanding the existing annotations because they were designed for predicting function(s) for new proteins, and thus lack the necessary mechanisms to integrate the partial knowledge depicted by the broader descriptions. As such, we propose a novel method for expanding function annotations, in which the correlations between different functions instead between proteins are studied. We explore Gene Ontology, a standardization of protein function descriptions using controlled vocabularies, to seek for the correlation patterns between various functions [1]. Based on the correlation profiles, we determine whether yet uncharacterized functions that a protein might have, or more specific function(s) could be annotated to the protein. There is a recent study [15] that attempts to use the partial function knowledge in Gene Ontology(GO) of genes and their expression profiles for further characterizing their functions (i.e., annotating to deeper GO terms). However, the researchers only analyzed the parent-child relationship and identified 14 learnable function terms among the significantly expressed genes.

Protein sequence data are one of the most abundant and reliable sources for inferring protein functions. Besides the homology search, the default method in many sequence databases such as SWISS-PROT, the protein sequence signature data achieved in InterPro [12], Prosite [7], PRINTS [2] and Pfam [3] provide alternative information and ways for inferring protein functions. The function of a protein might be obtained if it shares some signatures with the proteins of known functions. Nevertheless, this procedure can become complex when there is a many-to-many relationship in which one signature is conserved for many proteins or one sequence has a number of signatures. Obviously, a data mining tool that is able to efficiently learn and integrate high-dimension characteristics of an object is desired for gathering, analyzing and evaluating the information from the diverse sources. Decision tree algorithm stands out to be one of the best learners and is integrated into the Spearmint system [8] for characterizing proteins according to multiple signature matches. Our proposed approach with a decision tree as the learner adopts the categorization system inherent in the GO hierarchy. It differs from other function prediction methods in several aspects. First, it aims at expanding current annotations, and hence the proteins of partial knowledge become the targets to predict. Second, it is based on the correlation between the functions of proteins, i.e. the function attribute of proteins other than the attributes (e.g. transcriptions) used in the previous studies. Third, because of the nature, it fully utilizes the partial function knowledge that proteins already have. Hence, it is expected to have more power and prediction accuracy than an *ab initio* prediction algorithm that does not use the prior information. Base on the correlation patterns between GO function classes, and the derived classifiers, we have analyzed human and yeast proteomes, and demonstrated that high true positive ratios for most predictions can be achieved. Meantime, we obtain the results for the relationships between learnability and GO function levels.

2 Methods

None of the biological functions acts in isolate. Instead many functions often work in a concerted way to perform a common task. On the other hand, a single protein may play an important role in different cellular processes and biochemical cascades. Such multiple function attributes of a protein can be used to identify several related functions because they share a common element (the protein). Based on the activity of the protein, it is also possible to deduce which pairs of functions are more likely co-activated or co-repressed under the protein state.

2.1 Correlations Between Different Functions in Gene Ontology

GO is used in this study to explore the correlations between functions. It provides a comprehensive source of structured knowledge of gene product function in terms of "molecular function", "biological process" and "cellular function". The function categorizations are represented with nodes (or called terms, classes) in a directed acyclic graph in which one node may have more than one parent node. Protein function knowledge annotated to the deepest GO nodes gives the most specific function description(s).

We often have rather abstract (broad) function knowledge for some proteins, which are annotated to one of few nodes at high levels of GO hierarchy [9]. Prior to further wet-lab function studies, expanding the current annotation is necessary. According to the GO structure, we define two types of correlations between different GO terms. As shown in Fig. 1a, two types of correlations are defined: inherited (e.g. parent-child) and non-inherited (e.g. between siblings) correlation. GOi has two functionally more specific child nodes (GOi_1 and GOi_2). Its relation with GOj is non-inherited although they may be involved in regulating a same or similar biological process. A novel protein with function GOi will be judged whether it has the function GOj or the functions describing the child nodes (GOi_1 or GOi_2) by a protein signature matching.

2.2 Data-Driven Discovery of Rules for Protein Function Classification Using Sequence Signatures

For the data mining process, given a dataset or training dataset of protein sequences with known functions, the goal is to establish the correlation between each pairs of GO function nodes by inducing a classifier(s) of high discriminating power for distinguishing this correlation from others using protein signature profiling. Then, the function for a novel protein sequence can be assigned if it fits a partition established by the classifier (or alternative classifiers). The schematic framework for the proposed algorithm is illustrated in Fig. 1b.

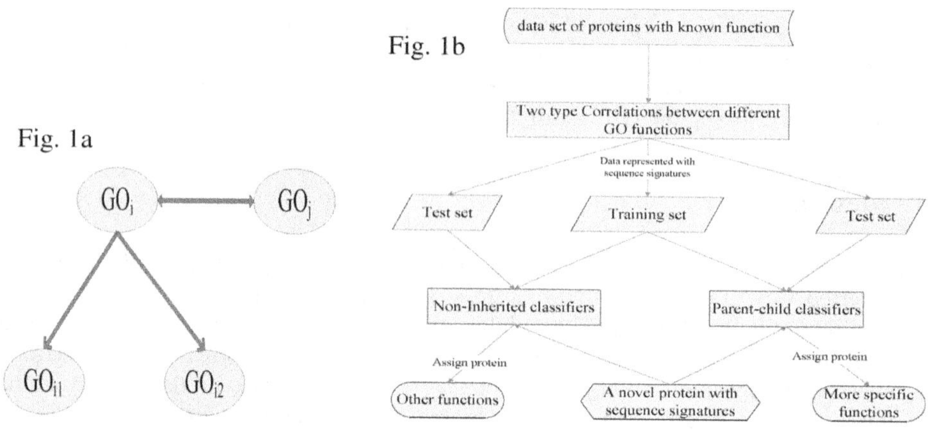

Fig. 1. (a): Two types of GO term correlations. (b): Algorithmic flow chart for expanding function knowledge.

Data Representation and Decision Tree Rlgorithm

Most data mining algorithms aim to characterize instances with a reduced subset of a finite set of attributes. In this study, protein sequences (in addition to its function attributes) are described by a vocabulary of sequence signatures from the databases integrated by InterPro such as Prosite, PRINTS or Pfam. The data formulations are very similar to Spearmint system. Suppose that a set of proteins contain a total of N signatures (any protein may lack some of the signatures). A protein is thus encoded by an N-bit binary profile where the i-th bit is 1 if the corresponding signature is present in the protein, and zero otherwise. Note that each binary profile is also associated with a label that identifies the function of the protein.

We use the CART decision tree algorithm for building signatures-based classifiers. To illustrate the procedure, an example is given in Fig. 2, compared with Spearmint system. The employed decision tree algorithm classifies the proteins of the training dataset into positive and negative domains based on a reduced subset of protein signatures. The resulting tree consists of a root node, internal nodes and leaf nodes. The internal nodes, obtained from learning all the sequences documented in some secondary databases like Pfam or Prosite, capture significant signature features for partitioning protein samples, whereas the root node depicts the most broad annotation that all the trained proteins share, and leaf nodes give the sample partitions that the tree achieves. The proposed decision tree algorithm differs from Spearmint system in two aspects. First, we use the GO terms as function annotation (Fig. 2b) while Spearmint uses SWISS-PROT keywords (Fig. 2a), which can be seen from the leaf nodes. Second, we use existing function knowledge to divide classification space to induce a set of similar proteins, while Spearmint selects the proteins belonging to a same InterPro domain as training samples to learn a classifier. By predicting a novel protein from the root node (its current annotation) to a leaf node (its newly assigned function), we expand the function knowledge of the protein.

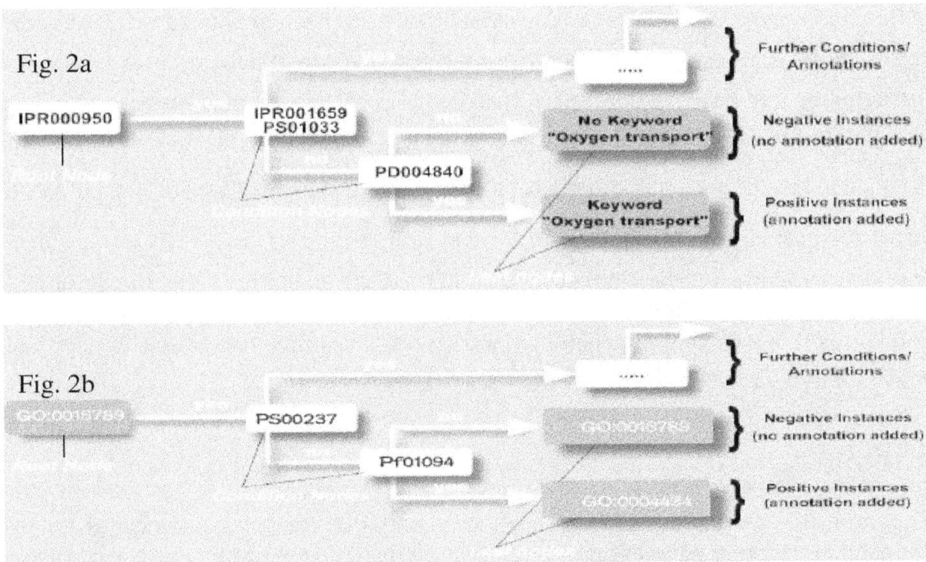

Fig. 2. (a): Decision tree that classifies proteins belonging to IPR000950. The Spearmint prediction is: "Add the Keyword "Oxygen transport", if the protein belongs to IPR000950 (InterPro), but neither to IPR001659 nor to PS01033 (PROSITE) and if it has a hit to PD004840 (ProDom)". (b): Decision tree that classifies proteins belonging to GO:0016789. The prediction rule is "Add the another annotation GO:0004434, if the protein has the annotation GO:0016789 and not belong to PS00237 (PROSITE), but has a hit to Pf01094 (Pfam)".

Local Classifiers for Two Types of GO Term Correlations

Decision tree learning for non-inherited relationship: we define the correlation between two GO terms (e.g. g_i and g_j in Fig. 1a) as non-inherited relationship if there is at least one protein whose annotation covers the annotations of two child proteins. First, we set g_i as the reference for defining the classification space and all the proteins annotated to it are used for the training dataset. An *M* matrix of protein signature attributes is constructed, where each row is an N-bit binary vector for a protein and each column corresponds to a signature. The matrix is then associated with a label vector whose elements are defined as g_{ij} if the corresponding protein (to a row in the protein signature matrix) has both annotations of g_i and g_j or $\overline{g_{ij}}$ if the corresponding protein has g_i but not g_j annotation. Proteins labeled with g_{ij} are positive instances, and the remaining are negatives. Once a decision tree classifier is built by learning a training dataset, it is used to annotate a novel protein from g_i to g_j. Then, g_j is taken as the reference and a classification space is defined as above, followed by constructing the protein signature profiles and the label vector, building the classifier and making decision for a novel protein in a similar way.

Decision tree learning for parent-child relationship: Suppose that GO terms g_{i_1}, g_{i_2}, ……, g_{i_n} are child nodes of g_i and $n \geq 2$. Classifiers trained for this type of relationship are able to acquire a more specific annotation(s) described by one or more child terms. Obviously, only proteins that have been annotated to child nodes are useful for learning the classifier. Here, we take the parent node g_i as the reference for defining the classification space that includes all the proteins annotated. For a given child node g_{i_h} $(1 \leq h \leq n)$, a protein is labeled with g_{i_h} (positive instance) if it is annotated to the node and otherwise $\overline{g}_{i_h}$ (negative instance). Then, a decision tree is learnt for pushing the annotation of a novel protein from g_i to g_{i_h}. Note that there may be n classifiers to be learned for n child nodes. Generally, a specific classification space and an attribute matrix are constructed for each learnable correlation. A correlation is learnable only when both positive and negative instances are more than λ, a criterion used for selecting learnable correlations.

3 Results

The numeric experiment is to explore the potential of the proposed method for expanding the existing annotations. By dividing a dataset into training and test dataset, we are able to obtain an unbiased estimate of accuracy that a classifier can achieve. For this purpose, we adopt a 5-fold cross-validation in which the positive instances and negative instances are divided randomly into *n* parts of roughly equal size, respectively. A random combination of a positive part and one negative part constitutes a test dataset, and the remaining data are used as the training dataset. One cycle of an *n*-fold cross-validation produces $n \times n$ pairs of training and test datasets, and also results in $n \times n$ classifiers. We use precision (=*TP*/(*TP*+*FP*)), where *TP* and *FP* are true and false positives, respectively), and recall (=*TP*/(*TP*+*FN*)), where *FN* are false negatives), as the measure for classification performance since we are most concerned on how a classifier is right in its positive predictions.

3.1 Performance from Human Proteome

In order to get the most accurate and comprehensive function annotation data, we use all the 9071 manually edited human proteins in GOA (Gene Ontology Annotation). GOA is a project sponsored by the European Bioinformatics Institute that aims to provide accurate assignments of gene products to the Gene Ontology (GO) database. In the first stage of this project, GO assignments have been applied to a dataset covering human proteome by using both electronic mapping and manual curation [4]. The download data are of direct annotations which are associated to the most specific nodes in GO. In GO, a gene annotated to a node is also annotated to its ancestor nodes of the same branch from the child node to root node. This upward propagation results in "general annotations", based on which the correlations between GO terms (nodes) are analyzed.

Non-inherited relationship: there are totally 71628 non-inherited relationships containing at least 10 proteins. For a numeric demonstration, we focus on the "Molecular Function" (MF) branch. The number of relationships is 2208, including 239 GO nodes if the criterion of 10 proteins (i.e. λ >10) is applied. Fig. 3a illustrates the relationship between classification performance and level of classification space (i.e. defined by g_i), after removal of low-performance classifiers with precision < 50%. As a result, half of the remaining classifiers have a precision >75%. In the range from level 2 to 9, precision rises slowly. The reason for the variation is that the more homogeneous the proteins are, the easier it is for the classifiers to learn the pattern. In Fig. 3b, the red line shows the non-inherited relationship classification sensitivity and the level of classification space, similar with precision, more homogeneous the proteins are, the more proteins are correctly predicted by the classifiers.

Parent-child relationship: we identify 316 learnable parent nodes with a total of 842 child nodes by using $\lambda > 10$. As illustrated in Fig. 3a, half of the classifiers have precision >81%. Similar to the non-inherited relationship, precision rises slowly with the level of parent nodes increasing. Compared to the previous study [15], a much larger of function classes are analyzed, and the proposed algorithm reaches a deeper level (up to the 10th GO hierarchy versus the 7th in the previous study). We also estimate the precisions for the three dimensions of GO: molecular function: 87%; biological process: 72%; and cellular component: 77%, respectively. It appears that protein signatures are more informative for revealing "molecular function" than other two dimensions, which might also be the reason of only dimension of "molecular function" being investigated previously [14]. The blue line in Fig. 4a shows that the sensitivity of classifiers learned from parent-child relationship performs better than non-inherited relationship.

3.2 Performance from Yeast Proteome

Protein signatures diverged gradually between different organisms during the evolutionary trajectories, and hence can be good molecular features for reversely constructing a phylogeny [16]. The protein signature profiles for human are much more complex than yeast that may have better performance for function knowledge expansion. The function annotation data for yeast are derived from SGD [5]. Fig. 3b gives the performances of the classifiers trained on the yeast data. For non-inherited relationship, precision increases with the nodes of deeper levels being used, which is the same as observed in human data, but precision is higher, with half of the classifiers

Fig. 3a

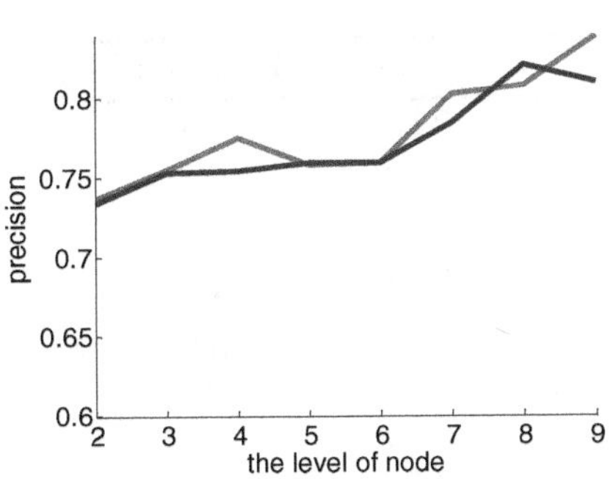

Fig. 3b

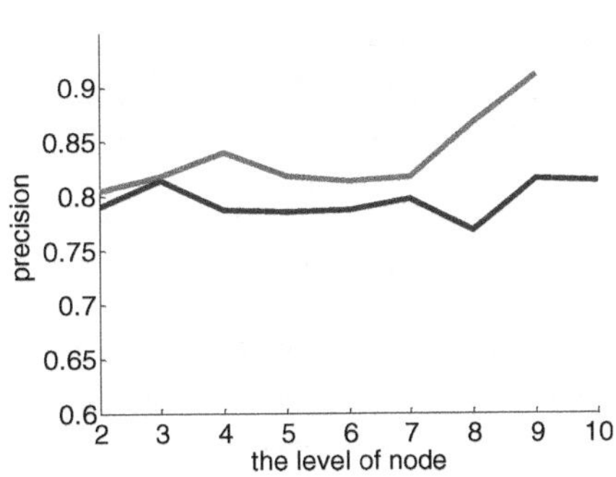

Fig. 3. (a) The relationship between classification performance and the level of nodes for human proteins. (b) The relationship between performance and the level of nodes for yeast proteins. Red – non-inherited relationship; Blue – parent-child relationship.

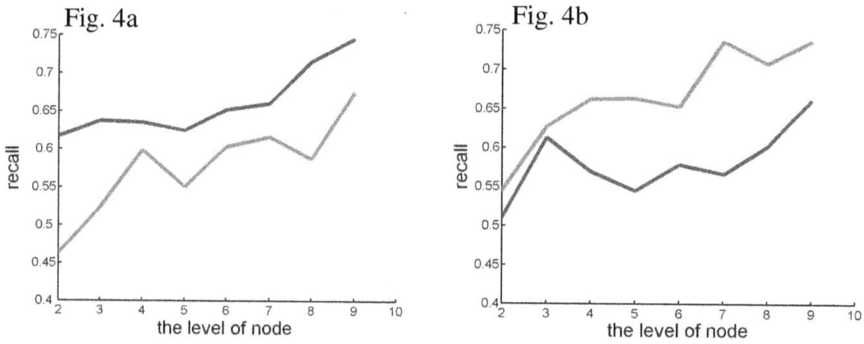

Fig. 4. (a) The relationship between classification sensitivity and the level of nodes for human proteins. (b) The relationship between classification sensitivity and the level of nodes for yeast proteins. Red -- non-inherited relationship; Blue -- parent-child relationship.

having a value > 83%. However, for parent-child relationship, the trend is not so clear, possibly due to the fact that the classifiers might have achieved the nearly optimal performance so that further improvement is not obvious. Another limiting factor for the function classes of deeper levels is smaller sample sizes. Half of the parent-child classifiers have a precision > 76%. The classifiers' sensitivity performance is illustrated in Fig. 4b.

We only find one work related to function expanding [15]. In that work, they used gene expression data to further predict gene function in Gene Ontology and used neural network to train classifiers. They implemented the method on the small serum dataset consisting of 517 differentially expressed genes, and only 14 learnable parent nodes were selected. Compared to their work, we use sequence signature instead of gene expression data, and the samples are 9071 compared to their 517 genes. By this way, more learnable parent nodes (316 parent nodes) are selected. The average precision of14 classifiers is 87.1%, 12 of which are also learnable in our work. Precision of the 13

Table 1. The learnable parent nodes and their performance based on either gene expression (Precision 1) or sequence signatures (Precision 2)

Parent node	Precision 1	Precision 2
GO:0030003, cation homeostasis	0.95	0.88
GO:0000279, M phase	0.96	0.77
GO:0046907, intracellular transport	0.75	0.70
GO:0009653, morphogenesis	0.73	0.51
GO:0042592, homeostasis	0.93	0.90
GO:0009058, biosynthesis	0.73	0.80
GO:0008202, steroid metabolism	0.95	0.80
GO:0006350, transcription	0.99	0.90
GO:0006351, transcription, DNA-dependent	0.85	0.41
GO:0006464, protein modification	0.77	0.78
GO:0007600,sensory perception	0.95	0.79
GO:0009607, response to biotic stimulus	0.90	0.76

classifiers are illustrated in Table 1. Compared to the precision for the three branches: molecular function: 87%; biological process: 72%; cellular component: 77%, their result is little better than ours. We think that the reason that their result is better is the smaller dataset. Since they had to use the differentially expressed genes, their dataset is only about 1/40 of ours in size. Consequently, the classifiers are easier to learn and the result is better. In comparison, we use much more learnable parent nodes. More importantly, our proposed algorithms can also consider non-inherited relationships. Naturally, we can further predict gene function both vertically and horizontally.

4 Discussion

Fragmentary knowledge for a large number of proteins as documented in public databases calls for studies for computationally expanding the current annotations. Though most *ab initio* function prediction approaches can accrue the number of current annotations by predicting the functions for totally unknown proteins, it is even more important to develop an efficient method(s) to expand the existing function annotations because as demonstrated in several studies including the present study, integrating the partially knowledge of a protein into a function prediction model is one of the most economical and promising ways to acquire the most accurate and specific function knowledge for a large number of proteins. This study describes such an efficient computation algorithm based on GO. We first define two types of correlations between function terms in GO hierarchy. Then, we identify the most important protein signatures for any pair of protein that follows either type of correlations in order to build a classifier for predicting a partially known protein. Next, a protein function expansion is realized by the prediction(s) of the classifier for the protein. This study suggests that protein signatures are informative and can be better features than gene transcriptions for function recognition. Protein signature data have advantages over gene expression data because they can be accurately characterized. In addition, use of sequence signatures rather than full-length sequences reduces the false-positive rate induced by multi-domains of proteins. Spearmint system uses 800 Swiss-Prot keywords to predict protein function automatically and achieves very small error rates for predictions. The next effort of Spearmint system is to further compile the annotations in GO, although analysis of more than 20 thousands function classes in GO can be formidable if without an efficient computational algorithm(s). Our results from analysis of protein signatures might provide some insights for large-scale function further annotations.

Acknowledgements

This work was supported in part by the National Natural Science Foundation of China (Grant Nos. 30170515, 30370798, 30571034 and 30570424), the National High Tech Development Project of China, the 863 Program (Grant No. 2003AA2Z2051), the Heilongjiang Province Department of Education Outstanding Overseas Scientist grant (Grant No. 1055HG009), National Science Foundation of Heilongjiang Province (Grant Nos. ZJG0501, GB03C602-4, and F2004-02) and Innovative Fund of Harbin Medical University Graduate Student.

References

1. Ashburner, M., et al.: Gene Ontology: Tool for the Unification of Biology. The Gene Ontology Consortium. Nat Genet. (2000)25-29
2. Attwood, T. K., et al.: PRINTS and Its Automatic Supplement, prePRINTS. Nucleic Acids Res. (2003)400-402
3. Bateman, A., et al.: The Pfam Protein Families Database. Nucleic Acids Res. (1994)138-141
4. Camon, E., et al.: The Gene Ontology Annotation (GOA) Database: sharing knowledge in Uniprot with Gene Ontology. Nucleic Acids Res. (2002)262-266
5. Dwight, S. S., et al.: Saccharomyces Genome Database (SGD) Provides Secondary Gene Annotation using the Gene Ontology (GO). Nucleic Acids Res. (1996) 69-72
6. Enright, A. J., et al.: Protein Interaction Maps for Complete Genomes Based on Gene Fusion Events. Nature. 402 (1998)86-90
7. Falquet, L., et al.: The PROSITE Database, Its Status in 2002. Nucleic Acids Res. 30 (1997) 235-238
8. Kretschmann, E., et al.: Automatic Rule Generation for Protein Annotation with the C4.5 Data Mining Algorithm Applied on SWISS-PROT. Bioinformatics. (1998) 920-926
9. Lagreid, A., et al.: Predicting Gene Ontology Biological Process from Temporal Gene Expression Patterns. Genome Res. (1998) 965-979
10. Letovsky, S., Kasif, S.: Predicting Protein Function from Protein/protein Interaction Data: a Probabilistic Approach. Bioinformatics. 19 Suppl (1998) 197-204
11. Marcotte, E. M., et al.: Detecting Protein Function and Protein-protein Interactions from Genome Sequences. Science. (1998)751-753
12. Mulder, N. J., et al.: InterPro, Progress and Status in 2005. Nucleic Acids Res. (1998) 201-205
13. Pellegrini, M., et al.: Assigning Protein Functions by Comparative Genome Analysis: Protein Phylogenetic Profiles. Proc Natl Acad Sci U S A. (1997)4285-4288
14. Tao, T., et al.: A Study of Statistical Methods for Function Prediction of Protein Motifs. Appl Bioinformatics. (1998) 115-124
15. Tu, K., et al.: Learnability-based Further Prediction of Gene Functions in Gene Ontology. Genomics. (1997) 922-928
16. Yang, S., et al.: Phylogeny Determined by Protein Domain Content. Proc Natl Acad Sci U S A. (1998) 373-378
17. Zhou, X., et al.: Transitive Functional Annotation by Shortest-path Analysis of Gene Expression data. Proc Natl Acad Sci U S A. (1997) 12783-12788

Identifying the Modular Structures in Protein Interaction Networks

Yanen Li, Feng Lu, and Yanhong Zhou

Hubei Bioinformatics and Molecular Imaging Key Laboratory, Huazhong University of Science and Technology,
Wuhan, Hubei 430074, China
yhzhou@hust.edu.cn

Abstract. Identifying the modular structures in Protein Interaction Networks (PINs) is crucial to the understanding of the organization and function of biological systems. We propose a new module definition taking into account the degree of a subgraph instead of a vertice, and design a corresponding agglomerative algorithm, which is implemented into a computational tool called ModuleSpider, to recognize modules within a network. The application of ModuleSpider to the yeast core protein interaction network identifies 97 simple modules which are biologically meaningful according to the GO annotation of proteins in the modules. In addition, the results of comparison analysis demonstrate that ModuleSpider modules show stronger biological relevance. Further, the ModuleSpider is able to construct the interaction web of modules, which provides insights to the high level relationships of different functional modules. ModuleSpider is freely available upon request to the authors.

1 Introduction

Accumulating evidence suggests that biological systems are composed of individual, interacting modules [1,2,3,4]. Recently, high-throughput techniques, such as yeast two-hybrid screens [5] and affinity purification using mass spectrometry [6], accelerated the discovery of the complete Protein Interaction Networks (PINs) in several organisms. The complete dataset of interactions, for the first time, enables us to investigate the uncharted territories of the modular structures in PINs, leading to a deeper understanding of the organization, function and evolution of cellular systems.

Much effort has been put in the identification of biological relevant modules. According to the description of the connectivity of proteins, the available approaches can generally be dichotomized into two categories. One kind of approaches considers the PIN as a weighted graph by quantifying the connection between proteins. For example, Pereira-Leal et al. weighted a protein interaction based on the number of experiments supporting the interaction [7]. Arnau et al. weighted the distance between two proteins by the length of the shortest path between them [8]. However, in the traditional clustering methods to find functional modules within PINs, their strategies usually generate many identical distances and lead to a *tie in proximity* [9] problem. Meanwhile, Asthana et al. treated protein interactions probabilistically to predict

D.-S. Huang, K. Li, and G.W. Irwin (Eds.): ICIC 2006, LNBI 4115, pp. 757 – 767, 2006.

protein complex membership [10], which, unfortunately, ignores the dependence of types of evidence used.

The other kind of broadly adopted approaches employs binary representation of protein interactions without accounting for the quality or quantity of evidence that supports each interaction. Under this assumption, many researches focused on identifying cliques, where every pair of nodes tends to interact with each other, in the network [11,12,13,14,15]. However, the complete subgraph rarely appears in reality; thus the clique-finding methods inevitably neglect the peripheral edges, and the subgraphs which are not fully connected but truly biologically relevant. Another approach to identify modules based on the binary topology is to iteratively divide the network into smaller and smaller subnetworks, and recognize modules during the division. Girvan and Newman [16] proposed the concept of edge betweenness, which is defined as the number of shortest paths between all pairs of vertices that run through the edge. Edges between modules tend to have more shortest-paths running through them than do edges inside modules, and thus have higher betweenness values. The deletion of edges with high betweenness can separate the network, while keeping the module structure intact. Based on this principle, Girvan and Newman proposed a divisive algorithm (G-N algorithm) to construct a tree to find module structures in an unweighted network. As the original G-N algorithm does not include a clear definition of module, it cannot formally tell which parts of the tree are modules. Radicchi et al. [17] combined the G-N division process with two new module definitions and proposed a new self-contained algorithm to identify modules from a network. However, the module definitions they proposed may not be suitable for finding modules in biological networks because of several limitations within them.

In this paper, we extend the concept of degree from vertices to subgraphs and propose a new quantitative definition of a module in a network. By combining this new module definition with the edges generated by the G-N algorithm, we propose a new agglomerative algorithm to identify modules in a network. A computational tool, called ModuleSpider, is developed based on this approach. ModuleSpider was applied to the yeast core protein interaction network from the Database of Interacting Proteins (DIP) [18]. Most modules identified are rich in significant co-occurrence of proteins of related biological functions, and exhibit lower P-values than the modules defined by Radicchi et al. We also compare the modules obtained by ModuleSpider with the quasi-cliques identified by the approach proposed by Bu et al. Further, ModuleSpider maintains the relationships between adjacent modules resulting in a network of modules. This higher-level interaction network between modules allows a system-level understanding of the relationship between modules representing different biological processes and functions.

2 Method

2.1 Definition of the Module

Many possible module definitions have been proposed [17,19,20]. Generally, a module in a network is a subnetwork that has more internal edges than external edges. A module definition must be able to capture the essence of this intuition. Radicchi et al.

extended the degree definitions to the vertices in an undirected graph and proposed two module definitions: strong modules and weak modules. Given an undirected graph, they defined the indegree of a vertex in a module as the number of edges connecting it to other vertices in the same module and the outdegree of a vertex as the number of edges connecting it to the vertices outside the module. For a strong module, each vertex has higher indegree than outdegree; and for a weak module, the sum of indegree is greater than the sum of outdegree of all vertices in the module. However, these definitions do not appear to follow the general understanding of a module precisely. Firstly, the determination of a strong module can be strongly influenced by the degree of a single vertex, which would neglect highly-connected peripheral nodes with high possibility of being members of the module. Secondly, in the weak module definition, the edges inside a subgraph have been counted twice, which will result in considering a subgraph, in which nodes are loosely grouping together, as a weak module (see Fig. 1). Thirdly, Radicchi et al. only test their module definitions on social networks, which are different from the biological networks. For example, the biological networks show strong disassortativity while most social networks do not [21]. Thus, the definitions of strong and weak modules may not be suitable for identifying biologically meaningful modules in biological networks.

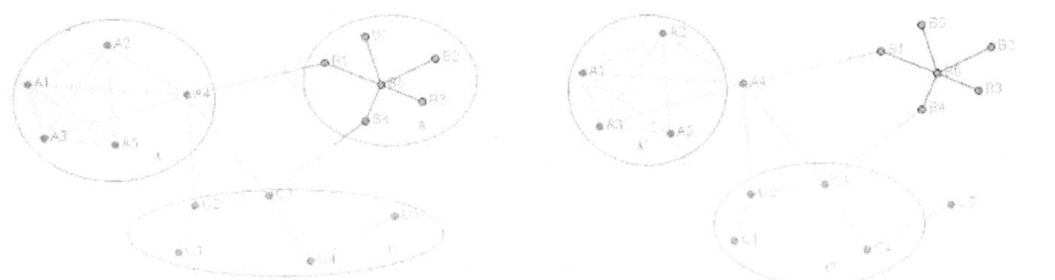

Fig. 1. Examples of the limitations of strong and weak module definitions. In the left panel, all three subgraphs are not strong modules because each subgraph has at least one vertex (A4, B4, C3) that does not have more indegree than outdegree. Only one strong module, Module A' at the right panel can be identified by removing the highly-connected node from the Module A. In the weak module definition, even if a subgraph (C' at the right panel) has more external edges than internal edges (4 versus 3), it may still be considered to be a module.

To overcome these limitations, we extend the concept of degree from the individual vertex to the subgraph in order to characterize the connectivity of a subgraph in a graph (a graph is identical to a network in this paper):

Definition of degrees of a subgraph. Given a graph G, let E denote the edge set and E_{ij} denote the edge connecting node i and node j (E_{ij} and E_{ji} are identical in an undirected graph and only counted once in the formulas below), then, for a subgraph $V \subset G$, the indegree of V is defined as:

$$K^{in}(V) = \sum E_{ij} \text{ ,.} \tag{1}$$

where both node i and node $j \in V$ ($i \neq j$). The outdegree of V is defined as:

$$K^{out}(V) = \sum E_{pq} \, . \tag{2}$$

where node $p \in V$ and node $q \notin V$, or node $q \in V$ and node $p \notin V$.

Definition of a module. Given a graph G, a subgraph $V \subset G$ is a module if

$$K^{in}(V) > \alpha K^{out}(V) \, . \tag{3}$$

where α is the module strictness tuning coefficient. To further distinguish different levels of modules in the network, we define the complex module and the simple module as follows.

Definition of complex and simple modules. A module is a complex module if it can be separated into at least two modules by removing edges inside it using the G-N algorithm. Otherwise, it is a simple module.

This module definition directly captures the general understanding of the module concept and it is very flexible. Similar to the approach of Radicchi et al, this definition is based on the relationship between internal and external links, rather than relying on internal links only [11]. Also, this module definition can be tuned to a stricter one by increasing the coefficient α, otherwise to a looser one by decreasing α. Interestingly, when $\alpha = 0.5$, definition (3) is equal to the weak module definition proposed by Radicchi et al. Furthermore, this definition, based on the connectivity of subgraphs, makes it easy to define the adjacent relationship between modules.

Definition of adjacent modules. Given two modules $U, V \subset G$, U and V are adjacent if $U \cap V = \varnothing$ and there are edges directly connecting vertices in U and V .

According to the adjacent module definition above, we can easily deduce that: Given two modules $U, V \subset G$, If U and V are adjacent, the subgraph $W = U \cup V$ is also a module. This deduction suggests that merging two adjacent modules will generate a complex module. Also, the definition of the complex module and the simple module implies that the separation of a simple module into two subgraphs using the G-N algorithm can at most generate one module.

2.2 Algorithm

We propose a new agglomerative algorithm to identify the simple modules of smallest size within a PIN, and implemented it into a computational tool, ModuleSpider.

Similar to conventional agglomerative algorithms, our agglomerative algorithm initially puts each vertex into a singleton subgraph. Then, the subgraphs are gradually merged to find the simple modules in the network. There are two important properties of our algorithm - the occurrence of merging and the order of merging.

As mentioned above, the order of deletion based on the betweenness in the G-N algorithm reflects the relative relationship between edges inside modules and edges between modules in the network. The later the edge is deleted, the more likely it is an edge inside a module. In ModuleSpider, the edge deletion order generated by the G-N algorithm is reversed and used as the merging order in the agglomerative algorithm. By gradually adding edges to the subgraphs in the reverse order deleted edges, ModuleSpider assembles the singleton subgraphs into simple modules. As a key goal of this approach is to identify simple modules of smallest size, only the mergence between two non-modules is allowed. And, once a simple module is identified, it will not be merged with other subgraphs. This merging scheme generates compact simple modules and prevents merging with loosely connected subgraphs. The agglomerative algorithm is summarized as follows:

1. The G-N algorithm [16] is run on the network until no edges remain, and the order of edge deletion is obtained.
2. An edge list is created in the reverse order of edge deletion in Step (1).
3. Initialize each vertex as a singleton subgraph with no edges.
4. Pop-up an edge from the top of the edge list.
5. If the edge connects vertices in the same subgraph, add it to the subgraph.
6. If the edge connects vertices in two different subgraphs:
   1) If the two subgraphs are both non-modules, Merge them, retain the edge, and:
      a. If the merged subgraph is a ModuleSpider simple module, set it as a simple module.
      b. Otherwise, go to Step (7).
   2) Otherwise, discard the edge and go to Step (7).
7. Repeat steps (4) – (7) until no edges are left in the edge list.

The computational complexity of the ModuleSpider algorithm is $M^2N + M \approx O(M^2N)$, where M is the number of edges and N is the number of vertices.

3 Results

3.1 Dataset

The dataset used in this study is the yeast core protein interaction network downloaded from the DIP database [18] (version ScereCR20060116). This interaction dataset is generated by filtering the large high-throughput protein interaction data using two different computational methods, the Expression Profile Reliability Index and the Paralogous Verification Method, to improve reliability of the interaction data [22]. After removing all self-connecting links, the final core protein interaction network includes 2554 yeast proteins and 5728 interactions.

3.2 Simple Modules Identified in the Yeast Core Protein Interaction Network

Using the strictness tuning coefficient $\alpha = 1.0$, ModuleSpider has found 192 compact simple modules, whose sizes range from 2 to 467 proteins, in the yeast core protein interaction network. All 192 simple modules together include 2158 of the 2554 proteins in the network. 97 modules (totally 1933 proteins) whose sizes are larger than 3 proteins are identified.

Table 1. Top 15 ModuleSpider modules with lowest P-value GO terms

Module No.	GO term	Cluster freq. [a]	Genome freq. [b]	P-value [c]
294	Nucleocytoplasmic transport	30/43	112/7291	1.17E-44
318	Golgi vesicle transport	28/32	144/7291	6.28E-44
339	Actin cytoskeleton organization and biogenesis	31/53	105/7291	2.76E-43
340	Some biogenesis and assembly	37/52	249/7291	1.47E-42
281	Proteolysis	25/28	164 /7291	1.94E-38
337	rRNA processing	28/38	170/7291	7.41E-38
342	Cellular morphogenesis	40/117	160/7291	2.85E-36
236	Hydrogen ion homeostasis	12/12	23/7291	9.71E-31
243	mRNA metabolism	17/18	188/7291	1.72E-26
122	mRNA 3'-end processing	10/10	21/7291	3.92E-26
244	Regulation of cell growth	9/11	11/7291	2.22E-24
110	Cyclin catabolism	8/8	12/7291	5.38E-23
304	Meiosis I	11/13	58/7291	6.20E-22
123	ATP biosynthesis	8/8	20/7291	3.20E-21
269	Peroxisome organization and biogenesis	9/11	38/7291	1.54E-19

[a] The number of proteins in the module annotated with the GO term out of the module size. [b] The number of proteins in the module annotated with the GO term out of the total number of proteins in the yeast genome. [c] The probability that the co-occurrence of proteins with this GO term had occurred in a cluster of this size by chance.

To substantiate whether the modules found by ModuleSpider are biologically meaningful, we first obtain the GO annotation for each protein in the network downloaded from the *Saccharomyces Genome Database* (SGD) [23]. Manual inspection of the annotations shows that most modules are rich in proteins of similar functions. For example, all 12 proteins of module 236 belong to the hydrogen ion homeostasis; all 10 proteins in module 122 are related to mRNA 3'-end processing; all 8 proteins of module 123 (ATP1, ATP2, ATP4, ATP5, ATP6, ATP7, ATP17 and ATP18) are subunits of the mitochondrial ATP synthase complex, which is crucial for ATP synthesis; 9 out of 11 proteins in module 244 are involved in the regulation of cell growth, and when using the SGD GO Term Finder to designate these proteins to a common biological process (BP) category, it is indicated that only 11 out of 7291 proteins in the whole SGD repository belong to this category. These findings suggest that proteins identified within a ModuleSpider module are closely related in similar biological pathways and functions.

To further evaluate the biological relevance of ModuleSpider modules, we estimate the statistical significance of GO BP term co-occurrence using the SGD GO Term Finder. The results show that most modules (95%) demonstrate significant GO terms enrichment beyond what would be expected by chance. If the P-value cutoff 1.82E-5 is used based on the Bonferroni correction, and if the modules with too large size (> 300 proteins) are removed, there are 79 ModuleSpider modules left. The average lowest P-value of GO terms for all 79 modules is 3.18E-7, ranging from 1.01E-5 to 5.93E-44. We list a part of compact simple modules with lowest P-value GO terms in Table 1.

3.3 Comparison with Other Module Definitions

To test the effectiveness of our module definition, we apply the same agglomerative algorithm, from the same dataset as ModuleSpider to detect strong and weak modules defined by Radicchi et al. There are 41 strong modules obtained from the yeast core protein interaction network with their sizes ranging from 3 to 300 proteins. All 41 strong modules include only 236 of the 2554 core proteins. At the same time, 135 weak modules are identified with their sizes ranging from 3 to 300 proteins. There are total 1597 proteins in these weak modules. Similarly, we use the SGD GO Term Finder to quantify the statistical significance of the GO BP term co-occurrence in the strong and weak modules obtained. There are 35 strong modules that show significant GO terms co-occurrence (P-value < 1.82E-5). However, the average lowest P-value of GO terms of the 35 strong modules is 6.65E-7, which is higher than that of the 79 simple modules identified by ModuleSpider (3.18E-7). For weak modules, 106 out of 135 weak modules show significant GO terms co-occurrence. The average lowest P-value of GO terms for these weak modules is 5.29E-7, still less significant than that of the 79 ModuleSpider simple modules. Fig. 2 shows the distribution of lowest P-value GO terms for modules obtained by different module definitions. There are fewer high P-value modules and more low P-value modules in simple module groups obtained by ModuleSpider than in other two module groups. For example, there are 7 (9.0%) ModuleSpider simple modules with P-value lower than 1.00E-35, but there is only 1 (0.9%) weak module and no strong module with such low P-values.

We also make a comparison between the modules obtained by ModuleSpider and the quasi-cliques obtained by the spectral analysis approach proposed by Bu et al [15]. We still use the GO Term Finder to quantify the statistical significance of the GO BP term co-occurrence of two kinds of modules, respectively. 44 of the 48 quasi-cliques show significant lowest GO BP term P-values, and the average P-value is 4.19E-8, which is lower than that of the 79 ModoleSpider modules. However, the major limitation of Bu's approach is that the requirement of the quasi-clique is too strict; such stringency makes it hard to detect subgraphs whose nodes are not densely connected with each other but are truly biologically meaningful, such as the Arp2/3 complex in yeast [24]. A careful observation of the 48 quasi-cliques also validates the conclusion drawn above. In these 48 quasi-cliques, there is no quasi-clique consisting of less than 10 proteins, which, obviously contradicts our awareness of the formations of functional complexes in reality. Further, if we only consider the ModuleSpider modules whose sizes are no less than 10 proteins, the average P-value of the 25 ModoleSpider modules obtained is 1.64E-8, which is more significant than that of the corresponding 48 quasi-cliques (4.19E-8). Finally, to find all modules in a protein interaction network, the ModuleSpider algorithm is more efficient than Bu's spectral analysis approach, which has been proved to be NP-Complete.

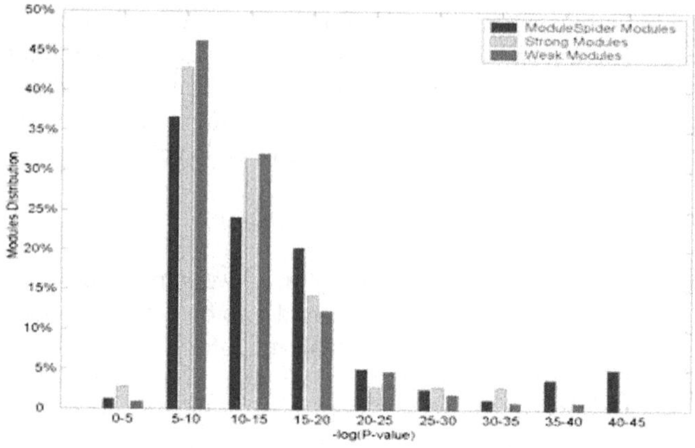

Fig. 2. Distribution of different types of modules according to their lowest P-value of designated GO terms. Modules whose sizes are less than 3 or larger than 300 are not counted.

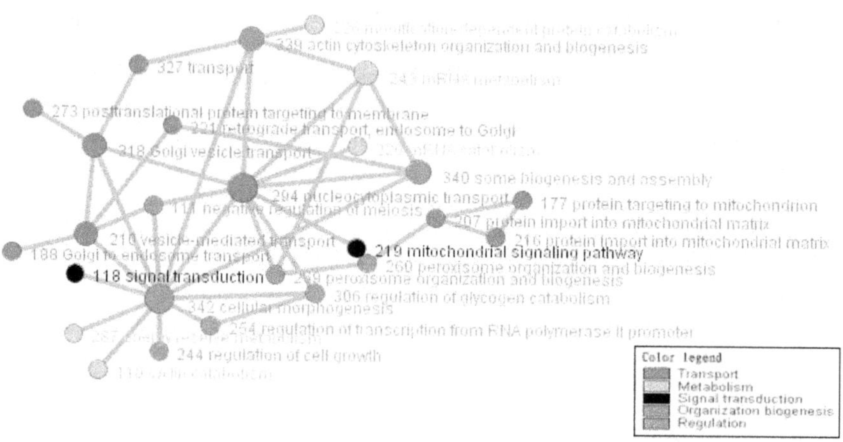

Fig. 3. A subsection of the interconnected module network. Each node denotes a module and each edge denotes a connection between two modules. Modules representing similar biological processes are shown by the same color.

3.4 Interaction Between Modules

In order to gain insights into how the ModuleSpider modules relate to each other within the cellular system, we assemble an interconnection network of the 97 simple modules identified by ModuleSpider from the yeast core protein interaction network. The network of modules is constructed as follows: for each adjacent module pair, the edge that is deleted last by the G-N algorithm is selected from all the edges that connect two modules to represent the link between them. The network of modules obtained is highly connected, which suggests that the cell is a complex web of highly interconnected functional modules. Fig. 3 is a subsection of this module network. In Fig. 3, the edges reflect the relationships between modules; the size of the nodes reflects the hub status of modules - the bigger the node is, the more edges connecting the module to other modules. Although these relative relationships are based on the structural information in the network, many of them are biologically meaningful. In Fig. 3, highly connected modules are usually connecting to closely related modules and to other highly connected modules (hubs). For instance, Module 294 (nucleocytoplasmic transport) serves as a hub for related modules including Module 243 (mRNA metabolism), Module 219 (mitochondrial signaling pathway) and Module 269 (peroxisome organization and biogenesis). At the same time, this module also connects with other hubs such as Module 342 (cellular morphogenesis), Module 340 (some biogenesis and assembly) and Module 318 (Golgi vesicle transport). The network of modules provides a panoramic viewpoint of the relationship between different biological processes and functions represented in each module.

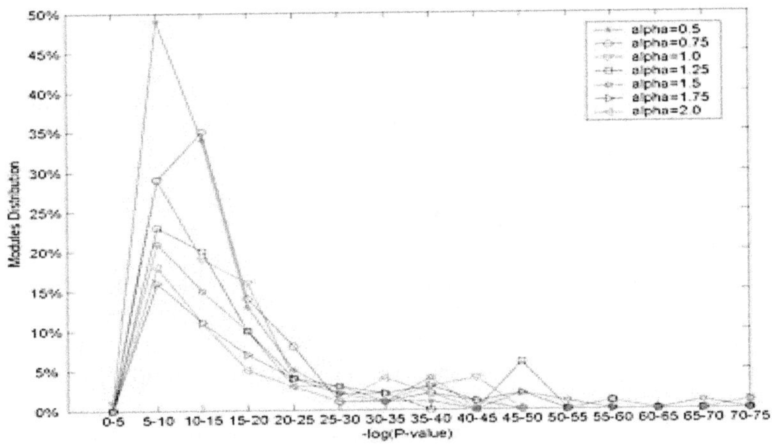

Fig. 4. Distribution of ModuleSpider modules according to the lowest P-value of designated GO terms, using 7 α values. Modules whose sizes are less than 3 or larger than 300 are not counted.

4 Discussion

The results of the ModuleSpider modules are obtained when the coefficient $\alpha = 1.0$. In order to analyze how different α values affect the identification results, we test the GO BP term co-occurrence of the ModuleSpider modules using the same agglomerative algorithm with different α values (from 0.5 to 2.0, with a step of 0.25). Fig. 4 shows the Distribution of ModuleSpider modules using 7 α values; Table 2 presents some statistics of the modules identified under different α values. Statistics in Table 2 indicates that when $\alpha = 1.25$, modules identified by ModuleSpider exhibit lowest average P-vaule of GO term co-occurrence; and Fig. 4 also shows a peak of module accumulation within the [1.0E-45, 1.0E-50] bin compared with that of other 6 α values. Thus, it is suggested that more simple modules with high biological significance can be recognized by ModuleSpider when $\alpha = 1.25$.

Table 2. Statistics of ModuleSpider modules identified using 7 different α values

α	Significant Modules	Ave P-value	P-value <1.0E-20
0.5	106	5.29E-07	8.5% (9)
0.75	92	3.89E-07	15.2% (14)
1.0	79	3.18E-07	17.7% (14)
1.25	69	2.54E-07	23.2% (16)
1.5	61	4.18E-07	24.6% (15)
1.75	49	2.58E-07	30.6 (15)
2.0	46	2.77E-07	26.1% (12)

5 Conclusion

In this paper, we extend the degree concept from the vertex to the subgraph and propose a new definition of module within a network based on the degree definition of the subgraph. A new agglomerative algorithm is designed to assemble simple modules from protein interaction networks, using the relative order of edges generated by the G-N algorithm as the merging order. The new approach has been implemented in the ModuleSpider program.

Application of ModuleSpider to the DIP yeast core protein interaction network identifies modules significantly rich in proteins of similar functions. These results imply that modules identified based on network structural information are biologically meaningful. Comparison with the strong and weak module definitions of Radicchi et al. shows that the ModuleSpider modules obtained are more biologically significant. ModuleSpider also casts superiorities upon the approach proposed by Bu et al.

Furthermore, the ModuleSpider can construct the network of modules conveniently. Manual evaluation of the module connections suggests that the interactions between modules are of biological significance.

In conclusion, the ModuleSpider algorithm and computational tool provides an excellent approach to decipher the modular structures of biological networks, which will deepen our understanding of the organization, function, and evolution of cellular systems.

Acknowledgement

This work was supported by the National Natural Science Foundation of China (Grants No. 30370354 and 90203011) and the Ministry of Education of China (Grants No. 20050487037 and 505010).

References

1. Barabasi, A. L., Oltvai, Z.N.: Network Biology: Understanding The Cell's Functional Organization. Nat.Rev.Genet., 5 (2) (2004) 101-113
2. Hartwell, L. H., Hopfield, J. J., Leibler, S.: From Molecular to Modular Cell Biology. Nature, 402 (6761) (1999) C47-C52

3. Ravasz, E., Somera, A. L., Mongru, D. A.: Hierarchical Organization of Modularity in Metabolic Networks. Science, 297 (5586) (2002) 1551-1555
4. Rives, A.W., Galitski, T.: Modular Organization of Cellular Networks. Proc. Natl. Acad. Sci. U. S. A., 100 (3) (2003)1128-1133
5. Uetz, P., Giot, L., Cagney, G.: A Comprehensive Analysis of Protein-protein Interactions in Saccharomyces Cerevisiae. Nature, 403(6770) (2000) 623-627
6. Gavin, A. C., Bosche, M., Krause, R.: Functional Organization of The Yeast Proteome by Xystematic Analysis of Protein Complexes. Nature, 415 (6868) (2002) 141-147
7. Pereira-Leal, J. B., Enright, A. J., and Ouzounis, C. A.: Detection of Functional Modules from Protein Interaction Networks. Proteins, 54 (1) (2004) 49-57
8. Arnau, V., Mars, S., Marin, I.: Iterative Cluster Analysis of Protein Interaction Cata. Bioinformatics, 21(3) (2005) 364-378
9. MacCuish, J., Nicolaou, C., MacCuish, N. E.: Ties in Proximity and Clustering Compounds. J. Chem.I nf. Comput. Sci., 41 (1) (2001) 134-146
10. Asthana, S., King, O. D., Gibbons, F. D.: Predicting Protein Complex Membership Using Probabilistic Network Reliability. Genome Res., 14 (6) (2004) 1170-1175
11. Spirin, V., Mirny, L.A.: Protein Complexes and Functional Modules in Molecular Networks. Proc. Natl. Acad. Sci. U. S. A., 100 (21) (2003) 12123-12128
12. Bader, G. D., Hogue, C. W.: An Automated Method for Finding Molecular Complexes in Large Protein Interaction Networks. BMC.Bioinformatics, (2003) 42
13. Gagneur, J., Krause, R., Bouwmeester, T. et al.: Modular Decomposition of Protein-protein Interaction Networks. Genome Biol. 5(8) (2004) R57
14. Zhang, C., Liu, S., Zhou, Y.: Fast and Accurate Method for Identifying High-quality Protein-interaction Modules by Clique Merging and Its Application to Yeast. J. Proteome. Res. 5(4) (2006) 801-807
15. Bu, D., Zhao, Y., Cai, L.: Topological Structure Analysis of The Protein-protein Interaction Network in Budding Yeast. Nucleic Acids Res. 31 (9) (2003) 2443-2450
16. Girvan, M., Newman, M. E.: Community Structure in Social and Biological Networks. Proc. Natl. Acad. Sci. U. S. A., 99 (12) (2002) 7821-7826
17. Radicchi, F., Castellano, C., Cecconi, F.: Defining and Identifying Communities in Networks. Proc.Natl.Acad.Sci.U.S.A., 101 (9) (2004) 2658-2663
18. Xenarios, I., Salwinski, L., Duan, X. J.: DIP, The Database of Interacting Proteins: A Research Tool for Studying Cellular Networks of Protein Interactions. Nucleic Acids Res. 30 (1) (2002) 303-305
19. Newman, M. E.: Fast Algorithm for Detecting Community Structure in Networks. Phys. Rev. E. Stat. Nonlin. Soft. Matter Phys., 69 (6 Pt 2) (2004) 066133
20. Wasserman, S., Faust, K.: Social Network Analysis. Cambridge Univ. Press, Cambridge, UK (1994)
21. Newman, M.E., Park, J.: Why social networks are different from other types of networks. Phys. Rev. E. Stat. Nonlin. Soft. Matter Phys. 68 (3Pt 2) (2003) 036122
22. Deane, C. M., Salwinski, L., Xenarios, I. et.al.: Protein Interactions: Two Methods for assessment of The Reliability of High Throughput Observations. Mol.Cell Proteomics, 1 (5) (2002) 349-356
23. Cherry, J. M., Adler, C., Ball, C.: SGD: Saccharomyces Genome Database. Nucleic Acids Res., 26 (1) (1998) 73-79
24. Robinson, R.C., Turbedsky, K., Kaiser, D.A.: Crystal Structure of Arp2/3 complex. Science, 294 (5547) (2001) 1679-1684

An Analysis of Gene Expression Relationships Between Periodically Expressed Genes in the Hela Cells

Yun Xiao[1], Xia Li[1,2,*], Shaoqi Rao[1,3,*], Juan Wang[1], Yun Zhang[1], and Lei Du[1]

[1] Department of Bioinformatics, Harbin Medical University, Harbin 150086, P.R. China
wintarcxy@yahoo.com.cn, Lixia6@yahoo.com, raos@ccf.org, wangjuanmary@yahoo.com.cn, chaosapollo@163.com, dulei@hrbmu.edu.cn

[2] Biomedical Engineering Institute, Capital University of Medical Sciences, Beijing 100054, P.R. China

[3] Departments of Cardiovascular Medicine and Molecular Cardiology, The Cleveland Clinic Foundation, 9500 Euclid Avenue, Cleveland, Ohio 44195, USA

Abstract. The availability of increasingly accumulated time-series expression profiles has provided the data structures to study the temporal relationship between genes. The aims of the study were to explore the modes of relationship between the genes periodically expressed in human HeLa cell cycle, and to reversely engineer the dynamic networks of gene transcription. We also studied the phase-specific properties of the genetic relationships by decomposing the whole network into the sub-networks according to the cell cycling phases. The results demonstrated that the gene-gene relationships within a same phase or between different phases followed different modes.

1 Introduction

A cell cycle is often composed of four stages: G1, S, G2 and M phase. In normal tissues, cell division in vivo is precisely regulated by a cascade of events that determine when a cell is going to divide. Therefore, a proper regulation of the cell division cycling is essential for the growth and development of all organisms. Understanding these underlying regulation mechanisms is of great importance for studying many disease mechanisms where abnormal cellular behaviors are involved, most notably some cancers. There are many known and yet unknown molecular determinants that control the cell cycle or lead to the transition of normal cells into cancerous cells when mutated. These interferences can occur in any of the cell cycle phases: G1, G2, S and M or boundaries. It is one of the focuses in current cell biology and human genetics to understand the dynamic behaviors of these molecular interplays that regulate the cell cycle in both normal and cancer cells.

DNA microarray technology [1] has enabled researchers to simultaneously monitor the expression levels of thousands of genes. In the experiments of time-series gene

* Corresponding authors.

D.-S. Huang, K. Li, and G.W. Irwin (Eds.): ICIC 2006, LNBI 4115, pp. 768 – 773, 2006.

expression, a comprehensive transcriptional profile of a gene can be obtained to capture its temporal trends and characteristics, by measuring its activities at different time points in a cell cycle. Up to now, the genome-wide transcriptional program during the cell cycle has been investigated in a wide range of the organisms. Some previous studies (e.g. [2]) focused on elucidating the gene relationships based on global correlation pattern over the whole time-course, and identifying the clusters of genes whose expression levels simultaneously rise and/or fall. Obviously, the global clustering algorithm is unable to distinguish the detailed dynamic facets of a gene relationship in the cell cycling. Therefore, in this study, a local clustering algorithm [3] is proposed to characterize four modes of temporal relationships: correlated, time-shifted, inverted and inverted time-shifted. We applied this method to analyze a HeLa cell cycle dataset, and revealed the relationships between the periodically expressed genes in the HeLa cells.

2 Materials and Methods

2.1 Gene-Expression Datasets

In this study, we used the expression dataset of HeLa cells [4] which were synchronized by three different methods, collected from five independent microarray experiments. The 1134 clones (representing 874 genes) that showed the most significant periodic variations during the cell cycle became our studied targets. Here, we only analyzed two datasets which were synchronized by double thymidine block methods (Thy2 and Thy 3).

2.2 Data Preprocessing

The expression data were normalized using a "Z score" method [5], so that for each gene the average expression ratio was 0 and the standard deviation was 1.

2.3 Local Clustering

The normalized expression level at time point i (or j, i, j= 1, 2, …, n) for gene x (or y) was denoted as x_i (or y_j). A "score matrix" was used for measuring the similarities between the expressions for genes x and y, and was defined as:

$$M(x_i, y_j) = x_i y_j \,. \tag{1}$$

Two summary matrices $E_{i,j}$ and $D_{i,j}$ were computed, with the initial conditions: $E_{0,j} = 0$, $E_{i,0} = 0$, $D_{0,j} = 0$, and $D_{i,0} = 0$.

$$E_{i,j} = \max(E_{i-1,j-1} + M_{i,j}, 0) \,. \tag{2}$$

$$D_{i,j} = \max(D_{i-1,j-1} - M_{i,j}, 0). \tag{3}$$

An overall maximal *Score* value was found by comparing the two maxima for matrices $E_{i,j}$ and $D_{i,j}$, representing the match *Score* for the two genes. If the maximum is off-diagonal in its corresponding matrix, the two genes have a time-shifted relationship. A maximal value from matrix $D_{i,j}$ indicates the two genes have an inverted relationship, and otherwise paralleled (i.e. the maximal *Score* is from $E_{i,j}$).

2.4 Statistical Significance Determined by Permutation Tests

A permutation statistic was used to determine the significance of an observed relation. One hundred sets of random expression profiles were generated by shuffling the normalized expression levels at different time points (e.g. interchanging their column (time point) locations). The P-value is the probability that the observed match *Score* is larger than any *Score* derived from the random profiles. The smaller P-value relates to the more significant *Score*. We set the critical value that corresponds to the 0.1% level of significance (P-value = 0.001). A statistically significant relationship was claimed if the match *Score* was greater than the critical value.

3 Results

3.1 Four Modes of Relationships

A total of 740 significant relationships were identified. Figure 1 shows the basic temporal characteristics of four relative modes (from analysis of the Thy3 dataset).

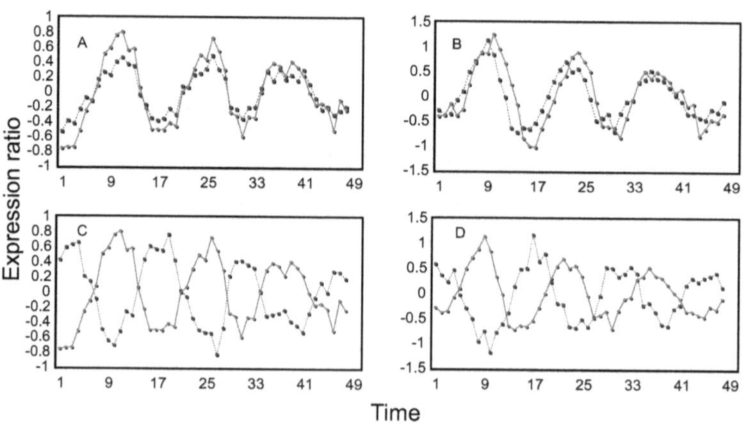

Fig. 1. Four expression relationships were found by a local clustering algorithm. (a) Simultaneous (correlated) relationship. (b) Time-shifted relationship. (c) Inverted relationship. (d) Inverted time-shifted relationship.

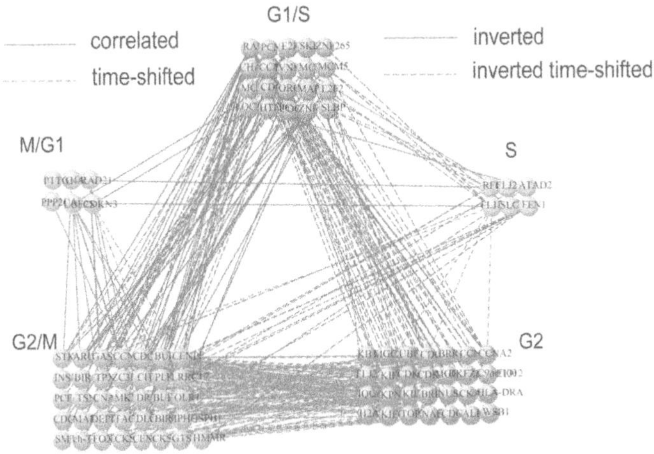

Fig. 2. Between-phase network. All expression relationships between different phases are shown. Each point represents a gene. Color is used to indicate the phase of the gene. Red solid and red dotted links indicate correlated and time-shifted relationships, respectively. Blue solid and blue dotted links indicate inverted and inverted time-shifted relationships, respectively.

3.2 Construction of the Relation Networks

To provide a picture view of the expression relationships, we built a network by globally clustering the HeLa cell cycle datasets. Then, we decomposed the whole network into two kinds of subnetworks: within-phase networks and between-phase networks.

Between-phase Networks
Figure 2 shows the between-phase network. Obviously, there is a triangle in this network, suggesting that the between-phase networking often occurred between the genes in the phase pairs of G1/S-G2, G2-G2/M and G2/M-M/G1. Figure 3A shows the more detailed between-phase sub-networks derived by direct peeling off the whole network. In G1/S-G2 sub-network most relationships are inverted time-shifted and in G1/S-G2/M most relationships are also inverted time-shifted. However, in G2-G2/M most relationships are time-shifted.

Within-phase Networks
We further constructed the within-phase networks (Fig 3B). To our surprise, we did not find any blue edge (i.e. inverted or inverted time shifted modes) in these sub-networks. Only positive relationships (i.e. correlated or time shifted modes) were observed in these within-phase networks.

4 Discussion

In each between-phase sub-network, some kind of relationships was predominant. For example, the majority relationships in G1/S-G2 and G1/S-G2/M sub-networks were inverted time-shifted. In G2-G2/M sub-network, the time-shifted relationship was the

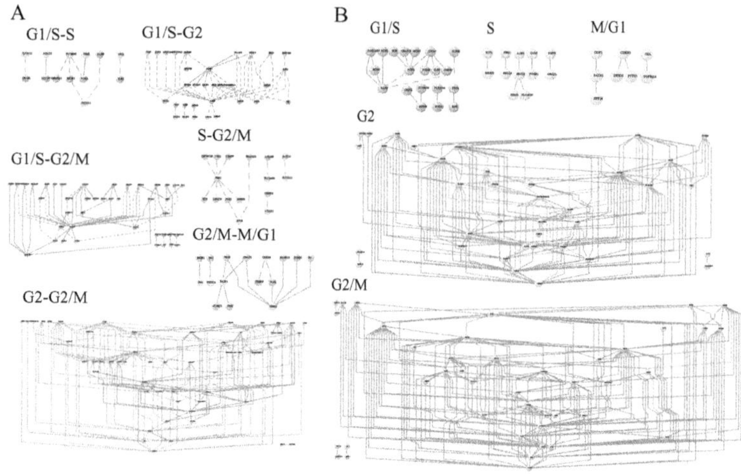

Fig. 3. Between-phase sub-networks and within-phase networks. A) Between-phase sub-networks constructed by peeling off the whole network shown in Figure 2. B) Within-phase sub-networks per phase.

leading mode. In the within-phase networks, the majority were of the correlated relationships. Although the reasons for these patterns are largely unknown, they may reflect the inherent characteristics of the gene-gene interplays for the cell cycling. Integrating the functional knowledge of the genes (e.g. using Gene Ontology) may help elucidate the detailed molecular mechanisms for the cell cycling.

The further studies might be to probe deeply the unknown space of the built networks, for example, to find the hub genes. These important genes should play an important role(s) for phase transitions. We found that gene RAMP played a hub role in G1/S, G1/S-G2 and G1/S-G2/M sub-networks. RAMP is the gene that encodes RA-regulated nuclear matrix-associated protein. The protein, a serine/threonine-rich protein, is associated with the nuclear matrix protein, and is translocated from the nucleus to the cytoplasm during mitosis and cytokinesis. It has a significant role in the proliferation of the human embryonic carcinoma cells[6]. The transcriptional activity of gene RAMP is highly correlated with the cell proliferation rate of NT2 (human embryonic carcinoma cell line) cells. It has been demonstrated that overexpression of RAMP induces a transient increase in the proliferation rate of NT2 cells. The present study also indicates that RAMP may be an important hub in the gene networks for the HeLa cell cycling, which deserves a further verification with wet-lab experiments.

5 Conclusion

In summary, we found four modes of gene-gene relationships among periodically expressed genes that regulate the Hela cell cycling. We constructed different kinds of gene networks to capture the dynamic and temporal properties of genetic regulations. The results demonstrated that the gene-gene relationships within a same phase or between different phases were inclined to follow different modes.

Acknowledgements

This work was supported in part by the National Natural Science Foundation of China (Grant Nos. 30170515, 30370798, 30571034 and 30570424), the Heilongjiang Province Department of Education Outstanding Overseas Scientist grant (Grant No. 1055HG009), and National Science Foundation of Heilongjiang Province (Grant Nos. ZJG0501, GB03C602-4, and F2004-02)

References

1. Duggan, D.J., Bittner, M., Chen, Y., Meltzer, P. , Trent, J.M.:Expression Profiling using cDNA Microarrays. Nat Genet, Vol. 21. (1999) 10-14
2. Kim, S., Dougherty, E.R., Bittner, M.L., Chen, Y., Sivakumar, K., Meltzer, P. , Trent, J.M.: General Nonlinear Framework for the Analysis of Gene Interaction via Multivariate Expression Arrays. J Biomed Opt, Vol. 5. (2000) 411-424
3. Qian, J., Dolled-Filhart, M., Lin, J., Yu, H. , Gerstein, M.:Beyond Synexpression Relationships: Local Clustering of Time-shifted and Inverted Gene Expression Profiles Identifies new, Biologically Relevant Interactions. J Mol Biol, Vol. 314. (2001) 1053-1066
4. Whitfield, M.L., Sherlock, G., Saldanha, A.J., Murray, J.I., Ball, C.A., Alexander, K.E., Matese, J.C., Perou, C.M., Hurt, M.M., Brown, P.O. , Botstein, D.:Identification of Genes Periodically Expressed in the Human Cell Cycle and Their Expression in Tumors. Mol Biol Cell, Vol. 13. (2002) 1977-2000
5. Cheadle, C., Vawter, M.P., Freed, W.J. , Becker, K.G.:Analysis of Microarray Data using Z score Transformation. J Mol Diagn, Vol. 5. (2003) 73-81
6. Cheung, W.M., Chu, A.H., Chu, P.W. , Ip, N.Y.:Cloning and Expression of a Novel Nuclear Matrix-associated Protein that is Regulated During the Retinoic Acid-induced Neuronal Differentiation. J Biol Chem, Vol. 276. (2001) 17083-17091

Analysis of Sib-Pair IBD Profiles Using Ensemble Decision Tree Approach: Application to Alcoholism

Yang Jiang[1], Qingpu Zhang[2], Xia Li[1,2,*], Lei Du[1], Wei Jiang[1], Ruijie Zhang[1], Jing Li[3], and Shaoqi Rao[1,*]

[1] Department of Bioinformatics, Harbin Medical University, Harbin, 150086, China
jybackup@163.com, zzqp2002@126.com, lixia6@yahoo.com, {dulei, jiangwei, Ruijie Zhang}@ems.hrbmu.edu.cn, raos@ccf.org
[2] Department of Computer Science, Harbin Institute of Technology, Harbin, 150051, China
[3] Harbin Institute of Blood and Tumor, Harbin, 150010, China
ljcross2008@126.com

Abstract. There is currently a great interest in using single-nucleotide polymorphisms (SNPs) in genetic linkage and association studies because of the abundance of SNPs as well as the availability of high-throughput genotyping technologies. Here, we apply a novel ensemble decision approach to extract the relevant SNPs for alcoholism using sib-pair IBD profiles of pedigrees. The results indicate that ensemble decision tree is a promising algorithm for mining genetic markers for complex genetic diseases.

1 Introduction

Single nucleotide polymorphisms (SNPs) can be used as genetic markers that, in combination with other technologies, will help reveal the genetic basis of complex human diseases and enable medicines to be targeted to those patients likely to benefit. Alcoholism is a complex disorder in which multiple genes may contribute to the risk. To address this complexity, the Collaborative Study on the Genetics of Alcoholism (COGA) researchers designed a large-scale family study and collected multiple alcoholism and alcoholism-related phenotypes from the participants.

Current linkage and association methods have proven to be extremely successful in identifying genetic loci responsible for the development of diseases shown to have simple Mendelian modes of inheritance. However, most inherited diseases do not follow simple Mendelian patterns. The etiologies of these diseases are considered to be complex in nature, consisting of the interactions of multiple genetic loci as well as possible environmental factors. It is important to investigate how our current methodologies perform in the realm of complex diseases. For the time being, there have been a number of approaches for extracting alcoholism relevant SNPs from the Genetic Analysis Workshop 14 (GAW14) COGA data. For example, genetic algorithm-support vector machine hybrid [1] and sliding windows [2] have been used to mine alcoholism relevant SNPs.

* Corresponding authors.

D.-S. Huang, K. Li, and G.W. Irwin (Eds.): ICIC 2006, LNBI 4115, pp. 774–779, 2006.

In this study, we extend our ensemble decision approach [3] for mining alcoholism relevant SNPs using the genome-wide SNP data generated for GAW14. Using the dbSNP [4] and the OMIM databases, we identified the SNPs close to some genes that are associated with alcoholism.

2 Methods

2.1 Defining the Phenotypic Attribute of a Sib Pair and Features to Be Mined

We extend our ensemble decision approach to sib-pair analysis of pedigrees. First, we define the phenotypic attribute of a sib pair, the affection status of a sib pair for alcoholism. For the binary trait, there are three possible attributes, of which two attributes are chosen to be the phenotypes for learning: both sibs in a sib pair are affected; and no sibs in a sib pair are affected. The genetic features are defined to be the estimated proportions of alleles shared IBD by the sib pair at the SNP positions, provided by the GENIBD of the SAGE package [5].

2.2 Mining Alcoholism Relevant SNPs

In this study, we use the Genetic Analysis Workshop 14 COGA real data sets to demonstrate the behaviors and properties of the proposed method for mining alcoholism relevant SNPs. For computational convenience, we perform the same analysis procedures separately for each investigated chromosome.

In order to build up multiple pairs of training sets $\{L_d\}$ and test sets $\{T_d\}$, we introduce n-fold cross validation (CV), in which the data of each class is randomly divided into n roughly equal parts. For example, there are two classes, in each class the data are randomly divided into n non-overlapping subsets of roughly equal size, denoted as $D_i(i=1,2,\cdots 5)$ and $N_i(i=1,2,\cdots 5)$. A random combination of D_i and N_i constitutes a test set and the rest of the subsets are used as the training set. The n-fold CV resampling produces $n \times n$ pairs of training sets and test sets. We repeat the cross-validation M times and obtain $n \times n \times M$ pairs of training and test sets.

Sib-pair IBD profiles can be described as an $n \times p$ matrix, $\mathbf{X} = (x_{ij})$, where x_{ij} is the IBD value for the ith sib pair at the jth SNP locus. Our proposed ensemble decision method is a supervised-learning approach based on a recursive partition tree. The procedures of tree building are as follows. First, we split the dataset into two subsets, training set and test set. Then, a binary tree is grown on training set. By a recursive partition algorithm. The search for feature SNPs starts at the root of the tree and proceeds to its leaves. At each internal node, a decision is made with regard to the choice of a feature SNP and a threshold value (cutoff) such that the class impurity is reduced to a minimum when a branch is made by an induction rule. After the optimal bifurcation is made, the samples are divided into two non-overlapping subsets (two child nodes). For each subset, the same process is conducted successively until a leaf is reached or stopping criteria for tree growth are satisfied. Here the impurity is assessed by Gini inequality index [3]. For each grown tree, one subset of SNPs is extracted and is tested on the sister test set. This process for feature selection is repeated on each pair of training sets, $\{L_d\}$ $(d = 1,2,\ldots,m)$ and test sets, $\{T_d\}$ $(d = 1,2,\ldots,m)$, which consequently forms the decision forest that is composed of an ensemble of SNP feature subsets, $G_1, \ldots, G_d, \ldots, G_m$, $G_d = \{g_1^d, g_2^d, \ldots, g_k^d\}$, or called SNP forests.

To test whether an extracted subset of SNPs (Tree G_d) is able to significantly distinguish the affection status of a sib pair using the sister test sample（T_d）, we propose a χ^2 statistic:

$$\chi^2 = \frac{\left[\left|n_{00} n_{11} - n_{01} n_{10}\right| - n/2\right]^2 n}{(n_{00} + n_{01})(n_{10} + n_{11})(n_{00} + n_{10})(n_{01} + n_{11})} \tag{1}$$

where n= $n_{00} + n_{01} + n_{10} + n_{11}$ and n_{00}, n_{01}, n_{10} and n_{11} are the frequencies for true negative, false positive, false negative and true positive, respectively. A Chi-squared test was carried out on a 2×2 table to see whether the extracted subset of SNPs (Tree G_d) is able to significantly distinguish the affection status of a sib pair using the sister test sample（T_d）. This statistic follows an asymptotic Chi-squared distribution with one degree of freedom.

Based on the constructed SNP forests, we make an ensemble decision for selecting an alcoholism relevant SNP. Whether a SNP feature is relevant to the disease depends on the magnitude of its relevance intensity (or called an ensemble vote), FV.

For a particular SNP feature g_k, define:

$$\mathrm{FV}(g_k) = \mathrm{F}(G_1, G_2, \ldots G_m) = \frac{\sum_d \omega_d \mathrm{I}(g_k, G_d)}{\sum_d \omega_d}, \tag{2}$$

where I (g_k, G_d) is an indicator function:

$$I(g_k, G_d) = \begin{cases} n_d^k & g_k \in G_d \\ 0 & otherwise \end{cases}. \tag{3}$$

A weight, ω_d, can be a measure for the classification performance of G_d, for example, $\omega_d = \chi_d^2$ or set $\omega_d = 1$ for an equal weight for all the SNP feature subsets.

We resort to a permutation approach to generate the null distribution of FV, for which we randomly assign a phenotypic attribute to a sib pair. A critical value FV_{β}^{0} for a specified significance level, β (e.g. 0.05 in this study) is then obtained. A SNP feature is selected if its $FV \geq FV_{\beta}^{0}$ (one-tailed).

3 Results

In this study, we perform a 5-fold cross-validation 20 times. Totally, we extract 344 significant SNPs at the 5% significance level, of which 166 are related to alcoholism (supported by the existing knowledge). The chromosomal distributions for the signifi-cant SNPs are shown in Fig.1. However, we focus our attentions on chromosomes 1, 4, and 7 (the results as shown in Table 1) because previous studies defined some regions on these chromosomes that may harbor several genes for alcoholism [6].

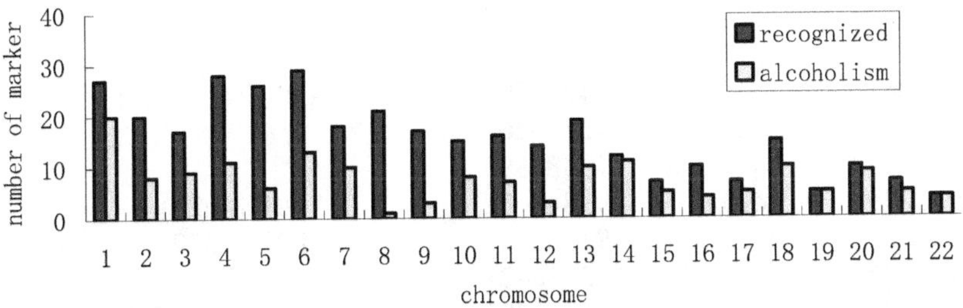

Fig. 1. The brown bar represents the SNPs we recognized with 5% significance level and the yellow bar represents the recognized SNPs related to alcoholism

In large-scale genome-wide association studies, 'false positives' can be a thorny problem. To see if the proposed ensemble decision approach can well control the number of 'false positive', we apply this method to the 100 simulated replicates for GAW14. The results from analyzing the simulated data demonstrate that we have successfully recognized all the simulation 'answers' or the SNPs closest to an answer, suggesting that the ensemble decision approach is robust and powerful in identifying the true loci for alcoholism. Nevertheless, the existing knowledge about the genes related to alcoholism in biomedical databases (e.g. NCBI GENE database) is very limited. Therefore, we view the results derived from this data analysis as exploratory, which might have support from the existing knowledge pool (i.e. in the manner of 'knowledge validation') or lacking of such a support waiting for further wet-lab experiment to verify their involvements in the pathogenic pathways to alcoholism.

Table 1. List of the significant SNPs on chromosomes 1, 4 and 7 (P<0.05)

SNP Name	Band	Genetic Position	Alcoholism gene region
rs1171466	1p34.2b	39215299	GRIK3(1P34-P33)
rs1441523	1p34.3d	34069870	
rs1408946	1p34.1e	42392561	
rs1418490	1p36.23a	7002001	FRAP1 (1p36.2)
rs1399437	1p36.13c	18227689	
rs1599169	1p36.31b	4349628	
rs320039	1p33b	48793370	AKR1A1(1P33-P32)
rs2896880	1p32.2b	56453092	
rs1933392	1p33d	46803463	
rs1079130	1p33b	49115058	
rs728719	1p32.2a	57482150	
rs2077402	1p13.2b	111186557	CEPT1(1P13.3)
rs1390474	1p31.3c	64748336	ADH5P2(1P31.1)
rs2388956	1p31.1b	80644475	
rs1354061	1p31.3a	64467663	CRYZ(1P31-P22)
rs1775986	1p22.3b	86587517	
rs3782	1p22.3c	85446633	
rs066751	4q25f	112347992	ADH5(4q21-25)
rs1349629	4q21.21b	75685262	
rs1604153	4q21.22a	78382893	
rs951299	4q23a	99843682	ADH6(4q23)
rs720327	4q22.1	91866437	ADH4(4q22)
rs725386	4p15.33c	12928421	
rs2310396	4p15.1f	31912292	PI4K2B(4P15.2)
rs1113428	4p12a	44643625	GABRA2(4p12)
rs1349629	4q21.21b	75685262	DCK(4q13.3-q21.1)
rs1604153	4q21.22a	78382893	
rs293441	4p13.3c	69495689	
rs961489	4p13.2c	66079872	
rs1005959	7q31.2a	112780691	CHRM2(7q31-q35)
rs1012709	7q32.2b	129770032	
rs967919	7q22.1h	102532434	TRRAP(7q21.2-22.1)
rs717352	7q22.2a	103367208	
rs1375670	7q22.1a	95057063	
rs2887978	7q21.11e	80899643	
rs722471	7q21.3d	94572400	
rs878903	7p15.1c	29559251	NPY(7p15.1)

4 Discussion

We extend ensemble decision approach to a wide range of areas, in this paper, we apply it to mine alcoholism relevant SNPs. In summary, some potential candidate regions on chromosomes1、4、and 7 linked with alcoholism susceptibility loci were found. These findings are consistent with previous reports [7]. Caution should be taken in interpretation of the SNP feature selection for alcoholism. Many causes can contribute the relevance of a SNP to alcoholism, for example, linkage disequilibrium between loci and gene-gene interactions, which require further genetic analysis to be clarified. For the time being, interactions are frequently at the center of interest in single-nucleotide polymorphism (SNP) association studies. When interacting SNPs are in the same gene or in genes that are close in sequence, such interactions may suggest which haplotypes are associated with a disease. The results show our approach delivers useful results and gives a competitive tool that facilitates the further study of global-view SNP relevance network(s) for alcoholism.

Acknowledgements

This work is supported in part by the National Natural Science Foundation of China (Grant Nos. 30170515 and 30370798), and National Science Foundation of Heilongjiang (No.D0212, F2004-02 and ZJG0501).

References

1. Gong, B.S., Guo, Z., Li, X. et al.: A Genetic Algorithm-Support Vector Machine Hybrid for Predicting Affection Status of a Sib Pair Based on Genome-wide SNP IBD Prfiles.2004 international Genetic Analysis Workshop14 conference, 14(20) (2004) 852-858
2. Li, C.X.: Analysis of sib IBD Profiles and Genomic Context for Identification of the Relevant Molecular Signatures for alcoholism.2004 international Genetic Analysis Workshop14 conference, 14(20) (2004) 845-851
3. Li, X., Rao, S., Wang, Y., Gong, B.: Gene Mining: a Novel and Powerful Ensemble Decisionapproach to Hunting for Disease Genes using Microarray Expression Profiling. Nucleic Acids Res, 32 (2004) 2685-2694
4. Brookes, A.J., Lehvaslaiho, H., Siegfried, M., Boehm, J.G., Yuan, Y.P., Sarkar, C.M., Bork ,P., Ortigao, F.: HGBASE: a Database of SNPs and Other Variations in and around Human Genes. Nucleic Acids Res. 28 (2000) 356-360
5. S.A.G.E.: Statistical Analysis for Genetic Epidemiology. Cork, Ireland (2003)
6. Almasy, L., Borecki, B.: Exploring Genetic Analysis of Complex Traits Trough the Paradigm of Alcohol Dependence: Summary of GAW11 Contributions. Genet Epidemiol 1999, 17 (1999) S1-S24
7. Zinn-Justin, A, Abel, L.: Genome Search for Alcohol Dependence using the Weighted Pair-wise Correlation Linkage Method: Interesting Findings on Chromosome 4. Genet Epidemiol 1999, 17 (1997) S421-S426

Association Research on Potassium Channel Subtypes and Functional Sites

Peng Wu[1], Xia Li[1,2,*], Shaoqi Rao[1,3,*], Wei Jiang[1], Chuanxing Li[1], and Jie Zhang[1]

[1] Department of Bioinformatics, Harbin Medical University, Harbin 150086, P.R. China
wpwpwp_@163.com, lixia6@yahoo.com,
{jiangwei, licx}@ems.hrbmu.edu.cn
[2] Department of Computer Science, Harbin Institute of Technology, Harbin, 150051, China
[3] Departments of Cardiovascular Medicine and Molecular Cardiology, The Cleveland Clinic Foundation, 9500 Euclid Avenue, Cleveland, Ohio 44195, USA
rao@ccf.org

Abstract. Potassium channels, a goup of membrane proteins and found in virtually all cells, are of crucial physiological importance for understanding cell biology and channelopathy. Studying the association of channel subtypes with their functional sites, and mining the functional sites mostly relevant to channel subtypes, can not only enhance molecular classification of potassium channels but only provide some clues for the functional targets. Here, we proposed to use functional sites profiles as features, where the functional sites are often used to characterize ion channels in biomedical research domains. We employed a novel integrated decision method to recognize the functional sites subset that is mostly relevant to the differentiation of potassium channel subtypes.

1 Introduction

Potassium channels, a group of special proteins in cytolemma, play critical roles in some important cellular processes such as the production of cell transmembrane current, and resting potential maintenance. The proteomic investigation of the potassium channels has been a focus in biomedical domains in recent years. All potassium channels implement their biological functions via some relevant functional sites, for example, kalium ion binding site, cuprum ion binding site etc. Inhibition of these sits can cause the loss of ion channel integrated activity leading to a list of disorders called channelopathy. Among these disorders, the channelopathy associated with the disruption of potassium channel is one of the most common types. Although biologists are able to localize the underlying functional sites for ion channels in laboratory by using X-ray crystal diffraction and nuclear magnetic resonance technique, these experiments are both too time-consuming and costly. Alternatively, with the rapidly accumulated biomedical data for ion channels, bioinformatics approaches such as various machine learning algorithm may provide the promises and were proposed to resolve the functional characterizations using the sequence

* Corresponding authors.

D.-S. Huang, K. Li, and G.W. Irwin (Eds.): ICIC 2006, LNBI 4115, pp. 780 – 785, 2006.

conserved regions [1]. Typical algorithms include h-function algorithm, Γ-function algorithm, Θ-function algorithm, Markov's chain model, vectorized sequence-couple model, and discriminant function algorithm. With more and more functional sites predicted and confirmed so far, how to make full use of these available data, and to establish the coherence and relevance between potassium channel subtypes and the functional sites become an immediate task in the proteomic studies of channelopathy.

Up to the present, no report for analyzing the relevance between potassium channels and their functional sites has been seen in literature. Here, we presented a bioinformatics approach to analyze and to evaluate the strength of relevance between the profiles of functional sites and ion channels. An ensemble approach based on decision trees [2] was employed to extract an optimal feature site subset(s) that was to precisely classify the studied potassium channel subtypes. In addition to identification of a functional site subset to achieve the maximal prediction accuracy, we also identified the functional sites mostly relevant to the ion channel subtype discrimination. An application of this approach to analyze the functional sites data from PDB [3]and PDBsite [4,5] is given to demonstrate its performance for mining relevant features based on the distribution of functional features over the decision forest using an ensemble voting approach.

2 Method

2.1 Constructing Functional Sites Profiles for Potassium Channels

First, we extracted channel subtypes and 3D structure information of potassium channels from 3D Protein Structure database (PDB). Then these data were annotated to the database of Protein's Functional site (PDBsite) to retrieve the corresponding information for functional sites. As a result, we constructed the functional sites profiles for potassium channels, which were linked with the phenotypic dimension of channels.

The potassium channel functional site profile can be described by an $m \times p$ matrix, $C = (c_{ij})$, where c_{ij} describes the existence for the j th functional site (s_j) on the i th sample C_i. When the subtype of an ion channel sample is known, the data for the sample consist of a vector of functional site profile, $C_i = (c_{i1} \ldots c_{ip})$ and a category label (y_i). Suppose that the studied channel samples belong to K categories $\omega_1, \omega_2, \ldots \omega_k$. Define a class label, y_i, to be an integer from 1 to *K*. Let n_k be the number of samples in the k th category.

2.2 Constructing the Partition Tree

An n-fold cross-validation (CV) resampling technique was employed to build up pairs of training sets, $\{L_d\}$ (d=1, 2,..., n),and test sets, $\{T_d\}$ (d=1, 2,..., n), for learning and testing, respectively. For each resampling constructed pair, L_d and T_d, a recursive partition tree is grown on the training set L_d with the class impurity as the criterion for tree splitting and the built tree was then tested on the holdout set T_d. Feature site selection proceeded in a manner that the best site was sought so that impurity was minimized at a node (say *t*), from which a new bifurcation was attempted. Hence, the process of growing a tree parallels the process of recognizing future sites. When tree growth was stopped, a subset of feature sites, S_d, was obtained from the particular training set, L_d.

The above feature site subset building algorithm was applied to the multiple training sets and test sets, generated by the n-fold cross-validation resampling technique. Consequently, a set of feature subsets were obtained by inducing the decision trees, denoted { S_1, $S_2, \cdots S_n$ }. For a feature functional site, whether it is relevant to channel subtypes depends on the magnitude of its relevance intensity (or called an ensemble vote), FV. For a particular feature site s_k, define:

$$\mathrm{FV}(s_k)=\mathrm{F}(S_1, S_2, \ldots S_n)=\frac{\sum_d \varpi_d I(s_k, S_d)}{\sum_d \varpi_d}, \tag{1}$$

where FV $\in$ [0, 1], and I(s_k, S_d) is an indicator function:

$$I(s_k, S_d)=\begin{cases}1 & \mathrm{s_k} \in S_d \\ 0 & \text{otherwise}\end{cases}. \tag{2}$$

A weight, ϖ_d, can be a measure for the classification performance of S_d, or for simplicity, ϖ_d = 1 for an equal weight for all the feature subsets. To determine the significance of a candidate subset of functional sites, we resorted to a permutation approach [2] to simulate the null distribution of *FV* and to identify the significance cut-off value, denoted as FV^*.

3 Data

We obtained the data on 62 functional sites for 45 potassium channels, by first retrieving 66 potassium channels from the Protein's 3D Structure Database (PDB) and then annotating to the PDBsite database where knowledge about protein functional sites is stored. The extract potassium channels belong to three subtypes: voltage-gated, calcium-gated and inward rectifier potassium channel. To better characterize the functional sites, we designed three types of profiles, each being a 45-by-62 matrix with a row a channel and a column a functional site. In the first profile (or called existence profile), the functional features were coded as defined by equation (3), i.e. the binary values for existence or absence of a functional site. The feature in the second profile (or called intensity profile) was an integer that corresponds to the repeated number of the feature site that a channel has. By using this information, we

$$c_{ij} = \begin{cases} 1 & \text{if site exists} \\ 0 & \text{otherwise} \end{cases} \tag{3}$$

expected to capture the accumulated effect of a functional site. Finally, to reveal the randomness of a feature site, we construct a random profile, in which each element with a non-zero value in the first and second profiles was substituted by a random value from 1 to 10. We applied the tree-based ensemble method to the three profiles for mining the functional sites that are most likely relevant to the partition of the three subtypes for potassium channels.

4 Results and Discussion

4.1 Two Classes Problem and Significant Level

We developed successfully a novel tree-based ensemble method to resolve classification problem of two-class sample and obtained satisfactory result. For example, we applied this ensemble method to gene expressional profile datasets for mining feature genes which related to disease. Here, we extended the application of this method to multi-class sample. The strategy is changing multi-class sample to two-class sample by taking out one class as the exist-alone-class in turn, and incorporates other classes as a new one. With this strategy, a three-class sample profile was changed into three two-class sample datasets (called voltage-gated, calcium-gated and inward rectifier datasets). Applying 4-fold cross-validation (CV) resampling technique to these datasets to generate training sets and test sets. Finally, we obtained 320pairs of training and test sets in all.

4.2 Effect of Site's Intensity and Subtype-Specific Phenomena

Compared the results of existence profile and intensity profile, we found that the sites identified from intensity profile had distinct difference and convergence in distribution.See Fig.1 (a) for example, there were 5 sites' frequency upon 0.7, while other sites' frequency below 0.05. It's easier for us to make estimate of subtype-related-sites. The reason of this phenomenon was that not only the existence but also the intensity of functional sites relevant to the subtype of the potassium channel.

During the compare of the sites identified from existence profile and random profile, we found that they had the similar distribution. The reason of this phenomenon was that the random intensity had no relevancy to the subtype of the channel. This phenomenon also confirmed our hypothesis that the intensity of sites had relevancy to the subtype of the channel.

Compared the results of Voltage-gated dataset and calcium-gated dataset, see Fig.1 (b), we found the sites identified from Voltage-gated dataset had same distribution among existence profile, intensity profile and random profile. It means the relevancy of site's intensity and the subtype of the channel has Subtype-specific.

(a)

(b)

(c)

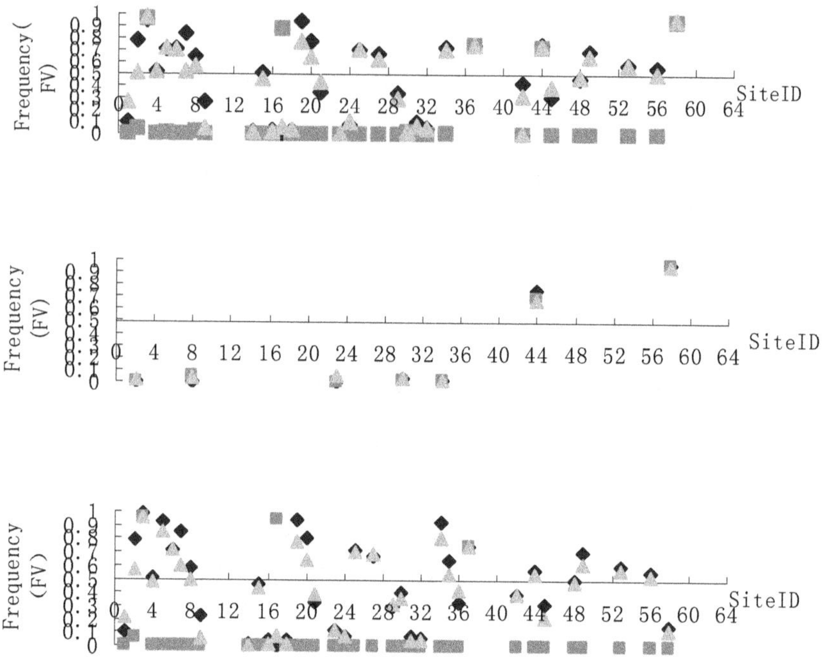

Fig. 1. a) The result of the voltage-gated dataset. b) The result of the calcium-gated dataset. c) The result of the inward rectifier dataset. Blue diamond, red square and green triangle means the site mining from the existence profile, the intensity profile and the random profile respectively.

Here, we choose $FV^* = 0.5$, the frequency of feature site more than FV^* was identified as the sites that had most relevant to the subtype of the potassium channel.

The results had been listed in table1. In table 1, the site of number 3, Copper Iron binding site, has the highest frequency in the result, which means that it has most relevant to the subtype of potassium channel. Luckily, this result had been confirmed by experiment. In 2003, F.J. Morera and D. Wolff reported that copper ion was one kind of important inhibitor of voltage-gating potassium channel; it inhibits the potassium channel by changing the gating kinetic properties of the channel [6]. In this paper, we applied a tree-based ensemble method to mine the functional site and got a set of functional site that most relevant to the subtype of potassium channel. By contrasting the result of the three profiles, we also found that not only the existence but also the intensity of the functional sites had relevant to the subtype of the potassium channel. In future, we'll try our best to research the similarity of these functional sites in structure and sequence.

Table 1. The functional site that most relevant to subtype of potassium channel from the intensity profile

Dataset	Site ID	FV	Notation	Site ID	FV	Notation
voltage-gated dataset	44	0.73	Metal atom in 562 binding site	58	0.947	Residues making water-mediated hydrogen bonds with ribose
	37	0.75	EGL Binding site			
	17	0.88	GOL binding site	3	0.963	Copper Iron binding site
calcium-gated dataset	44	0.69	Metal atom in 562 binding site	58	0.969	Residues making water-mediated hydrogen bonds with ribose
inward rectifier dataset	37	0.75	EGL binding site	3	0.963	Copper Iron binding site
	17	0.94	GOL binding site			

Acknowledgements

This work was supported in part by the National Natural Science Foundation of China (Grant Nos. 30170515, 30370798, 30571034 and 30570424), the Heilongjiang Province Department of Education Outstanding Overseas Scientist grant (Grant No. 1055HG009)， National Science Foundation of Heilongjiang Province(Grant No. ZJG0501, GB03C602-4, F2004-02) and the Innovative Fund of Harbin Medical University Graduate Student.

References

1. Chou, K.C.: Analytical Biochemistry, 233 (1996) 1-14
2. Li, X., Rao, S., Wang, Y., Gong, B.: Gene Mining: A Novel and Powerful Ensemble Decision Approach to Hunting for Disease Genes Using Microarray Expression Profiling. Nucl Acids Res, 32 (2004) 2685–2694
3. Berman, H.M., stbrook, J., Feng, Z., Gilliland, G., N.Bhat, T., Weissig, H., Shindyalov, I.N.,Bourne, P.E.: The Protein Data Bank Nucleic Acids Research, 28 (2000) 235-242
4. Morera, F.J., Wolff, D.: External Copper Inhibits the Activity of The Large-Conductance Calcium- and Voltage-sensitive Potassium Channel from Skeletal Muscle. Journal of Membrane Biology, 192 (2003) 65-72
5. Ivanisenko VA, Pintus SS, Grigorovich DA, Kolchanov NA. PDBSite: a database of the 3D structure of protein functional sites. Nucleic Acids Res., (2005) V33, D183-D187.
6. Ivanisenko VA, Pintus SS, Grigorovich DA, Kolchanov NA. PDBSiteScan: a program for searching for active, binding and posttranslational modification sties in the 3D structrues of proteins. Nucleic Acids Res., (2004)W549-W554.

Nonequilibrium Model for Yeast Cell Cycle

Yuping Zhang[1], Huan Yu[1], Minghua Deng[1], and Minping Qian[1,2,⋆]

[1] School of Mathematical Sciences and Center for Theoretical Biology, Peking University, Beijing 100871, P.R. China
[2] qianmp@pku.edu.cn

Abstract. In the living cells, molecules including proteins, DNAs, RNAs and so on, with interactions between them cooperate as networks that govern various cellular functions. In this paper, a stochastic model with trigger mechanism is proposed based on what are known about the genes and proteins controlling the cell cycle of budding yeast. With respect to the biological observations, it looks more natural and understandable than deterministic dynamical model and our former stochastic model. Our model vividly describes that the protein interaction network goes through the biological pathway and forms an endless loop.

1 Introduction

The emergence and development of many high-throughput data-collection techniques, such as microarrays [1], protein chips or yeast two-hybrid screens [2], automated RTPCR and 2-D gel electrophoresis help us to simultaneously obtain the status of a cell's components and determine how and when these molecules interact with each other. It presents an opportunity to infer the real gene networks from experimental observation. Meanwhile it stimulates quantitative understanding of biological systems and functions from their components and interactions. Recently, a considerable amount of attention has been paid to the quantitative modeling and understanding of the budding yeast cell cycle regulation [3,4,5,6,7,8,9,10,11,12].

In particular, Li *et al.* [5] introduced a deterministic Boolean network model and investigated its dynamic and structural properties. In Li's work, the network was both dynamically and structurally stable. Seven global attractors of the dynamics were identified but only one could be interpreted as a biological stationary state, and the corresponding attracting trajectory as a biological pathway. We further advanced a stochastic model [13] based on Li's model, in which only the biological stationary state and the pathway were conserved under a wide range of noise level. We used a pseudo potential function to describe the dynamic landscape of the system and the biological pathway was considered as a valley in the landscape [14,15]. Unfortunately, those former works were all built upon equilibrium systems without concerning of extern signals such as cell size. In fact, the biological cell cycle is an endless progress accompanying with positive entropy production and thus a nonequilibrium system. In this work, we introduced a nonequilibrium stochastic model for yeast cell cycle.

⋆ Corresponding author.

D.-S. Huang, K. Li, and G.W. Irwin (Eds.): ICIC 2006, LNBI 4115, pp. 786–791, 2006.

2 Nonequilibrium Stochastic Model

Our nonequilibrium stochastic model is based on the updated protein interaction network of Li *et al.* [5], which is depicted in Fig. 1. In the network, Cln3 activates a pair of nodes, SBF and MBF. SBF and MBF stimulate the transcription of G1/S genes, including those of Cln2 and Clb5. The S phase cyclin Clb5 initiates DNA replication, after which the transcription factor complex Mcm1/SFF is turned on, which stimulates the transcription of many G2/M genes including the gene of the mitotic cyclin Clb2. The cell will exit from mitosis and divide into two after Clb2 is being inhibited and degraded by Cdc20, Cdh1 and Sic1. The cell (or two cells: the mother and the daughter) now comes back to the stationary G1 state, waiting for the signal for another round of division.

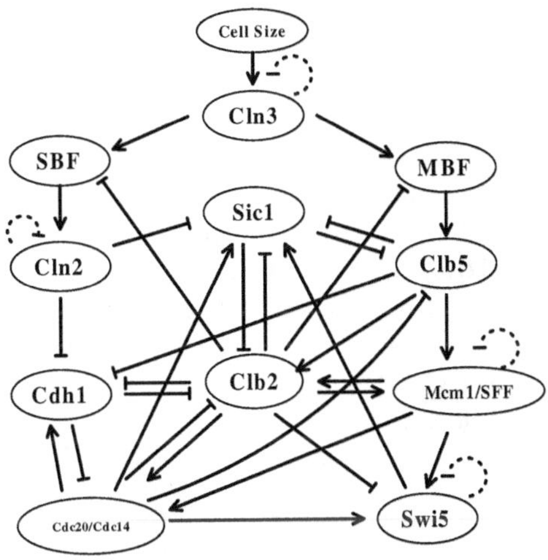

Fig. 1. The cell-cycle network of the budding yeast. Each node represents a protein or a protein complex. Arrows are positive regulation, "T"-lines are negative regulation, dotted "T"-loops are degradation.

The evolution of the network has the Markov property. That is, the statistical behavior of the cell state at the next step is determined by the cell state at the present step. In our model[13], the states of 11 nodes in the network shown in Fig. 1, namely, Cln3, MBF, SBF, Cln2, Cdh1, Swi5, Cdc20, Clb5, Sic1, Clb2, and Mcm1 are represented by variables $(s_1, s_2, ..., s_{11})$, Each node i has only two states: $s_i = 1$ and $s_i = 0$, representing the active state and the inactive state of the protein i, respectively. We define the transition probability of the Markov chain as follows:

$$\begin{aligned} &P_r(s_1(t+1), ..., s_{11}(t+1)|s_1(t), ..., s_{11}(t)) \\ &= \prod_{i=1}^{11} P_r(s_i(t+1)|s_1(t), ..., s_{11}(t)), \end{aligned} \tag{1}$$

where

$$P_r(s_i(t+1) = \sigma_i | s_1(t), ..., s_{11}(t)) = \frac{\exp(\beta(2\sigma_i - 1)\sum_{j=1}^{11} a_{ij}s_j(t))}{\exp(\beta\sum_{j=1}^{11} a_{ij}s_j(t)) + \exp(-\beta\sum_{j=1}^{11} a_{ij}s_j(t))},$$

if $\sum_{j=1}^{11} a_{ij}s_j(t) \neq 0, \sigma_i \in \{0,1\}$;
and

$$P_r(s_i(t+1) = s_i(t) | s_1(t), ..., s_{11}(t)) = \frac{1}{1+e^{-\alpha}}, \tag{2}$$

if $\sum_{j=1}^{11} a_{ij}s_j(t) = 0$ and $\{s_1(t), ..., s_{11}(t)\} \neq \{0,0,0,0,1,0,0,0,1,0,0\}$;
and

$$P_r(s_{cln3}(t+1) = 0 | S(t) = \{0,0,0,0,1,0,0,0,1,0,0\}) = \frac{1}{1+e^{-\gamma}}, \tag{3}$$

We define $a_{ij} = 1$ for a positive regulation of protein j to protein i and $a_{ij} = -1$ for a negative regulation of protein j to protein i. If the protein i has a self-degradation loop, $a_{ii} = -0.1$. The positive constants α, β and γ are temperature-like parameter characterizing the noise in the system [16]. Especially, γ plays a trigger role, so $0 < \gamma < \alpha$ and the last equation act as an extern cell size signal.

The Markov chain is ergodic since it consists of finite states and is irreducible. The steady-state probability distribution can be estimated by the simulation of stochastic processes as follows:

$$\pi_n = \frac{\#n}{\sum_{m=0}^{2047} \#m}, \tag{4}$$

where $\#m$ denotes the number of times the state m is being visited in the simulate iterations.

3 Results

For state m and n, the pure probability flux from m to n is denoted as $\pi_m p_{mn} - \pi_n p_{nm}$, where p_{mn} is the transition probability from m to n. Fig. 2 is an example of the probability flux among all 2048 states. There is a distinct difference between Fig. 2 and the Fig.3 in [13]. The arrow representing the probability flux from the stationary G1 state to the excited G1 state (the START of the cell-cycle) is a real part of the cycle instead of a dashed line. States of proteins on different stage of cell cycle can be seen clearly in Fig. 3, which fit the known biologically knowledge very well. The result demonstrates that the stationary G1 state is not a steady state, but a metastable state. Other states in the system are more unsteady metastable states. Qian *et al.*([17]) proved that irreversible

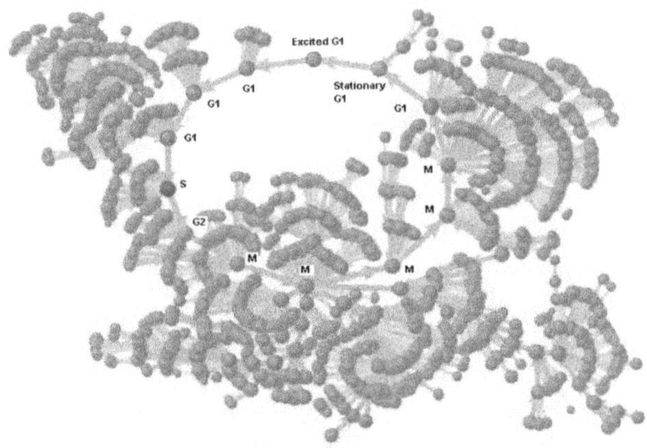

Fig. 2. The probability flux. Each node represents one of the 2048 states. The size of a node reflects the stationary distribution probability of the state. If the stationary probability of a state is larger than a given threshold value, the size of the node is in proportion to the logarithm of the probability. Otherwise, the node is plotted with the same smallest size. The nodes on the biological pathway (the loop in the figure) are denoted with different characters: Stationary G1, Excited G1, G1, S, G2 and M. The arrows reflect the pure probability flux (only the largest flux from any node is shown). The width of an arrow is in proportion to the logarithm of the probability flux it carries. For a node, only the largest flux from it is shown. The simulations were done with $\alpha = 3$, $\beta = 3$ and $\gamma = 1$.

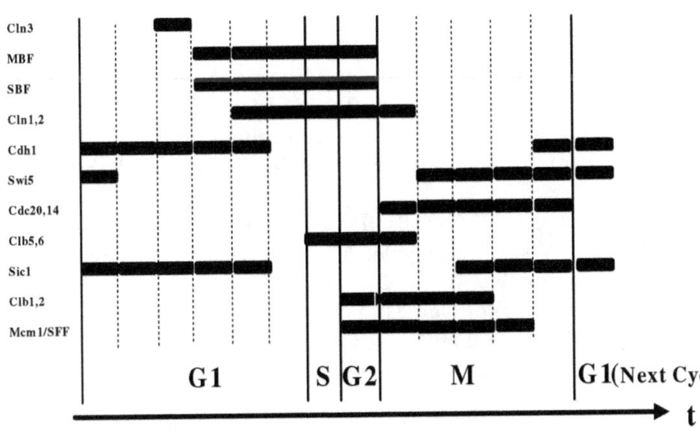

Fig. 3. The states of proteins on different stage of cell cycle. Black bar means activation.

Markov chain will inevitablely lead to a loop. Such a system will generate positive entropy. That is also the property of alive system, such as cell cycle system. Qian *et al.*([17]) gave the definition of entropy production rate as

$$e_p = \frac{1}{2} \sum_{i,j \in S} (\pi_i p_{ij} - \pi_j p_{ji}) \ln \frac{\pi_i p_{ij}}{\pi_j p_{ji}} \tag{5}$$

We computed the entropy production rate when $\beta = 3$, $\alpha = 3$ and $\gamma = 1$, and the value is about 12.2.

4 Conclusion

In conclusion, we introduced a nonequilibrium stochastic model for the yeast cell cycle network by considering extern signals effects (such as cell size) in the model. A real alive life system must be an open system. The loop existing in real cell cycle can be seen clearly in the result of our model. Because of negative entropy production rate from external environment (absolute value of which equals the positive entropy production rate of the internal system), the system can maintain cycling.

Acknowledgments

This research is supported by the grants from National Natural Science Foundation of China (No. 30570425, No. 90208022) and the National Key Basic Research Project of China (No. 2003CB715903), and supported in part by Microsoft Research Asia (MSRA).

References

1. Zweiger, G.: Knowledge Discovery in Gene-Expression-Microarray Data: Mining the Information Output of the Genome. Trends Biotechnol **17**(11) (1999) 429–436
2. Ito, T., Chiba, T., Ozawa, R., et al.: A Comprehensive Two-hybrid Analysis to Explore the Yeast Protein Interactome. Proc Natl Acad Sci U S A **98**(8) (2001) 4569–4574
3. Chen, K.C., Csikasz-Nagy, A., Gyorffy, B., et al.: Kinetic Analysis of a Molecular Model of the Budding Yeast Cell Cycle. Mol Biol Cell **11**(1) (2000) 369–391
4. Cross, F.R., Archambault, V., Miller, M., et al.: Testing a Mathematical Model of the Yeast Cell Cycle. Mol Biol Cell **13**(1) (2002) 52–70
5. Li, F., Long, T., Lu, Y., et al.: The Yeast Cell-Cycle Network is Robustly Designed. Proc Natl Acad Sci U S A **101**(14) (2004) 4781–4786
6. Chen, H.C., Lee, H.C., Lin, T.Y., et al.: Quantitative Characterization of the Transcriptional Regulatory Network in the Yeast Cell Cycle. Bioinformatics **20**(12) (2004) 1914–1927
7. Chen, K.C., Calzone, L., Csikasz-Nagy, A., et al.: Integrative Analysis of Cell Cycle Control in Budding Yeast. Mol Biol Cell **15**(8) (2004) 3841–3862
8. Cross, F.R., Schroeder, L., Kruse, M., et al.: Quantitative Characterization of a Mitotic Cyclin Threshold Regulating Exit from Mitosis. Mol Biol Cell **16**(5) (2005) 2129–2138
9. Futcher, B.: Transcriptional Regulatory Networks and the Yeast Cell Cycle. Curr Opin Cell Biol **14**(6) (2002) 676–683

10. Murray, A.W.: Recycling the Cell Cycle: Cyclins Revisited. Cell **116**(2) (2004) 221–234
11. Ingolia, N.T., Murray, A.W.: The Ups and Downs of Modeling the Cell Cycle. Curr Biol **14**(18) (2004) R771–R777
12. Tyers, M.: Cell Cycle Goes Global. Curr Opin Cell Biol **16**(6) (2004) 602–613
13. Zhang, Y., Qian, M., Ouyang, Q., et al.: Stochastic Model of Yeast Cell Cycle Netwok (in appear)
14. Ao, P.: Potential in Stochastic Differential Equations: Novel Construction. Journal of Physics A: Mathematical and General (3) (2004) L25–L30
15. Zhu, X.M., Yin, L., Hood, L., et al.: Calculating Biological Behaviors of Epigenetic States in the Phage λ Life Cycle. Funct. Integr. Genomics **4**(3) (2004) 188–195
16. Albeverio, S., Feng, J., Qian, M.: Role of Noises in Neural Networks. Physical Review. E. Statistical Physics, Plasmas, Fluids, and Related Interdisciplinary Topics **52**(6) (1995) 6593–6606
17. Jiang, D., Qian, M.: Mathematical Theory of Nonequilibrium Steady States. Springer (2004)

Tissue Classification Using Gene Expression Data and Artificial Neural Network Ensembles*

Huijuan Lu[1,2], Jinxiang Zhang[3], and Lei Zhang[1]

[1] Institute of Computer Applications, China Jiliang University, 310018 Hangzhou, China
hjlu8@cjlu.edu.cn
[2] College of Computer Science, Zhejiang University, 310027 Hangzhou, China
zjx@zjei.net
[3] Department of Computer Science, Zhejiang Education Institute, 310012 Hangzhou, China
leizhang4662@sina.com

Abstract. An important challenge in the use of large-scale gene expression data for biological classification occurs when the number of genes far exceeds the number of samples. This situation will make the classification results are unstable. Thus, a tissue classification method using artificial neural network ensembles was proposed. In this method, a feature preselection method is presented to identify significant genes highly correlated with tissue types. Then pseudo data sets for training the component neural network of ensembles were generated by bagging. The predictions of those individual networks were combined by simple averaging method. Some data experiments have shown that this classification method yields competitive results on several publicly available datasets.

1 Introduction

Microarray experiment technology allows gene expression levels to be measured for the entire genome simultaneously. Such experiments collect enormous amounts of gene expression data that clearly reflects underlying biological processes. Array methodologies have led to a tremendous acceleration in the rate at which gene expression pattern information is accumulated [6], [13], [14]. In the medical field, determination of cancer type and stage is often crucial to the optimal treatment of patients [10], [12]. However, poorly differentiated cancers can be difficult to diagnose by routine histopathology and morphology. Normal cells can evolve into malignant cancer cells through a series of mutations that control the cell cycle, apoptosis, and genome integrity. Diagnostic classification of cancers from their gene-expression profiles has become a novel promising method. In this field, the major tasks of bioinformatics research are to identify gene sets that contributed to cancer classification and find optimal classification methods for diagnosis. These tasks are also called *feature selection* and *supervised classification*, two commonly addressed problems in machine learning. But the number of genes in microarray data set is very large, about 2000 to

* This work was supported by the Zhejiang Provincial Major Scientific and Technological Project (No.2005C21028) and Zhejiang Provincial Natural Science Foundation of China (No.Y109456).

D.-S. Huang, K. Li, and G.W. Irwin (Eds.): ICIC 2006, LNBI 4115, pp. 792 – 800, 2006.

20000, it is impossible to search all the possible feature subsets for global optimal results. Furthermore, the sample sizes are limited that will make the classification results unstable. So microarray techniques have led to several important challenges for pattern recognition. In recent years, several mathematical approaches have been developed to tissue classification using microarray data, including *voting* scheme [10], *unsupervised clustering* [2], *support vector machine* [17], *Bayesian regression models* [19] and some hybrid approaches [8], [22] . Ben-Dor et al.[3] and Dudoit et al. [7] have compared performances of several traditional classification approaches that applied in this area. These classification methods include *k*-NN, *Classification And Regression Trees* (CART) and so on.

One of the basic characteristics of *Artificial Neural Networks* (ANNs) is that they can discover interrelationships within large sets of data. In complex biological systems, characterized by an abundance of data, ANNs allow pattern recognition, which would otherwise prove difficult. They have been chosen as a primary analysis tool in medical studies. Some research groups [12], [15], [20] reported tumor classification by multilayered backpropagation neural networks respectively. Hornik et al. showed that feedforward artificial neural networks with one hidden layer can approximate any functions in any accuracy. However, until now there is no rigorous theory indicating how to do such error-free approximation. Therefore whether an artificial neural network based application will be successful or not is almost fully determined at present by the who is the user. In general, the more experiences the user has on artificial neural networks, the more chances the application will have in gaining success. Unfortunately, in real-world applications the users are often those with little knowledge on neural computing. Therefore artificial neural networks techniques sometimes do not lead to ideal results in the microarray application [7]. In the beginning of the 1990s, Hansen and Salamon [11] showed that the generalization ability of an artificial neural networks system can be significantly improved through ensembling artificial neural networks, i.e. training several artificial neural networks and combining their predictions. Since artificial neural network ensembles work remarkably well and are easy to be used, they are regarded as a promising methodology that can profit not only experts in artificial neural network research but also engineers in real-world application. The techniques have been successfully applied to many real-world domains such as handwritten digit recognition, scientific image analysis, face recognition, OCR, seismic signals classification, breast cancer diagnosis, and in-vitro fertilization treatment [22].

In this paper, based on the recognition of the power of artificial neural network ensemble, a novel classification approach for gene expression microarray data analysis was proposed. In this approach, significant genes for classification were selected by Wilcoxon test. Then learning data sets for each member of neural network ensembles were generated by *convex pseudo-data* (CPD) methods. The predictions of those individual networks were combined by simple averaging method.

2 Methods

For our purpose, gene expression data on p genes for n mRNA samples may be summarized by an $n \times p$ matrix $\boldsymbol{X}=(x_{ij})$, where x_{ij} denotes the expression level of gene (variable) j in mRNA sample (observation) i. The expression levels might be either absolute (e.g. oligonucleotide arrays) or relative with respect to the expression levels of a suitably defined common reference sample (e.g. cDNA microarrays). When the mRNA samples belong to known classes (e.g. follicular lymphoma), the data for each observation consist of a gene expression profile $\boldsymbol{x}_i=(x_{i1},x_{i2},\cdots,x_{ip})$ and a class label y_i, i.e., of predictor variables $\boldsymbol{x}_i$ and response y_i. For K classes, the class labels y_i are defined to be integers ranging from 0 to K-1.

A predictor or classifier for K tumor classes partitions the space $\boldsymbol{X}$ of gene expression profiles into K disjoint and exhaustive subsets, such that for a sample with expression profile $\boldsymbol{x}_i=(x_{i1},x_{i2},\cdots,x_{ip})$, the predicted class is k, that is, one of the values between 0 to K-1. Here, K is set to 2 for we will only discuss two label classification problems. Predictors are built from past experience, i.e. from observations which are known to belong to certain classes. Such observations comprise the learning set (LS) $\mathbf{L}=\{(\boldsymbol{x}_1,y_1), \cdots,(\boldsymbol{x}_{n\mathrm{L}},y_{n\mathrm{L}})\}$. Predictors may then be applied to a test set (TS), $\mathbf{T}=\{\boldsymbol{x}_1, \cdots,\boldsymbol{x}_{n\mathrm{T}}\}$, to predict for each observation $\boldsymbol{x}_i$ in the test set its class $\hat{y}_i$.

2.1 Gene Selection

The intrinsic problem with classification from microarray data is that sample size n is much smaller than the dimensionality of the feature space, i.e. the number of genes p. Many genes are non differentially expressed across the samples and irrelevant for phenotype discrimination. Dimensionality reduction of the feature space has been performed by many authors, see for example Golub et al. [10], Ben-Dor et al. [3] and Dudoit et al. [7], among others. It drastically eases the computational burden and for many problems improves class prediction due to the reduced amount of noise. Our feature selection is based on scoring each individual gene g, with $g \in \{1,\cdots,p\}$ according to its strength for phenotype discrimination. We use a nonparametric method that is based on ranks and was presented by Park et al. [16]. It is in fact equivalent to the test statistic of Wilcoxon's two sample test,

$$s(g)=\sum_{i\in N_0}\sum_{j\in N_1} I(x_j^{(g)}-x_i^{(g)}\le 0) \tag{1}$$

where $x_i^{(g)}$ is the expression value of gene g for individual i and N_m represents the set of the sample indices $\in\{1,\cdots,n\}$ having two different class labels in $m\in\{0,1\}$ (i.e. belonging to two different classes in this case). $I(.)$ denotes the indicate function, equaling 1 if the condition in parentheses is true and 0 otherwise. The score function can be interpreted as counting for each individual having response (class label) value zero, the number of instances with response one that have smaller expression values, and summing up these quantities. Viewing it as Wilcoxon's test statistic, it allows ordering of the genes according to their potential significance. It captures to what extent a gene g discriminates the response categories and it is easy to notice that both values near the minimum score zero and the maximum score n_0n_1 indicate a differentially expressed, informative gene. The quality measure $q(g)=\max(s(g),n_0n_1-s(g))$, thus gives the highest values to those genes whose expression levels have the best strength for phenotype discrimination. We then simply take the $p^*<<p$ genes with the highest values of $q(g)$ as our top features and restrict the boosting classifier to work

with this subset. The number of predictor variables is a tuning parameter whose optimal value varied across different datasets. A formal choice of p is possible via cross validation on the learning data or by determining the correct null distribution by bootstrap methods and a decision on significance levels as in Park et al. [16]. Many more variable selection criteria for gene expression data have been proposed in the literature. We think that our approach based on Wilcoxon's test statistic is more suitable in the context of gene expression data than the t-statistic used in Dudoit et al. [7] or the TNoM score [3] which corresponds to counting the number of errors made by the best stump, a decision tree with two terminal nodes. The situation is similar to the trade-off between t-, Wilcoxon- and sign-test. It is known from robustness theory that the t-test is highly sensitive to outliers and (even small) deviations from the normal distribution, whereas the sign-test (TNoM score) wastes useful information about the magnitude of gene expression levels. A good compromise is the Wilcoxon test which has nearly optimal power properties over a large class of data-generating distributions. We selected p^* with highest $q(g)$ value as our predictors' feature subset..

2.2 ANN Construction

In our work, the component neural network has same topology. Each member of ensemble is a multilayered feedforward with error backpropagation neural network. $O_j^{(l)}$ denotes the output signal of the jth neuron in the lth layer and $w_{ij}^{(l)}$ the connection weight coming from the ith neuron in the (l-1) layer to the jth neuron in the lth layer. $b_j^{(l)}$ is the bias of the jth neuron in the lth layer. $f(.)$ is the transfer function. Thus, the relationships between different layers can be represented by

$$O_j^{(l)} = f(\sum_i w_{ij}^{(l)} O_i^{(l-1)} + b_j^{(l)}) . \tag{2}$$

In this application, the topology of each neural network in the ensemble adopted three layer structures. The number of neurons in the input layer was equal to the number of genes p^* used for diagnosis of each sample. The number of neurons in the hidden layers is 5. Despite increasing this number can reflect more complicated problems, but the accuracy of our application is not increased remarkably and it will add burden to training time. Our research is only limited to two-class classification, the output layer adopt only one node, it gave an output between 0 (not this category) and 1 (this category). Purelin function, i.e., $f(y) = y$ was used as the transfer function between input layer and hidden layer. Logsig function, i.e., $f(y) = 1/(1+e^{-y})$ was used as the transfer function between hidden layer and output layer. For training these neural networks, batch gradient descent approach with momentum was adopted. The number of the epochs for backpropagation training algorithm equals 1000. The learning rate is 0.7 and the Momentum is set to 0.3.

2.3 Artificial Neural Network Ensembles

The major feature of microarray data is the numbers of samples are limited, not more than a few dozens. This situation will lead to single neural networks hard to be trained. In general, a neural network ensemble is constructed in two steps, i.e., training a number of component neural networks and then combining the component predictions. As for training component neural networks, the most prevailing approaches are *Bagging* and *Boosting*. Bagging which based on *bootstrap sampling* was proposed by Breiman [4]. It generates several learning sets from the original training set and then trains a component neural network from each of those training sets. Boosting was proposed by Schapire [18] and improved by Freund et al. [9]. It generates a series of component neural networks whose training sets are determined by the performance of former ones. Training instances that are wrongly predicted by former networks will play more important roles in the training of later networks. There are also many other approaches for training the component neural networks. Bagging and Boosting have been found to be powerful classification techniques that can highly increase accuracy on various problems, especially in higher dimensions. Since the speeds of neural network training always converge slowly, Bagging is used more frequently.

In our work, we create perturbed learning data sets based on convex pseudo data (CPD). Each perturbed learning sets is generated by repeating the following n_L times:

Select two instances $(\boldsymbol{x},y)$ and $(\boldsymbol{x}', y')$ at random from the learning set **L**.

Select at random a number v from the interval $[0,d]$, $0 \le d \le 1$, and let $u = 1 - v$. The new instance is $(\boldsymbol{x}'', y'')$ where $y'' = y$ and $\boldsymbol{x}'' = u\boldsymbol{x} + v\boldsymbol{x}'$.

As in Bagging, multiple altered learning set $\mathbf{L}_b$, of the same size as the original learning set **L**, are generated and used to train the components of ensembles. Because the performance is improved in the initial turns, we only generated 5 data sets for each original learning sets by above algorithm. As for combining the predictions of component neural networks, *simple averaging* was used, that is, the outputs of each component neural networks were averaged as the finial output. To gain the best performance, several values of d were tried, d=0.1,0.25,0.5,0.75,1 and the values of d with the smallest test set error rate were retained. Note that when d values is 0, CPD reduced to bagging.

3 Results

3.1 Datasets

We explored the performance of our classification techniques on two publicly available datasets:

1. *Leukemia Dataset.*This dataset[10] contains gene expression levels of n=72 patients either suffering from acute lymphoblastic leukemia (ALL, 47 cases) or acute myeloid leukemia (AML, 25 cases) and was obtained from Affymetrix oligonucleotide microarrays.
2. *Colon Dataset.*In this dataset, expression levels of 40 tumor and 22 normal colon tissues for 6500 human genes are measured using Affymetrix technology. A selection of 2000 genes with highest minimal intensity across the samples has been made by Alon et al. [2], and these data are publicly available at http://microarray.princeton.edu/oncology/.

3.2 Independent Tests and Cross Validations

We first describe the results on the leukemia data set [10]. In this study, the data were already divided into a training set of 38 mRNA samples and a test set of 34 mRNA samples. The observations in the two set came from different labs and were collected at different times. The test set comprises a broader range of samples, including samples from peripheral blood as well as bone marrow, from childhood AML patients, and from laboratories that used different sample preparation protocols. When training the classifiers and for the pre-selection of genes, we first reduced the set of available genes to the top p*=50, 100 and 200 genes as ranked in terms of Wilcoxon score based Park et al. procedure [16]. To compare our results (ANN ensembles) we also applied the three other traditional discriminant analysis procedures single artificial neural network (ANN), 1-NN and CART. The results are given in table 1.

Table 1. Classification rates by the four methods for the leukemia data set with 38 training samples (27 ALL, 11 AML) and 34 test samples (20 ALL,14 AML). Given are the number of correct classification out of 38 and 34 for the training and test samples respectively. The d values are reported in parentheses.

P*	ANN ensembles	ANNs	1NNs	CARTs
50	34 (0.05)	33	33	33
100	33 (0.05)	33	33	33
200	33 (0.5)	33	33	33

Among these results, all the 34 test samples using 50 genes and ANN ensembles were classified accurately, which is performed better than that of Golub et al. in the same number of genes. To further evaluate the performances of the classifiers, we applied the cross validation tests to the above two datasets. In these tests, we randomly selected one third of the original data sets as a test set and remaining ones as a training set. In the principle comparison, for each learning set/test set run, the top p*=50, 100 and 200 genes with largest Wilcoxon's scores are selected using the learning sets. Next, predictors are constructed using the learning set and test set error rates are obtained by applying the predictors to the test sets. For CPD, several values of the parameters d are examined: d=0.1, 0.25, 0.5, 0.75 and 1. All above procedure is repeated 150 times.

Table 2. Comparison of accuracy rates with four classification methods

P*	ANN ensembles (%)	ANNs (%)	1NNs (%)	CARTs (%)
leukemia dataset				
50	98.03±2.63 (0.05)	97.75±3.26	93.88±4.67	94.96±4.65
100	97.50±2.77 (0.05)	97.36±3.03	94.93±4.35	94.55±5.19
200	97.31±2.56 (0.5)	97.03±3.07	92.64±5.70	94.26±5.36
Colon dataset				
50	85.90±5.62 (0.5)	83.11±6.13	80.70±6.29	78.53±7.63
100	85.87±6.59 (0.5)	83.27±7.00	78.90±6.92	76.70±7.66
200	86.22±6.49 (0.75)	84.70±7.08	76.47±7.34	77.07±7.67

The average error rates are recorded in the table 2. The *d* values used in CPD are reported in parentheses. From the table above, one can see that the accuracy rates predicted by ANN ensembles are higher than that predicted by other classifiers.

3.3 ROC Curves

Estimates of classification accuracy give only a partial insight on the performance of a method. In our evaluation, we treated all errors as having equal penalty. In a clinical setting, one often prefers to punish misclassifications asymmetrically, since false negative error, i.e., classifying a tumorous tissue as normal can be fatal, whereas false positive errors, i.e., predicting a normal tissue as a tumor may be less serious since in this case additional tests will be carried out. Receiver Operator Characteristics (ROC) curves illustrate how accurate classifiers are under asymmetric losses, by plotting the tradeoff between false positives and false negatives. Each point on the two dimensional ROC curve corresponds to a particular probability $\beta \in [0,1]$ that was used as a threshold for positive (tumorous) classification. The (x,y) coordinates of each point are then the fractions of negative and positive samples that are classified as positive with this particular threshold β. In general, ROC curves are between these two extremes, with a larger area under the ROC curve (AUROC) indicating a better performance. In the ideal case, the ROC curve goes through (0,1), the upper left corner of the plot, that is, AUROC equals to 1. We selected top 50 genes for tests in this data set. The AUROC of neural network ensemble and single neural network are 0.890 and 0.851 respectively. It demonstrated that the sensitivity of neural network ensembles is better than that of single neural networks.

4 Conclusion

The artificial neural network ensemble is a recently developed technology, which has the ability of significantly improving the performance of a system where a single artificial neural network is used. In this paper, we proposed a tissue classification

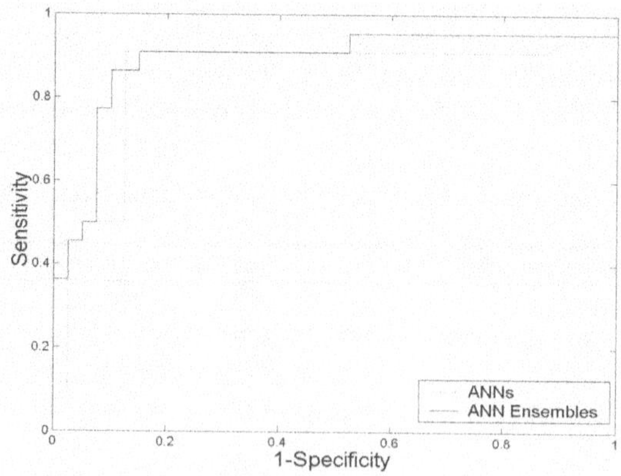

Fig. 1. ROC curves for neural network ensembles and neural networks applied on colon data set

method using DNA microarray data, which utilize artificial neural network ensemble to identify different cancer subtypes. The experiments demonstrated that this method will improve the performances of classification. Its accuracy of classification is better than single neural networks, 1NNs and CARTs. The results showed the feasibility of cancer classification based on neural network ensembles and microarray data. In this paper, we only consider the model for two class type discriminant. Extension to more than two categories using neural network ensembles needs further research. Some previous reports, e.g. [1] presented some hints to resolve this problem.

References

1. Anand ,R., Mehrotra ,K., Mohan, C.K., Ranka, S.: Efficient Classification for Multiclass Problems Using Modular Neural Networks. IEEE Transactions on Neural Networks,Vol.6, (1995) 117-124
2. Alon, U., Barkai, N., Notterman, D., Gish, K., Ybarra, S., Mack, D., Levine, A.J.: Broad Patterns of Gene Expression Revealed by Clustering Analysis of Tumor and Normal Colon Tissues Probed by Oligonucleotide Arrays. Proceedings of the National Academy of Sciences, Vol. 96, (1999) 6745-6750
3. Ben-Dor, A., Bruhn, L., Friedman, N., Nachman, I., Schumme,r M., Yakhini, Z.: Tissue Classification with Gene Expression Profiles. Journal of Computational Biology ,Vol. 7, (2000) 559-583
4. Breiman, L.: Bagging Predictors. Machine Learning, Vol. 24, (1996) 123-140
5. Breiman, L.: Using Convex Pseudo-Data to Increase Prediction Accuracy. Technical Report 513, Statistics Department, U.C. Berkeley, USA, 1998
6. DeRisi, J.L., Iyer, V. R., Brown, P.O.: Exploring The Metabolic and Genetic Control of Gene Expression on A Genomic Scale. Science, Vol. 278, (1997) 680-686
7. Dudoit ,S., Fridlyand, J., Speed ,T.: Comparison of Discrimination Methods for the Classification of Tumors using Gene Expression Data. Journal of the American Statistical Association, Vol. 97, (2002) 77-87
8. Fort ,G. , Lambert-Lacroix, S.: Classification Using Partial Least Squares with Penalized Logistic Regression. Bioinformatics, Vol 21, (2005) 1104-1111
9. Freund, Y. ,Schapire, R.E.: A Decision-Theoretic Generalization Of On-Line Learning And An Application To Boosting. Journal of Computer and System Sciences, Vol. 55, (1997) 119-139
10. Golub, T.R., Slonim, D.K., Tamayo, P., Huard, M., Gaasenbeek, M., Mesirov, J.P., Coller, H., Loh, M.L., Downing, J.R., Caligiuri ,M.A., Bloomfield ,C.D., Lander, E.S.: Molecular Classification of Cancer: Class Discovery and Class Prediction by Gene Expression Monitoring. Science, Vol. 286, (1999) 531-537
11. Hansen ,L.K. and Salamon, P.: Neural Network Ensembles. IEEE Transactions on Pattern Analysis and Machine Intelligence, Vol. 12, (1990) 993-1001
12. Khan ,J., Wei, J.S., Ringner, M., Saal, L.H., Ladanyi, M., Westermann ,F., Berthold, F., Schwab, M., Antonescu, C.R., Peterson, C., Meltzer, P.S.: Classification and Diagnostic Prediction of Cancers Using Gene Expression Profiling and Artificial Neural Networks. Nature Medicine, Vol. 7, (2001) 673-679
13. Khan J., Simon R., Bittner M., Chen Y., Leighton S.B., Pohida T., Smith P.D., Jiang Y., Gooden G.C., Trent J.M., Meltzer P.S.: Gene Expression Profiling of Alveolar Rhabdomyosarcoma with cDNA Microarrays. Cancer Research , Vol. 58.,(1998) 5009-5013

14. Lockhart, D.J., Dong, H., Byrne, M.C., Follettie ,M.T., Gallo, M.V., Chee, M.S., Mittmann, M., Wang, C., Kobayashi, M., Horton, H., Brown, E.L.: Expression Monitoring by Hybridization to High-Density Oligonucleotide Arrays. Nature Biotechnology, Vol. 14, (1996) 1675-1680
15. O'Neill, MC. and Song, L.: Neural Network Analysis of Lymphoma Microarray Data: Prognosis Diagnosis Near-perfect. BMC Bioinformatics, Vol. 4, (2003) 13
16. Park ,P., Pagano ,M., Bonetti ,M.A.: Nonparametric Scoring Algorithm for Identifying Informative Genes from Microarray Data. Pacific Symposium on Biocomputing, Vol. 6, (2001) 52-63
17. Ramaswamy, S., Tamayo, P., Rifkin, R., Mukherjee, S., Yeang, C.H., Angelo ,M., Ladd, C., Reich, M., Latulippe, E., Mesirov, J.P., Poggio, T., Gerald ,W., Loda, M., Lander, E.S., Golub, T.R.: Multiclass Cancer Diagnosis Using Tumor Gene Expression Signatures. Proceedings of the National Academy of Sciences, Vol. 98., (2001) 15149-15154
18. Schapire, R.E.: The Strength Of Weak Learnability. Machine Learning, Vol.5, (1990) 197–227
19. West, M., Blanchette, C., Dressman, H., Huang, E., Ishida, S., Spang, R., Zuzan, H., Olson, JA Jr, Marks ,J.R., Nevins ,J.R.: Predicting the Clinical Status of Human Breast Cancer by Using Gene Expression Profiles. Proceedings of the National Academy of Sciences, Vol. 98, (2001) 11462-11467
20. Xu, Y., Selaru ,F.M., Yin, J., Zou, T.T., Shustova ,V., Mori, Y., Sato, F., Liu ,T.C., Olaru ,A., Wang, S., Kimos ,M.C., Perry, K., Desai, K., Greenwald, B.D., Krasna, M.J., Shibata, D., Abraham, J.M., Meltzer, S.J.: Artificial Neural Networks and Gene Filtering Distinguish between Global Gene Expression Profiles of Barrett's Esophagus and Esophageal Cancer. Cancer Research, Vol. 62., (2002) 3493-3497
21. Yeung, K.Y., Bumgarner, R.E., Raftery, A.E. :Bayesian Model Averaging: Development of an Improved Multi-Class, Gene Selection and Classification Tool for Microarray Data.Bioinformatics, Vol. 21, (2005) 2394-2402
22. Zhou, Z.H.,Chen, S.F.: Neural Network Ensemble. Chinese Journal of Computers, Vol.25, (2002) 1-8

Author Index

Lecture Notes in Bioinformatics

Vol. 4115: D.-S. Huang, K. Li, G.W. Irwin (Eds.), Computational Intelligence and Bioinformatics, Part III. XXI, 803 pages. 2006.

Vol. 4075: U. Leser, F. Naumann, B. Eckman (Eds.), Data Integration in the Life Sciences. XI, 298 pages. 2006.

Vol. 4070: C. Priami, X. Hu, Y. Pan, T.Y. Lin (Eds.), Transactions on Computational Systems Biology V. IX, 129 pages. 2006.

Vol. 3939: C. Priami, L. Cardelli, S. Emmott (Eds.), Transactions on Computational Systems Biology IV. VII, 141 pages. 2006.

Vol. 3916: J. Li, Q. Yang, A.-H. Tan (Eds.), Data Mining for Biomedical Applications. VIII, 155 pages. 2006.

Vol. 3909: A. Apostolico, C. Guerra, S. Istrail, P. Pevzner, M. Waterman (Eds.), Research in Computational Molecular Biology. XVII, 612 pages. 2006.

Vol. 3886: E.G. Bremer, J. Hakenberg, E.-H.(S.) Han, D. Berrar, W. Dubitzky (Eds.), Knowledge Discovery in Life Science Literature. XIV, 147 pages. 2006.

Vol. 3745: J.L. Oliveira, V. Maojo, F. Martín-Sánchez, A.S. Pereira (Eds.), Biological and Medical Data Analysis. XII, 422 pages. 2005.

Vol. 3737: C. Priami, E. Merelli, P. Gonzalez, A. Omicini (Eds.), Transactions on Computational Systems Biology III. VII, 169 pages. 2005.

Vol. 3695: M.R. Berthold, R.C. Glen, K. Diederichs, O. Kohlbacher, I. Fischer (Eds.), Computational Life Sciences. XI, 277 pages. 2005.

Vol. 3692: R. Casadio, G. Myers (Eds.), Algorithms in Bioinformatics. X, 436 pages. 2005.

Vol. 3680: C. Priami, A. Zelikovsky (Eds.), Transactions on Computational Systems Biology II. IX, 153 pages. 2005.

Vol. 3678: A. McLysaght, D.H. Huson (Eds.), Comparative Genomics. VIII, 167 pages. 2005.

Vol. 3615: B. Ludäscher, L. Raschid (Eds.), Data Integration in the Life Sciences. XII, 344 pages. 2005.

Vol. 3594: J.C. Setubal, S. Verjovski-Almeida (Eds.), Advances in Bioinformatics and Computational Biology. XIV, 258 pages. 2005.

Vol. 3500: S. Miyano, J. Mesirov, S. Kasif, S. Istrail, P. Pevzner, M. Waterman (Eds.), Research in Computational Molecular Biology. XVII, 632 pages. 2005.

Vol. 3388: J. Lagergren (Ed.), Comparative Genomics. VII, 133 pages. 2005.

Vol. 3380: C. Priami (Ed.), Transactions on Computational Systems Biology I. IX, 111 pages. 2005.

Vol. 3370: A. Konagaya, K. Satou (Eds.), Grid Computing in Life Science. X, 188 pages. 2005.

Vol. 3318: E. Eskin, C. Workman (Eds.), Regulatory Genomics. VII, 115 pages. 2005.

Vol. 3240: I. Jonassen, J. Kim (Eds.), Algorithms in Bioinformatics. IX, 476 pages. 2004.

Vol. 3082: V. Danos, V. Schachter (Eds.), Computational Methods in Systems Biology. IX, 280 pages. 2005.

Vol. 2994: E. Rahm (Ed.), Data Integration in the Life Sciences. X, 221 pages. 2004.

Vol. 2983: S. Istrail, M.S. Waterman, A. Clark (Eds.), Computational Methods for SNPs and Haplotype Inference. IX, 153 pages. 2004.

Vol. 2812: G. Benson, R.D. M. Page (Eds.), Algorithms in Bioinformatics. X, 528 pages. 2003.

Vol. 2666: C. Guerra, S. Istrail (Eds.), Mathematical Methods for Protein Structure Analysis and Design. XI, 157 pages. 2003.

www.ingramcontent.com/pod-product-compliance
Lightning Source LLC
Chambersburg PA
CBHW071403230926
51846CB00001B/8
* 9 7 8 3 5 4 0 3 7 2 7 7 6 *